ENERGY

Volume III: Nuclear Energy and Energy Policies

ENERGY

Volume I
by S. S. Penner and L. Icerman

Demands, Resources, Impact, Technology, and Policy, 1974

Volume II
by S. S. Penner and L. Icerman

Non-nuclear Technologies, 1975

Volume III
edited by S. S. Penner

Nuclear Energy and Energy Policies, 1976

Contributors

Keith A. Brueckner*   (pp. 349–454)
A. Hochstim**   (pp. 150–236)
J. P. Howe**†   (pp. 1–149, 455–523, 551–558)
Larry Icerman**‡   (pp. 524–550, 668–690)
S. S. Penner**   (pp. 455–523, 551–581, 600–690)
W. B. Thompson*   (pp. 237–348)

University of California, San Diego
   Energy Center and Department of Physics
  **Energy Center and Department of Applied Mechanics and Engineering Sciences*
†*On leave from Wayne State University, Detroit, Michigan*
‡*Present address: Department of Technology and Human Affairs,*
   Washington University, St. Louis, Missouri

Michael Z. Nagel and R. J. Cerbone   (pp. 582–599)
General Atomic Company, San Diego, California

Bruce J. West   (pp. 150–236)
Physical Dynamics, Inc., San Diego, California

ENERGY

Volume III
Nuclear Energy
and Energy Policies

One of a Three-Volume Set of Lecture Notes

Edited by S. S. Penner

Energy Center and
Department of Applied Mechanics and Engineering Sciences
University of California, San Diego
La Jolla, California

1976
Addison-Wesley Publishing Company, Inc.
Advanced Book Program
Reading, Massachusetts

London · Amsterdam · Don Mills, Ontario · Sydney · Tokyo

ISBN 0-201-05564-3
ISBN 0-201-05565-1 pbk.

Reproduced by Addison-Wesley Publishing Company, Inc., Advanced Book Program,
Reading, Massachusetts, from camera-ready copy prepared by the authors.

Manufactured in the United States of America.

ABCDEFGHIJ-HA-79876

CONTENTS

CONTENTS

Page

Page

GLOSSARY OF SYMBOLS

A = ampere, Angstrom unit (10^{-8} cm), or nuclear mass number, depending on the context

a.c. = alternating current

amu = atomic mass unit

atm = atmosphere

$\vec{B}$ = magnetic field intensity

B_ϕ = toroidal component of $\vec{B}$

B_θ = poloidal component of $\vec{B}$

barn = 10^{-24} cm^2

bbl = barrel

Btu = British thermal unit

c = vacuum velocity of light $\simeq 3 \times 10^8$ m/sec

$^\circ$C = degrees Centigrade

cal = calorie

Ci = curie

cm = centimeter

d = day

D = diffusion coefficient, dollars, deuterium, or deuteron, depending on the context

D_b = "banana" diffusion coefficient

D_{class} = classical diffusion coefficient

D_{PS} = Pfirsch-Schlüter diffusion coefficient

d.c. = direct current

e = magnitude of the charge of the electron = 1.6021×10^{-19} Coulomb

e^+ = positron

E or $\mathcal{E}$ = energy

$\vec{E}$ = electric field

e_f = electric charge density or safety factor, depending on the context

$\vec{E}_i$ = electric field of the ion

esu = electrostatic units

ev = electron volt

(e) or subscript e = equivalent energy or electrical energy, depending on the context

f = particle distribution function

$\vec{F}$ = force

$^\circ F$ = degrees Fahrenheit

f_+ or f_- = particle flux

$f_e(\vec{v}, \vec{r}, t)$ = single particle electron distribution function

$f_i(\vec{v}, \vec{r}, t)$ = single particle ion distribution function

ft = foot

g = gram

$\vec{g}$ = relative velocity

G = Gauss

gal = gallon

Gw = gigawatt = 10^9 watts

h = hour or thermal power per unit area, depending on the context

H = heat transfer coefficient or enthalpy, depending on the context

hp = horsepower

Hz = Hertz

in. = inch

j = joule

$\vec{j}$ = electric current density

J = second adiabatic invariant

k = Boltzmann's constant = 1.3804×10^{-23} joule/$^\circ$K = 1.38×10^{-16} erg/$^\circ$K

$\vec{k}$ = wave vector

$\overline{k}$ = rate constant for fusion reaction

K = electron thermal conductivity

$°K$ = degrees Kelvin

k_D = Debye wave number

K_{eff} = electron thermal conductivity (effective)

k_D = Debye wave number

kcal = kilocalorie = 10^3 calories

kev = 10^3 ev

kg = kilogram = 10^3 grams

kG = kilogauss = 10^3 Gauss

kj = kilojoule = 10^3 joules

km = kilometer = 10^3 meters

kt = kiloton = 10^3 tons

kv = kilovolt = 10^3 volts

kw = kilowatt = 10^3 watts

ℓ = density scale or fractional probabilistic neutron losses, depending on the context

L = characteristic length or connection length, depending on the context

$\mathcal{L}$ = Lundquist number

$\mathcal{L}_f$ = probability of fast neutron containment in a fission reactor

$\mathcal{L}_t$ = probability of thermal neutron containment in a fission reactor

ℓ_o = collisionless screening number

lb = pound

lbm = pound mass

m = meter or electron mass, depending on the context

M = ion mass or Mach number, depending on the context

m_e = mass of the electron

m_n = mass of the neutron

m_p = mass of the proton

Mev = 10^6 ev

mi = mile

min = minute

Mj = megajoule = 10^6 joules

mph = mile(s) per hour

mrem = millirem = 10^{-3} rem

mt = metric ton

Mw = megawatt = 10^6 watts

n = newton or neutron, depending on the context

N_e = electron concentration

N_i = concentration of species i or ion concentration, depending on the context

N_θ = thermal speed

N_θ^e = electron thermal speed

N_θ^i = ion thermal speed

p = pressure or proton, depending on the context

p^{br} = power density of the bremsstrahlung radiation

p^{cyc} = power density of the cyclotron radiation

p^{sync} = power density of the synchrotron radiation

p_B = magnetic field pressure

p_e = electron plasma pressure

p_N = ion plasma pressure

P_s = specific power of a fission reactor

$p_\parallel$ = component of the pressure tensor parallel to $\vec{B}$

$p_\perp$ = component of the pressure tensor perpendicular to $\vec{B}$

ppm = parts per million

psi = pounds per square inch

psia = pounds per square inch (absolute)

psig = pounds per square inch (guage)

$Q = 10^{18}$ Btu or energy released from a nuclear reaction, depending on the context

r = toroidal coordinate

R = major radius of torus, radius of curvature of field lines, or conversion of breeding ratio, depending on the context

R_L = Larmor radius

rad = radian or unit of measure of energy of ionizing radiation absorbed per unit mass (kg), depending on the context

rem = Roentgen-equivalent man

rpm = revolutions per minute

S = entropy

SCF = standard cubic feet

sd = standard day or stream day

sec = second

STP = standard temperature and pressure

SWU = separative work unit

t = ton = 2,000 pounds

T = temperature (kev) or tritium or triton, depending on the context

T^e or T_e = electron temperature

t_d = doubling time

T^i or T_i = ion temperature

$t_{\frac{1}{2}}$ = half life for a first-order nuclear reaction

Tw = terrawatt = 10^{12} watts

(th) or subscript t = thermal energy

v = volt

$\vec{v}$ = velocity

$\bar{v}$ = average velocity of a particle

$\vec{v}_D$ = drift velocity

v_p = phase velocity

$v_{\parallel}$ = component of velocity parallel to $\vec{B}$

$v_{\perp}$ = component of velocity perpendicular to $\vec{B}$

w = watt

W = power density

W^{br} = power density due to bremsstrahlung

W^{cyc} = power density due to cyclotron radiation

w_{ℓ} = thermal power per unit length

W^{nuc} = power density from nuclear fusion

W_p = present worth in dollars

$W^{rad}_{(line)}$ or $W^{rad}_{(recomb)}$ = power density from line or recombination radiation

W_y = worth in year y in dollars

y = year

Z = nuclear charge number or normalized constant in a probability distribution function, depending on the context

α = magnetic field line coordinate, energetic helium nuclei emitted in a nuclear reaction, or ratio of nuclei emitting γ radiation to nuclei fissioning after combination with a neutron or alpha particle, depending on the context

β = magnetic field line coordinate, electrons emitted from atomic nuclei, fraction of delayed neutrons or ratio of magnetic to kinetic pressure, depending on the context

γ = instability growth rate or photons emitted by atomic nuclei, depending on the context

$\delta(x)$ = Dirac delta function

δE_J = perturbation in energy for fixed J

δf = perturbed particle distribution function

ϵ = inverse aspect ratio or fast fission effect, depending on the context

$\epsilon(\omega, \vec{k})$ = dielectric coefficient

η = electrical resistivity or average neutron yield per neutron-nucleus reaction, depending on the context

θ = toroidal coordinate

ι = rotational transform

λ = mean free path or reaction rate constant, depending on the context

λ_D = Debye wave length

μ or μm = micron = 10^{-4} cm

μ = magnetic moment

μCi = microcurie = 10^{-6} curie

μsec = microsecond = 10^{-6} sec

ν = collision frequency or neutron yield of a fission reaction, depending on the context

ν_e, ν_{eN} or ν_{ei} (depending on the context) = electron collision frequency

ν_i or ν_{ii} (depending on the context) = ion collision frequency

$\bar{\nu}$ = average neutron yield from a fission reaction or average electron collision frequency, depending on the context

ξ = fractional average energy loss per scattering collision between a neutron and a nucleus

ρ = mass density

σ = electric conductivity

$\sigma(\underline{a}, b)$ = cross section for reaction between particles $\underline{a}$ and b

$\sigma_{a, c, \text{ or } \gamma}$ = cross section for capture of a neutron by a nucleus, followed by emission of a gamma ray

σ_f = cross section for a neutron-induced fission reaction

σ_s = cross section for scattering of neutrons over all angles

$\Sigma_{T, f, c, s}$ = macroscopic cross section for the respective neutron reactions

$\sigma_T(E)$ = total cross section for all outcomes of a reaction between a neutron with energy E and a nucleus

τ = confinement time for a fusion reaction

τ_B = bounce period

τ_H = hydrodynamic period

τ_η = resistive period

$\overline{\tau}$ = average half life of emitter of delayed neutrons

ϕ = toroidal coordinate or electric potential, depending on the context

Φ = electrostatic potential or magnetic flux, depending on the context

Ψ = magnetic flux function or magnetic surface label, depending on the context

ω_B = bounce frequency

ω_c = Larmor frequency

ω_c^e = electron gyrofrequency

ω_c^i = ion gyrofrequency

PREFIXES

milli or m = 10^{-3}

micro or μ = 10^{-6}

nano or n = 10^{-9}

pico or p = 10^{-12}

kilo or k = 10^{3}

mega or M = 10^{6}

giga or G = 10^{9}

tera or T = 10^{12}

TABLE OF ACRONYMS

ABCC = Atomic Bomb Casualty Commission

ACRS = Advisory Committee on Reactor Safeguards

AEC = Atomic Energy Commission

AISI = American Iron and Steel Institute

ANL = Argonne National Laboratory

APS = American Physical Society

BEIR = biological effects of ionizing radiation

BHP = biological hazard potential

BNL = Brookhaven National Laboratory

BWR = boiling water reactor

CB = containment building

CFR = Code of Federal Regulations

CI = containment integrity

CRBR = Clinch River breeder reactor

CTR = controlled thermonuclear reactor

CVCS = chemical and volume control system

DBA = design basis accident

DOL = Division of Licensing (and Regulation)

DWS = demineralized water supply

ECCS = emergency core cooling system

EHCS = electrical heating control system

EIR = environmental impact report

EIS = environmental impact statement

EPA = Environmental Protection Agency

ERDA = Energy Research and Development Administration

ESF = engineered safety features

ESS = engineered safety systems

FBR = fast breeder reactor

FEA = Federal Energy Administration

FSAR = final safety analysis report

FFTF = fast fuel test facility

GCFR = gas (helium) cooled fast (breeder) reactor

HTGR = high temperature gas (helium cooled) reactor

IAEA = International Atomic Energy Agency

ICRP = International Commission on Radiation Protection

IHX = intermediate heat exchanger

LASL = Los Alamos Scientific Laboratory

LMFBR = liquid metal (cooled) fast breeder reactor

LOCA = loss of coolant accident

LWBR = light water breeder reactor

LWR = light water reactor

MPC = maximum permissible concentration

NASA = National Aeronautics and Space Administration

MSBR = molten salt breeder reactor

NCAR = National Center for Atmospheric Research

NCRP = National Council on Radiation Protection (and Measurements)

NRC = Nuclear Regulatory Commission

NRTS = National Reactor Testing Station

NSSS = nuclear steam supply system

NTIS = National Technical Information Service

OPEC = Organization of Petroleum Exporting Countries

ORNL = Oak Ridge National Laboratory

OTA = Office of Technology Assessment

PAHR = post accident heat removal (system)

PARR = post accident radioactivity removal (system)

PCRV = prestressed concrete reactor vessel

PFR = prototype fast reactor

PNL = Pacific Northwest Laboratories

PRT = pressure relief tank

PSAR = preliminary safety analysis report

PV = pressure vessel

PWR = pressurized water reactor

RHRS = residual heat removal system

RRV = rupture relief valve

RTPR = reference theta pinch reactor

SIS = safety injection system

UNSCEAR = United Nations Committee on Effects of Atomic Radiation

USAEC = United States Atomic Energy Commission

USERDA = United States Energy Research and Development Administration

WPS = waste processing system

Table 1.2-2 Units, conversion factors, energy consumption. [*]

$1 \text{ joule} = 10^7 \text{ erg } (= 10^7 \text{ dyne-cm}) = 6.24 \times 10^{12} \text{ Mev} = 6.24 \times 10^9 \text{ Bev} = 1.0$ newton-m $= 0.736 \text{ ft-lb} = 0.24 \text{ cal} = 0.949 \times 10^{-3} \text{ Btu} = 2.78 \times 10^{-4}$ wh $= 3.73 \times 10^{-7} \text{ hph}, 2.78 \times 10^{-7} \text{ kwh} = 2.38 \times 10^{-10} \text{ ton of TNT}$ equivalent $= 1.22 \times 10^{-13}$ of the fusion energy from the deuterium in 1 m^3 of seawater $= 1.11 \times 10^{-14}$ g of matter equivalent $= 1.22 \times 10^{-14}$ of the fission energy of 1 kg of U-235 equivalent $= 6.7 \times 10^{-23}$ of the average daily input of solar energy at the outside of the atmosphere of the earth $= 5.8 \times 10^{-32}$ of the daily energy output from the sun.

1 metric ton of coal $\simeq 27.8 \times 10^6$ Btu (1 metric ton = 1 mt $\simeq$ 2,200 lb).
1 bbl of petroleum $\simeq$ (5.60 to) 5.82 (or more) $\times 10^6$ Btu (1 bbl = 42 gallons).
1 SCF of natural gas $\simeq 10^3$ Btu.
1 cord of wood $\simeq 1.95 \times 10^7$ Btu (1 cord = 128 ft^3).

9,500 Btu (th)/kwh$_e$ at 36% conversion efficiency.

$1 \text{ Q} = 10^{18} \text{ Btu} = 1.05 \times 10^{21} \text{ joule} = 2.93 \times 10^{14} \text{ kwh(th)} = 1.22 \times 10^{10} \text{ Mwd(th)}$
$= 3.35 \times 10^7 \text{ Mwy(th)} = 1.7 \times 10^{11}$ bbl of petroleum equivalent; $1 \text{ y} = 8.76 \times 10^3$ h.

Coal conversion to oil: $\geqslant$2 bbl/mt ($\geqslant$42% energy-conversion efficiency).

Energy consumption estimates: [**]

USA (1970) -- 0.07 Q/y [2×10^8 people, 11.7 kw(th)/p].
(2000) -- 0.16 Q/y [3×10^8 people, 17.8 kw(th)/p].
(2020) -- 0.3 Q/y [4×10^8 people, 25 kw(th)/p].
WORLD (1970) -- 0.24 Q/y [4×10^9 people, 2 kw(th)/p].
(2000) -- 2.1 Q/y [7×10^9 people, 10 kw(th)/p].
(2050) -- 6 Q/y [10×10^9 people, 20 kw(th)/p].

Note: See footnotes to this table on the following page.

*Table abbreviations used: bbl = barrel; Btu = British thermal unit; cal = calorie; d = day; g = gram; h = hour; hp = horsepower; kg = kilogram; kw = kilowatt = 10^3 watt; m = meter; Mw = megawatt = 10^6 watt = 10^3 kw; p = person; SCF = standard cubic foot, corresponding to the gas volume at a pressure of 14.73 psi (= 1 atmosphere) and a temperature of 60°F; w = watt; y = year; the symbol (e) or the subscript e identify electrical energy; the symbol (th) or the subscript t identify thermal energy. The subscript e is also used occasionally in place of the phrase "equivalent energy"; the particular meaning attached to e should generally be clear from the context.

**Representative estimates from various sources; forecasts to the year 2000 and beyond are uncertain by factors of 2 or more.

Contents of Volume I, 1974

Contents of Volume II, 1975

PREFACE

Volume III of this series of lecture notes deals with nuclear energy technologies and with energy policy. It differs from the preceding volumes in the following important respect: the contents include contributions by separately identified authors and thus may not show the same coherent level of exposition which was hopefully achieved in the earlier volumes. The subject matter under discussion is, however, so diverse that multiple authorship appeared to be essential in order to achieve authoritative coverage. The Editor has made every effort to assure continuity in exposition and style.

The reader will find the discussion on nuclear fission-energy (except for the Appendix) to be largely descriptive and generally suitable for use in a lower division course. On the other hand, the remaining material in Chapter 21, as well as the exposition of fusion energy, are addressed to upper division physical science majors. The discussion of energy policies is predicated on the assumption that the reader has attained some sophistication concerning the analysis of energy problems.

I am happy to acknowledge the excellent typing job done by Kay Hutcheson, who prepared this entire volume.

In order to assist the reader in the use of this book, we have reproduced Table 1.2-2 of Volume I on units, conversion factors, and energy consumption. The reader should note that reference numbers again refer to individual Sections.

S. S. Penner, Editor

CHAPTER 21

NUCLEAR FISSION ENERGY

21.1 Introduction

In this chapter, we describe the production of useful energy from nuclear fission. We will begin with a summary of applicable nuclear phenomena and a brief history of nuclear reactors. Our principal concerns are brief schematic descriptions of the main types of nuclear power plants, illustrative analyses of how the designs of reactors arise, some unique requirements for reactor materials, key issues in nuclear safety, and environmental effects and problems in the storage of radioactive materials. The objective is to examine characteristic aspects of the subject rather than to learn how to design operating nuclear power plants. Our purpose will be served if we succeed in setting important and interesting problems and issues in reasonable perspective. In some respects, engineering designs are works of art and deserve appreciation as such.

The development of nuclear reactors began in earnest in 1942. During World War II, the scientists and engineers who developed the means of producing fissile plutonium and uranium and weapons were confident that nuclear

fission would be an important potential source of energy for peaceful applications. As soon as the war ended, many of these people in the U.S. became politically active[1] and strongly influenced congressional opinion and legislation
in order to place future developments of nuclear-energy utilization in civilian
rather than military hands. Through the first Atomic Energy Act of 1946, the
Congress of the U.S. charged the Atomic Energy Commission (AEC) with the
development of nuclear power reactors, as well as all other applications of
nuclear energy.

Eclipsed for some time by military demands, the development of nuclear fission reactors for peaceful purposes nevertheless occupied the attention
of many scientists and engineers who were involved in nuclear reactor development in the post-war period. Under government control, the development of
commercial nuclear power plants gathered additional momentum in 1952-53.
In 1955, the first Atoms for Peace Conference was held in Geneva, Switzerland,
under the auspices of the United Nations. Here, a large amount of information
on the development of nuclear power was publicly released for the first time by
several nations. The Atomic Energy Act of 1957 defined the role of the U.S.
Atomic Energy Commission in developing and regulating the use of nuclear
fission reactors for the production of commercial electrical power. New legislation passed by Congress during the later part of 1974 became effective in February 1975 and incorporated governmental authority for the development of application of nuclear energy (also weapons) in a new Energy Research and Development Administration (ERDA), while delegating the responsibility and authority
for regulation of nuclear power production to a new Nuclear Regulatory Commission (NRC).

[1]Alice Kimball Smith, A Peril and a Hope, The Scientists' Moment in America
    1945-47, The University of Chicago Press, Chicago, Ill., and London, U.K.,
    1965.

The status of the development of nuclear fission power plants in the U.S. as of December 1974 is shown in Fig. 21.1-1. Twenty-four countries, in addition to the U.S., are currently using or scheduling the construction of somewhat over 200 installations to produce 112 gigawatts (Gw_e) = 112,000 Mw_e of electrical power.

As of the end of 1973, the planned production by 1980 in the U.S. of approximately 232 Gw_e represented over 20% of our anticipated total electrical generating capacity and over four times the existing 1972 capacity of approximately 5.5 Gw_e. By the end of 1974, several of these planned installations were postponed because of the scarcity and high cost of capital to finance the construction. However, the current need to reduce the consumption of oil is stimulating national attention on the solution of the particular problems which cause delays of 10 to 12 years in the construction of nuclear power plants, as compared with typical construction times of 6 to 7 years for fossil-fuel plants. The issues to be resolved include the finding of sites, granting of licenses, regulation of prices, production of over 7×10^5 tons of U_3O_8, as well as the raising of the required capital of many hundreds of billions of dollars.

At the present time, a relatively small number of very active people is stimulating opposition to the use of nuclear fission power by overstating the likelihood of possible hazards. In several states, legislation is being promoted to declare moratoria on the construction of nuclear power plants. Thus, the future progress of the field is subject to large antithetical forces, the resolution of which will require extensive knowledge both of energy technology and of the psychological needs of people. Objective knowledge of the benefits and risks associated with alternative means of energy production and use must be gained by all those who wish to make intelligent judgements on U.S. and World energy options.

Fig. 21.1-1 A 1975 projection for approximate locations of nuclear power reactors in the United States during the period 1975 to 1985. Oper-
able in 1975: ■, 55 units, 37 Gw_e; under construction: ▲, 73 units,
74 Gw_e; planned: ●, 90-100 units, 105-110 Gw_e. Total prospective
U.S. capacity in 1985: 190 to 210 Gw_e. Total prospective world
capacity in 1985: 372 Gw_e. Planned beyond 1985: 13 Gw_e. Companies in the U.S. are supplying reactors as follows: Westinghouse 36%, General Electric 36%, Combustion Engineering 15%,
Babcock and Wilcox 13%. This information is similar to that published in Nuclear News 18, 63-75 (1975) and is based on AEC data.

21.2 Fission, Science and Engineering

Meitner and Frisch[1] announced the discovery of nuclear fission six years after Chadwick's discovery of the neutron in 1932 had opened up the fields of neutron physics and chemistry. Study of the chronology of the following key events leading to discovery of the fission reaction may strengthen our appreciation of the relation of scientific discoveries to our lives and culture.

The discovery of uranium oxide in pitchblende was made by O. Klaproth (Germany) in 1789. The atomic hypothesis was proposed by John Dalton (England) in 1808. The laws of conservation of energy were formulated by G. Leibnitz (Germany) in 1693, by B. Thompson (USA and Germany) in 1798, by J. R. Mayer (England) in 1842, and by H. von Helmholtz (Germany) in 1847. The periodic law for correlating the properties of the elements was developed by L. Meyer (Germany) in 1864 and by D. Mendeleeff (Russia) in 1869. Identification of electrons was accomplished by J. J. Thompson (England) in 1897. Radium was isolated from pitchblende by P. and M. Curie (France) during 1898. Observations of α-, β-, and γ-rays in radioactive decays were first reported by E. Rutherford (England) in 1899. The equivalence of mass and energy was proposed by A. Einstein (Germany) during 1905. The nuclear atom was postulated by E. Rutherford (England) during 1911. The discovery of the neutron was announced by W. Bothe and H. Becker (Germany), I. Curie and F. Joliot (France), and J. Chadwick (England) in 1932. Neutron reactions and diffusion of neutrons were first studied by E. Fermi, L. Szilard and many others (Europe, USA) in 1932. The discovery of the fission reaction by L. Meitner, O. Frisch, F. Strassmann,

---

[1]See S. Glasstone, Source Book on Atomic Energy, D. Van Nostrand Company, Inc., 250 Fourth Avenue, New York, N.Y., 3rd edition, 1967.

and O. Hahn (Germany) occurred during 1939. Determinations of the number of neutrons per fission were made by H. von Halban, F. Joliot, and L. Kowarski (France) in 1939. Measurements of delayed neutrons from fission were performed by R. Roberts, N. Meyer, L. Hafstad, and P. Wang (USA) also in 1939. Transmutation of U-238 to Pu 239 was accomplished by G. T. Seaborg, J. Kennedy, A. Wahl, and E. Segre' (USA) during 1940. The first critical, chain-reacting assembly was completed by E. Fermi and W. Zinn (USA) in 1941.

Although the listed advances in knowledge involve basic scientific contributions, our discussion of nuclear power plants falls properly in the domain of technology. The relation between science and technology is complex and has been discussed and clarified by studies of two contemporary historians of science, Derek J. de Solla Price[2] and Thomas H. Kuhn.[3] De Solla Price recognizes science and technology as two distinct human activities. Kuhn emphasizes that science is defined more by its customs than by logic. Science and technology are each carried on by subcultures with their own defining and guiding examples or paradigms.[3] Interaction between the subcultures occurs largely through sharing of techniques and movement of people, as will be illustrated by the development of this chapter. During the period 1940-1945, the war-time urgency of plutonium production for use in nuclear weapons introduced scientists and the techniques of nuclear physics and chemistry into the world of technology. Engineers learned the science and helped create the technology. Some of the scientists returned to scientific pursuits after the war; some became primarily technologists and assisted with the birth of nuclear engineering.

---

[2] Derek J. de Solla Price, "Is Technology Historically Independent of Science? A Study in Statistical Historiography," Technology and Culture VI, No. 41965, pp. 555-568 (1965).

[3] Thomas H. Kuhn, The Structure of Scientific Revolutions, 2nd. ed., University of Chicago Press, Chicago, Ill., 1970.

21.3 The Energetics and Products of Fission[†]

Nuclear-fission occurs, for example, following absorption of a neutron by fissile nuclei of uranium (U-233, U-235) and plutonium (Pu-240). A representative fission reaction is the following:

$$\,^{1}_{0}n + \,^{235}_{92}U \rightarrow \,^{236}_{92}U^{*} \rightarrow \,^{A_1}_{Z_1}F_1{}^{*} + \,^{A_2}_{Z_2}F_2{}^{*} + \nu\,^{1}_{0}n + Q. \qquad (21.3\text{-}1)$$

The meaning of the symbols appearing in Eq. (21.3-1) will be made clear in the following discussion. When the uranium-235 nucleus captures a neutron ($\,^{1}_{0}n$), a transition occurs to an excited state of the compound nucleus (an excited nuclear state is identified by an asterisk), which may subsequently release its excitation energy in several ways. The energy acquired by the compound nucleus equals the sum of the kinetic and binding energies of the captured neutron. For heavy nuclei with odd mass numbers (e.g., U-235), the excitation energy associated with neutron capture is greater than the threshold or activation energy for separation into two excited fission products $F_1{}^{*}$ and $F_2{}^{*}$ with atomic numbers Z_1 and Z_2 and mass numbers A_1 and A_2, respectively; at the same time, ν neutrons and the kinetic energy Q are released. The thermal energy Q represents the total thermal energy of all of the products formed by the reaction. The primary fission products lose excitation energy over a considerable period of time by emitting beta (β) and gamma (γ) rays and a few neutrons. A nuclide emitting a beta ray increases its atomic number (or nuclear charge) by one unit and becomes a new element. A fraction α of the excited uranium-236 loses energy by γ-emission to become the relatively stable nucleus U-236. The following

---

[†] See also Chapter 2, Section 2.14, in Volume 1 of this series.

approximate energies in Mev are associated with the reaction products: $Q_1 + Q_2$ = 167 for the fission fragments F_1 and F_2; Q_n = 5 for the neutrons, Q_γ = 13 for gamma rays (photons), Q_β = 7 for beta rays (electrons), and Q_o = 11 for the neutrinos, which are particles having no charge or rest mass. Of the total fission energy of 203 Mev per atom of U-235, that associated with neutrinos escapes completely, leaving 192 Mev to be degraded to heat in the reactor and transferred to useful processes. Since combustion of a carbon atom with oxygen yields about 4 ev, it follows that the relative masses of fossil (C^{12}) and nuclear (from U^{235}) fuels required for equivalent energy production are in the approximate ratio

$$\left[\frac{192 \times 10^6 \text{ (ev/atom)}}{235 \text{ (g/mole)}} \middle/ \frac{4 \text{ (ev/atom)}}{12 \text{ (g/mole)}} \right] = 2.45 \times 10^6 .$$

The sum of the mass numbers of the fission products and of the ν neutrons emitted equals the mass number of the excited nucleus. The change in actual mass times the square of the velocity of light equals the energy released. The potential energy shows a double hump with separation of fragments. Unsymmetrical fission may relate to the potential function. The distribution of yield with mass number is shown in Fig. 21.3-1. The laws of conservation of energy and momentum require that the lighter fragments have greater kinetic energy. On the average, Q_1 = 99 Mev and Q_2 = 68 Mev. Reference to Fig. 21.3-1 suggests the following ordering for the dominant atomic mass numbers of the fission fragments that are expected to be formed: $73 < A_1 < 115 < A_2 < 162$. In the presence of fast neutrons with additional energy, the range of observed fission-product masses is widened above that occurring when only thermal neutrons are present (cf. the two sets of curves shown in Fig. 21.3-1).

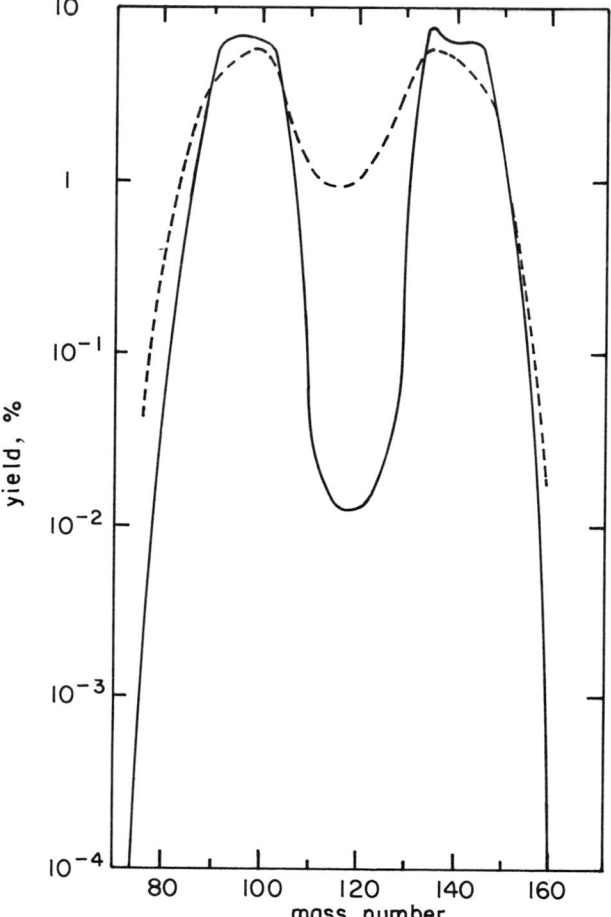

Fig. 21.3-1 Yield of nuclides vs. mass number for fission of U-235 with
          thermal (solid curve) and fast neutrons (dotted curve). Very few
          fissions yield nuclides of equal mass, although increased energy
          added by the neutron increases the yield of symmetrical fission.
          Data are taken from Reactor Physics Constants, Argonne Na-
          tional Laboratory, Report ANL5800 (1963).

21.4 Energy of Prompt Neutrons Produced by Fission

The energy of the neutrons produced by fission ranges from 0.1 to 10 Mev with an average value of about 2 Mev. In the center of mass system, a Maxwell-Boltzmann distribution law probably represents the number of primary neutrons occurring within a given energy interval, thus suggesting that the excitation energy is partitioned over many nucleons. For U-235, the number of neutrons, $dN(E)$, with energy in the interval dE at E is given fairly well by the relation

$$dN(E)/dE = AE^{\frac{1}{2}} \exp(-E/E_o) , \tag{21.4-1}$$

where $E_o = 1.29$ Mev and $A = 0.77 \nu$ (Mev)$^{-3/2}$. This distribution function is plotted in Fig. 21.4-1.

21.5 Neutron Yield

In 1939, Bohr and Fermi[1] suggested that the number of neutrons produced by fission, ν, was greater than 2. This surmise was soon verified experimentally by H. von Halban, F. Joliot and L. Kowarski.[2] Since ν is greater than 2, the number of neutrons produced by fission is sufficient to supply not only the neutrons required to sustain the chain reaction but also to replace the neutrons lost by capture reactions. Of equal importance for the successful im-

---

[1] H. D. Smyth, Atomic Energy for Military Purposes, U.S. Government Printing Office, 1946; this report may be obtained from the National Technical Information Service, Springfield, Va. 22154.

[2] See S. Glasstone, Source Book on Atomic Energy, p. 485, D. Van Nostrand and Co., Inc., Princeton, N.J., 3rd edition, 1967.

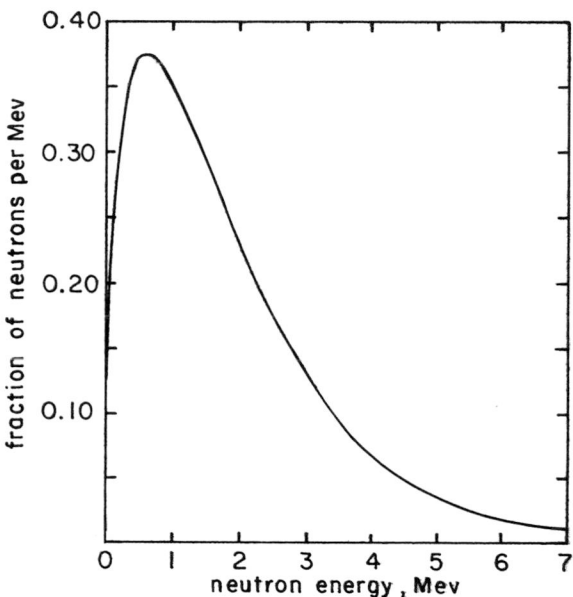

Fig. 21.4-1 A plot of the number of neutrons in an energy interval around E
            emitted during fission caused by thermal neutrons; calculated from
            Eq. (21.4-1). Although the most probable neutron energy is ap-
            proximately 0.6 Mev, the average energy is near 2 Mev.

plementation of controlled nuclear-power production is the emission of a small

fraction, β,* of the neutrons after a delay of several seconds. These delayed

neutrons are emitted by such nuclides as isotopes of bromine and iodine which

carry an excess of neutrons and energy and for which the probability of neutron

emission exceeds that of beta decay. Relevant data, including the yields of neu-

trons, are listed in Table 21.5-1. It should be noted that the constant A in the

distribution function dN(E) for neutron energies [see Eq. (21.4-1)] is determined

by the requirement that the integral of the function over energy equals ν.

---

*The use of β as a symbol for the fraction of delayed neutrons should be dis-
tinguished in context from its use to label electrons (beta particles) emitted
from an excited nucleus.

Table 21.5-1 Neutronic data for fissile and fertile nuclides; reproduced from Ref. [1] in Section 21.7.

Nuclide	Neutron energy	σ_f, barn	α	$\bar{\nu}$	η	β, %	$\bar{\tau}$, sec
U-233	0.253 ev at $\bar{v}=2200m/s$	527 ± 10	0.098 ± 0.002	2.54 ± 0.04	2.51	0.30	16.7
	1 Mev	1.95	0.03	2.998	2.94	- - -	- - -
U-235	0.0253 ev	582	0.19	2.43	2.049	0.75	14.1
	1 Mev	1.25	0.078	2.79	2.60	- -	
U-238	0.0253 ev	- - -	($\sigma_c = 2.73$)	- - -	- - -	- - -	- - -
	2 Mev	0.58		2.6	- -	- -	- - -
Pu-239	0.0253 ev	746	0.38	2.89	2.08	0.23	14.4
	0.3 ev	3300	0.77	2.9	1.64	- - -	- - -
	1 Mev	1.8	0.0267	3.12	3.04	- - -	- - -
Th-232	0.0253	- - -	($\sigma_c = 7.45$)	- - -	- - -	- - -	- - -
	2 Mev	0.010		2.54	- - -	- - -	- - -

Legend: ev = electron volt; $\bar{v}$ = average velocity of neutrons; σ_f = cross section for fission; 1 barn = 10^{-24} cm^2; σ_c = cross section for capture; α = ratio of cross section for capture to that for fission; $\bar{\nu}$ = average yield of neutrons per fission; η = average yield of neutrons per neutron absorbed in the fissile nuclide; β = fraction of delayed neutrons; $\bar{\tau}$ = average or mean half life for emission of de-layed neutrons.

21.6 Breeding and Conversion

Of essential importance to the scientific and engineering feasibility of obtaining useful fission energy are the reactions of neutrons with fertile nuclides. These are U-238 [half life for α emission $= t_{\frac{1}{2}}(\alpha) = 4.51 \times 10^6$ y] which constitutes 99.3% of natural uranium, and Th-232 [$t_{\frac{1}{2}} \simeq 1.41 \times 10^{10}$ y] which is the only thorium isotope sufficiently stable to exist in nature. The reactions involved and the transmutation of the products by beta emission may be written as follows:

$$^{238}_{92}U + n \rightarrow {}^{239}_{92}U \xrightarrow[2.35 \text{ min}]{\beta} {}^{239}_{93}Np \xrightarrow[2.3 \text{ d}]{\beta} {}^{239}_{94}Pu \ [t_{\frac{1}{2}}(\alpha) = 2.4 \times 10^4 y], \quad (21.6\text{-}1)$$

$$^{232}_{90}Th + n \rightarrow {}^{233}_{90}Th \xrightarrow[22 \text{ min}]{\beta} {}^{233}_{91}Pa \xrightarrow[27 \text{ d}]{\beta} {}^{233}_{92}U \ [t_{\frac{1}{2}}(\alpha) = 1.62 \times 10^5 y]. \quad (21.6\text{-}2)$$

There are many additional capture reactions that must be considered in calculating neutron and other material balances in nuclear reactors. The reactions listed in Eqs. (21.6-1) and (21.6-2) permit the production, called conversion or breeding, of valuable fissile nuclides that may be used to fuel nuclear reactors. The term conversion refers to the production of another fissile material, e.g., Pu, by the fission of U; the term breeding refers to producing more fissile material by neutrons from its own fission.

21.7 Basic Neutronic Data

Probabilities, yields and time constants for nuclear reactions are among the basic data that are used in the design and performance analysis of a nuclear reactor. The cross sections for neutron reactions depend on neutron energy or velocity. Some useful data are listed in Table 21.5-1. The total cross section for all neutron-nucleus interactions is labeled σ_T and is made up of cross sections giving the probability of scattering (σ_s), capture (σ_c or σ_a or σ_γ), fission (σ_f), and other possible processes. A group at the Brookhaven National Laboratory measures, collects, and systematizes (using applicable nuclear theory and statistical analysis) and, in collaboration with user groups, maintains a library of standardized cross sections and other nuclear data. This information is available in tabular and computerized form for application in nuclear reactor calculations. The former AEC publication[1] Neutron Cross Sections contains some of this information and is often referred to as the "Barn Book." The curves in Fig. 21.7-1 are taken from this reference and show how σ_f and α vary with neutron energy for $_{94}Pu^{239}$. Since $\eta = \nu/(1+\alpha)$, we note that the neutron yield per neutron absorbed varies strongly with energy. The yield of neutrons is most favorable for either low energy, slow (thermal) or for high energy, fast neutrons. For these reasons, practical reactors fall into two classes, thermal (neutron) and fast (neutron) reactors.

21.8 Neutron Scattering and Slowing Down

The cross sections for the fission of U-233, U-235 and Pu-239 are relatively large for neutrons having energies in the range of molecular energies at room temperature. For this reason, assemblies of fissile and moderating materials in which neutrons are slowed with relatively little capture by collision with nuclei of a moderator, such as water or graphite, require a relatively small (critical) mass of fissile material for maintaining the chain reaction. A

---

[1] D. J. Hughes et al, Neutron Cross Sections, BNL325, supplements and second edition, U.S. Atomic Energy Commission, 1960. Available from National Technical Information Service, Springfield, Va., 22151.

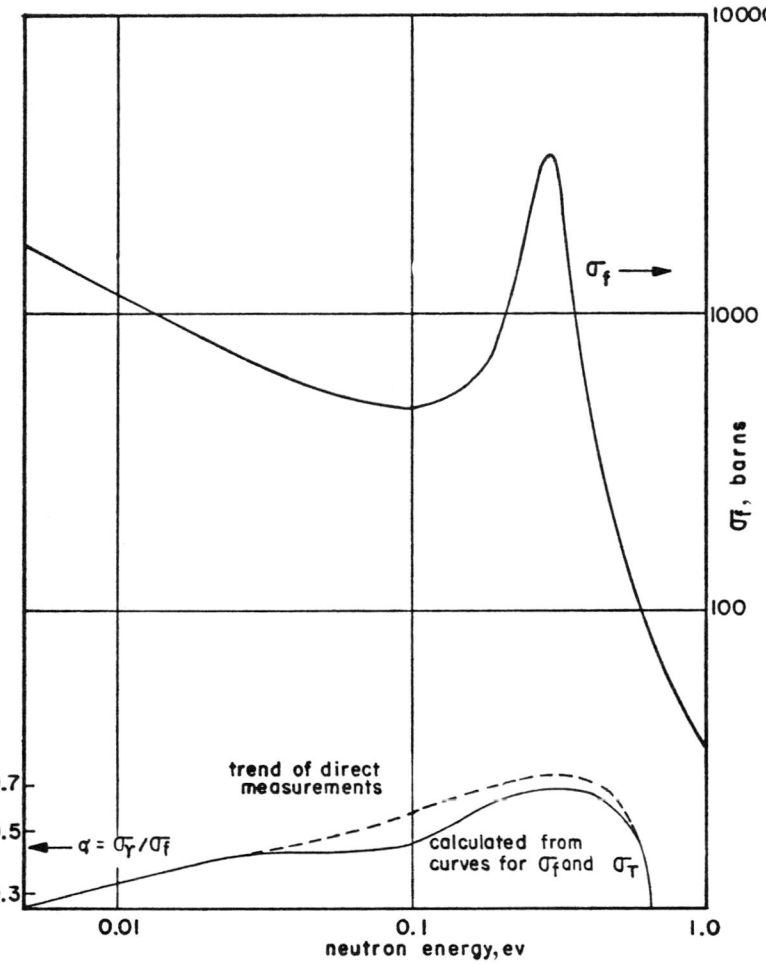

Fig. 21.7-1 A plot representing experimental determinations of the total cross
section for a neutron of given energy with Pu-239, showing resonance absorption at 0.270 ev. Near this resonance, the ratio of
capture to fission becomes relatively large, as is shown in the
lower curve, resulting in a low yield of neutrons per neutron absorbed. This loss of neutrons limits the utilization of plutonium
and uranium in thermal neutron reactors. U-233 is a better fuel
for such reactors. The curves are adapted from Ref. [1] of Section 21.8.

figure of merit for a moderator is given by the product of the scattering cross section (σ_s) and the average of the logarithmic energy loss per collision (ξ), divided by the absorption cross section. Some of these quantities are listed in Table 21.8-1 for useful moderators. Also tabulated are data on the physical and chemical effects resulting from the displacement of atoms by fast neutrons.

In calculating the neutronic behavior of assemblies, it is convenient to introduce a quantity called the macroscopic cross section (Σ), which is defined as the cross section per atom (σ) times the number of nuclei of a given species per cm^3. Although this quantity may be thought of as the cross section of 1 cm^3 containing the nuclei in question, it is more appropriately considered to be a reciprocal mean free path, as shown below, or average distance traveled by a neutron between reactions in the medium in question. Since the probability of a neutron traveling a distance x in a medium having N atoms per cm^3, without being scattered, may be written as $p(x) = A \exp(-xN\sigma_s) = A \exp(-x\Sigma_s)$, where A is a normalization constant, then the mean free path may be expressed as follows:

$$\frac{\int_0^\infty xP(x)dx}{\int_0^x P(x)dx} = \frac{A \int_0^\infty xe^{-x\Sigma_s}dx}{A \int_0^\infty e^{-x\Sigma_s}dx} = \Sigma_s^{-1} \; .$$

Thus, Σ_s^{-1} is a mean free path. The units of Σ_s are $(cm^2/atom) \times (atom/cm^3)$ = cm^{-1}.

Thermal neutron reactors are reactors in which neutrons are slowed and moderated to thermal energy by the moderator and have a Maxwellian velocity spectrum of the type illustrated in Fig. 21.8-1. In fast (neutron) reactors, neutrons are slowed down as little as possible, although they lose energy by

Table 21.8-1 Properties of moderators that are used or that have been extensively evaluated.

Material	The number of moderating (lighter) atoms per cm^3 at room temperature	σ_s, barns	σ_a, barn	ξ	$\Sigma_s \xi$ (slowing-down power)	$\Sigma_s \xi / \Sigma_a$ (moderating ratio)	Temperatures (°C) at melting / boiling / critical conditions	Comments
H_2O	3.3×10^{22}	44.4 (5 at 1 Mev)	0.66	0.927	1.36	70	0 / 100 / 374	reversible radiation induced decomposition into H_2 and $\frac{1}{2}O_2$ easily tolerated in power reactors.
D_2O	3.3	10.5	0.0011	0.510	0.18	2.1×10^4	0 / 100 / 374	reversible radiation induced decomposition into H_2 and $\frac{1}{2}O_2$ easily tolerated in power reactors.
$C_{18}H_{14}$ (terphenyl)	3.4	25	0.33	0.9	0.715	62	30 to 80 / 315 / ---	irreversible radiation and thermally induced decomposition tolerable below 400°C but requires purification and make-up.
$ZrH_{1.6}$	3.6	44.4	0.67	0 if E<0.14 ev; 1 if E>0.14	1.47	---	--- / --- / ---	hydrogen escape is tolerable below 650°C; decomposition pressure is 1 atm at approximately 850°C
Be	1.24	6.1	0.009	0.207	0.16	150	1350 / --- / ---	the reaction n + ^{7}Be → 2He leads to gas bubbles and swelling; atom displacements by fast neutrons results in anisotropic crystallite growth and cracking.
BeO	7.5	9.9	0.009	0.2	0.16	183	2350 / --- / ---	changes similar to those in Be limit usefulness to fluence < 10^{22} n/cm^2 at temperatures between 800 and 1000°C. Replacement too expensive for commercial reactors.
Graphite	11.3	4.7	0.0045	0.158	0.063	170	--- / 3905* / ---	anisotropic distortion of crystallites leads to distortion and cracking. Useful up to fluence of 2×10^{22} n/cm^2 in temperature range 600 to 1400°C. Replacement costs tolerable in commercial reactors.

Legend: σ_s = scattering cross section; σ_a = absorption cross section; ξ = logarithm of the fractional energy loss per collision; Σ_s = σ_s multiplied by σ_s, the number of atoms of scatterer per cm^3 multiplied by σ_s; Σ_a = the number of atoms of absorber per cm^3 multiplied by σ_a; σ_s = the number of atoms of

*The vapor pressure reaches one atmosphere above the solid at the indicated temperature.

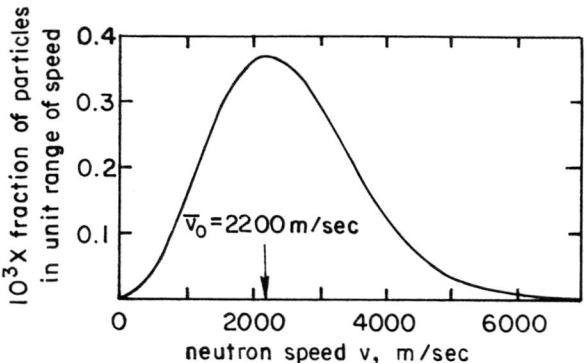

Fig. 21.8-1 Plot of a Maxwellian distribution of numbers of neutrons having velocity in an interval dv at v for temperatures $298.15^{\circ}K$ ($20^{\circ}C$).

elastic and inelastic collisions; they are absorbed while making random collisions that may reduce their energies to a few kilovolts.

21.9 Nuclear Fission Reactors

Fissile, fertile, structural, cooling, scattering, or moderating, control and shielding materials may be arranged in a very large number of ways to allow fission, breeding, transfer of heat, control, and safe operation of nuclear reactors for the production of power and nuclides. Design and design analysis, the heart of engineering, encompass the ways of conceiving, optimizing, constructing, and operating useful systems in terms of sets of explicit requirements or criteria. We may gain appreciation for this field of engineering by inspecting a number of the design concepts that are now in use or under development. Some general considerations and historical highlights will add to our perspective.

21.10 Early Reactors

The first chain-reacting pile was assembled by Fermi, Zinn and many coworkers in the West Stands of the University of Chicago athletic field during the late Fall of 1942. It was stacked with purified graphite blocks and four-inch diameter right cylinders of purified UO_2 and uranium metal arranged on a lattice. It had no provisions for cooling. Today we would call a construction of this type a critical assembly. Many of the basic parameters required for quantitative calculations of the neutronic properties of this and other assemblies were determined by using neutron counters and measuring the radioactivity induced in metal-foil neutron absorbers, the distribution of neutron flux, the rate of increase or decrease of flux, and the changes in "reactivity" caused by introducing materials into the assembly. Measurements to determine the operating characteristics of nuclear reactors are performed to this day for each reactor more or less in this manner.

The second reactor was put in operation at what is now the Oak Ridge National Laboratory in 1943. It consisted of a pile of four-inch square graphite logs. Every other graphite log had a channel bearing a uranium metal slug that was canned in aluminum, over which passed a stream of air moved by downstream blowers. This so-called X-10 pile produced a few megawatts of heat and enough Pu for a determination of chemical properties. Extensive experiments (e.g., measurements of temperature and of power coefficients of reactivity) were made to define the performance characteristics of heat-producing assemblies. During 1944, Zinn and coworkers at the University of Chicago put into operation a low power, D_2O moderated, natural uranium metal reactor at the site in the Argonne Forest of Cook County, Illinois, where the Argonne National Laboratory is now located.

The du Pont Company began construction of reactors for the production of plutonium for nuclear weapons in 1943 at the Hanford site on the Columbia River in the State of Washington. This work was performed under contract to the Manhattan District of the U.S. Corps of Engineers. The first production reactor was started in September 1944. By the summer of 1945, these reactors produced enough plutonium for experiments on weapon design and for the atomic bomb used at Hiroshima, Japan. After the war, the production of Pu was expanded at Hanford, Washington and, through the construction and operation of D_2O-moderated reactors, at a Savannah River site in Georgia. Many tons of weapons-grade plutonium have been produced.

During World War II, several European, British and Canadian scientists and engineers started the construction of a D_2O-moderated natural uranium metal reactor at Chalk River, Ontario, Canada. This facility was first operated in 1947 and has provided neutron flux ranging to 10^{14} n/cm^2-sec for many experiments and tests with reactor components. A D_2O-cooled critical assembly, or zero-energy, experimental pile, ZEEP, represented an early phase in this project. Dramatic early World-War II history is associated with the movement of several scientists and a store of D_2O from Europe to Canada. [1]

For some years, fission-power technology followed different routes in the USA, UK, and France. Now there is considerable convergence in technology. At the close of World War II, in spite of pressures to produce nuclear weapons, many of those involved in the development of nuclear reactors in the USA wanted to emphasize instead the development of nuclear-energy technology for peaceful applications. These research workers pursued significant developments in the war-time laboratories that became known as the Argonne National Laboratory (ANL) near Chicago, Illinois, the Oak Ridge National Laboratory

[1]Margaret Gower, _Independence and Deterrence_, Volume II, MacMillan Co., Cambridge, England (1974).

(ORNL) at Oak Ridge, Tennessee, and the Los Alamos Scientific Laboratory (LASL) at Los Alamos, New Mexico. Studies were also initiated at industrial laboratories, especially at the General Electric Company, the Westinghouse Electric Corporation, and the North American Aviation Company. Most of the basic ideas now used or under development in reactor designs were formulated at the National Laboratories between 1942 and 1950 and are named in the next section. A listing of important reactors and their design characteristics is given in Table 21.10-1.

21.11 Power Reactor Concepts

The gas-cooled reactors built and operated in the UK produced the first commercial electrical power. The British chose to build dual-purpose nuclear plants using natural uranium from which they obtained plutonium for weapons as well as heat to generate electrical power.

In the U.S., nuclear reactors were used only to produce weapons-grade plutonium during the early post-war years. A large capacity to enrich the U-235 content of uranium by gaseous diffusion was also created. Shortly after the war, a major military effort on nuclear reactors for submarine propulsion was begun. Thus, while national defense programs diverted effort from the development of commercial nuclear-powered systems, these defense priorities led to commercialization of water-cooled reactors with enriched uranium fuel.

In France, gas-cooled natural uranium reactors received early emphasis, whereas the principal development effort has been rededicated to the development of fast breeder reactors since the early sixties.

At the 1955 Atoms for Peace Conference, at which most countries unveiled their fission-reactor developments, Russian representatives described a water-cooled, high-pressure, graphite-moderated, enriched uranium-alloy-

Table 21.10-1 A condensed tabulation of important reactor designs.

Name Location	Purpose	Fissile Fuel Fertile Fuel	Fuel Compound	Moderator-Reflector	Coolant and Temperature	Cladding	Vessel and Primary Circuit	Control	Shield
Hanford Washington, USA	Pu production for weapons	U-235 (0.7115%) U-238	U metal (0.1% Fe, Al)	graphite	H_2O at <90°C	Al	Al tubes, steel pipes	boron-steel	graphite, steel, cellulose
Savannah River Georgia, USA	Pu production for weapons	U-235 (0.7115%) U-238	U metal (0.1% Fe, Al)	D_2O	H_2O at <90°C	Al, Zircaloy	Al tubes, steel pipes	boron-steel	
Magnox UK and France	Pu and power	U-235 (0.7115%) U-238	U metal (0.1% Fe, Al)	graphite	CO_2 at 400°C	alloy of Mg-Al-Be (magnox)	steel	boron-steel	concrete
Candu Canada, India	power	U-235 (0.7115%) U-238	U metal (0.1% Fe, Al)	D_2O	H_2O at 300°C, terphenyl at 400°C	zircaloy	Al calandria, zircaloy tubes, steel piping	boron-steel	concrete
PWR	power	U-235 (1 to 4%) U-238	UO_2	H_2O (100 atmos.)	H_2O at 300°C	zircaloy	steel, with stainless liner	Ag-In-Cd or B_4C in steel tubes	H_2O and steel
BWR		U-235 (1 to 4%) U-238	UO_2	H_2O (67 atmos.)	H_2O at 285°C	zircaloy	steel, with stainless liner	B_4C in steel tubes	H_2O and steel
HTGR, USA HTR, UK AVR, Germany	power and heat	U-235 (1%)* Th-232 (4.5%)	UC_2 ThC_2	graphite	He at 50-100 atmos. and 700-900°C	pyrocarbon and SiC	prestressed concrete, steel lined	B_4C in inconel	graphite, concrete
LMFBR, USA PFR, UK and Phenix France BR, Russia	power Pu	Pu (10 to 30%) U-238	U, PuO_2	none	Na at 600°C	AISI-316 (stainless steel)	AISI-316 steel vessels and piping for sodium	B_4C in AISI-316	steel, concrete
GCFR, USA Germany	power U-233, Pu	Pu (10 to 30%) Th-232 U-238	U, PuO_2	none	He at 100 atmos. and 700°C	AISI-316 (stainless steel)	prestressed concrete, steel lined	B_4C in inconel	steel and graphite
MSBR (E) ORNL	power U-233	U-233 Th-232	U, Li, Be, Zr-fluoride	graphite	Li, Be, ZrF at 600°C		hasteloy-N	B_4C in hasteloy-N	
Naval	propulsion	U-235	Zr-U alloy UO_2	H_2O	H_2O	zircaloy	steel	Hf	H_2O and steel

*Supplied as 93% enriched U.

fueled power plant that had produced some electricity. Since 1955, both water-moderated and fast reactors have been built in the U.S.S.R. In Sweden, heavy- and light-water-moderated reactors have been emphasized.

We shall now present a brief summary of reactor concepts developed at the AEC National Laboratories.

ANL (1) The fast-breeder concept, uranium metal, fuel-cooled with molten sodium; development of the Experimental Breeder Reactors I and II (EBR-I, II) located at the National Reactor Test Site (NRTS) near Idaho Falls, Idaho; the EBR-II is now (1975) operating. (2) The pressurized-water-reactor concept; this technology was transferred to the Westinghouse Naval Reactor Division and other companies beginning in 1946 and led to the commercial PWR. (3) The boiling water reactor concept was tested in 1953 in EBWR and BORAX (transient) at the NRTS; the technology was further developed by the General Electric Atomic Product Division and other companies.

ORNL (1) Research and testing reactors of the swimming-pool type; enriched uranium-aluminum alloy plates clad with aluminum and water-cooled with natural or forced convection; marketed as research reactors by several companies. (2) Partial development of a helium-cooled, beryllium-oxide-moderated, high-temperature power-reactor concept. (3) Aqueous homogeneous breeder and power-reactor concept leading to a homogeneous reactor experiment (HRE) in 1957; uranyl sulfate dissolved in heavy water and a ThO_2 slurry blanket were tested and found to be infeasible because of corrosion of the zirconium alloy vessel. (4) The molten-salt-breeder reactor (MSBR) was conceived and an experiment initiated (MSRE), which is still in progress.

LASL (1) A homogeneous experimental and research reactor "Water Boiler" was designed to use uranyl nitrate in light water; this reactor has a small critical mass and has been marketed as a research reactor by the North American Aviation Company. (2) Small fast reactors with plutonium alloy fuel and mercury cooling.   (3) The molten plutonium (alloy) fast-reactor concept was evolved and experimental work done (LAMPRE).

BNL   The molten bismuth and uranium alloy reactor-concepts were developed.

We shall confine our attention to the five important U.S. reactor types, as follows: light-water-moderated and cooled reactors (LWR) including the pressurized-water reactor (PWR) and the boiling-water reactor (BWR); high temperature gas(helium)-cooled, graphite-moderated reactors (HTGR); fast neutron-breeder reactors (FBR) including the liquid metal (sodium) cooled (LMFBR) and the gas(helium)-cooled [GCF(B)R].

Important omissions include heavy-water-moderated, light-water-cooled, natural uranium reactors of the type developed and used in Canada, India, Argentina, and elsewhere; molten-salt-fueled and cooled, graphite-moderated, Th-U-233 cycle breeders (MSBR) under development at ORNL; light-water-moderated and cooled thorium-blanket breeders (LWBR) under development by the Westinghouse Company; CO_2-cooled, graphite-moderated reactors.

21.12   The Origin of Light-Water-Moderated Nuclear Power Reactors (LWR)

Water-cooled and -moderated reactors have become the principal nuclear source of heat and steam for central-station use in commercial electrical generating plants of the U.S.   Experience with production reactors has shown that clean water will provide the fairly high required rates of heat transfer. For naval propulsion, both water and sodium systems have been developed to transfer heat to steam generators.   However, it has proved to be difficult to make reliable sodium-to-water heat exchangers or steam generators.   Unusually detailed care is demanded in the manufacture of this kind of equipment to avoid small leaks and consequent severe corrosion and embrittlement of the materials of construction by sodium hydroxide. By contrast, dependable water-to-water heat exchangers or steam generators are much more easily made using well known and carefully controlled manufacturing methods and tests. Thus, pressurized water-moderated and cooled reactors have proved to be suitable as steam supplies for submarine propulsion. In large measure, this technology has determined the early and continuing path followed in the development of commercial reactors.

During the period 1945-1953, the National Laboratories were operated by contractors for the AEC and a few industrial laboratories became more active in proposing novel concepts (e.g., some of those listed in Table 21.10-1) for the development of nuclear steam supplies for commercial application. The utility of pressurized-water technology, coupled with the success achieved by Admiral Rickover in managing the development of a nuclear navy since about 1954, led to the construction of a prototype nuclear power plant by the Westinghouse Electric Company at Shippingport, Pennsylvania. The Duquesne Power and Light Company operated the plant, which began generating about 60 megawatts of electrical power in 1961. During 1964, the power level was increased to about 150 Mw_e. This type of reactor is described in detail below and became the commercial reactor of the Westinghouse Electric Company. The first commercial demonstration plant was the Yankee plant at Rowe, Massachusetts.

In the 1952 to 1954 time period, the General Electric Company made its entry into the commercial nuclear energy business with a different type of light-water reactor. Also cooled and moderated with water, the GE reactor allows water to boil at the top of the nuclear core. After passing through a separating system, the dry steam is led directly to a turbine. This boiling water reactor (BWR), unlike the PWR, does not interpose a heat-transfer loop between the reactor and the steam-turbine loop. We shall now examine the designs of the PWR and BWR nuclear-power systems in greater detail.

21.13 Pressurized Water Reactors (PWR)

At the end of 1974, there were approximately 24 PWRs producing electricity and another 100 under construction or on order. The thermal energy output of single reactors has increased from 392 megawatts (Mw_t) by the Yankee reactor at Rowe, Massachusetts, in 1958 to approximately 4000 Mw_t for the plants that are currently on the drawing boards. A single large unit will furnish about 1300 Mw_e, which is enough electrical power for 0.7 to 1.3×10^6 people.

The principal reactor components are illustrated in the following figures. Figure 21.13-1 shows schematically the essential components of the primary and secondary reactor loops, while Fig. 21.13-2 is a somewhat more detailed drawing of the cooling system. Water at approximately 146.7 atm (2200 psi) is circulated through a loop, one leg of which is the reactor, while the second is a steam generator. Coolant temperatures range from near $280^{\circ}C$ at the reactor inlet to approximately $320^{\circ}C$ at the outlet. Steam is raised in the second loop at 37 to 40 atm (500 to 600 psi) and 260-280$^{\circ}C$. Some of the energy content of the steam is converted to mechanical work in a turbine which drives an electrical generator. An efficient design optimizes energy utilization subject to requirements and constraints which include the following: the amount and rate of electrical energy production; control of output according to demand; safety and health of operators and of the public; cost of capital, operations, maintenance and fuel processing; impact on the environment; and other parameters. In the process of design, the engineer seeks to optimize some objective value or set of values. Often, too few factors are included among the values considered. Hopefully, we will learn to include in the optimization all pertinent conditions for human welfare.

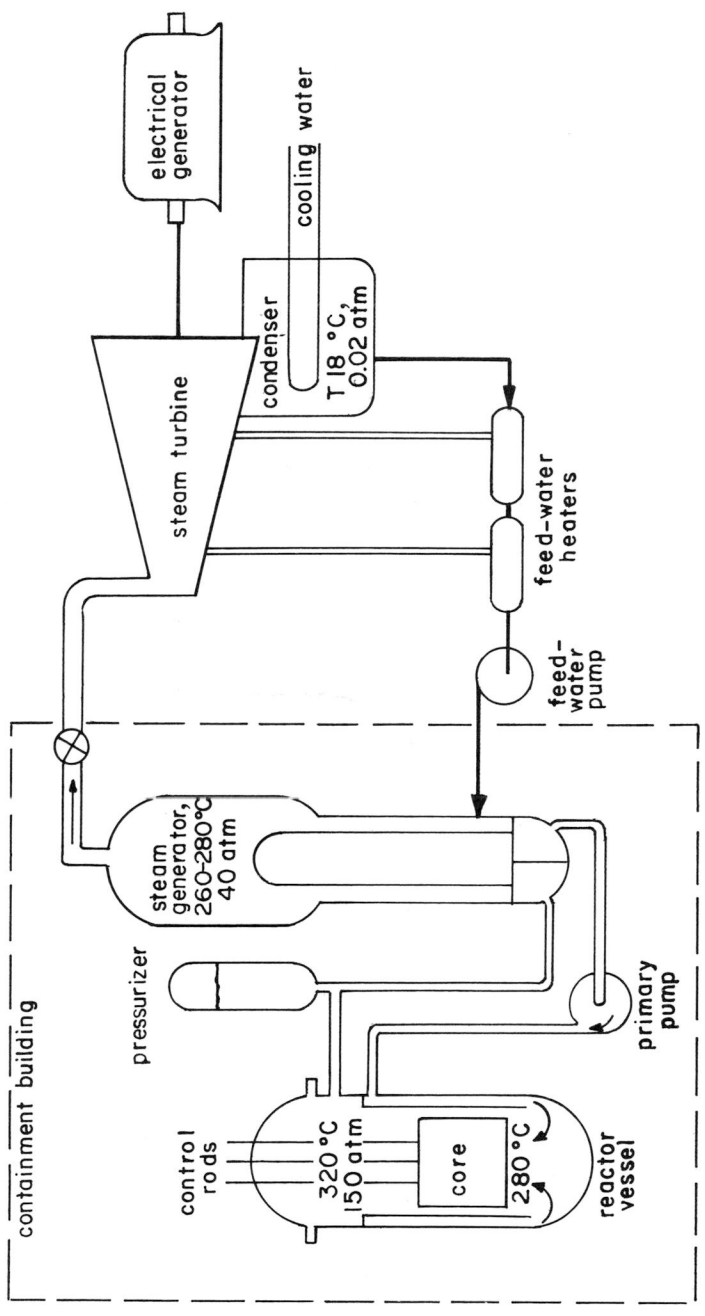

Fig. 21.13-1 Representation of the essential elements of the cooling- and power-generating systems of a pressurized-water reactor; operating temperatures and pressures are indicated.

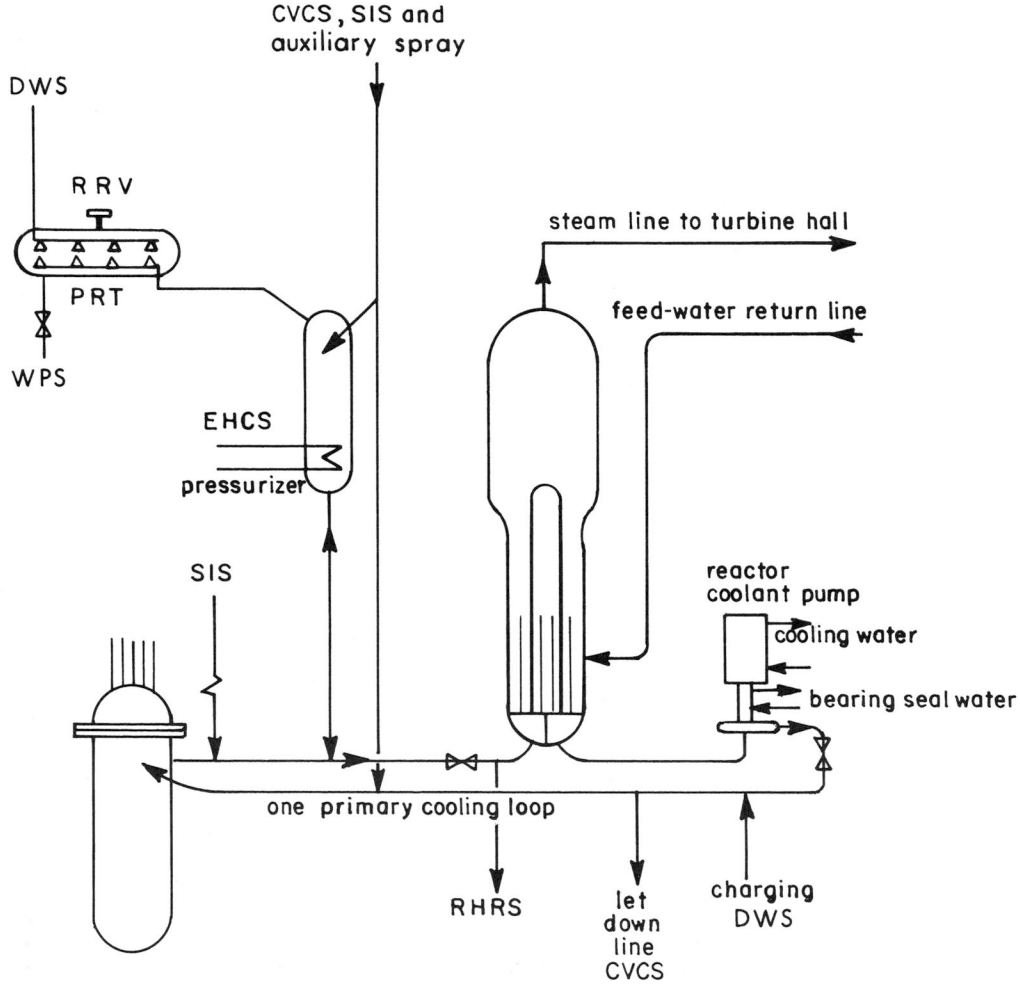

Fig. 21.13-2 Schematic diagram showing a few essential features of the pressurized-water primary cooling system of the PWR. Explanation of symbols: CVCS = chemical and volume control system, SIS = safety injection system, RHRS = residual heat removal system, WPS = waste processing system, DWS = demineralized water supply, EHCS = electrical heating control system, PRT = pressure relief tank, RRV = rupture relief valve. Many valves and control loops are omitted.

A. General Arrangement of a PWR Nuclear Steam-Supply System

Figure 21.13-3 shows some reactor details. The huge steel pressure vessel is nearly 10 m (30 feet) high, over 3 m (10 feet) in internal diameter and has walls that are 20 to 25 cm (8 to 10 in.) thick. The centrally located reactor core, where most of the nuclear reactions take place, is 2 to 4 m (6 to 12 ft) in diameter and of comparable height. Control rods are inserted into the reactor from above and are moved up and down by electric motors mounted at the top closure. Baffles direct the incoming cooling water (which enters just above the core) downward, thus keeping the vessel at the inlet temperature. Below the baffles, the cooling water enters the core, flowing upward inside a core barrel that, just above the top of the core, connects with outlet nozzles from which the heated water flows to the steam generators.

Figure 21.13-4 shows a plan view indicating the radial locations of essential components. Water in the core scatters and slows the neutrons which activate the fission reaction. The water surrounding the core also scatters or reflects neutrons, thereby reducing leakage from the core. The core barrel serves to confine the upward flow of heated water. A thick, steel thermal shield surrounding the barrel absorbs gamma radiation and some neutrons. The downward flowing water removes gamma-ray heat from the shield and cools the pressure vessel. Some neutrons will reach the pressure vessel, thereby altering its mechanical properties.

Figure 21.13-5 indicates the layout of the steam-raising system or nuclear steam-supply system (NSSS). It also shows such other essential components of the cooling system as the pressurizers. These pressurizers are electrically-heated steam producers. By controlling the electrical power input, the pressure in the constant volume system may be regulated in spite of

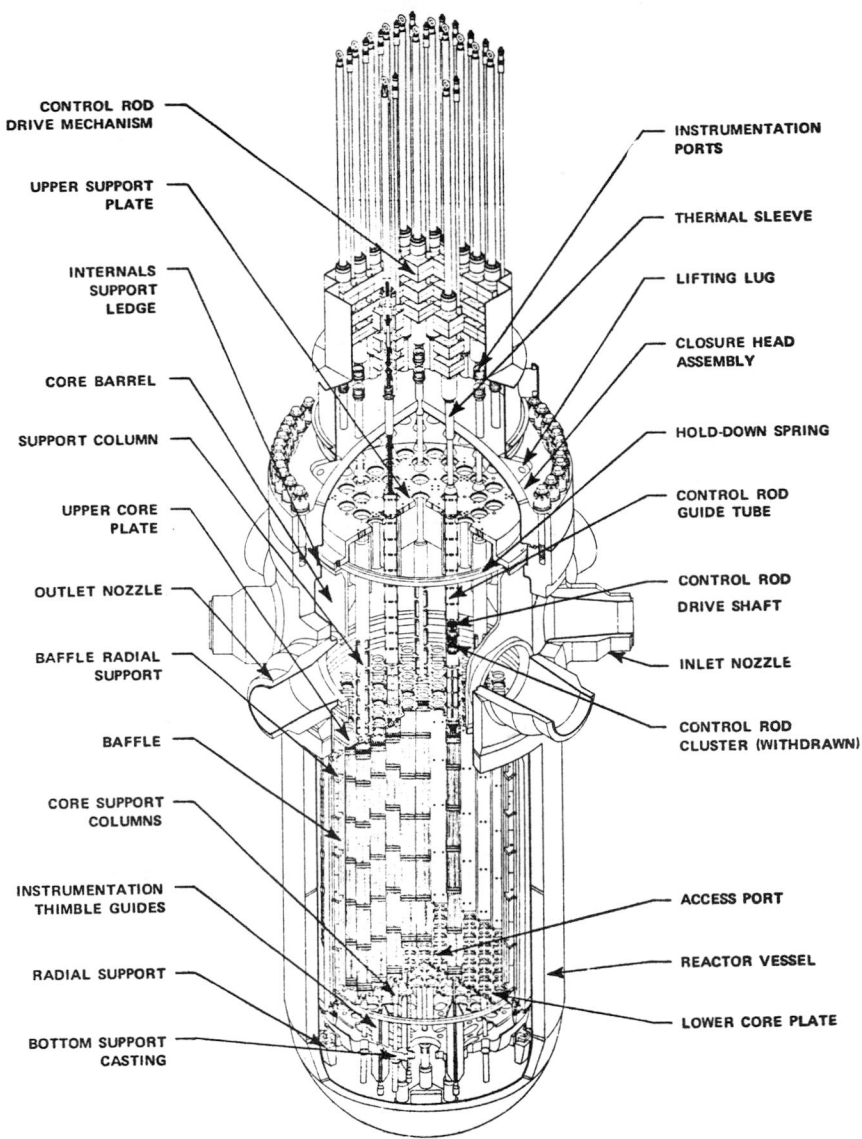

CONTROL ROD
DRIVE MECHANISM

UPPER SUPPORT
PLATE

INTERNALS
SUPPORT
LEDGE

CORE BARREL

SUPPORT COLUMN

UPPER CORE
PLATE

OUTLET NOZZLE

BAFFLE RADIAL
SUPPORT

BAFFLE

CORE SUPPORT
COLUMNS

INSTRUMENTATION
THIMBLE GUIDES

RADIAL SUPPORT

BOTTOM SUPPORT
CASTING

INSTRUMENTATION
PORTS

THERMAL SLEEVE

LIFTING LUG

CLOSURE HEAD
ASSEMBLY

HOLD-DOWN SPRING

CONTROL ROD
GUIDE TUBE

CONTROL ROD
DRIVE SHAFT

INLET NOZZLE

CONTROL ROD
CLUSTER (WITHDRAWN)

ACCESS PORT

REACTOR VESSEL

LOWER CORE PLATE

Fig. 21.13-3 Cutaway of reactor vessel and internals. The vessel is over
10 m in height and the walls are 20-25 cm thick. Reproduced
with permission from the Westinghouse Electric Corporation.

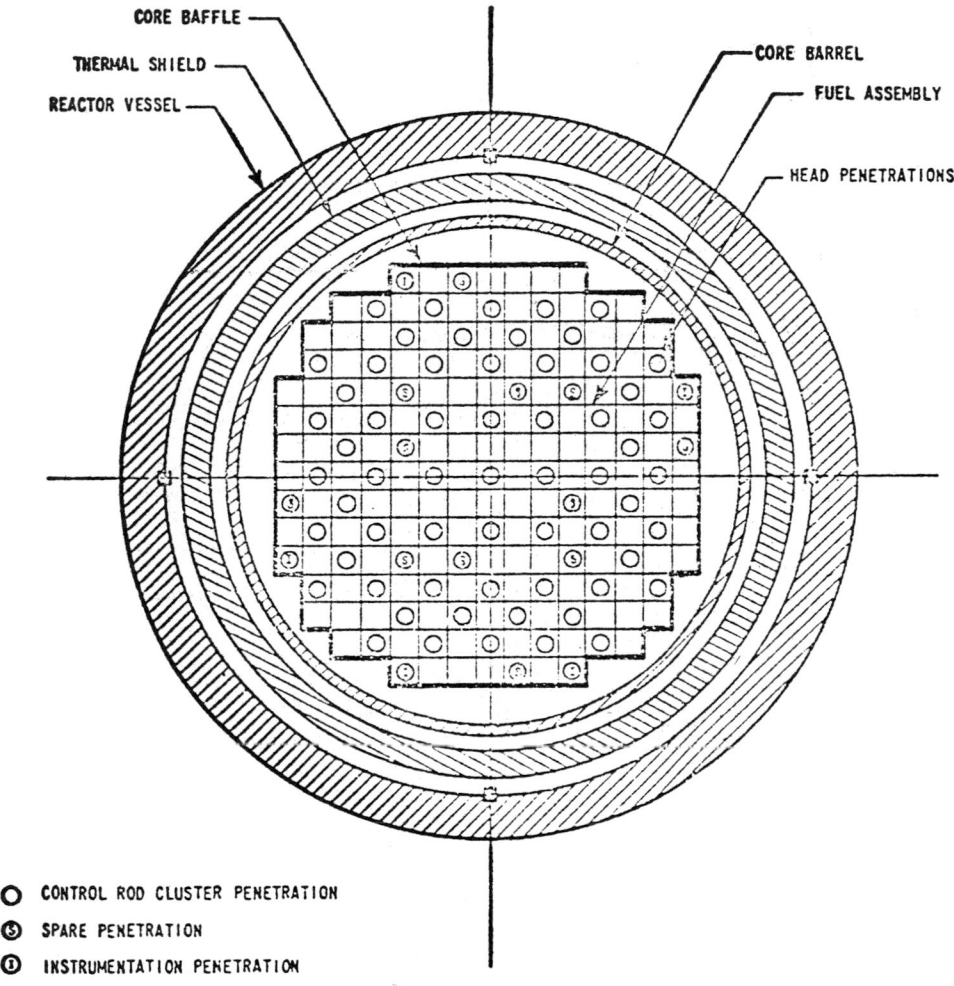

CORE BAFFLE
THERMAL SHIELD
REACTOR VESSEL
CORE BARREL
FUEL ASSEMBLY
HEAD PENETRATIONS

O CONTROL ROD CLUSTER PENETRATION
Ⓢ SPARE PENETRATION
Ⓘ INSTRUMENTATION PENETRATION

Fig. 21.13-4 Cross section of a four-loop PWR core showing 193 fuel assemblies centrally located within the thick walled pressure vessel,
the thermal shield and the core barrel. The barrel supports the
core and separates downward flow of incoming cooling water that
cools the vessel and shield from the upward flow that removes
heat from the core. Reproduced with permission from the
Westinghouse Electric Corporation.

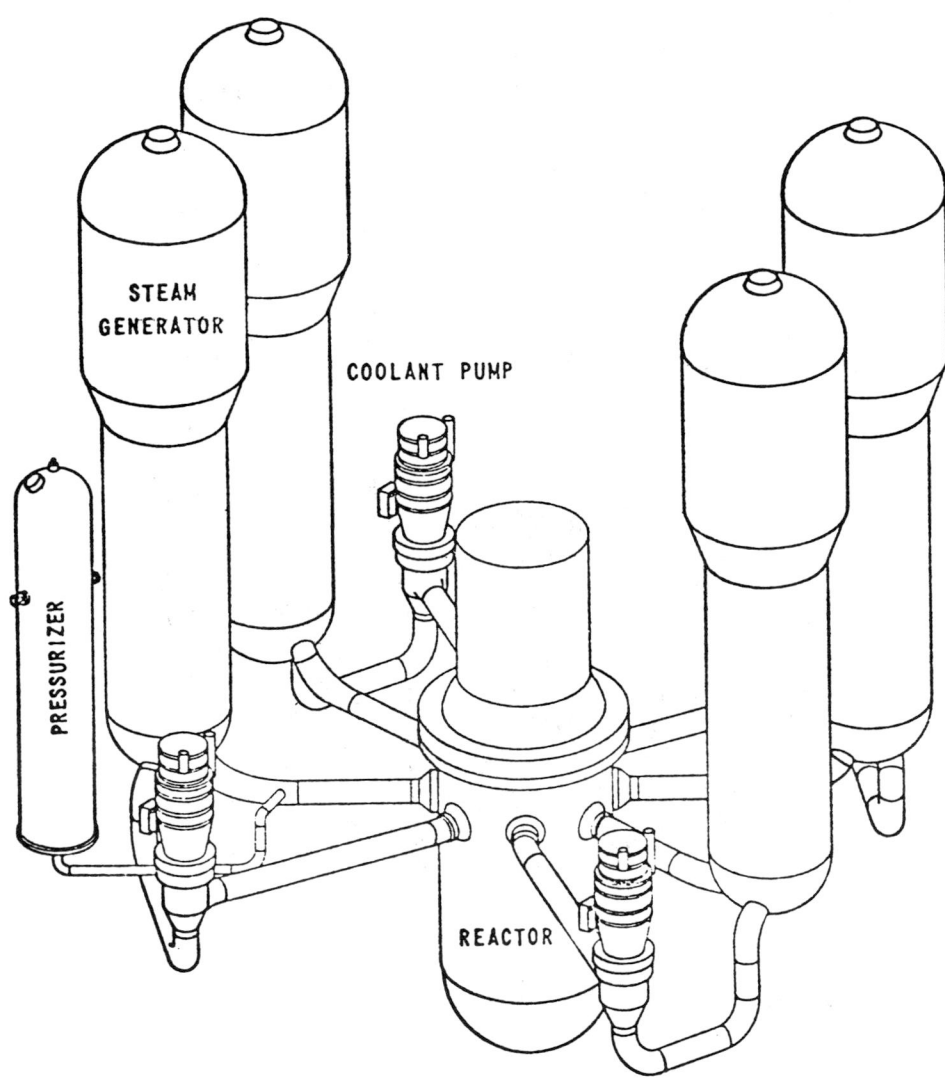

Fig. 21.13-5 A simplified diagram of a four-loop nuclear steam-supply system (NSSS) illustrating the relative sizes and locations of the reactor vessel, the steam generators, the pumps that circulate
pressurized cooling water and the pressurizers. Reproduced
with permission from the Westinghouse Electric Corporation.

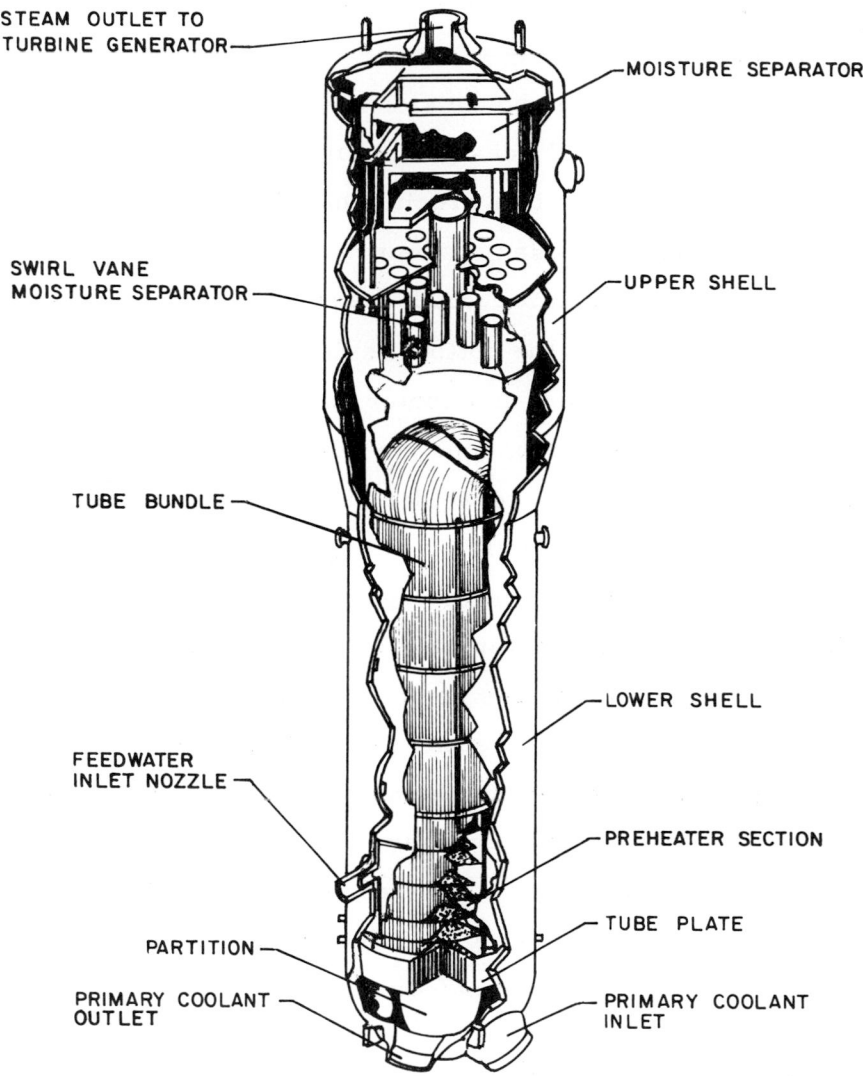

STEAM OUTLET TO
TURBINE GENERATOR

MOISTURE SEPARATOR

SWIRL VANE
MOISTURE SEPARATOR

UPPER SHELL

TUBE BUNDLE

LOWER SHELL

FEEDWATER
INLET NOZZLE

PREHEATER SECTION

PARTITION

TUBE PLATE

PRIMARY COOLANT
OUTLET

PRIMARY COOLANT
INLET

Fig. 21.13-6 Cutaway of a typical steam generator showing several essential
features. The large number of tubes (each making a hairpin loop
from the hot to the cooled sides) is required in order to minimize the temperature drop from the pressurized water to the
generated steam. The large volume at the top of the steam
generator provides the required moisture separation. Reproduced with permission from the Westinghouse Electric Corporation.

temperature fluctuations which occur in the relatively incompressible water elsewhere in the loop.

Figure 21.13-6 indicates how a steam generator is constructed. This equipment must also be carefully designed to minimize inefficiencies in the transfer and use of heat, which is a crucial consideration for the LWR because of temperature limitations imposed by the properties of water and the limiting value of the maximum pressure for which the pressure vessel has been designed. Water from the reactor enters the bottom cavity of the exchanger. A partition divides this space into a hot inlet and cooler outlet sections. A thick, perforated steel disc, called a tube sheet, tops the inlet and outlet cavities. Hairpin-like loops of steel-boiler tubing are welded at each end into the tube sheet and carry the high pressure water, while providing the large surface that is required for the transfer of heat to the steaming region. Feed water is pumped from a condenser through feed-water heaters and enters just above the tube sheet. Above the tubes is a more voluminous region containing swirl-inducing and other baffles and chambers which are carefully placed to rid the steam of water droplets. A steam pipe passes to the turbine and is welded to the nozzle at the top.

Careful and detailed engineering are required to handle both steam and water, provide high efficiency, good control, safety, reliability, proper chemical composition of fluids, control of radio-nuclides and other substances, limit corrosion, and, in general, assure safe and economical plant performance.

B.   Reactor Core and Fuel Elements

As we saw in Figs. 21.13-3 and 21.13-4, the reactor core is made up of square elements that contain and locate the nuclear fuel as required for the neutron chain reaction, for transfer of heat from the fuel to the flowing coolant and for control of nuclear reactions by adjusting the position of neutron absorb-

ers within the core. Qualitatively we may summarize the considerations, discussed in Appendix 21A, that determine size, shape and composition of core
components as follows. The total thermal power required from the core determines the rate of fissioning, the total surface area, the length and hence diameter of the cylindrical rods. The balancing of the rate of production, absorption, diffusion and leakage of neutrons determines relationships between the
number of atoms per unit volume of all constituents, the length and diameter
of the core. Changes in this balance due to changes in temperature and consumption of materials determine the amount of neutron absorbers introduced.
The strength required to maintain the position of all items in the core determines the effective thickness of tubing and other items. Finally, in the process
of design, these several requirements are adjusted to optimize plant efficiency
subject to restraints on cost.

In Fig. 21.13-7, we attempt to show the complete structural details of
the nuclear fuel assemblies. Of 225 positions defined by grid spacers with fif-
teen holes on a side, 205 carry cylindrical fuel rods (see the cutaways in
Fig. 21.13-7). These rods are about 1 cm in diameter and 2 to 4 m long.
Their walls are about 0.05 cm thick and made of zircaloy, which is a zirconi-
um-base alloy containing 2 to 4% Sn and a few tenths of a percent of Nb to control corrosion. The tubes contain up to 10^5 kg of UO_2 that has been formed by
pressing and sintering into right cylindrical pellets having about 10% porosity.
After insertion of the pellets, the tubes are usually filled with He, closed by
welding at each end, and carefully tested for leaks by using mass-spectrometric
helium leak detectors. In the larger reactors, 59 of the 193 fuel assemblies
carry control rods. Five assemblies carry instrumentation sensors. Spare
penetrations are also provided.

Figure 21.13-8 shows four fuel assemblies in plan view, two with control rods, one without, and one with an instrumentation thimble. Fuel elements

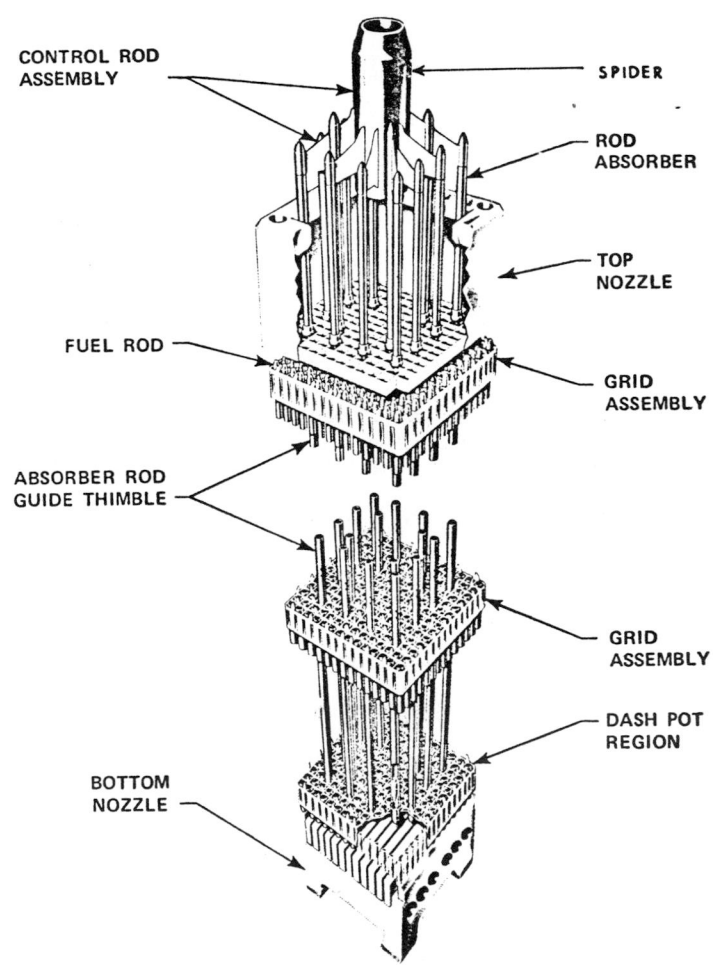

Fig. 21.13-7 A cutaway isometric view of a cluster of fuel and control rods. The cutaway fuel rods reveal the location of the 16 movable control rods that project from the spider and drive rods at the top down through the assembly into individual dashpot snubbers at the bottom. The fuel-rod locations are shown by the stubs in each gridplate. Reproduced with permission from the Westinghouse Electric Corporation.

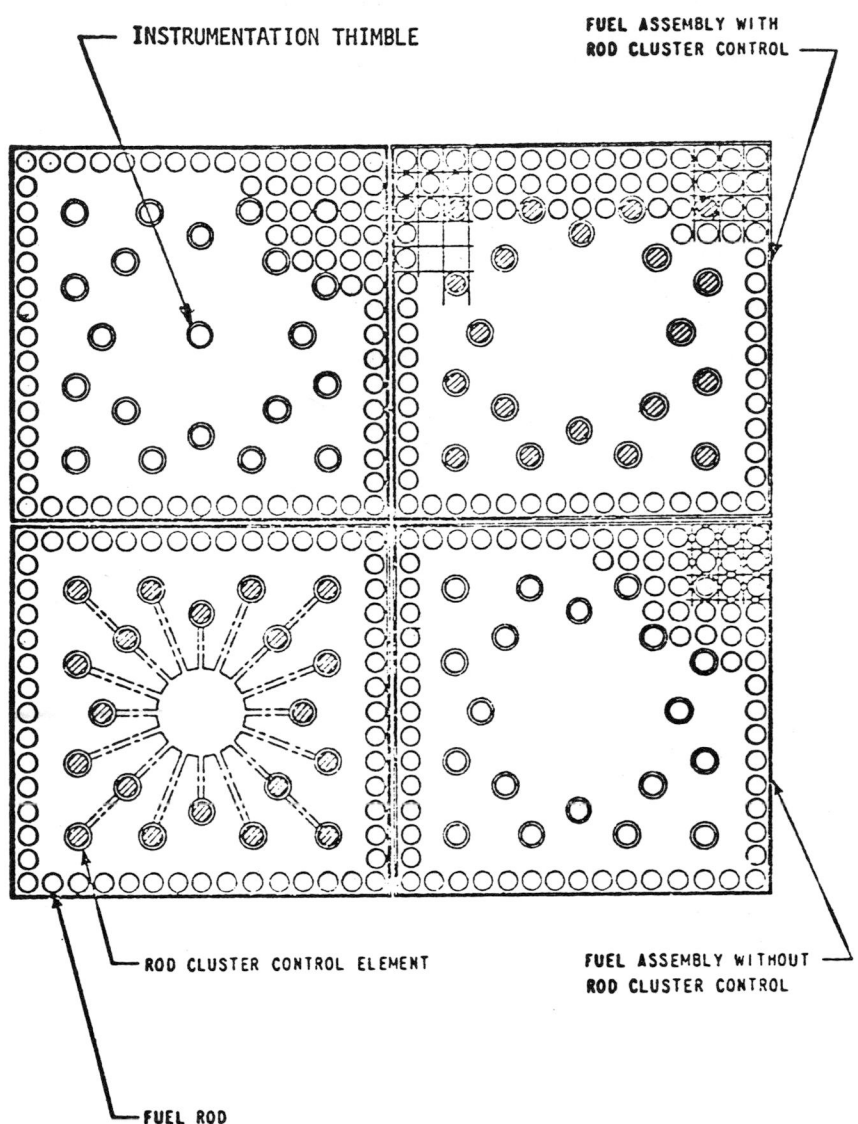

INSTRUMENTATION THIMBLE

FUEL ASSEMBLY WITH
ROD CLUSTER CONTROL

ROD CLUSTER CONTROL ELEMENT

FUEL ASSEMBLY WITHOUT
ROD CLUSTER CONTROL

FUEL ROD

Fig. 21.13-8 A cross section of fuel assemblies with and without control-rod cluster. Most of the fuel rods are omitted. Their locations are clear from the lattice arrangements of the supporting grids. Reproduced with permission from the Westinghouse Electric Corporation.

of this type give the required service, amounting to about 3 years in a reactor. After 2 to 3% of the fuel has been fissioned, the accumulated fission products absorb neutrons, while the remaining fuel no longer contributes enough heat or neutrons for the required level of power.

After removal from the reactor, the fuel elements are cooled under 6 to 10 m of water, which also provides adequate shielding of gamma radiation from fission products. About 90 days later, the elements may be shipped to storage or to a reprocessing plant where fission products are isolated. The management and storage of these radioactive materials will be discussed in Chapter 24.

C. Control Rods

Stainless steel tubes filled with boron carbide or an alloy of silver, indium, and cadmium are neutron absorbers that may be inserted into the appropriate positions pictured in Figs. 21.13-7 and 21.13-8. The upper ends of these rods are attached to a spider and a drive rod and, as already mentioned, are positioned by electric drives mounted just above the pressure vessel. Signals from sensors are processed through computer-logic and decision systems and used to determine the positions and movements of the control rods. Fail-safe devices and systems, redundancy, diversity to prevent common-mode failures, and limitations on rates of change of absorber strengths (rod positions) are rigorously used to prevent error or failure in control or shutdown. For example, if the electrical power fails, all control rods "fall" into the core. Because fresh fuel supplies more neutrons than exposed fuel, a boron-containing salt is dissolved in the cooling water to "poison" or reduce the initial neutron reactivity of the core. This boron "burns out" thus partially compensates for the reactivity changes of the fuel.

D. Reactor Containment Structure and Pressure Suppression;
    Engineered Safety Features (ESS)

Public safety requires that only extremely limited amounts of radioactive materials may be released from a nuclear power plant. To contain accidental releases from the primary circuit, a gas-tight containment building is
placed around the NSSS. In a conservative design, containment must be assured
for the worst imaginable failure that is consistent with physical laws. This situation, which is called a design basis accident (DBA), will be analyzed further
in Chapter 24. Although highly improbable, rupture of the primary pressure
system is taken to be the DBA. Thus, in addition to being gas-tight during normal operation, the building must be capable of containing and condensing, at
less than 3 atm pressure, all of the steam flashed from the 147 atm, 320°C primary system. Figure 21.13-9 shows how steam escaping from the NSSS would
be largely condensed on passage through a heat exchanger containing ice. Alternatively, the steam may be led through pools or sprays of water. This shell
is further protected from outside damage by a substantial concrete building.

E. Emergency Core Cooling System (ECCS)

The accidental rupture of the primary cooling system (i.e., the DBA)
results in loss of both the cooling and moderating functions provided by the confined water. Although fission ceases promptly, the power produced by decaying
fission products is high and corresponds immediately after occurrence of the
DBA to slightly over 10% of operating power, dropping to 8% in 10 seconds and
to 5% in slightly under 40 seconds. To remove this heat, water must again be
forced through the core in a shut-down condition. This prompt cooling function
is provided by high and low pressure injection pumps and by the accumulator
shown in Fig. 21.13-9. Prolonged cooling is provided by elevated water-storage

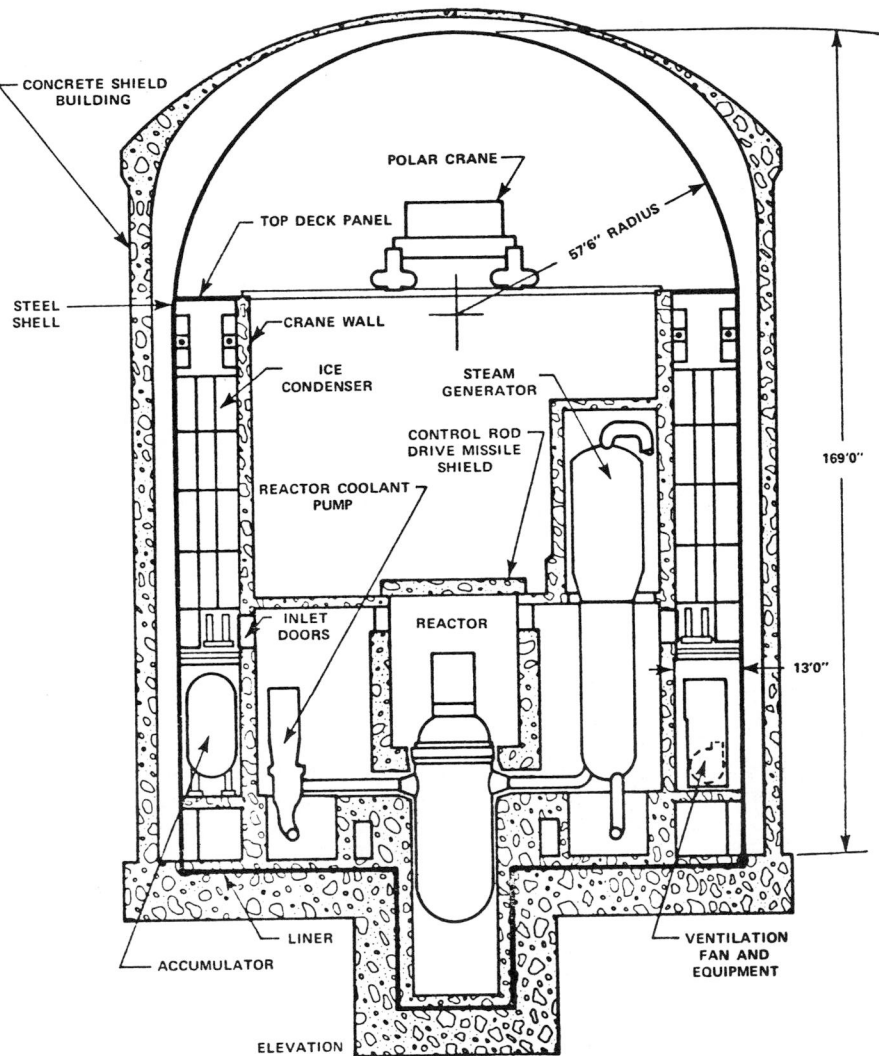

Fig. 21.13-9 A vertical cross section through a PWR containment structure
             showing the location of ice-cooled condensers and passageways
             through which steam would flow if a major break occurred in the
             primary cooling system. This arrangement is one of a few op-
             tions for limiting the pressure inside the containment structure
             in case of an accident. Reproduced with permission from the
             Westinghouse Electric Corporation.

tanks which are not shown in Fig. 21.13-10. The DBA has been assumed to involve a complete break in the piping of the primary loop.

The hypothetical accident and the provisions for emergency core cooling (ECCS) have been the subject of considerable study and debate, including several months of congressional hearings.[1]

The essential technical questions relate to (i) the timing for reinjection of water in view of the fact that, if not cooled, the temperature rises several hundred degrees centigrade in a few seconds; (ii) the reestablishment of normal convective heat transfer from the fuel rods after the film of steam has been swept away from the overheated rods; (iii) the extent of exothermic chemical reaction between zirconium in the cladding and water, which further endangers the integrity of the fuel. Tests on full length bundles of electrically heated rods show that, if water is reinjected within about twenty seconds after the break, normal convective heat transfer will be reestablished. Under these circumstances, the maximum temperature of the cladding does not exceed $1175^{\circ}C$ ($2100^{\circ}F$) and the reaction rate between zirconium and steam does not exceed safe limits.

F. Plant Layout

The balance of the plant outside the containment shell includes the steam turbine, steam-condenser cooling system, return-feed-water pumping and treatment apparatus, control room, fuel-handling and storage facilities, auxiliary

---

[1] Federal Register (36 F.R. 22774) National Archives and Records Service, General Services Administration, Wash. D.C. 20405. See "Report to the American Physical Society by the study group on light-water reactor safety." Rev. Mod. Phys. 47, Supp. No. 1 Summer 1975.

power and water supplies, and many other features associated with the monitoring and control of effluents, proper introduction and return of cooling water from
a river or ocean, evaporative cooling towers, etc. Although plant safety, cost
and efficiency of construction and operation dictate the proper arrangement for
these components, there are many options. As we proceed, we will infer guidelines on the engineering designs associated with these plant features.

21.14 Boiling Water Reactors (BWR)

S. Untermeyer, W. H. Zinn and others at the Argonne National Laboratory developed and proved the feasibility and stability of the BWR. Heat-transfer coefficients were verified using electrically heated rods. The stability of
reactor cores during boiling and changes in moderator density were demonstrated with steady and transient reactor tests (EBWR and BORAX, respectively) at
the National Reactor Testing Station (NRTS) in Idaho. Workers at the General
Electric Company adopted the basic design concept developed at ANL, built an
experimental BWR at Vallecitos, California, and then designed and constructed
a commercial demonstration BWR electrical generation plant at Dresden, Illinois. This plant went on line in 1959. By the beginning of 1975, 3 BWRs were
producing electricity. In addition, 23 BWRs were being built or are on order.
The BWR reactor assembly is shown in Fig. 21.14-1.

The BWR concept was developed because of desirable reduction in pressure
and plant costs by elimination of the large temperature drops and expense associated with the steam generators of the PWR. Also, BWR pressure vessels may
have thinner walls. Disadvantages arise mainly from radioactivity in the cooling water and in the steam circulating through the turbine and other parts of the
heat-power loop, thus potentiating larger releases to the atmosphere. Furthermore, in the event of extensive release of fission products from fuel to the

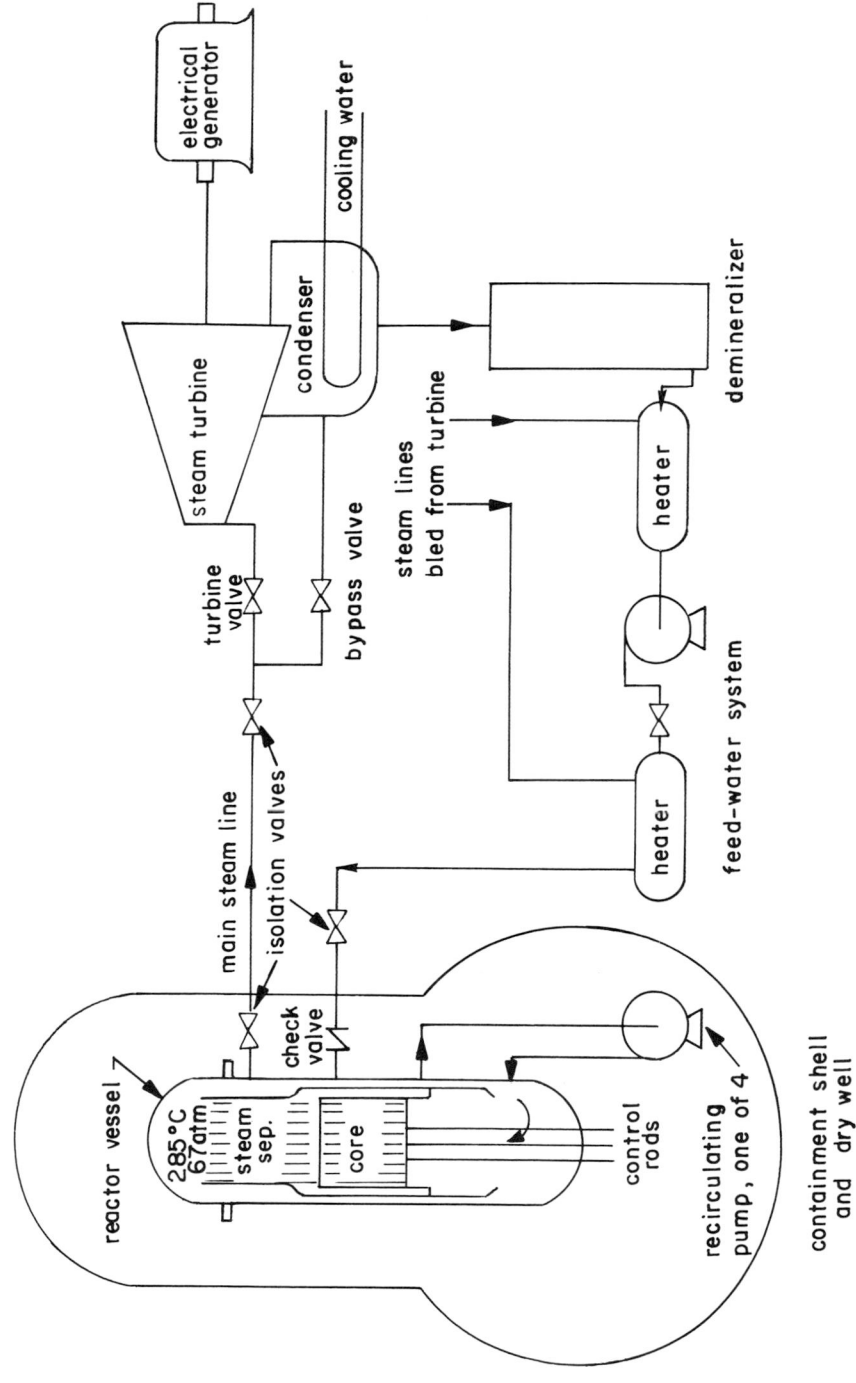

Fig. 21.14-1 Schematic diagram of the cooling and steam-power systems used in the BWR. The means for isolating the reactor system from the steam-power system should be noted.

cooling systems, one barrier has been removed between this region and the containment vessel or building that surrounds all reactors.

A. BWR Steam-Power Cycles

With steam generation in the reactor-pressure vessel, three different loops have been devised in order to improve the rate and extent of heat transfer and the cycle efficiency, as is indicated in Fig. 21.14-2. The loop at the top (a) represents a natural convection system. The flow of water through the core is enhanced by a chimney above the core where a mixture of the hottest liquid and steam bubbles provides a low density leg that is displaced upward by denser incoming water. Figure 21.14-2 (b) portrays a system in which a pump induces forced circulation and water flow through the core without the need of a chimney. Forced circulation is necessary for currently used power levels. Figure 21.14-2(c) shows a dual-cycle arrangement in which heat is extracted from circulating water in order to supply steam to the turbine at intermediate pressure levels. Figure 21.14-3 shows the essential components of the steam-power system.

B. General Arrangement

Figure 21.14-3 shows the location of the reactor core, the steam raising apparatus, means for directing water and steam flows, and the control rods inside the reactor vessel. Moisture separation and production of steam with acceptable quality, at steady pressure and flow rates, must be accomplished above the reactor. The downward action of gravity directly on the control rods is not fail-safe. However, a hydraulic accumulator supplies the required perpetual force to drive the control rods upward when needed. Thus, although the internal pressure of about 67 atm (1000 psi) in the BWR is lower than in the PWR,

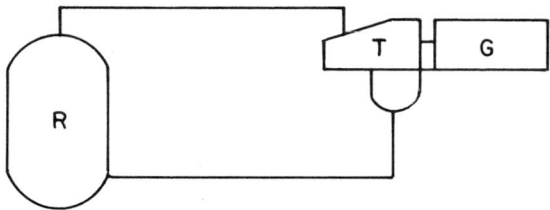

(a) Natural circulation, single-cycle turbine.   Convection causes flow through
the reactor.   Steam from the turbine condenses and returns to the bottom of the
reactor core.

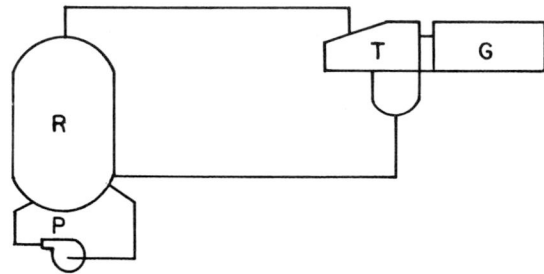

(b) Forced circulation, single-cycle turbine.   A separate pumped-loop circulating water through the core is required for large systems.   The pumped circulation entrains a large flow through the reactor core by means of injectors.

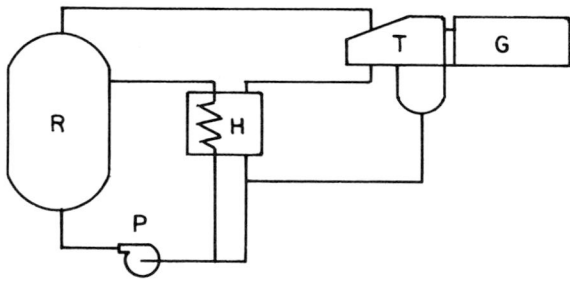

(c) Forced circulation, dual-cycle turbine.   Steam is produced both at the top
of the reactor and in a steam generator placed in the circulating loop.

Fig. 21.14-2 Boiling water reactor cycles; R = reactor, T = turbine, G = elec-
              tric generator, C = steam condenser, P = pump, H = heat ex-
              changer.

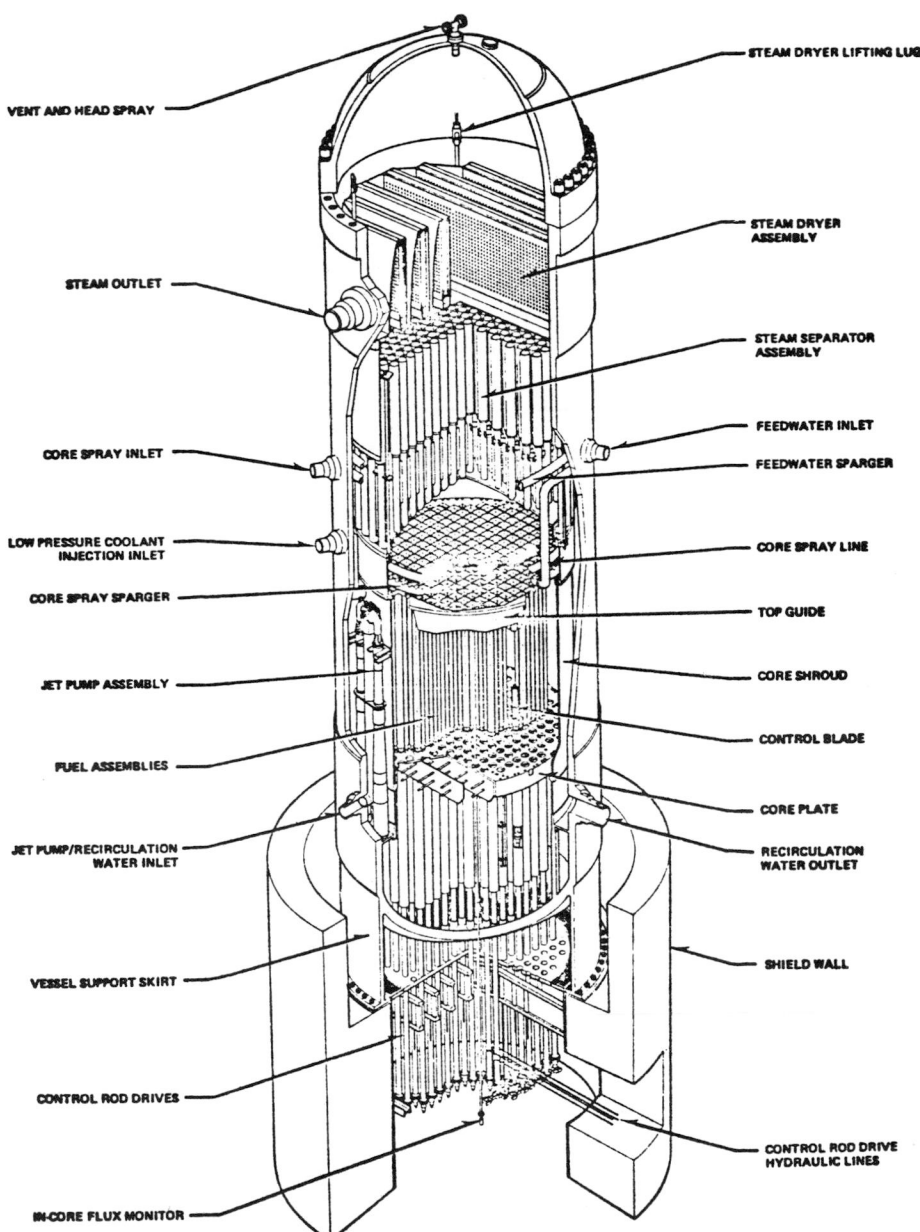

VENT AND HEAD SPRAY

STEAM OUTLET

CORE SPRAY INLET

LOW PRESSURE COOLANT
INJECTION INLET

CORE SPRAY SPARGER

JET PUMP ASSEMBLY

FUEL ASSEMBLIES

JET PUMP/RECIRCULATION
WATER INLET

VESSEL SUPPORT SKIRT

CONTROL ROD DRIVES

IN-CORE FLUX MONITOR

STEAM DRYER LIFTING LUG

STEAM DRYER
ASSEMBLY

STEAM SEPARATOR
ASSEMBLY

FEEDWATER INLET

FEEDWATER SPARGER

CORE SPRAY LINE

TOP GUIDE

CORE SHROUD

CONTROL BLADE

CORE PLATE

RECIRCULATION
WATER OUTLET

SHIELD WALL

CONTROL ROD DRIVE
HYDRAULIC LINES

Fig. 21.14-3 Isometric illustration of the essential features of a BWR reactor
             assembly equipped for recirculation of water through the core.
             Reproduced with permission from the General Electric Company.

the vessel must be larger. The BWR weighs 860,000 lbs at the 1,000 Mw_e power level and is 5 m in diameter, 16.5 m high and 14 cm thick. The recirculating water is directed upward through the core while steam above the core passes through separators into a steamdrum. Feed water is entrained and directed downward into the recirculating flow. The emergency core cooling is activated if the steam-power loop fails.

C. Reactor Core and Fuel Elements

The BWR and PWR cores and the fuel rods are similar. There are also some differences. Typically, 49 uniformly spaced bundles of rods are held by grids enclosed in zircaloy boxes. These and their end fittings direct the recirculating flow along the rods to assure uniform heat transfer. The cruciform blades (control elements) may be inserted between fuel boxes. The blades are stainless steel tubes filled with compacted B_4C. Control elements regulate the power level and determine the distribution of neutrons and heat. Control elements are withdrawn as fuel burns out or are inserted when new fuel is added. A typical fuel assembly is shown in Fig. 21.14-4, a lattice in Fig. 21.14-5.

D. Reactor Containment and Safety Features

The engineering safety features for a BWR are different from those of the PWR. Pressure suppression is accomplished (see Fig. 21.14-6) by leading the steam from a relatively low pressure shell surrounding the reactor vessel into a pool of water in a large annular pressure shell below the reactor. A concrete building surrounds the BWR and limits escape of gas through filters connected to stacks. Water sprays may be turned on to cool and flood the core if the water-steam circulation loop is breached.

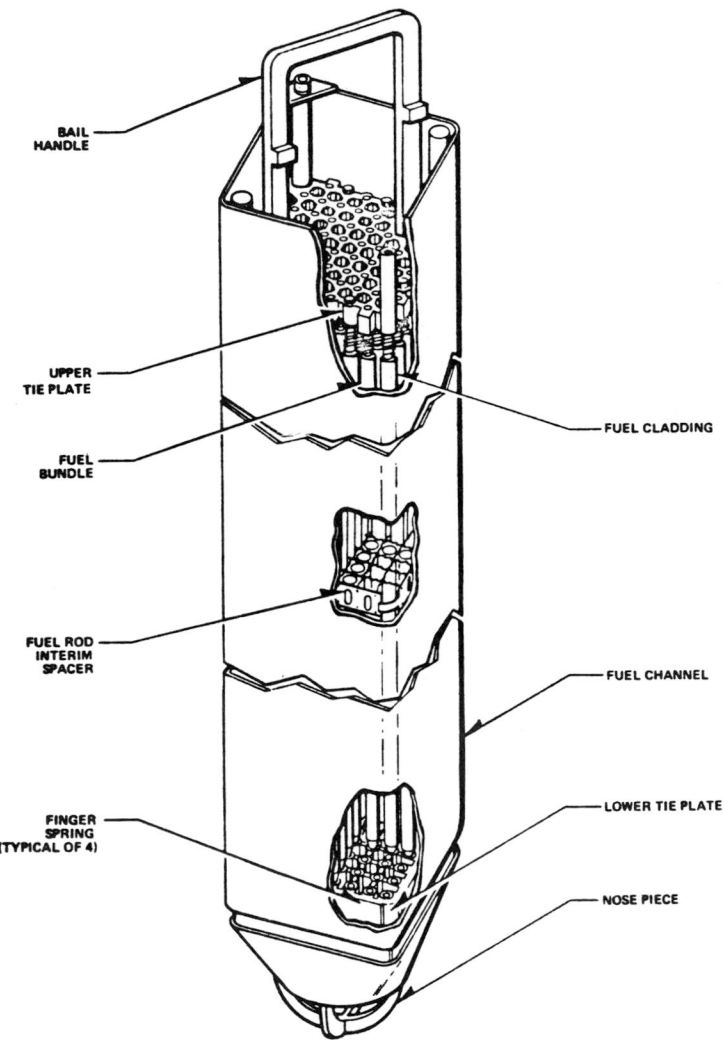

Fig. 21.14-4 Typical fuel assembly containing 49 positions for rods inside a
square zircaloy box or duct. The lengths range from 2 to 4 m,
depending on reactor size. The boxes are approximately 24.8
cm × 24.8 cm square. Reproduced with permission from the
General Electric Company.

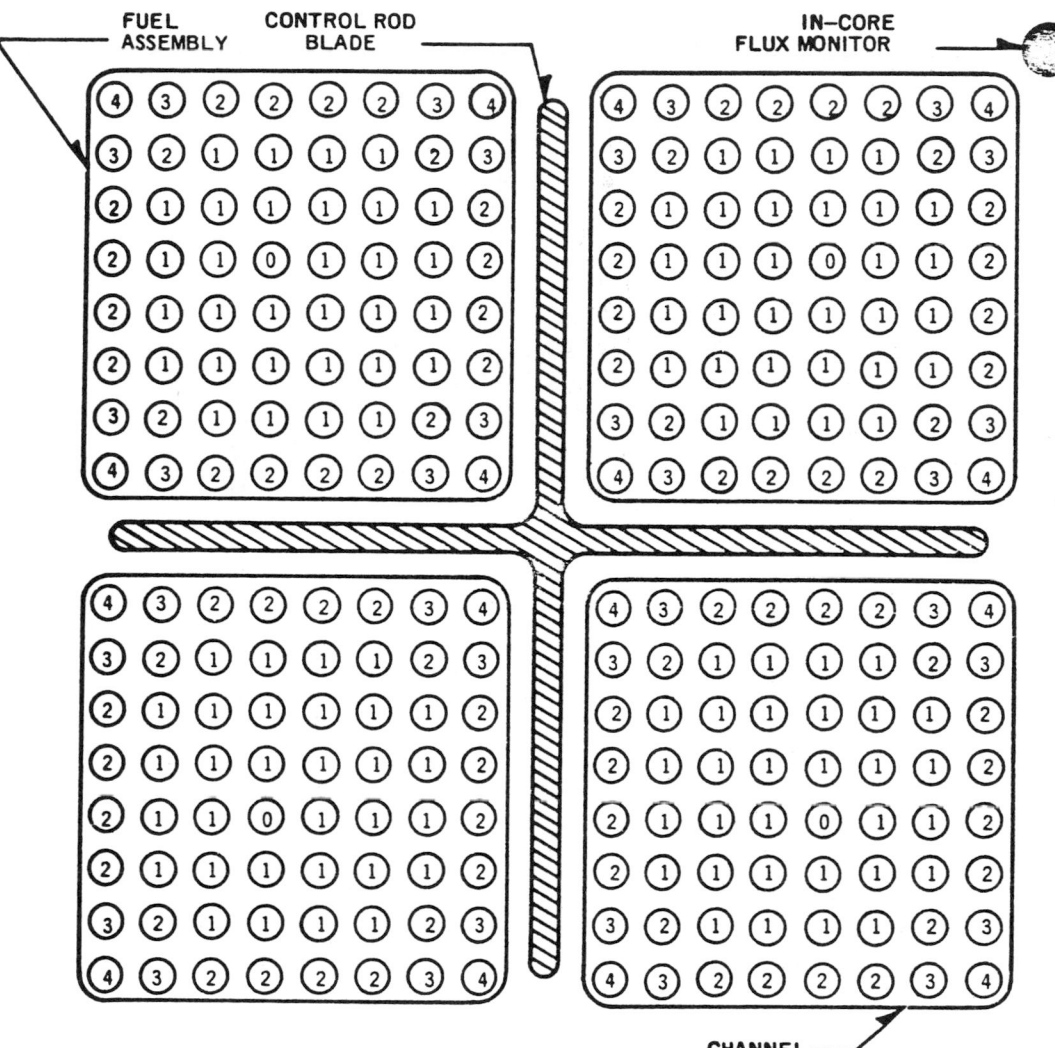

Fig. 21.14-5 A diagram representing cells in the lattice of a reactor fuel
loading. As fissile material is consumed, control blades are
moved. Rods containing uranium of different enrichment are
located so as to make heat generation as uniform as possible.
Approximately one third of the core is replaced at each loading
and enrichments altered as required for optimum use; 1, 2, 3, &
4 indicate locations of four different rod enrichments in the fuel
assembly; 0 indicates a water-containing rod. Reproduced with
permission from the General Electric Company.

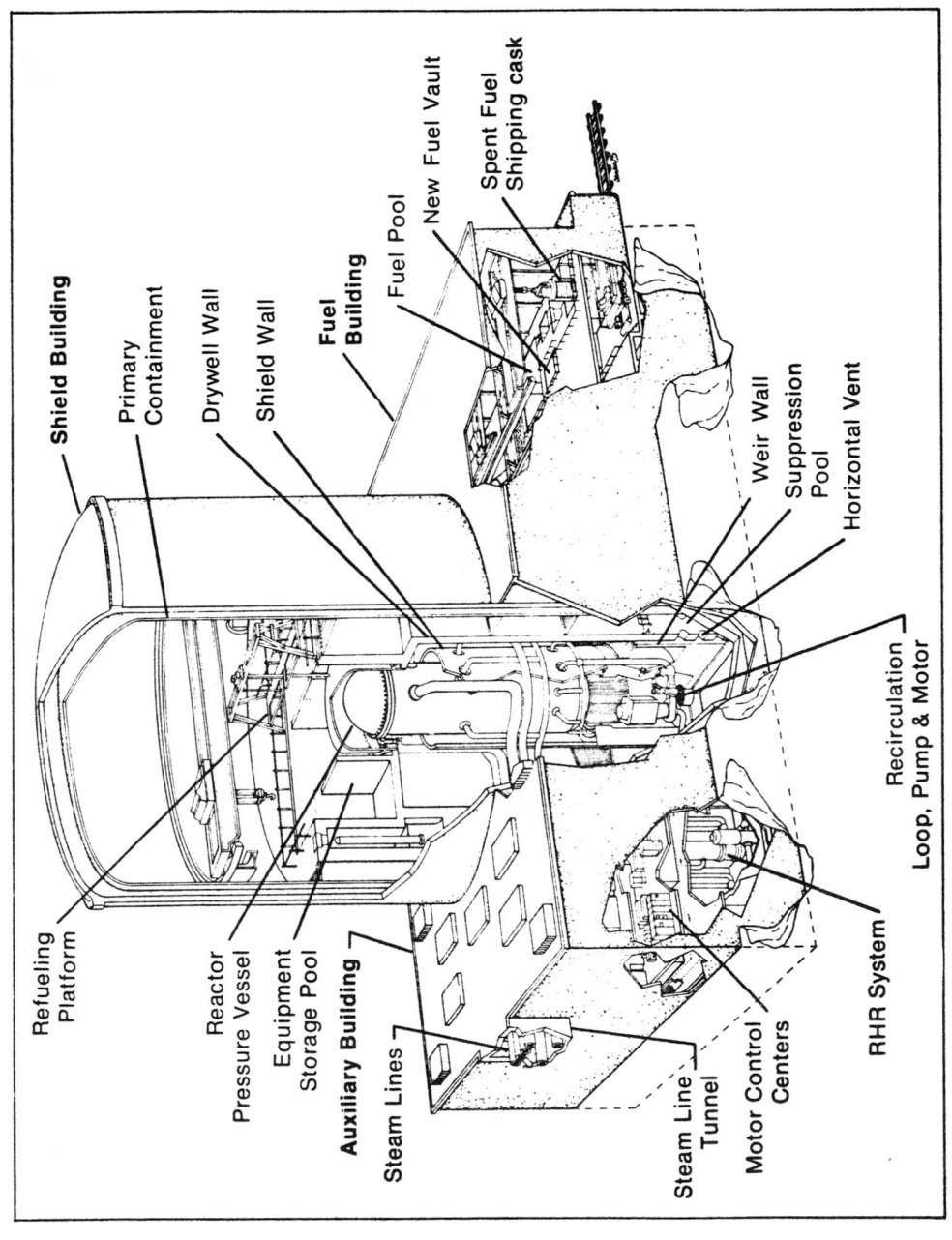

Fig. 21.14-6 An illustration of the containment-structure, reactor vessel, engineered safety features, and service equipment for a Mark III General Electric BWR; RHR = residual heat removal system. Reproduced with permission from the General Electric Company.

21.15 High-Temperature-Gas(Helium)-Cooled Reactors (HTGR)

The use of helium to transfer heat permits the use of graphite and other materials at elevated temperatures. These materials provide thermal and mechanical stability and the transfer of heat at temperatures up to 850°C and probably up to $1,000^\circ$C. This feature affords higher heat-power efficiency, the option of steam or gas turbines for power production, and the use of heat for chemical processing. The reactor produces $^{233}_{92}$U from fertile Th. If combined with FBRs, enrichment of ^{235}U will be unnecessary. Because of these two features, an HTGR requires less uranium and separative work than an LWR. However, unless $^{233}_{92}$U is produced by supplemental methods, highly enriched uranium (usable in weapons) is required as fuel. A unit of separative work (SWU) refers to the energy required to double the ^{235}U content in one kg of uranium (output). It is related to the entropy of mixing.

A. General Arrangement

The general arrangement of the HTGR is illustrated in Fig. 21.15-1, which shows an isometric cutaway view of a complete plant. Figure 21.15-2 shows an isometric, three-quarter, vertical section of the core and pressure vessel. The reactor core and the heat-exchange loops are seen to be located within an integrated structure. The core is composed of hexagonal graphite blocks which contain the fissile and fertile materials and passages for the helium coolant. The heat-exchanger steam-generators, helium circulators, auxiliary heat exchangers and circulators, as well as the control rods and core support, are contained within the structure in cavities. The thick-walled, reinforced concrete vessel is called a prestressed concrete reactor vessel

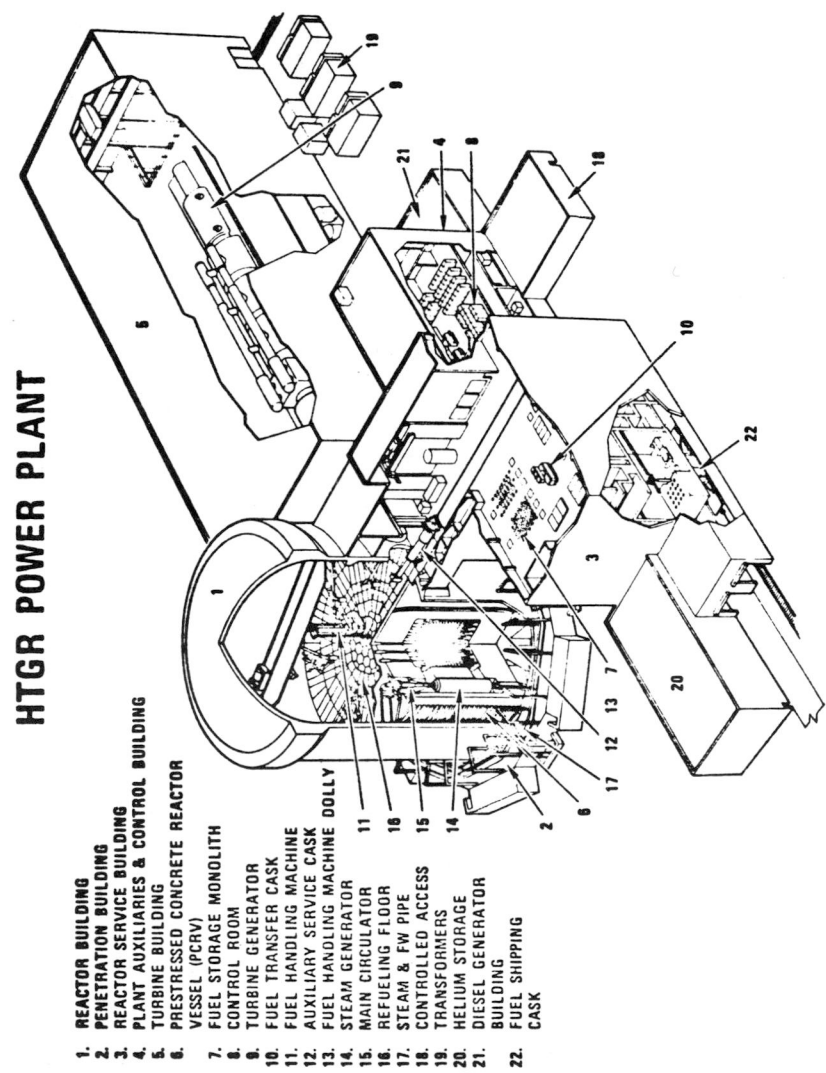

HTGR POWER PLANT

1. REACTOR BUILDING
2. PENETRATION BUILDING
3. REACTOR SERVICE BUILDING
4. PLANT AUXILIARIES & CONTROL BUILDING
5. TURBINE BUILDING
6. PRESTRESSED CONCRETE REACTOR VESSEL (PCRV)
7. FUEL STORAGE MONOLITH
8. CONTROL ROOM
9. TURBINE GENERATOR
10. FUEL TRANSFER CASK
11. FUEL HANDLING MACHINE
12. AUXILIARY SERVICE CASK
13. FUEL HANDLING MACHINE DOLLY
14. STEAM GENERATOR
15. MAIN CIRCULATOR
16. REFUELING FLOOR
17. STEAM & FW PIPE
18. CONTROLLED ACCESS
19. TRANSFORMERS
20. HELIUM STORAGE
21. DIESEL GENERATOR BUILDING
22. FUEL SHIPPING CASK

Fig. 21.15-1   Representation of a complete HTGR power plant showing the helium-cooled, graphite-moderated reactor in the cylindrical building flanked by the fuel-service building in the right-hand foreground, the steam-turbine-electrical-generator hall in the right-hand background, and the plant-control room included in the angle between.   Reproduced with permission from the General Atomic Company.

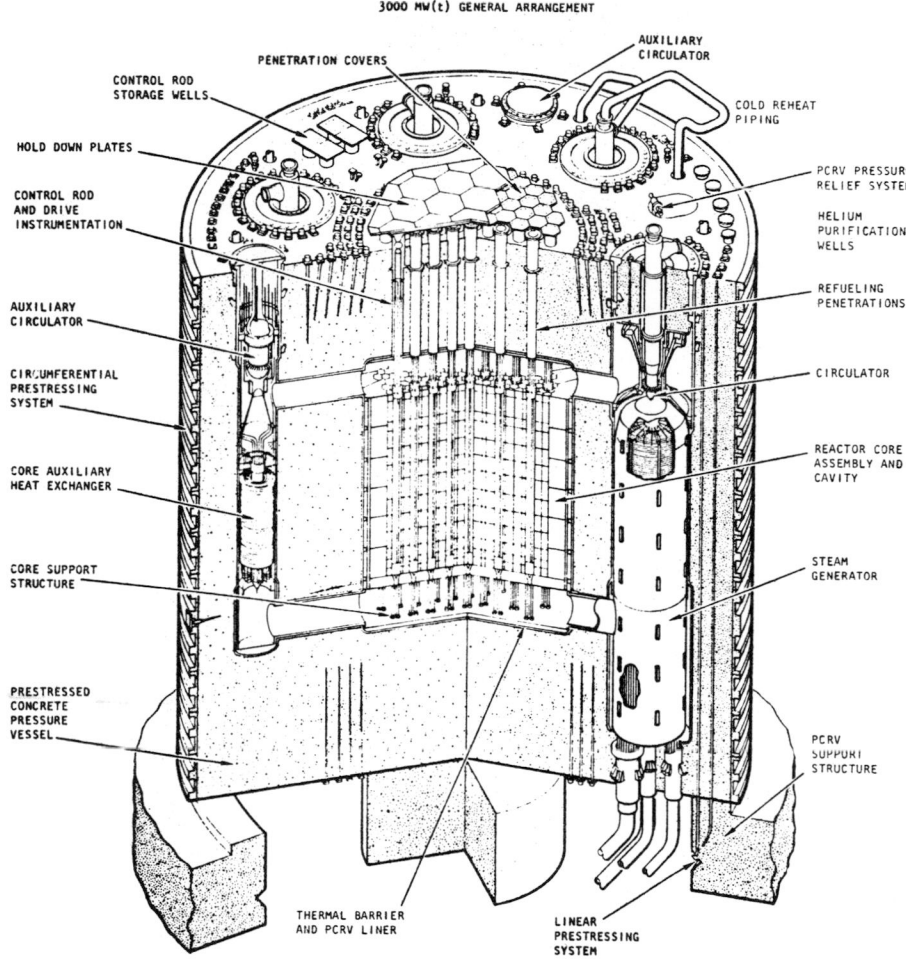

3000 MW(t) GENERAL ARRANGEMENT

PENETRATION COVERS

AUXILIARY
CIRCULATOR

CONTROL ROD
STORAGE WELLS

COLD REHEAT
PIPING

HOLD DOWN PLATES

PCRV PRESSURE
RELIEF SYSTEM

CONTROL ROD
AND DRIVE
INSTRUMENTATION

HELIUM
PURIFICATION
WELLS

AUXILIARY
CIRCULATOR

REFUELING
PENETRATIONS

CIRCUMFERENTIAL
PRESTRESSING
SYSTEM

CIRCULATOR

CORE AUXILIARY
HEAT EXCHANGER

REACTOR CORE
ASSEMBLY AND
CAVITY

CORE SUPPORT
STRUCTURE

STEAM
GENERATOR

PRESTRESSED
CONCRETE
PRESSURE
VESSEL

PCRV
SUPPORT
STRUCTURE

THERMAL BARRIER
AND PCRV LINER

LINEAR
PRESTRESSING
SYSTEM

Fig. 21.15-2 Isometric, three quarter vertical section of an HTGR showing
the major features of the primary-cooling and steam-generating
system or NSSS. In normal operation, helium is circulated downward through the core and through the four loops having steam
generators. Reproduced with permission from the General
Atomic Company.

(PCRV). The PCRV is stressed in compression by numerous strong tendons that provide the containing forces. The cavities are lined with a gas-tight, water-cooled, steel membrane. Helium at about 47 atm (700 psi) is circulated through the reactor and steam generators. Penetrations and seals are provided for the control-rod drivers, steam lines, helium fill, sampling lines, and instrumentation conduits. A helium-purification system is enclosed within the cavity of the PCRV. Larger penetrations with bolted covers are provided for the removal of components.

The PCRV itself is located inside a containment shell capable of holding the helium at a pressure of about two atmospheres.

Fuel removed from the core by remote handling machines may be transferred through access ports to a fuel-storage building in which cooling is provided for residual radioactivity. In the storage facility, spent radioactive fuel blocks (after cooling) may be loaded into large, shielded, shipping containers for transfer to a chemical reprocessing plant located elsewhere.

The largest building in the complex is the hall housing the turbine and generator (see Fig. 21.15-1). This structure and its machinery are similar to those used in a modern, efficient steam power plant. High pressure, super heated steam at approximately $530^{\circ}C$ and 160 atm (2400 psi) is fed to the high pressure turbine.

B. HTGR Fuel Elements

Figure 21.15-3 shows the hexagonal prismatic graphite blocks that make up the reactor core. A standard element is 36 cm (14.17 in.) across flats and 79 cm (31.22 in.) long. Fuel contained in molded carbon cylinders 5 by 1.6 cm (2 in. by 0.62 in.) is inserted into 132 full length, longitudinal holes, 1.59 cm (0.624 in.) in diameter. Coolant flows through 72 holes which

HTGR FUEL COMPONENTS

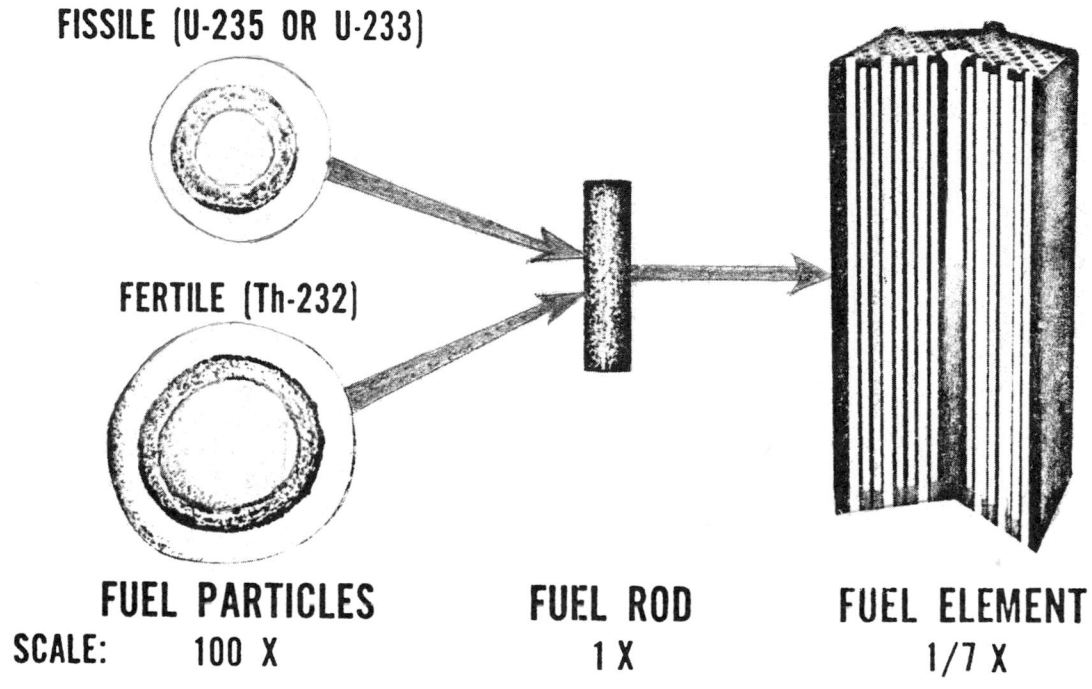

FISSILE (U-235 OR U-233)

FERTILE (Th-232)

FUEL PARTICLES

SCALE: 100 X

FUEL ROD
1 X

FUEL ELEMENT
1/7 X

Fig. 21.15-3 Illustration of the relative sizes and make-up of the components
of an HTGR core. At the right is the machined and drilled graphite block. Nearly two thirds of the longitudinal holes contain fuel
rods, the remainder being coolant channels. Each rod is made
up of the two kinds of particles bonded together by a carbonaceous
material. Reproduced with permission from the General Atomic
Company.

are 2.1 cm (0.828 in.) in diameter. To provide for control, appropriately located blocks having a central hole large enough to receive a rod with an outside
diameter of 8.26 cm (3.25 in.) consisting of a tube made of Incoloy 800 (i.e.,
a high-temperature, high-nickel, -chromium, -austenitic steel) and filled with
B_4C. In these elements having control channels there is room for only 76 fuel
holes and 43 coolant channels.

The number of blocks making up a reactor core depends on the required
power output. A standard design furnishing 1160 MW_e (3000 MW_t) is made up
of 3,944 fuel blocks stacked in 493 columns to give a core 8.45 m (27.7 ft) in
diameter and 6.34 m (20.8 ft) high. Hexagonal blocks not containing fuel, but
permitting controlled flow of helium, surround the core to provide a neutron
reflector with a mean thickness of 1.05 m (41.9 in.).

C. HTGR Fuel Particles

A unique feature of the HTGR is the method used in containing fissile
and fertile materials and fission products within the fuel elements. Fissile
particles are made by coating microspheres of uranium dicarbide ($UC_{1.86}$),
which are approximately 200 μm in diameter, with successive layers of porous
carbon, dense and extremely fine-grained impervious isotropic pyrocarbon,
dense and impervious silicon carbide, and with an outer layer of strong and impervious pyrocarbon. Each layer is 20 to 50 μm thick, giving a total thickness
of about 170 μm. The coatings are deposited by the pyrolytic decomposition of
a low molecular weight hydrocarbon or a gaseous organic silicon compound on
the particles in a bed that is fluidized by flowing argon. The processing conditions have been established for providing coatings with the required structure
and composition that give the needed properties and behavior, namely, retention of
fuel and fission products at the required temperatures of approximately 1350°C.

The designed life of the fuel is 4 to 6 years. During this service, bombardment by some 8×10^{21} energetic neutrons per cm^2 (fluence) displaces over ten percent of the carbon atoms in the coatings at least once and also alters the properties of the pyrocarbon. The multi-layered, coated spheres are called TRISO particles.

The effects produced by recoiling atoms caused by neutron collisions are interesting and important enough to warrant separate discussion. These phenomena have been studied intensively since 1942. In addition, the effects of recoiling fission products and their accumulation in fuel materials require special attention.

Fertile material in the form of thorium dioxide (ThO_2) is contained in particles with kernels as large as 500 μm in diameter, which are provided with porous and impervious pyrocarbon coatings having a combined thickness of approximately 160 μm. This type of coated particle is named BISO. Since fewer fission products accumulate and are rather well retained within these particles, an SiC coating is not considered to be essential.

The two kinds of particles are easily separated and purified for reuse of the unburned U-235 and the bred U-233. Burning in oxygen rids the fuel of carbon. Silicon carbide does not burn. Consequently, the smaller particles may be separated by sieving and then separately dissolved chemically.

The assembly of HTGR fuel particles into a fuel rod and subsequent construction of fuel elements are shown schematically in Fig. 21.15-3.

D. Prestressed Concrete Reactor Vessels

The PCRV adds greatly to the safety of nuclear reactors. Although requirements for its construction are exacting, it is erected directly in place. The high strength steel tendons are tensioned after construction. Since these tendons are located toward the outside of the vessel and at building temperature, they are well protected from corrosion or other exposure. The load, strain and stress of each tendon may be monitored and a tendon may be replaced if excessive elongation or failure occur. Because of ample redundance in construction, failure of a single tendon is not serious.

Tests show that a properly designed and constructed PCRV does not fail catastrophically but will instead crack and leak slowly if the internal pressure becomes excessively high (i.e., approximately twice the working pressure). The design basis accident (DBA) for this reactor is failure at one of the penetrations. The area for gas flow from the failed penetration may be limited in such a way that the time constant for the pressure drop will be at least 30 seconds, thus giving ample time for shutdown of the reactor and activation of auxiliary devices, with little rise in core temperature. Moreover, the massive graphite core has such a large heat capacity that, in the case of failure of the regular and of the back-up shutdown mechanisms, the rise in temperature would be sufficiently slow to permit the operator ample time for implementation of tertiary emergency shutdown methods. Similarly, leakage of water through a faulty steam generator may be limited to quantities that are insufficient to damage the graphite seriously or to produce hazardous amounts of H_2 and CO.

E.   Balance of Plant; Steam-Power Generation

The power generating system to which the NSSS supplies steam is often called the balance of plant. Its design has involved long and well developed power-engineering principles and practice. Integration of the two systems and overall optimization implies balancing heat and mass flow and optimizing operating characteristics, safety, and costs.

Figure 21.15-4 shows a schematic flow diagram and typical temperatures

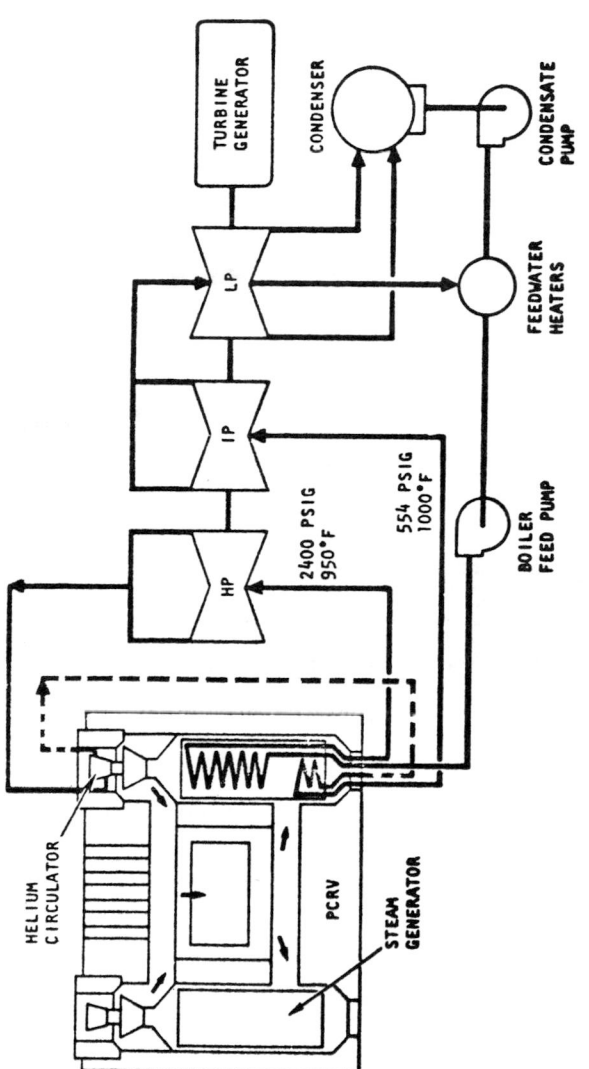

Fig. 21.15-4 A schematic steam-piping diagram for an HTGR power plant. Steam from the steam generator feeds the high pressure (HP) turbine and then the helium circulator. After being reheated, the steam drives the intermediate-pressure (IP) and low-pressure (LP) turbines. These and the feed-water-heating stages resemble those in a high efficiency fossil-fueled plant. Reproduced with permission from the General Atomic Company.

and pressures for the heat-power system, as well as the flow paths for helium, steam, and water.

A special feature of the HTGR is the method used in providing power for the main helium circulators. A disadvantage of gas cooling arises from the fact that the pumping power required to provide adequate velocity and mass flow for heat transfer is approximately 4% of the plant output, as compared with less than 1% for liquid cooling. By using steam from the outlet of the first high-pressure turbine to drive the main axial compressors, one may avoid taking all of the penalty from the plant output since this steam is not lost but is reheated (as shown on the dashed line of Fig. 21.15-4) and piped to the intermediate pressure turbine. Moreover, the stored energy in the steam system will maintain flow for some time after shutdown or loss of helium.

This steam-power cycle has an overall optimized efficiency of approximately 39%. Higher efficiencies are possible, but additional heat-exchange area and piping would cost more than the gain in output is worth.

Referring to Fig. 21.15-2, we note that auxiliary circulators are provided in the alternate outer cavities. When required, these axial compressors are driven by electrical power from an independent source to maintain helium circulation if the steam system fails. Thus, four independent primary circulators and four independent auxiliary circulators are available to assure helium flow. One loop with helium at a pressure of two atm is adequate for removal of the heat produced by the residual radioactivity after shutdown.

F. Gas-Turbine HTGR

Helium may be heated to sufficiently high temperatures to permit the use of closed-cycle gas turbines for power generation at efficiencies of about 36% if the heat is discharged directly. Up to 45% may be realized if heat from the high-temperature cycle is put through a second vapor-turbine cycle before discharge to a cooling system at ambient river or ocean temperatures.

Design studies show that gas turbines operating with helium at an inlet pressure of 80 atm (1200 psi) and a temperature of $825^{\circ}C$, producing approximately 270 Mw_e, are about the same size as the 50 Mw_e open-cycle turbines now in use. The turbine wheels for these machines are 5 to 6 feet in diameter. Consequently, the turbines, compressors and necessary heat exchangers may be fitted into cavities of the same size as are used for the steam system. Figure 21.15-5 illustrates this arrangement and indicates the flow path for hot helium to the turbine. There are heat exchangers, a regenerator or recuperator and the precooler (see Fig. 21.15-6). The precooler transfers heat from the working fluid. The power turbine provides an expansion leg which is approximately adiabatic (isentropic). The recuperator transfers heat from the hot gases leaving the turbine to the cooler gases flowing from the compressor to the reactor.

A national program, supported by the government, utilities and manufacturers is underway and is dedicated to develop, engineer, test, and produce a gas-turbine HTGR demonstration plant by about 1983. A similar international program is also current and involves groups in Germany, England and France.

21.16 Heat-Power Cycles

We shall now consider the principal factors that determine the efficiencies of nuclear fission reactors. We begin with an examination of energy-conversion cycles. The efficiency of a cycle using ammonia as the working fluid was considered in Section 14.7 of Volume II. Here we make graphical comparisons of the Rankine steam cycles used in the PWR and HTGR with the super-critical ammonia and Joule cycles proposed for use with the Gas-Turbine HTGR.

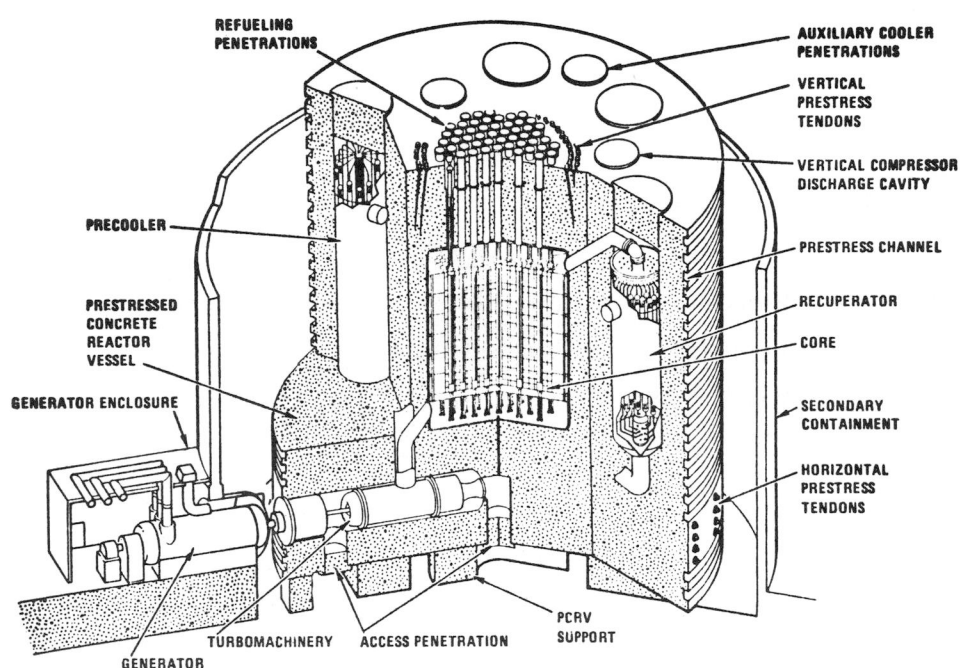

Fig. 21.15-5 A cutaway isometric diagram of a closed-cycle gas-turbine
HTGR showing how the heat exchangers, compressors and turbines may be arranged within a PCRV and made accessible for
maintenance. For one of four loops, ducts from the reactor to
the turbine and from the recuperator to the reactor are visible.
Ducts from the turbine to the precooler, compressor and on to
the recuperator are also located within the concrete. Reproduced with permission from the General Atomic Company.

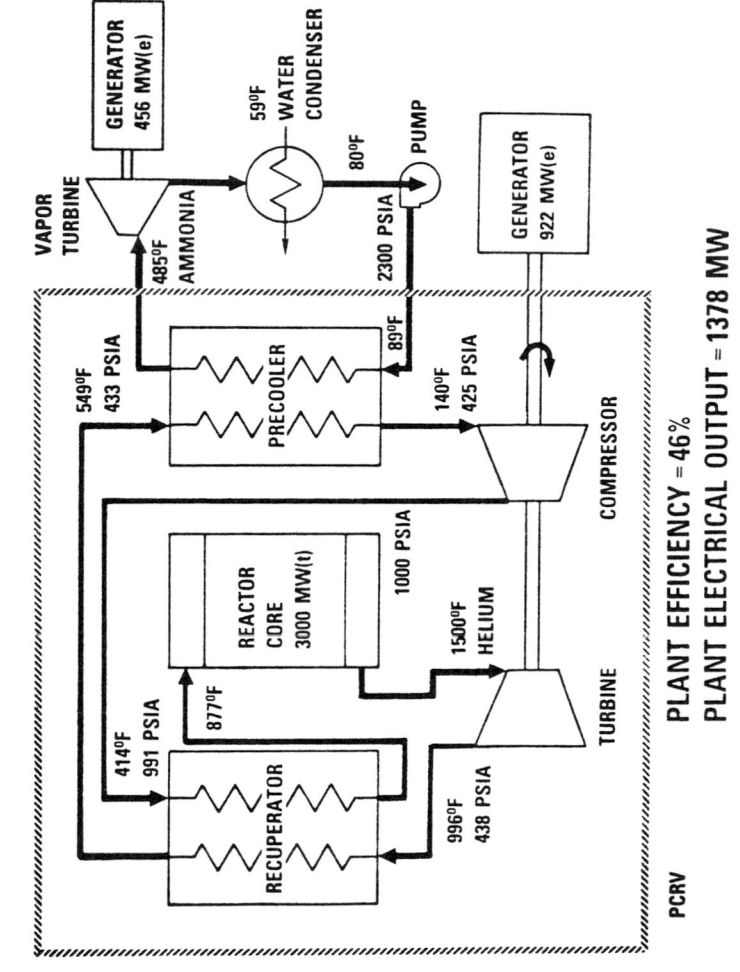

Fig. 21.15-6 Flow diagram for the closed-cycle gas-turbine HTGR system showing a combined supercritical ammonia heat-power cycle. Reproduced with permission from the General Atomic Company.

Figure 21.16-1 shows the approximate temperature-entropy diagram for these four cycles.

The cycle design and required equipment have evolved for the purpose of maximizing the useful work (which is represented by the enclosed areas in Fig. 21.16-1), subject to appropriate economic constraints on initial and maintenance costs. Major losses are associated with irreversible processes, i.e., primarily the temperature differences required for heat flow and the pressure drops for overcoming resistances to gas flow through the turbines that produce the work. Maximum conversion of heat to work occurs for small temperature and pressure drops. Maximum power demands larger gradients to achieve the required rates of flow. Within limits, increasing the area of heat exchangers and the number of expansion stages in the turbines (corresponding to smaller pressure drops per stage) increases efficiency at some cost. Thus, the process of design involves trade-offs in optimization.

A. Steam Rankine Cycles

Pinch points or bottlenecks for the transfer of heat to boiling water and steam occur near points a and a' in Fig. 21.16-1. The issue is illustrated in Fig. 21.16-2, which shows a plot of reactor-coolant enthalpy changes as a function of working-fluid temperature. We note that, in the neighborhood of point a, the driving force for heat flow will be a minimum under just these conditions that require a large amount of heat to be transferred to the working fluid. Moreover, to the right, the larger differences in temperature represent inefficiencies. One would prefer to transfer most of the heat from the coolant at its highest temperature.

Superheat, that is, raising the steam temperature well above the liquid-vapor equilibrium temperature, is not feasible in the case of the LWR. How-

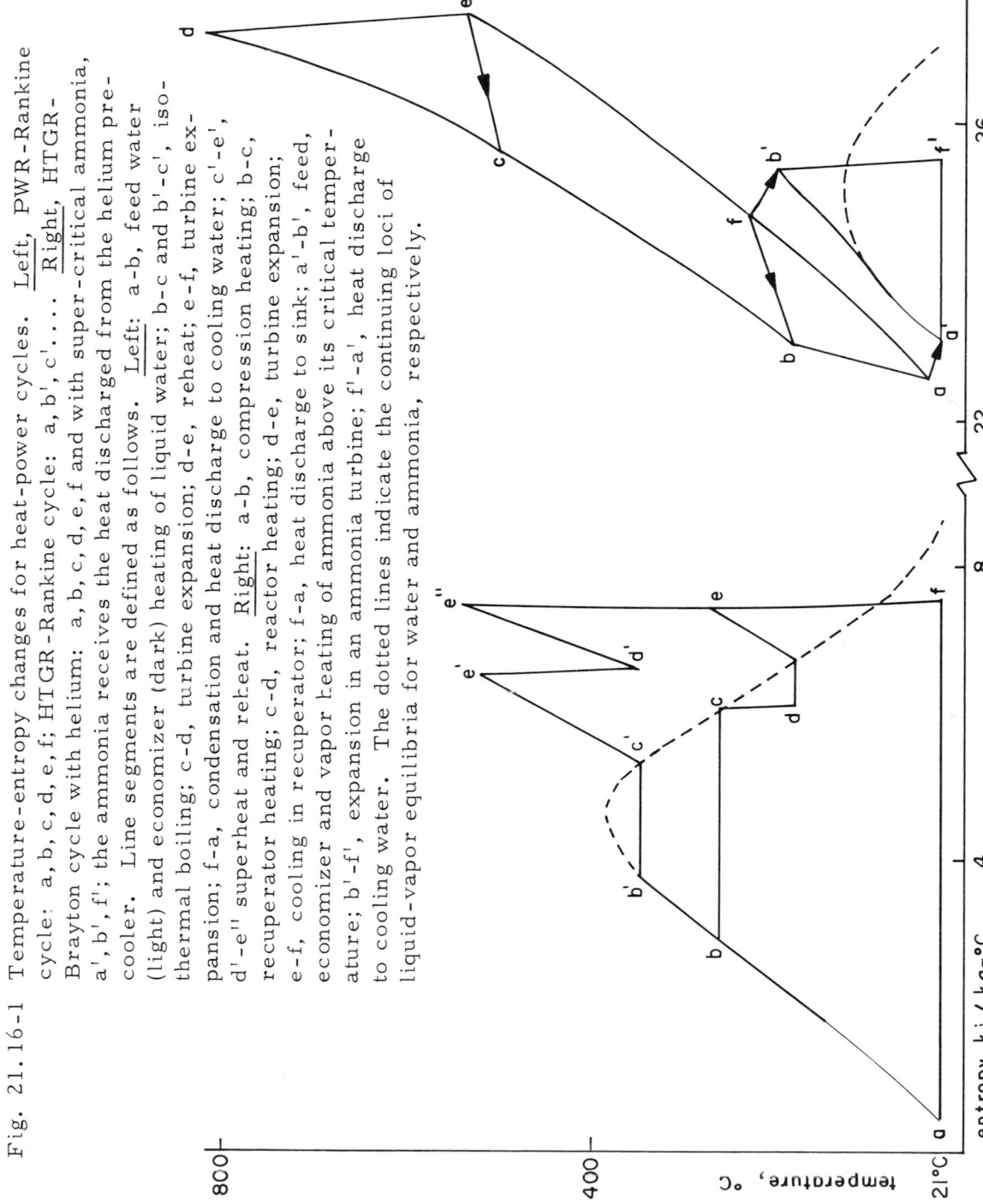

Fig. 21.16-1 Temperature-entropy changes for heat-power cycles. Left, PWR-Rankine cycle: a, b, c, d, e, f; HTGR-Rankine cycle: a, b', c', ... Right, HTGR-Brayton cycle with helium: a, b, c, d, e, f and with super-critical ammonia, a', b', f'; the ammonia receives the heat discharged from the helium pre-cooler. Line segments are defined as follows. Left: a-b, feed water (light) and economizer (dark) heating of liquid water; b-c and b'-c', iso-thermal boiling; c-d, turbine expansion; d-e, reheat; e-f, turbine ex-pansion; f-a, condensation and heat discharge to cooling water; c'-e', d'-e" superheat and reheat. Right: a-b, compression heating; b-c, recuperator heating; c-d, reactor heating; d-e, turbine expansion; e-f, cooling in recuperator; f-a, heat discharge to sink; a'-b', feed, economizer and vapor heating of ammonia above its critical temper-ature; b'-f', expansion in an ammonia turbine; f'-a', heat discharge to cooling water. The dotted lines indicate the continuing loci of liquid-vapor equilibria for water and ammonia, respectively.

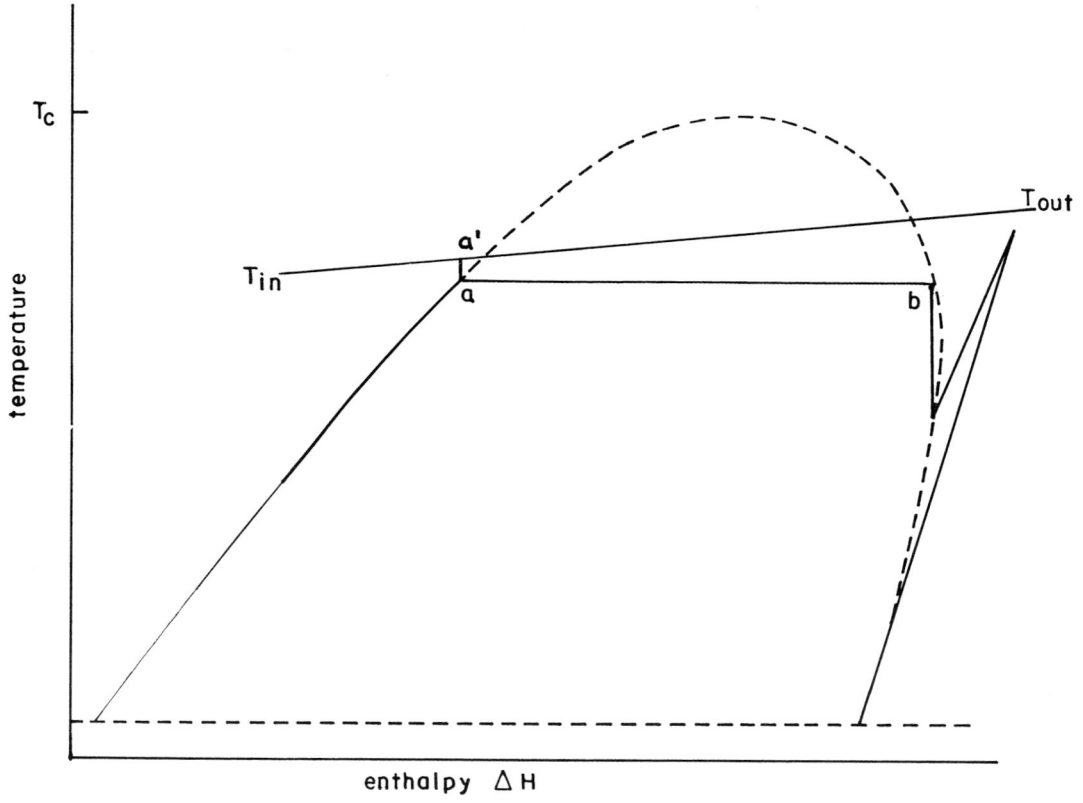

Fig. 21.16-2 A schematic plot of the enthalpy of a reactor coolant as it is
cooled from the reactor outlet temperature T_{out} to the reactor-
inlet temperature T_{in} by the transfer of heat in a steam generator to water and steam as isothermal vaporization occurs at the
boiler pressure between points a and b. The vertical distance
at _a_ represents the pinch point or minimum ΔT for the required
transfer of heat.

ever, helium from the HTGR may be used to heat steam well above the boiling temperature at the working pressure used (see Fig. 21.16-1).

Expansion of steam to produce work may be nearly isentropic. Actual efficiencies are approximately 80%. For the LWR, the steam is saturated, i.e., in equilibrium with liquid at the entrance to the turbine. Condensation occurs upon expansion and the turbine must accommodate water droplets that are contained in the gas. As is shown in Fig. 21.16-1, the expanded steam is reheated in a section of the steam generator and further expanded through a low-pressure turbine before it is condensed by cooling water.

In the HTGR steam cycle, the steam is superheated from the boiling temperature of 350°C to about 525°C before it is expanded to produce work in a high pressure turbine. This cycle resembles those used in conventional power plants. Referring back to Fig. 21.15-4, we recall that the power to circulate helium is also obtained from steam flowing from the high-pressure turbine through a turbine drive for the helium circulator. Following these two stages of expansion and cooling, the steam is reheated to about 550°C in the reheat section of the helium-heated steam generator and is then expanded through an intermediate section and a low-pressure turbine which converts useful enthalpy to work.

The condenser is a large heat exchanger in which the latent heat of condensation is transferred to cooling water. The line f-a in Fig. 21.16-1 indicates the isothermal change of phase from expanded steam to liquid water. This wide base of the T-S diagram for the Rankine cycle shows that the enclosed area, and thus the amount of useful work, are sensitive functions of the condensation temperature. We conclude that the cooling conditions are very important in power-plant design and, therefore, site selection must be carefully considered.

Feed-water pumping and heating represent the final step in completing the heat-power cycle. Relatively little power is required to pump the comparatively small volume of condensed water to the pressure of the boiler. In order to bring the temperature of the feed water to the boiling temperature, heat is fed to it in stages using steam either from low-temperature turbine stages or from an economizer section of the steam generator. By using heat from parts of the cycle at the lowest temperature that is just sufficient to heat the liquid one optimizes plant performance. Through this practice, the maximum possible amount of heat is transferred from the reactor coolant to the working fluid at the highest available temperature.

The piping diagram for an actual steam cycle is very much more complex than has been indicated in Figs. 21.13-1 or 21.15-4. Steam pipes lead from several locations in the low-pressure turbine stages to a corresponding number of feed-water heating stages. The complete cycle is composed of several sub-cycles. As a result, the system may be thought of as a network at the nodes of which the flows of enthalpy, work, entropy, and mass may be balanced or conserved, and the useful work maximized. The numerical details connected with these balances are conveniently incorporated into computer programs.

B.   Gas-Turbine Cycles

Although the final arrangement is very similar to the Rankine steam-cycle, the use of a non-condensing gas requires a thermodynamic cycle that looks different in the T-S plane. There are also several practical differences. The origin of the heat-power cycle may be ascribed to Joule. However, among mechanical engineers in the U.S., it is labeled the Brayton cycle.

As before, we may follow the cycle by starting with the input of heat to the working fluid, as indicated by the heavy line in the upper part of the cycle on the right-hand side of Fig. 21.16-1. As is shown in Fig. 21.15-6, the temperature of the helium is increased in the reactor from approximately 496°C to 815°C. An expected advantage of this direct cycle is the elimination of a heat-exchange step between the reactor coolant and the working fluid.

Expansion of the working fluid to produce useful work in the gas turbine is more nearly isentropic than in the steam turbine because the expansion ratio is only of the order of 2 to 2.4 instead of several hundred. The turbine blades may be shaped and sized to match the gas-dynamic requirements. As a result, the gas turbine has an efficiency of slightly over 90%.

Heat recovery occurs in the heat exchanger which is called a regenerator or recuperator. It is represented by the parallel sloping lines e-c and f-b of the Brayton cycle shown in Fig. 21.16-1; this stage is the analogue of the feed-water and economizer stages of the steam cycle. As may be seen, most of the enclosed area of the cycle is due to the length and spacing of these parallel legs. Therefore, the regenerator must have a large surface area and low resistance to the flow of both heat and gas. The relatively low difference in pressure between the two sides of the heat exchanger eases the task of meeting these requirements.

The precooler in a gas-turbine cycle is, as mentioned above, the analogue of the condenser in a steam cycle. In this heat exchanger, the enthalpy that is not convertible to work (by the Joule cycle) is transferred as heat to a sink. Although efficient operation demands that the temperature for discharging heat be as low as possible, inspection of Fig. 21.16-1 indicates that it must range from about 200°C to 21°C and that the area enclosed in the cycle is not very sensitive to these values. However, heat in this temperature range has considerable value for several purposes, in particular, for producing more work when using a properly chosen working fluid like ammonia.

The compressor pumps the helium through the cooler side of the regen-

erator and restores the working pressure. Although it corresponds to the feed-
water pump in a Rankine steam cycle, it is more nearly comparable to the gas
turbine in size and consumes an appreciable fraction of the power produced by
the turbine. The efficiency of the compressor is slightly less than that of the
turbine, just under 90%. Customarily, the compressor and turbine are on the
drive shaft which extends to the electrical generator, as is shown in Fig. 21.15-6.
It is feasible and at times convenient to separate the turbine into two stages, one
of which drives the compressor and the second of which supplies net power to
the generator.

Combined cycles, in which advantage is taken of the use of high-temperature heat in the Joule cycle and of the extra work obtainable from a condensing
fluid in the lower temperature range of the gas-turbine discharge, are finding
increasing application. The bottoming cycle a', b', f' shown in Fig. 21.16-1 has
been carefully chosen to avoid pinch points for heat transfer and to maximize
the enclosed area lying between the T-S line representing the transfer of heat
from the helium and the isothermal line representing condensation of a fluid at
a suitable cooling water temperature, at the same time permitting the pumping
of reasonable volumes of liquid to the vapor generator. A fluid with a critical
temperature in the same range as that of the precooler, at a reasonable critical
pressure, is desired. Of several fluids meeting these requirements, ammonia
appears to be the best because its energy density is relatively high. In fact,
only H_2O can provide higher energy densities than NH_3. The use of NH_3 permits the use of suitably sized piping, vapor turbines and feed pumps.

The supercritical heat-power cycle is almost completely analogous to the
Rankine cycle except that boiling does not occur. Above the critical temperature and pressure of a fluid, there is no change of phase and hence no isothermal range in the entropy- or in the enthalpy-temperature plane. Consequently,
as has already been emphasized, the heat-input curves of the working fluid may

be matched to the heat-discharge curve of the heat source and the efficiency and available work from the cycle maximized. Since the condenser temperature is below the critical temperature, the work required to pump the smaller volume of liquid back to the working pressure is minimized.

C. Heat Discharge; Cooling Systems

Both the power-engineering and the environmental aspects of the discharge of heat from power generating systems deserve emphasis. Rankine and supercritical cycles require low-temperature heat sinks, which are readily available near the seashore or on a river. In the U.S. and Europe, sites with adequate water for direct cooling are nearly exhausted. Many parts of the world already have insufficient water. The use of evaporative cooling towers reduces the intake of water by a factor of 10^4 compared to direct, once-through cooling; in evaporators, approximately $8 \times 10^4 \, m^3$ or metric tons (2.1×10^7 gallons) of water per day are required for a 1,000 Mw_e plant. In most contemporary power plants, once-through cooling is used because the cooling costs are reduced by as much as 0.4 mil/kwh. In the future, evaporative cooling towers or cooling ponds will be employed in a majority of the new power plants in the U.S. In addition, there may be a few sites, probably no more than 15% of the total, that will require direct transfer of heat to air in <u>dry</u> cooling towers. It is worth noting that cooling by the evaporation of water in a cooling tower, in a cooling pond, or by the direct use of river water, actually transfers heat to the air. A river into which heat is discharged equilibrates with the atmosphere within 1 to 10 miles downstream. Near the point of discharge, biota may suffer. Indeed, all heat released or received on earth is eventually equilibrated between the atmosphere and the oceans or else is radiated into space.

The use of an air-cooled heat exchanger to remove heat from the heat-

power loop increases the cost of energy, as compared to the cost of water cooling, in two ways: the cost is raised because of the required larger heat transfer
surface and because of the decreased efficiency associated with the increased
temperature of the discharged heat. Optimization forces a trade-off between
added equipment expense and decreased power output.

 The gas-turbine system, with somewhat inferior performance because
of higher discharge temperatures, requires relatively much smaller heat transfer surfaces than steam systems for two reasons: a higher ΔT is available and
the shape of the T-S cycle reduces the loss of useful work for a given increase
in the temperature at which heat is discharged. This difference is emphasized
in Fig. 21.16-3 through a scaled representation of optimized cooling-tower and
plant sizes for three different plants with top steam temperatures of $300^{\circ}C$ and
$520^{\circ}C$ and for a gas-turbine plant. The largest towers would be between 400 and
500 feet in height.

 Dry cooling is currently employed in a few fossil-fuel-fired plants in
Central Europe. We may expect limited applications in the future in the U.S.
The climatological impacts of the heat, up-drafts and air movements associated
with these installations have been evaluated to some extent (compare Chapter 6
in Volume I).

21.17 Process Heat

 An appreciable fraction of the total energy consumed is used as heat.
The potential environmental impacts of the heat evolved as the result of energy
use have been discussed in Chapter 6 of Volume I. The concept of a hydrogen
economy and its associated technology forms the subject of Chapter 11 in Volume II. Here, we are concerned with application of process heat in chemical
synthesis.

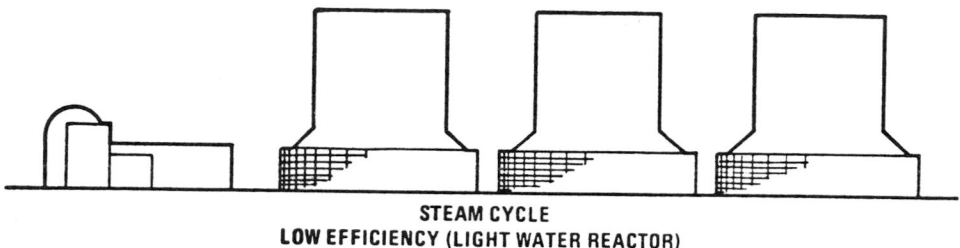

STEAM CYCLE
LOW EFFICIENCY (LIGHT WATER REACTOR)

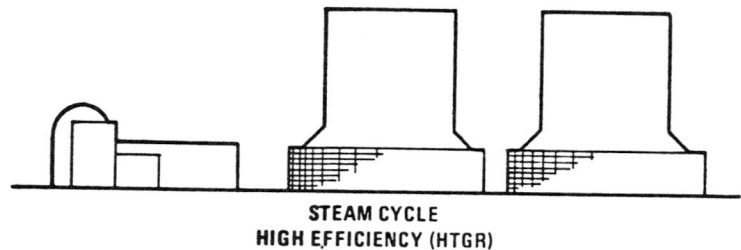

STEAM CYCLE
HIGH EFFICIENCY (HTGR)

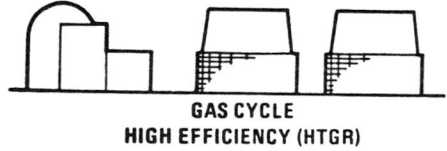

GAS CYCLE
HIGH EFFICIENCY (HTGR)

Fig. 21.16-3 Diagrammatic representation of the sizes of dry cooling towers for 3 types of nuclear power plants, each generating 1000 Mw_e. Reproduced with permission from the General Atomic Company.

Commercial production of hydrogen involves the following reactions of carbon or a hydrocarbon with water:

$$C(s) + H_2O(g) \rightarrow H_2(g) + CO(g), \quad \Delta H = 31.38 \text{ kcal at STP} \qquad (21.17\text{-}1)$$

or

$$CH_4(g) + H_2O(g) \rightarrow 3H_2(g) + CO(g), \quad \Delta H = 49.27 \text{ kcal at STP.} \quad (21.17\text{-}2)$$

Both reactions consume heat. The second reaction proceeds readily over a catalyst in a chemical reactor or reformer, in the temperature range from 800 to $1000^{\circ}C$. This process has been used predominantly for the production of hydrogen, with the heat usually derived from the combustion of a hydrocarbon. Recently, the cost for heat from clean fossil fuels has become greater than that from nuclear reactors. Consequently, there are economic reasons for considering the use of heat from reactors such as the HTGR for processes based on the chemical reactions described by Eqs. (21.17-1) and (21.17-2). In addition to heating the reactants, supplying steam and furnishing the enthalpy of reaction, the reactor may be used to provide electrical power for compression and movement of materials. Engineering development is underway in several countries. Groups in Japan (a country without indigenous supplies of fossil fuels from which to produce coke for steel making) are developing processes and equipment for employing hydrogen and carbon monoxide at 800 to $1000^{\circ}C$, which may then be used in the reduction of iron ore. In November 1974, the American Iron and Steel Institute announced the initiation of a similar program.[1] An HTGR will

---

[1] D. J. Blickwede, "The Use of Nuclear Energy in Steel Making," Nuclear News 17, 65-69, No. 18, October 1974.

readily supply 3000 Mw_t, which is equivalent to the heat and electrical requirements of a steel plant making approximately 4×10^6 tons of steel per year. Investigators in England, Germany and the U.S. are also supporting and collabo-
rating on similar developments in order to relieve the shortages, increasing
expense, and pollution associated with fossil-fuel use. Current projections
indicate a total demand for the process industries for heat which is equivalent
to one sixth to one quarter of that for utilities.

Waste heat at temperatures in the range of 30° to 250°C from electric
power production, and, perhaps also the process industries, may have economic
value. More extensive use of this heat will obviously conserve energy resources,
albeit at some expense because of greater use of material resources. In the cities of Stockholm, Sweden, and Moscow, U.S.S.R., warm water is being distributed from conventional power plants for heating buildings as far as 15 kilometers
(9 miles) away. Heat derived from a gas-turbine plant and operating without a
bottoming cycle could be used for desalting sea water, producing steam for industrial processes or heating and for other purposes, depending on whether the
value of the heat in a particular application is greater or less than that of the
electric power obtained in a bottoming cycle which lowers the discharge temperature.

21.18 Some Design Requirements

In the foregoing sections, we have portrayed essential features of the
fission reactors that will be used extensively during the next fifty years. Reactors planned in 1975 should be in operation by about 1984 and should have useful lives to the period 2014 to 2025. We expect that, before the year 2000, fission reactors, which are much more efficient in the utilization of resources and
heat, will have been fully developed and put into operation.

We shall now illustrate how the features of present and future reactors are determined by engineers in response to requirements and in accordance with accepted design principles.

A. Economic Considerations

Estimates of costs and returns on which commitment of capital and other resources may be based depend on the optimism of the estimators. A decision on the part of industry to commit a large amount of money is customarily made on the basis of comparisons among alternative investments and estimated returns on investments. An investment with anticipated return occurring far in the future may have lesser value than an investment with smaller but earlier return. Comparisons are made by discounting a future return to the present time (see pages 64-65 in Volume II for a discussion of the discounted cash flow method for estimating the rate of return on investment). The present worth, W_p, of a future return is related to future worth, W_y, after y years in the future by the compound interest formula

$$W_p(1 + r)^y = W_y ,$$

where r is the assumed rate of return. Often, 10% is taken as a needed rate of return. In most of the manufacturing industries, money now costs at least 16%. Thus, if a project will take 10 years to begin to pay, say, D dollars in the tenth year, the present worth of this return is only $D/(1.1)^{10} = 0.387$ D. On the other hand, the same return in five years is worth 0.62 D. Consequently, future returns must be large to be attractive. Such considerations have strongly affected the industrial support for technologies and entry into commercialization of nuclear energy. Subsidies of developments by government have served to overcome this obstacle for utilization of fission energy and in defense industries. Subsidies are being used again in the realization of fusion energy and the commercialization of solar energy. Military developments of fission reactors carried many of the front-end development expenditures with tax money.

At the present time, ERDA contracts with an industry to carry out a proposed development project at an estimated cost plus a fixed fee. The fee is often negotiated to represent a reasonable return on the value of the facilities which industry makes available to the project. These are mechanisms for co-operation between industry and government on long-range, high-cost developments. It is apparent that high interest rates or high costs of money have detrimental effects on energy technology, which must be projected 15 to 50 years into the future.

B. Design Requirements

The essence of good design is often intuitive and imaginative guessing, followed by accurate analysis, experimental measurements, and confirmation in field tests.

The formal process starts with the formulation of design requirements. Some of these constraints are given and some are formulated by engineers. The given requirements may be in the form of business, economic, or social incentives or objectives, such as the following: (i) production of a fixed amount of power on some schedule over a specified period of time at a total cost less than or equal to a number which is usually determined or projected from past experience; (ii) protection of the public from hazards to health, life, or property; (iii) minimum environmental impact or improvement of the environment; (iv) conservation of natural resources and prevention of harmful land use; (v) development of a new product or line of business; (vi) in some countries, the production of energy may be linked to the capability for producing weapons and other items for defense or aggression; (vii) enhancement of national prestige in technology; (viii) various, often tacit, values ranging from highly ethical to crassly selfish considerations.

It is the engineers' task to formulate a set of specific requirements that are accepted and approved by many independent parties, including vendors, buyers, operators, and regulators. Negotiations among the parties involved may be termed to be interactive and iterative.

The engineering process has many stages. In an early inventive or conceptual stage, the creative engineer may picture requirements and designs in his head. The novel parts of the design are successively reduced to equations, words and drawings. Standard or prior practice may suffice for portions of the expression of the concept or design. Following the conceptual stage, detailed "paper" studies are used to evaluate crucial questions concerning technical feasibility, as well as economic and social feasibility. Experiments and tests may be required to provide quantitative relations among engineering variables or to confirm spatial arrangements, durability, etc. Items of social and political feasibility are evaluated. Eventually, a design that may be priced, procured, constructed, and operated is produced, provided the criteria and requirements of each successive phase of development are met.

A partial list of types of organizations or parties involved in these projects includes vendors who manufacture apparatus and processes such as nuclear steam-supply systems, nuclear fuels, systems for fuel reprocessing, turbines, generators, instruments, pumps, valves, piping, structural materials, etc., and who may supervise initial operations; architect-engineers who plan the overall plant-design layout, the construction schedule, the site design and supervise construction; constructors who perform actual site work, including erection of buildings and installation of equipment; buyers such as utilities who may be owner-operators or vendors of services and who often use consultants to evaluate and compare, to give advice during negotiations, and to do additional architectural engineering; regulators, including the Nuclear Regulation Commission, the Federal Power Commission, State Public Utility Commissions, Environmental Protection Agency, as well as State, County, and City zoning and planning boards, State Occupational Safety and Hazards boards, the judicial system, and the public.

C. Systems, Components, and Materials

The final systems design is made to the satisfaction of all parties concerning the output, input, internal make-up, structure, and behavior of the system. Output includes not only energy to meet a scheduled demand but also economic gains or loses, environmental impact, and anticipated social and political effects or reactions. Input includes all components, materials, costs and skills.

In addition to the reactor, the following items are involved in the complete system: electrical subsystems, including the electrical load to be supplied through a transmission network, transmission lines, towers, transformers and switching gear substations, generators, sensors, control devices and loops for all interacting parts of the system; heat-power subsystems such as turbines, heat exchangers, reactors, fuel elements, control rods and drives, containment structures, safety systems, cooling towers, sensors and controls; fuel-cycle subsystems exemplified by fuel-handling, storage and shipping, fabrication, re-fabrication, reprocessing, isotope enrichment, fuel-element fabrication and mining and milling; materials such as U_3O_8, UO_2, $U_{1-x}Pu_xO_{2-\delta}$ $(0.1 < x < 0.3$; $\delta < 0.5)$, ThO_2, structural steels and related nickel-base alloys, zirconium alloys, boron carbide, hafnium, silver-cadmium alloys, graphite, pyrolytic carbons, water-system chemicals, concrete. Finally, environmental constraints must be satisfied with respect to site and locality, population distribution, water and water movement, meteorology (air movement and weather), biota, paths for transport and hold up of materials, other industries, human activities.

D.   Fuel Cycle

The plant output is used to feed an electrical energy distribution grid or loop. At the same time, the plant input is connected to fissile and fertile fuel loops or cycles which, in turn, are fed by raw materials from mines and from which materials are sent to storage. Figure 21.18-1 shows the flows and the component operations or processes in this cycle. More complete discussions of these operations may be found in such references as Sesonske. [1] Estimation of costs is a highly developed art. Although simple in principle, it is complex in detail because of the many components involved and because of financial difficulties associated with the escalation and inflation of rates and other changes in the charges for different kinds of labor, materials, interest, taxes, tax credits, etc. Price changes are also caused by technological advances and regulatory practices. The data listed in Table 21.18-1 allow some comparisons between fuel-cycle cost components for commercial LWRs and projected FBRs. Greatly reduced consumption of uranium, elimination of enrichment and the value of bred plutonium account for the much lower fuel-cycle costs for FBRs. The GCFR produces U-233, which may be more valuable than Pu.

[1]Alexander Sesonske, Nuclear Power Plant Design Analysis, USAEC Technical Information Center, Office of Information Services, 1973. Available from National Technical Information Service (NTIS), Springfield, Va. 22151.

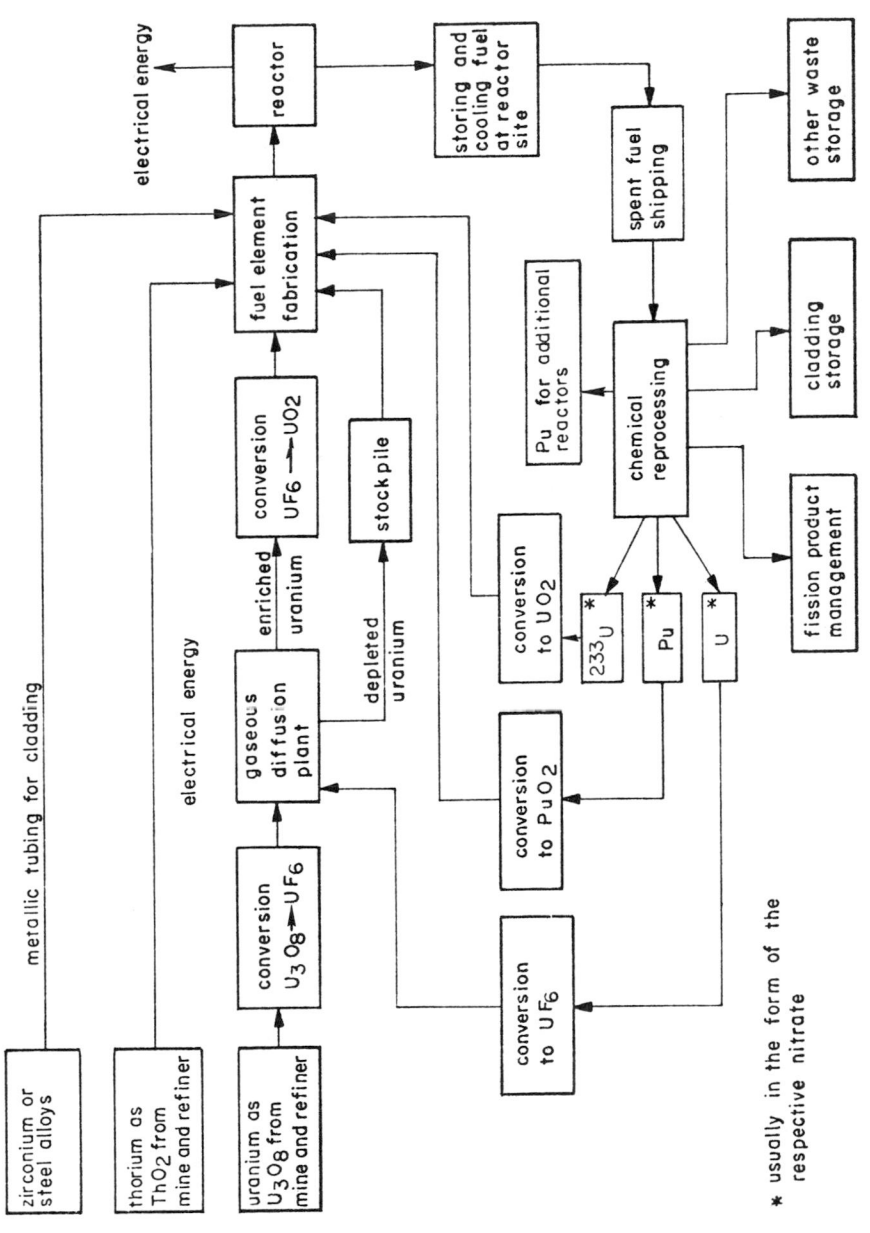

Fig. 21.18-1 Schematic representation of the nuclear-fuel cycle. Each box represents a stage in the processing of materials from their raw state to their reuse or management as radioactive wastes.

Table 21.18-1 Illustrative costs of feed materials, plant operations and money
invested in fuel-cycle plants and equipment.

Reactor type	LWR[1]	LMFBR[2]	GCFR[2]
material fed to system	UF_6	UO_2	UO_2
burn-up, Mwd/kg	30	150	100
costs or (credits), mills/kwh [*]			
feed material	2.74	0.03	0.03
isotope enrichment	1.58	-	-
fabrication of fuel	0.74	1.07	1.34
reprocessing, shipping and waste disposal	1.93	0.89	1.55
fuel converted or bred	(1.36)	(0.57)	(3.36)
capital invested in facilities at 16%/y with a 75% plant factor	3.97	1.14	1.35
total net cost for the fuel cycle	9.6	2.56	1.0

[*]Since Reichle gives costs projected to 1985 when the LWR plants now
planned will be operating, the FBR costs have also been adjusted from
1972 to 1985 by using the comparison escalation rates listed by Reichle.

[1]Adapted from L. F. Reichle, "Economics of Nuclear Power," 46th Annual Electric Utility Executives Meeting, Phoenix, Arizona, October
1975.

[2]Adapted from USAEC, "LMFBR Program Plan," Vol. 8, AEC Report
WASH 1108, 1972. Available from NTIS, Springfield, Va., 22151.

E. Energy Load or Demand

The desired output corresponds to the electrical energy that is fed into a transmission network. New plants are more efficient than old ones and, after they are properly shaken down, usually carry the main or base load in as constant or steady a fashion as is possible. Planning is based, in large measure, on experience. The demand history for a plant over its anticipated 30-year life is represented in Fig. 21.18-2. Plants are designed to follow electrical loads in varying degree. Scheduled, and occasionally unscheduled, shutdowns occur. Peak loads are supplied by low capital-cost equipment such as pumped-storage hydropower or fossil-fired, open-cycle gas turbines with steam-bottoming cycles. Storage batteries are being developed and evaluated for this service.

21.19 Illustrations of Design Analysis

The design of a nuclear energy system (power plant) often starts with the synthesis of a reasonably complete mathematical model in computer-code form. This model shows the essential performance features of the system. We start with a specification of the number density of nuclear species as a function of position in a reactor composed of fissile, fertile, scattering and absorbing materials. The nuclear reaction rates may then be calculated as functions of position and time. If a desired reaction rate is specified at each position, the problem may be inverted to determine the required density for each type of nuclide in any element of volume.

Similarly, given the fission rate at any position, the rate of heat generation and, therefore, removal is specified. The laws for heat transfer by conduction and convection may be expressed reasonably well mathematically and

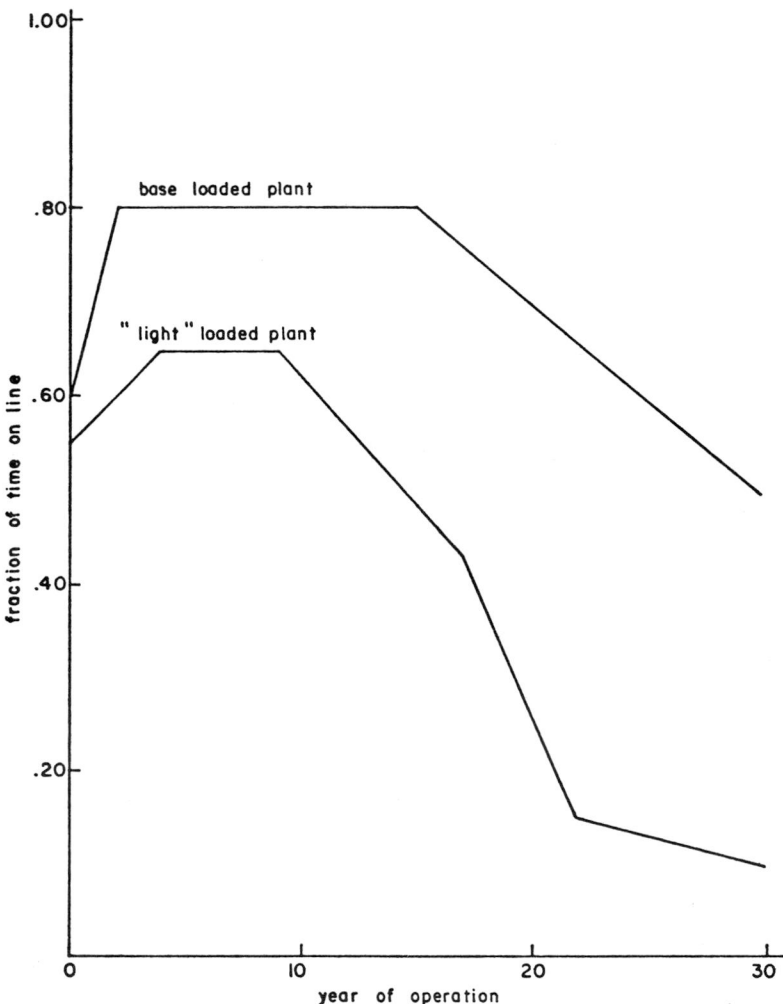

Fig. 21.18-2 Curves representing the load history of typical central station
power plants. Such curves are often used in economic analyses
of projected installations. Reproduced from USAEC Proposed
Final Environmental Statement: Liquid Metal Fast Breeder Program, WASH-1533, December 1974. Available from NTIS,
Springfield, Va. 22151.

may be related to temperature differences and gradients. Thus, mathematical descriptions of temperature and fluid flow are coupled to those of the nuclear reactions. Masses, pressure differences, temperatures, and temperature differences enter into relationships for strain, stress and strength within the reactor structure. As the result, all variables and parameters are related to performance. Optimum values may thus be calculated for all parameters. We present some simple illustrations of this type of analysis in Appendix 21A.

As may be appreciated from the foregoing remarks on the complete electrical-power systems, the list of design requirements on which an analysis would be based is too extensive to detail here. In the discussion of reactors, we emphasize the following items, among others, that must be included in a complete list: total electrical power, efficiency or heat rate, heat output of reactor, temperature of heat, mass and mass flow of fissile and fertile materials, size and number of fuel elements, fission-rate-neutron-flux control, conversion or breeding ratio, mass flow and velocity of coolant, upper limits on cost, capital construction, fuel cycle, operation and maintenance, interest rate, transmission, upper limits on release of radionuclides under all circumstances, upper limits on release of all effluents, aesthetic appearance.

21.20 Nuclear Safety, Environmental Protection and the Management of Radioactivity

In preceding sections, we have emphasized the requirement that the nuclear engineer must be concerned with the safe operation of nuclear power plants, the management of radioactive materials throughout the fuel cycle, and the control of harmful effluents, including heat. We shall now discuss the problems involved in the proper implementation of these guidelines.

Design requirements relating to public health, safety and the environment involve the interests and activities of utilities, manufacturers and constructors, individuals, the public, several levels and branches of government, financiers, insurance companies, and other institutions. Thus, in setting and meeting criteria, diverse customs and paradigms for decision making must be recognized. Conspicuous, in addition to engineering methods, are legal, financial and emotional approaches involved in the resolution of issues. Clearly, a large fraction of our population must acquire adequate perspective and assume the responsibilities and authority that are involved in the making of important decisions. The Code of Federal Regulations lists the rules for such decisions.

Safety engineering concerns all of the means by which either scheduled or unplanned activities or events connected with the production of energy from fission can harm people, property, surroundings and, in fact, the world. For historical and legal reasons, nuclear safety analyses dealing with accidents are performed separately from environmental analyses which are focused on the location or siting of plants and on the planned consequences involved in the disposal of unavoidable and irreducible effluents. Thus, evidence that the engineers' designs and the completed facilities will meet safety and environmental requirements is manifested through two series of documents, two legal procedures and different testing, surveillance and monitoring methods. The two series of documents are called safety-analysis reports and environmental statements, respectively. The two sets of considerations are clearly interrelated. Preparation of the reports, some of which have thousands of pages, requires contributions from diverse specialists over periods of three or more years and may cost several millions of dollars. Three additional years may elapse during review and decision concerning licensing of construction and operation of power and fuel processing plants.

The engineering basis for safe operation of each nuclear power plant is subjected to public review through hearings that are conducted by a 3-member hearing board in the locality of the proposed facility. Through this procedure and the judicial appeals process that may follow, adversaries have influenced the scope, basis and manner of decisions on matters of public safety and environmental effects. In addition, congressional hearings have been held to determine the adequacy and contents of the laws pertaining to nuclear safety.

In fact, the arguments about nuclear safety and the opposition to nuclear power have spread beyond the forums mentioned above and have taken the form of nationwide debates and, as mentioned earlier, public initiatives to regulate by laws the development and implementation of nuclear power.

Public debate has served to emphasize and distort important questions on the probabilities and consequences of disruptive accidents, the provisions for managing radioactive wastes, and the reliabilities of safeguards against diversion of potentially dangerous fissile and radioactive materials. In the debates between opponents and proponents of nuclear fission energy, the legitimate issues are probably less involved with problems of providing adequate safeguards in engineering designs than with complexities in construction and operation of power plants and the handling of radioactive materials by all of the diverse participants. In other words, competence and good intentions are more at issue than engineering principles.

Opponents of nuclear energy have offered an ultimate argument through the assertion that adequate controls of fissile and radioactive materials and the safety of nuclear-power and fuel-reprocessing plants can be achieved only with serious infringements on personal freedoms. On the other hand, the proponents of nuclear energy are confident that guarding the safety and health of humans will only require rational and intelligent behavior by individuals and groups. Thus, an important issue relates to societal structures that will prevail in the future. Incompetence, alienation, insurgency, and terrorism are inimical to a good society, not only to assurance of an adequate supply of essential energy. Both the proponents and opponents of nuclear power should properly assume a stable society in which dissent, debate and objective evaluation of societal alternatives are fostered. In designing, building, accepting and using nuclear-energy systems, thorough understanding and excellent judgement must prevail in order to balance and optimize the trade-offs between efficiencies and economies of scale, integration and standardization, and the stability and reliability associated with diversity, redundancy and adaptability.

APPENDIX 21A

ILLUSTRATIONS OF DESIGN ANALYSIS

21A.1 Introduction

In this Appendix, we comment on the theoretical basis for nuclear engineering-design analysis. The theories that describe and predict the behavior of neutron-chain reactors, flowing fluids and deformable solids have considerable beauty and power. Computation using fast, high capacity digital computers and highly developed computational codes and programs greatly facilitate design analysis. This analysis allows description of the plants and apparatuses that are the practical embodiment of imaginative concepts. The concepts are created intuitively by persons who have a thorough analytical understanding of physical phenomena and of practical human requirements.

The theoretical bases and the methods of engineering analysis of nuclear fission reactors have been published by a number of authors. Murray[1]

[1](a) R. L. Murray, Nuclear Energy: An Introduction to the Concepts, Systems and Applications of Nuclear Reactions, Pergamon Press, Inc., Oxford, U.K., 1975. (b) R. L. Murray, Introduction to Nuclear Engineering, Second Edition , Prentice Hall, Englewood Cliffs, N.J., 1961.

provides useful introductions to nuclear science and technology for undergrad-
uate students and non-specialists. Glasstone and Edlund[2] and Glasstone and
Sesonske[3] show how nuclear engineering arose from theory. Wigner and
Weinberg[4] have treated the physical theory of neutron chain reactors rigorously and elegantly. El-Wakil[5] emphasizes some of the power-engineering
aspects of nuclear energy. Sesonske[6] presents the methods of engineering
analysis now in use.

Here, we shall show how an accounting for neutron production, transport or diffusion and absorption leads to a description of the nuclear reactions
and heat production in a reactor. Finally, we shall attempt to show how size,
shape and the arrangement of a reactor fit the requirements and means for
removal of heat.

21A.2 Neutron Balance and Multiplication

A convenient way of accounting for neutrons in a thermal reactor, from
birth in fission to death by absorption or escape or leakage from the reactor,

[2]S. Glasstone and M. C. Edlund, The Elements of Nuclear Reactor Theory,
    D. Van Nostrand Company, Inc., 250 4th Avenue, New York, N.Y., 1952.

[3]S. Glasstone and A. Sesonske, Nuclear Reactor Engineering, D. Van
    Nostrand Company, Inc., 250 4th Avenue, New York, N.Y., 1963.

[4]E. P. Wigner and A. M. Weinberg, The Physical Theory of Neutron Chain
    Reactors, The University of Chicago Press, Chicago, Illinois, 1958.

[5]M. M. El-Wakil, Nuclear Energy Conversion, Scranton In Text Educational
    Publishers, Scranton, Pa., 1971.

[6]A. Sesonske, Nuclear Power Plant Design Analysis, USAEC Technical In-
    formation Services, NTIS, Springfield, Va. 22151, 1973.

is tabulated in Table 21A.2-1. Here, the numbers of neutrons are related or normalized to the numbers absorbed by fissile (e.g., ^{235}U) nuclei. The primary events determining the population of neutrons in a thermal reactor are birth in fission of ^{235}U due to absorption of a thermal neutron; birth in fission of ^{238}U due to fast neutrons directly from fission; random migration and loss or leakage at the edges of the reactor core; resonance absorption primarily in ^{238}U as the neutron is slowed down by collisions; random migration as a thermal neutron and leakage or absorption in structural materials and, lastly, capture by ^{235}U. If the number of neutrons born in this last process is greater, the same, or less for each successive generation, the population grows exponentially, remains steady or decreases, respectively. The ratio of neutron number density in succeeding generations is defined by the multiplication constant

$$k \equiv \eta \, \epsilon p f \mathcal{L}_f \mathcal{L}_t. \tag{21A.2-1}$$

The fractional change per generation is $k - 1$.

It is convenient to investigate the change in number density, n, with time in a simplified reactor that is sufficiently large to make the leakage vanishingly small. As the dimensions $\to \infty$, $\mathcal{L}_f = \mathcal{L}_t \to 1$ and $k \to k_\infty \equiv \eta \epsilon p f$. In this case, we may write

$$(1/n)dn/dt = (k_\infty - 1)/\tau, \tag{21A.2-2}$$

where τ = the average life of one generation. Clearly, if $k_\infty = 1$, the number density is constant or steady. This condition defines <u>criticality</u>. We recall (see Section 21.5) that a fraction β of the neutrons from fission is delayed.

Table 21A.2-1 Nuclear processes that determine the population of neutrons
in a thermal reactor.

Description of the processes producing and removing neutrons	Ratio of neutrons after the process to the original number	Number of neutrons remaining
One neutron is absorbed by a fissile nuclide. The number of neutrons produced by thermal fission per neutron absorbed = $\nu/(1 + \alpha) \equiv \eta$.	η	η
Some fast neutrons cause fission giving a "fast effect" and multiplying the neutron number by ϵ.	ϵ	$\eta \epsilon$
Moving randomly, some fast neutrons reach the edge of the reactor and "leak" out. The probability of not leaking is $\mathcal{L}_f$.	$\mathcal{L}_f$	$\eta \epsilon \mathcal{L}_f$
As they are moderated, some neutrons are lost by resonance capture by heavy nuclei. The probability of escaping resonance capture is p.	p	$\eta \epsilon \mathcal{L}_f p$
After being moderated, thermal neutrons may leak from the reactor as they move randomly. The thermal non-leakage probability is $\mathcal{L}_t$.	$\mathcal{L}_t$	$\eta \epsilon \mathcal{L}_f \mathcal{L}_t$
Some thermal neutrons are utilized for fission. The thermal utilization factor is f.	f	$\eta \epsilon p f \mathcal{L}_f \mathcal{L}_t$

These neutrons are emitted after fission by six fission products from ^{235}U
having known half lives. The longest half life is about 55.7 sec and the average
16.7 sec. If $(k_\infty - 1)$ is very slightly greater than 0, completion of a generation

will require over 55.7 sec and the population would increase slowly. Control
of the reactor by adjusting the absorption of neutrons in non-fissioning absorbers (control rods) could be accomplished deliberately by an operator observing a meter that indicates the output of a neutron counter or ionization
chamber filled with BF_3 gas $[(\eta, \alpha)$ $^{10}B \rightarrow$ $^4He +$ $^7Li]$. If $(k_\infty - 1)$ were allowed
to be significantly larger than zero, fewer of the delayed neutrons would be
required to replace those lost. Thus, the generation time would decrease
while the increase in population per generation would become larger. If
$(k_\infty - 1) = \beta \simeq 0.0075$, only prompt neutrons would be required for criticality;
this condition is labeled prompt critical. In power reactors, changes as large
as β in the multiplication constant k or the reactivity $[(k - 1)]$ are avoided by
physically limiting changes in the amounts and densities of absorber, moderator or fissile materials that can occur without a corresponding intrinsic
physical compensation, usually a decrease in k with temperature rise. That
is to say, the compensation must be independent of the operator and instrumentation.

The shortest generation time in an assembly of fissile materials is
associated with prompt criticality. In a thermal reactor, this time is that for
slowing down, migration in the moderator and capture of the neutron in the
fissile material, which requires several hundred collisions. During most of
its life, the neutron has a speed of 2200 to 3000 m/sec. Thus, the generation
time can be no shorter than 10^{-4} to 10^{-3} sec. In a fast reactor, far fewer collisions occur between birth and capture and the median neutron speed is greater
than 4 to 5×10^6 m/sec. Thus, the generation time can be as short as 10^{-6} sec.
In contrast, a nuclear explosive requires the largest excess reactivity and
shortest generation time that are physically attainable.

The complete kinetic equations for the time dependence of n may be
written as a set of linear, first-order differential equations involving the

populations and the rates of production and decay of the six delayed neutron emitters, which are accurately known. Solutions, e.g., by using the methods of the Laplace transformation to the frequency domain, provide the quantitative relations needed for design of a control system.

21A.3 The Diffusion of Neutrons

Neutrons move randomly in a reactor core. In LWRs, most of the collisions occur with the protons in water. On the average, a fast neutron loses approximately one half of its energy in each collision. The probability of a scattering collision is measured by Σ_s (as defined in Section 21.8) and is approximately equal to $N_H \sigma_{Hs}$ in LWRs, where N_H is the number of hydrogen atoms per cm^3. Here, σ_{Hs} varies from about 1 b for fission neutrons to 44.4 b for thermal neutrons. In water at $25°C$, $N_H \simeq 3.3 \times 10^{22}$ and the average distance between collisions of thermal neutrons is $1/\Sigma_s = [(3.3 \times 10^{22}/cm^3) \times (44.4 \times 10^{-24} cm^2)]^{-1} = 0.68$ cm. Over regions having dimensions appreciably larger than this distance, the behavior of large numbers of neutrons may be accurately modeled and described by diffusion theory.

For a diffusing species, Fick's law states that the current J or net flow in a given direction, e.g., the x-direction, crossing an area of 1 cm^2 is

$$J_x(x) = -D'(\partial n/\partial x),$$

where D' is the coefficient of diffusion and the minus sign signifies flow down the gradient. In three-dimensional vector notation,

$$\vec{J} = -D' \text{ grad } n.$$

In this relation, D' has the dimensions cm^2/sec.

For neutron diffusion, it is convenient to use a scalar flux $\phi = nv$, where v is the velocity of a group of neutrons. The product $(neutrons/cm^3)$ times (cm/sec) equals $(neutron/cm^2\text{-}sec)$ and is numerically equal to the number of neutrons in a volume v cm long with a cross-sectional area of 1 cm^2. By incorporating v in the diffusion coefficient, $D = D'/v$, Fick's law may be written as

$$\vec{J} = -D \text{ grad } \phi. \tag{21A.3-1}$$

In this relation, D has the dimensions of length and is related to a mean-free-path for scattering. By solving the equations of motion for two colliding particles, it can be shown that if scattering is isotropic or uniform in all directions and if the absorption is small compared to elastic scattering, then

$$D \simeq 1/3\Sigma_s. \tag{21A.3-2}$$

Collisions between neutrons and protons are nearly isotropic in a coordinate system centered on the center of mass of the colliding pair of particles. A more accurate value for D can be calculated from collision mechanics and is

$$D = 1/3(\Sigma_t - \Sigma_s\bar{\mu}), \tag{21A.3-3}$$

where $\bar{\mu}$ is the average of the cosine of the scattering angle, i.e., the angle between trajectories of the neutron before and after collision. As defined earlier, Σ_t is the total macroscopic cross section of the medium. The value of $\bar{\mu}$ for billiard-ball type collisions is $2/(3A)$, where A is the mass number of the scattering nucleus. Thus $\bar{\mu}$ becomes small compared to unity as A increases. Furthermore, if absorption is small compared to scattering, $\Sigma_t \simeq \Sigma_s$ and the approximate relation in Eq. (21A.3-2) holds.

The rate of any neutron-induced process may be calculated if the flux and the macroscopic cross section are known. Thus, the number of fissions per cm^3-sec in a reactor may be written as

$$R_f = \phi\Sigma_f. \qquad\qquad (21A.3-4)$$

In a finite reactor, the flux varies with position and, in some cases, with time. We may calculate this variation as follows. We consider a single direction, x. In a thin slab of thickness dx located at x, any change in the number of neutrons per unit volume must be due to three processes: (1) production by a source at x, s(x), chiefly fission; (2) losses at x, $\ell(x)$, due to absorption by any nuclide in dx; (3) the difference between flow in and out. The last contribution is equal to the difference in neutron current at each face of the thin slab. This change in current across the slab in the volume dx may be written as

$$J_x(x + dx) - J_x(x) = \frac{\partial J_x(x)}{\partial x}\, dx = -\frac{\partial}{\partial x} D \frac{\partial\phi}{\partial x}\, dx.$$

Adding the three contributions, the net change of neutrons per unit volume and time can be expressed as

$$\frac{\partial n}{\partial t} = \frac{\partial}{\partial x} D(x)\frac{\partial\phi}{\partial x} + s(x) - \ell(x).$$

In vector notation and using the vector $\vec{r}$ to signify a location with respect to a convenient coordinate system such as the three orthogonal axes x, y and z fixed in the reactor, the general three-dimensional relation becomes

$$\frac{1}{v} \frac{\partial \phi(\vec{r}, v, t)}{\partial t} = - \text{div } D(\vec{r}v)\text{grad } \phi(\vec{r}, v, t) + s(\vec{r}, v, t) - \ell(\vec{r}, v, t), \qquad (21A.3-5)$$

where we recognize that the quantities are functions of position and time for a group of neutrons having velocity v. Since the energy of a free neutron is kinetic, the velocity parameter can be mapped on the energy variable. The range of neutron energies from fast to thermal can be divided into intervals and an equation written for each interval giving a set of coupled equations. Furthermore, both s and ℓ may be expressed as linear functions of ϕ. Thus, a set of linear, partial, second-order differential equations, about which much is known, provides the means of calculating the neutron flux in fairly complex reactors (such as those described in Chapters 21 and 22) in which the amounts of fissile, moderating and other materials vary with position. The mathematical methods for solving such equations are well formulated and provide an extensive set of tools for either analytical or computer-programmed calculations.

21A.4 Nuclear-Reactor Criticality

In order to examine how these equations describe reactor behavior, we will solve two simple cases and then sketch how a complete calculation might be formulated. First, we consider a single group of thermal neutrons in a homogeneous, uniform, infinite slab of material having thickness H and made up of fissile and scattering materials. Since D is constant, the equation sim-

plifies to

$$\frac{1}{v}\frac{\partial \phi}{\partial t} = D\frac{\partial^2 \phi}{\partial x^2} + s - \ell.$$
   (21A.4-1)

Absorption of ℓ neutrons occurs within the slab.   Thus, $\ell = \Sigma_a \phi$, where Σ_a
is the total macroscopic absorption cross section.   We note that the neutron
source is a fission reaction which gives fast neutrons.   In the neutron balance
given in Section 21A.2, the number of thermal neutrons resulting from fission
per neutron absorbed by fissile nuclides is k_∞.   Thus,

$$\frac{1}{v}\frac{\partial \phi}{\partial t} = D\frac{\partial^2 \phi}{\partial x^2} + k_\infty \Sigma_a \phi - \Sigma_a \phi.$$
   (21A.4-2)

Inspection of Eq. (21A.4-2) shows that the first term on the right-hand side,
which represents the net flow of neutrons, will be into the thin slab if the curve
representing $\phi(x)$ is concave upward ($\partial^2 \phi/\partial x^2 > 0$) and out if it is concave downward.   This curvature of the flux function is called buckling.

A separation of variables may be made in the standard fashion for such
equations by letting

$$\phi(x, t) = \psi(x)T(t).$$

On substitution and dividing by ϕ, we obtain the following ordinary differential
equations for each member of the product:

$$\frac{1}{Dv}\frac{1}{T}\frac{dT}{dt} = \frac{1}{\psi}\frac{d^2\psi}{dx^2} + \frac{(k_\infty - 1)\Sigma_a}{D} = -B^2,$$

where B^2 is a real, positive constant of separation expressing the fact that the other two terms must be independent of each other. The two equations may be examined separately. The time dependence of ϕ can be written as

$$(1/T)(dT/dt) = -B^2 Dv + (k_\infty - 1)v\Sigma_a \qquad (21A.4-3)$$

and integrates directly to

$$T = T(0)\exp[(-B^2 Dv + (k_\infty - 1)v\Sigma_a)t]. \qquad (21A.4-4)$$

The equation for $\psi(x)$ is

$$(d^2\psi/dx^2) + B^2\psi = 0. \qquad (21A.4-5)$$

The boundary conditions for ψ may be examined by means of Fig. 21A.4-1, in which we have placed the origin of the coordinate system at the center of the slab. A general solution of Eq. (21A.4-5) has the form $\psi = A \sin m\pi x + B \cos m\pi x + Cx + D$, where m is an integer. However, with the chosen coordinates and in view of the physical necessity that ψ must vanish as $\pm x \to \infty$, only the cosine term is usable. The second necessary boundary condition requires that the current in the positive x-direction be the same on each side of the boundary at $x = H/2$ while the current in the negative x-direction vanishes for $x > H/2$. Solutions satisfying these conditions are

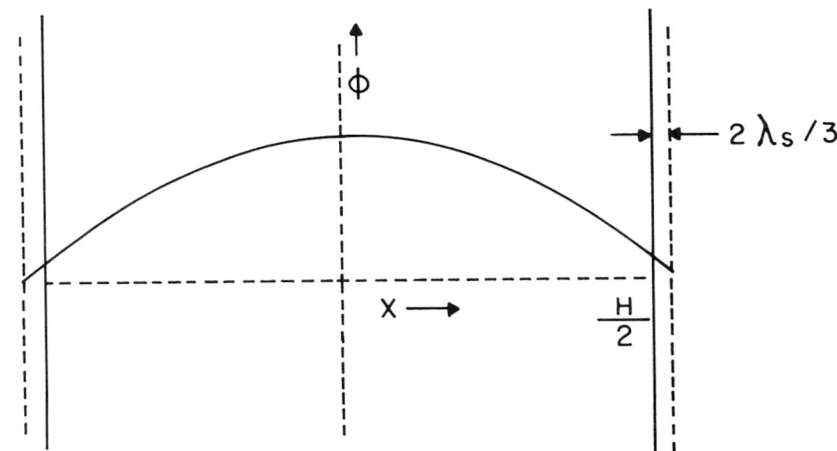

Fig. 21A.4-1 A plot of $\phi(x) = \phi \cos[\pi x/(H + \delta)]$ showing ϕ vanishing just outside
of the boundary of the slab.

$$\psi_m(x) = \psi_o \cos[m\pi x/(H + \delta)], \qquad\qquad (21A.4-6)$$

where $\delta = 2\lambda_s/3$; λ_s is the mean free path for scattering and is usually small
compared to H. For the purposes at hand, δ is neglected. Thus, there is a
set of eigenfunctions for Eq. (21A.4-5) and for each function an eigenvalue,
$B^2 = (m^2\pi^2/H^2)$, which is called the geometric buckling because of the shape
of the curve representing the flux, as portrayed in Fig. 21A.4-1. The complete solution of Eq. (21A.4-2) may then be represented by summing over all
eigenvalues, viz.

$$\phi(x, t) = \phi_o \sum_{m=1}^{\infty} \exp\{[-B_m^2 Dv + (k_\infty - 1)v\Sigma_a]\} \cos(m\pi x/H). \qquad (21A.4-7)$$

By inspecting Eq. (21A.4-7), we see that, in order for the flux to be independent of time, the argument of the exponential function must vanish. In general, this condition requires the lowest eigenvalue for the critical geometrical buckling,

$$B_c^2 = \pi^2/H^2 = (k_\infty - 1)\Sigma_a/D, \tag{21A.4-8}$$

thus establishing a relation between the geometrical buckling, which contains parameters related to size and shape, and what is called the <u>material buckling</u> that contains the parameters specifying neutron production, scattering and absorption. This general relation is called the <u>criticality equation</u> for a reactor. Noting that D and Σ have the dimensions, respectively, of length and reciprocal length, it is convenient to define a quantity

$$L^2 = D/\Sigma_a \simeq 1/3\Sigma_s\Sigma_a. \tag{21A.4-9}$$

For any assembly representing a reactor for which the geometrical buckling can be calculated and for which the flux may be modeled by a single group of neutrons, one can state the criticality condition in the following, slightly rearranged form:

$$k_\infty/(1 + L^2 B_c^2) = 1. \tag{21A.4-10}$$

Recalling the definitions given in Table 21A.2-1 and recognizing that we have only thermal leakage, we note that the effective neutron multiplication constant is

$$k_{eff} = n \epsilon pf/(1 + L^2 B_c^2);$$

(21A. 4-11)

therefore,

$$\mathcal{L}_t = 1/(1 + L^2 B_c^2).$$

(21A. 4-12)

As we assumed earlier, the larger the reactor relative to the magnitude of the mean-free-path for scattering, the greater the non-leakage probability.

Solution of partial differential equations of the type given in Eq. (21A. 4-7) are well known for a variety of geometries. The parallelepiped, the sphere and the cylinder are most familiar. For these three cases, the geometrical bucklings are, respectively, $B_c = \pi^2/A^2 = \pi^2/B^2 + \pi^2/C^2$, $B_c = \pi^2/R^2$, and $B_c = 2.405/R^2 + \pi^2/H$, where A, B, C, R, and H are the essential dimensions. The reactors that we have discussed are reasonably well represented as cylinders. The flux distribution is represented by the cosine functions of Eq. (21A. 4-6) parallel to the axis of such a reactor, except under the perturbing influence of a control rod, sensor, void or other irregularity. In a uniform bare cylinder, the steady state, radial flux distribution will be given by a zeroth-order Bessel function, which leads to the numerical constant 2.405. However, in practice, the density of fissile material and of the other parameters vary radially. As a consequence, results from numerical integration of sets of radial equations derived from Eq. (21A. 3-5) are used. In some instances, it is convenient to express $\phi(r)$ in a series of Bessel functions that satisfy the conditions at all boundaries. Thus, the results expressed in Eqs. (21A. 4-11) and (21A. 4-12) are readily extended to shapes other than a slab by inserting the appropriate geometrical buckling and the corresponding eigenfunctions.

The foregoing analysis is based on a single group of neutrons and illustrates the basic mathematics of criticality. However, the simple model does not represent the neutronic behavior of an actual reactor well. By adding a group of fast neutrons, we can illustrate the concept of multigroup analysis and also improve the model of a uniform, homogeneous reactor having no reflector.

We shall assume a time-independent neutron flux and set $d\phi/dt = 0$ in Eq. (21A.3-5). Since the reactor is uniform, D is independent of the spatial variable. Hence, the first term of the equation simplifies to

$$\text{divD grad } \phi = D\nabla^2\phi, \tag{21A.4-13}$$

where ∇^2 is the three-dimensional, second-order differential la Placian operator. In Cartesian coordinates, $\nabla^2 = (\partial^2/\partial x^2) + (\partial^2/\partial y^2) + (\partial^2/\partial z^2)$.

Cross sections for scattering and absorption may be assigned for each group. The source of thermal neutrons is the slowing down or moderation of the fast group. For our purposes, we will disregard neutrons that are in the process of losing energy. The source of fast neutrons is fission, which is induced by absorption of thermal neutrons. In this manner, the equations for the two groups are coupled.

The loss term is seen to be determined by the product of the respective total absorption cross sections for the assembly and the flux for each group. If the assembly is made up of slightly enriched uranium, water and structural materials, the total cross section for removal of neutrons from the fast group may be written as a sum of the absorption by each kind of nuclide in a given cm^3, i.e.,

$$\Sigma_f = N_U\sigma_{Uf} + N_H\sigma_{Hf} + N_O\sigma_{Of} + N_i\sigma_{if},$$

where the N and σ are the numbers of atoms per cm^3 and total (scattering and absorption) cross sections for each nucleus, respectively. The cross section for oxygen in the water is quite small. The last term, labeled i, represents the loss to such structural materials as the zircaloy cladding and is represented by a sum over the constituents of the alloy. Materials are selected and arranged so as to minimize losses. A similar expression for Σ_t represents the thermal group. Thus, the respective loss terms may be written $\ell_f = \phi_f \Sigma_f$ and $\ell_t = \phi_t \Sigma_t$.

The corresponding source terms may be found from the definition of k_∞. The factor p contained in k_∞ represents the neutrons (per thermal neutron absorbed by fissile material) that reach thermal energies, i.e., those that escape resonance capture in the slowing down energy range. Consequently, the source term for the thermal group is $s_t = p\Sigma_f \phi_f$.

To estimate the source of fast neutrons, we note that f is the fraction of neutrons absorbed by fissile material and that this absorption creates $\eta\epsilon$ fast neutrons. Now $f\eta\epsilon = k_\infty/p$ and thus the source term for the fast group may be written as $s_f = (k_\infty/p)\Sigma_t \phi_t$.

The neutron balance for each group at the steady state can be expressed as follows by using the symbol for the operator introduced in Eq. (21A.4-13) and the relations developed above for the sources and sinks:

$$D_f \nabla^2 \phi_f - \Sigma_f \phi_f + (k_\infty/p)\Sigma_t \phi_t = 0, \qquad (21A.4-14)$$

$$D_t \nabla^2 \phi_t - \Sigma_t \phi_t + p\Sigma_f \phi_f = 0. \qquad (21A.4-15)$$

Two principles lead directly to a solution for criticality. First, we note that geometrical considerations, namely, the boundary conditions discussed for Eq. (21A.4-5), require that both groups have the same buckling. Thus, we replace ∇^2 by B^2 as indicated by Eq. (21A.4-5). We also use a theorem without proof[1] which shows that the function ϕ, for a bare reactor, is separable into a product of two functions, one depending only on energy, the second only on space. Thus,

$$\phi_f = A_f \psi(r), \quad \phi_t = A_t \psi(r).$$

On substitution in Eqs. (21A.4-14) and (21A.4-15), we obtain

$$- D_f B^2 A_f \psi - \Sigma_p A_f \psi + (k_\infty/p)\Sigma_t A_t \psi = 0,$$

$$- D_t B^2 A_t \psi - \Sigma_t A_t \psi + p\Sigma_f A_f \psi = 0.$$

To represent a physical situation, these equations must have a non-trivial solution $\psi \neq 0$. We may then divide by ψ and further require that the determinant of the coefficients A_f and A_t be zero, viz.

$$\begin{vmatrix} (- B^2 D_f - \Sigma_f), & (k_\infty/p)\Sigma_t \\ \\ p\Sigma_f, & (- B^2 D_t - \Sigma_t) \end{vmatrix} = 0 \qquad (21A.4-16)$$

---

[1] Ref. [4] of Section 21A.1, p. 382.

or $(B^2 D_f + \Sigma_f)(B^2 D_t + \Sigma_t) - k_\infty \Sigma_f \Sigma_t = 0.$

To put the criticality condition in its usual and most convenient form, we divide by $\Sigma_f \Sigma_t$, incorporate the scattering or transport cross sections contained in the diffusion coefficients, and the total cross section in the quantities L^2 just as we did in Eq. (21A.4-10) and rearrange as before to write the criticality equation as

$$\frac{k_\infty}{(1 + B_c^2 L_f^2)(1 + B_c^2 L_t^2)} = 1. \qquad (21A.4-17)$$

The non-leakage probabilities are

$$\pounds_f = 1/(1 + B_c^2 L_f^2), \quad \pounds_t = 1/(1 + B_c L_t^2).$$

As before, B_c is given by the size and shape of the bare reactor and k_∞ and L by the appropriate constants for the nuclear reactions and numbers of nuclei per unit volume in the reactor. With appropriate choice of cross sections, plausible representations of simple, real assemblies result. However, the cross sections are necessarily averages over poorly specified ranges of energy and are really empirical constants rather than a priori representations of actual probabilities of reactions.

While the foregoing simplified theory illustrates the principal features of the physical theory of neutron-chain reactors reasonably well, it does not suffice for engineering calculations of reactors. Three kinds of physical factors require explicit representation in the equations: the spatial distribution of all of the reacting nuclei; the dependences of the flux function and cross

sections on energy over a complete range of neutron energies; and the inclusion
of time-dependent processes, in particular, for the six groups of delayed neutrons and changes of temperature. Each factor can be specified or selected by
the designer and included in detailed and complicated numerical calculations.

      The spatial problem includes such features as (1) the radial and, in
some cases, the longitudinal variation of the amount of fissile and other material, as initially loaded to flatten the power density and as depleted or burned
out by fission and capture; (2) the presence of a thick layer of moderator,
called a reflector, located around the core to reduce neutron leakage; (3) the
location of neutron-absorbing control rods; (4) positions at which fuel may be
absent to allow for insertion of sensors; and (5) the variation of atom densities
and cross sections with temperature. Two kinds of mathematical methods are
employed to represent the spatial aspects. The whole assembly may be divided
into zones or regions and solutions found that satisfy the boundary conditions
at each interface. For smaller variations, perturbation theories that are
highly developed for linear, second-order partial differential equations are
used. The mathematics of linear vector spaces and orthogonal functions supply the analytical tools.

      The dependence of the flux function and reaction-rate parameters on
energy has been handled in two ways. Beautiful theories of the slowing down
of neutrons, based on diffusion in energy space, have been developed and coupled
with the knowledge of nuclear cross sections based on theories of the nucleus.
A second approach that is more useful in contemporary nuclear engineering is
to divide the energy range from 10 Mev to thermal energy into several carefully
selected intervals, develop sets of cross sections from experimental measurements appropriately averaged over each interval, and work with sets of coupled
equations, one for each interval; these relations are analogous to the two Eqs.
(21A.4-14) and (21A.4-15)

The construction of multigroup equations is straightforward and shows how most of the required physical relationships may be combined for computation. Because the slowing down of neutrons by elastic collisions leads to an energy spectrum of the type $d\phi(E)/dE \propto 1/E$, logarithmic energy intervals weight the neutron groups more favorably than linear intervals. Accordingly, the range of a function called the neutron lethargy = $\log(E_o/E)$, where E_o can be set equal to 10 Mev ($10 \text{ Mev} \geq E > 0$), can be divided into G groups, where G ranges from 10 to 50 depending on the detail required, the numbers of spatial intervals needed and the capacity of the computer. Each group and its parameters may be labeled by assigning an index integer i or j ranging from 1 to G.

For each group, the loss and source terms may be expressed in mathematical language that is readily used in digital computation. The loss from each group is due to absorption and to scattering to groups of lower energy. The microscopic cross sections for each kind of nucleus are found from the data banks (see Section 21.7), which are summed over all nuclei located in the designated spatial interval and are then averaged over the energy interval of the group. Thus, the loss from the ith group at position $\vec{r}$ may be written as

$$\ell_i(r) = \phi(E_i, r, t)\left[\Sigma_a(E_i, r, t) + \sum_{j=1}^{G} \Sigma_s(E_i \rightarrow E_j, r, t)\right].$$

In this expression, we have separated the different processes, namely, absorption or capture and the scattering of neutrons of energy E_i to groups of lower energy j, $i \leq j \leq G$. We note, however, that the sum in the square bracket is merely the total cross section for neutrons of energy E_i, a quantity that has been measured directly for the types of nuclei that the designer may specify in the volume element located at $\vec{r}$. Therefore, we write

$$\ell_i = \phi_i \Sigma_t(i),$$

where the spatial and time variables are understood and the index i represents the energy parameter.

The source term is more complex because it is necessary to symbolize the contribution to the ith group of neutrons both directly from fission, if there is a contribution in the interval in question, and through scattering from each of the groups having energy greater than E_i. The spectra of fission neutrons and the cross sections for the scattering neutrons of energies E_j to E_i have been measured. This knowledge can be expressed in the source term as follows:

$$s_i = \sum_{j=1}^{i-1} \phi_j \Sigma_s(j \rightarrow i) + X_i \sum_{j=1}^{G} \phi_j \nu(j) \Sigma_f(j),$$

where the first summation over j gives the scattering into the ith group from each group having energy greater than E_i and the second summation adds the neutrons from fission having energy in the ith group. The factor X_i gives this fraction and these fission factors for each i are such that their sum over all groups equals unity. This way of writing the fission contribution recognizes that the yield of fission neutrons depends on the energy of the neutron causing fission. This term in X_i also contains the fast effect discussed in Section 21A.2.

Collecting the terms representing net flow (leakage) in or out, losses and sources, the rate of change of flux in an element of volume may be expressed for each group as in Eq. (21A.3-5). In greater detail,

$$\frac{1}{v}\frac{\partial \phi_i}{dt} = \text{div}(D_i \text{ grad } \phi_i) + \sum_{j=1}^{i-1} \phi_j \Sigma(j \rightarrow i) + X_i \sum_{j=1}^{G} \phi_j \nu(j)\Sigma_f(j) - \phi_i \Sigma_t(i).$$

$$(21A.4-18)$$

Thus, there are G simultaneous, coupled equations for the G functions ϕ_i. Although complex, the set of equations contains most of the necessary physical relationships governing the behavior of neutrons. Solutions can be obtained by machines.

The reactor designer works primarily with the steady-state equations obtained by equating to zero the term on the right of Eq. (21A.4-18). The resulting equation can be arranged as follows:

$$\nabla \cdot (D_i \nabla \phi_i) - \phi_i \Sigma_t(i) + \sum_{j=1}^{i-1} \phi_j \Sigma(j-1) = (X_i/\lambda) \sum_{j=1}^{G} \phi_j \nu(j)\Sigma_f(j), \quad (21A.4-19)$$

where ∇ is the symbol for a first-order differential operator for the coordinate system used to represent the reactor (most often the radial equation in cylindrical coordinates) and λ represents an eigenvalue. Often there are 3 or 4 radial zones, each containing an assigned number of nuclei per unit volume.

The numerical methods for solving equations of this type are well developed.[2] Programs for digital computations are available from user groups, for example, the Argonne National Laboratory of ERDA, and others.

Much is known about the general nature of solutions of the set of Eqs. (21A.4-10) to (21A.4-19). We saw, in connection with Eqs. (21A.4-14) and (21A.4-15),that, for a homogeneous region, the critical buckling of each steady-state

---

[2]Ref. [6] of Section 21A.1.

group must be the same. A set of ϕs may be assumed and, using tabulated cross section and neutron-yield data (often on tapes that are read by the computer), the summations may be done for the inhomogeneous term on the right-hand sides of Eqs. (21A.4-16) to (21A.4-19). Further, the second-order differential term is represented as a set of second-order finite differences on a finite set of points in the coordinate system. From these steps, a set of algebraic equations results that has a secular equation and a set of eigenvalues in a manner formally similar to Eqs. (21A.4-14) and (21A.4-15). In turn, numerical eigenfunctions are found that match values of flux and current at boundaries and that are better than the set first assumed. Following successive iterations and computer manipulation of matrices, convergent solutions may be obtained that are as accurate as desired.

The foregoing equations display in compact form the physics of neutron production, scattering, energy loss, absorption, and leakage. Before digital computers came into extensive use, much effort went into functional analysis of these and related mathematical representations. One way of approaching the analytical problem is to use representations of the cross sections and, in effect, replace summations with integrals. Then one may regard the set of equations as infinite and integrate over the entire range of lethargy to obtain an integro-differential equation that contains much but not all of the information provided by Eq. (21A.4-18). The scattering and slowing down portions of the problem may be treated accurately by analytical methods. Resonance absorption is more difficult to represent.

In the early days of neutron physics, before the discovery of fission, Fermi formulated an elegant theory of neutron diffusion and slowing down. [3] This formulation is analogous to the conduction of heat and cooling of a body.

---

[3] Ref. [2] of Section 21A.1.

It permits calculation of resonance absorption if the absorption is not too great. It also represents leakage and thermal utilization more realistically than the two-group formulation and was extensively used in the analysis of reactors.

Diffusion theory leads to poor physical representations of flux close to discontinuities or in regions in which there is strong variation of parameters with distance. Better mathematical descriptions of transport are available but have incompletely formulated mathematical tools for practical solutions. In one approach, transport equations are used to formulate the laws giving the distribution of neutrons as functions of energy, space and time. An equivalent approach is the Monte Carlo calculation in which a neutron is followed from birth through successive collisions, with instructions for the computer to assign probabilities for the various possible events.

The production of heat raises the temperature of the reactor, more in some regions than in others. The reaction probabilities and rates depend on temperature primarily through the density of atoms per unit volume of the materials. For moderated assemblies, the energy or velocity of thermal neutrons vary with temperature. In the resonance region, the change in cross section with relative velocities for nuclides and neutrons (Doppler effect), alters capture and fusion probabilities and is important in reactor control. Practicable reactors are constructed in such a way that k decreases with temperature for all possible modes of temperature variation.

Reactivity also changes with time: as fission products accumulate, fissile material is "burned up" and other changes occur. At the beginning of life, a power reactor must have enough absorbing material that can be gradually removed or burned out to make the reactor controllable.

Calculations are often made for "cold and clean" and "hot and dirty" conditions to determine the amount of control. In order that the excess reactivity at the beginning of life be safely within the range of the control system, boron is introduced as a "burnable poison" into many reactors. For the PWRs, soluble boron compounds are added to and removed from the cooling water. In the BWRs, control blades serve this purpose. Neutron absorptions for purposes of control reduce the number available to fertile materials and for fission. Thus, the control function increases the critical mass and decreases the production of fissile material.

21A. 5 Heat Transfer

The removal of heat from a reactor is intimately coupled with the neutronic design aspects. We may illustrate this topic by examining some of the estimates an engineer might do in his head or on the back of an envelope during the conceptual design phase.

The heat-transfer medium in practical reactors flows longitudinally or axially through a reactor and is confined transversely or radially. Axial flow of coolant influences other design features such as the mode of introduction and removal of fuel, of control elements and of instrumentation. These components are introduced and removed along the axis of coolant flow and, preferably, at the cool end.

The first requirement on coolant flow results from an overall heat balance that maintains a steady temperature throughout the reactor. The mass flow of coolant ($\dot{M}$) times the temperature rise (ΔT) and the average heat capacity (C) must equal the thermal power produced by the reactor. Since C will vary with temperature and pressure, an average $\overline{C}$ is required along the length of the coolant-flow path. The thermal power P_t of the reactor is related to the electrical power P_e by an efficiency η, as follows:

$$P_t = P_e / \eta = \dot{M}\overline{C}\Delta T.$$

In a power reactor, there will be some desired or allowable ΔT between inlet and outlet. Within allowable limits, each channel should have about the same exit temperature so that the average temperature of the mixed coolant is as close as possible to that at the exit of the hottest channel. Flow through each channel ($\dot{M}_c$) is adjusted according to the heat production expected for the channel. Because of approximate symmetry about the central axis, there will be a ring of channels producing about the same amount of heat. Usually only a few different radial zones (often three) suffice, each with its properly sized inlet orifice.

The overall flow in a 1,100 Mw_e PWR may be estimated by using some of the design parameters for the large PWR listed in Table 21A.5-1: electrical power = 1.1×10^9 w_e, efficiency = 0.325, thermal power = 3.4×10^9 w_t, ΔT = $T_{out} - T_{in}$ = $37°C$, specific heat for water = 5.3×10^3 w/kg-$°C$. The mass flow is $\dot{M}$ = $3.4 \times 10^9 /(37 \times 5.3 \times 10^3)$ = 1.7×10^4 kg/sec. An obvious require-ment on the number and size of coolant passages is that they permit the re-quired mass flow at a reasonable drop in pressure and a reasonable power consumption for pumping. For a channel near the center of a reactor producing maximum power, the neutron flux, fission rate and heat production will vary along the length of the channel in accord with Eq. (21A.4-6) and Fig. 21A.4-1. Except near the ends or other nonuniformities (e.g., control rods), the heat production per unit length is

Table 21A.5-1 Selected design parameters for a large PWR.

total thermal power	3.411×10^9 w
power generated in the fuel	3.34×10^9 w
maximum linear power	620 w/cm
approximate fuel loading in the core	87,700 kg
specific power	39 w/g
length of the fuel rod (H)	3.66 m
approximate number of rods	3.96×10^4
approximate outer diameter of the rods	1.04 cm
approximate diameter of the UO_2 pellets	0.92 cm
approximate fuel density	9.45 g/cm^3
approximate total surface area of the rods	4.8×10^7 cm^2
average heat flux Q	70 w/cm^2
ratio of peak-to-average heat flux or the hot-channel factor	2.80

$$W_{\ell} = (\pi W/H) \sin (\pi x/H), \qquad\qquad (21A.5-1)$$

where W is the total power of the channel of length H and x is measured from the entrance rather than the middle of the channel. The average power per unit length is W/H and the peak power at the midpoint is $\pi W/2H$ or 1.57 times the average value.

The heat content or enthalpy increase of the coolant is the integral of this function [Eq. (21A.5-1)] along the channel. Roughly, the enthalpy increase of the coolant is proportional to its temperature rise. Consequently, this temperature T_c is reasonably well represented by

$$T_c = T_{in} + (W/2\dot{M}\bar{C}H)[1 + \cos(\pi x/H)]   \qquad (21A.5-2)$$

A plot illustrating these relations is easily constructed and is shown in Fig. 21A.5-1. Also, we may sketch a curve for the central temperature of a fuel rod in the channel by adding a sine curve to T_c on the basis that the temperature difference between the center of the rod and the coolant is proportional to W_{ℓ}. Along the center line of the fuel element, the highest and limiting temperature occurs just downstream of its mid-point.

Next, we examine the flow of heat within and out of an element made of UO_2 used in an LWR, LMFBR or GCFR. The temperature at the center of the UO_2 is as high as permitted by the behavior of materials. Heat flows radially through poorly-conducting UO_2 down an increasing temperature gradient. At the interface between the UO_2 cylinder or pellet, contact resistance causes an abrupt drop in temperature. The drop through the metal cladding with relatively high conductivity is small. At the boundary between the cladding and coolant, there is a layer of relatively slowly moving coolant through which heat must flow primarily by conduction. In the body of the coolant, turbulent flow mixes the warmer and cooler coolants rapidly and, thus, transfers heat relatively rapidly by convection. Neutronic analysis (see Section 21A.4) permits calculation of the heat-production rate in each volume element.

Radial heat flow by conduction through the solid-fuel element may be analyzed with the help of Fig. 21A.5-2. In the cylindrically symmetrical ring of volume $2\pi r dr$, the calculated heat generation per unit volume is $q(r)$. At $\vec{r}$, the total heat that must flow through the cylindrical surface $2\pi r$ in circumference and of unit axial length is the integral of the above expression from

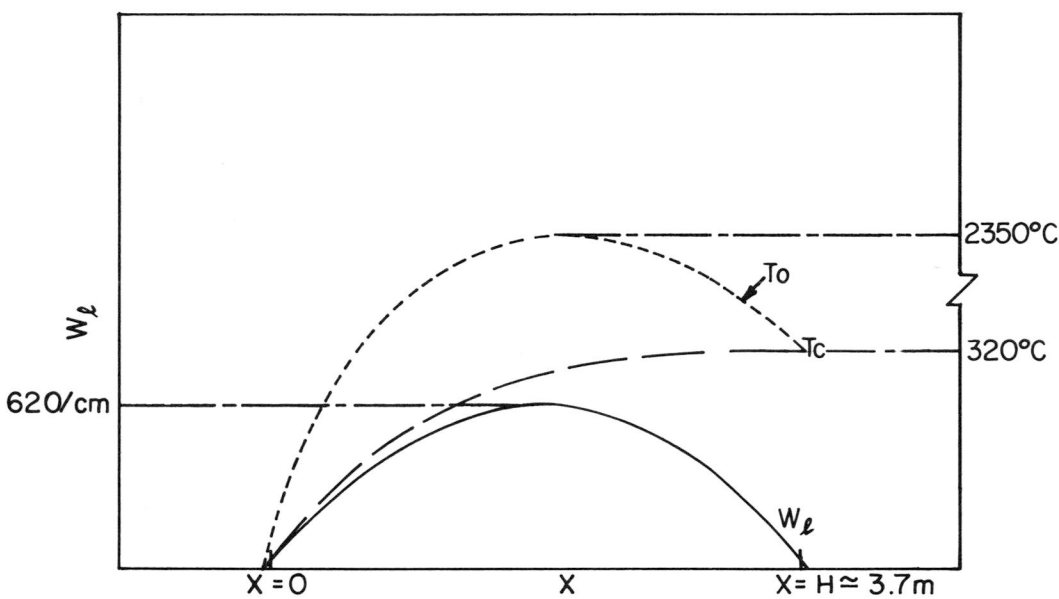

Fig. 21A.5-1 Schematic representation of power and temperature along a fuel channel. The power per unit length, W_ℓ is represented by a plot of Eq. (21A.5-1). The temperature of the cladding, T_c, is given by a plot of Eq. (21A.5-2) (longer dashed line, — —). The temperature at the center of a fuel rod T_o is represented by addition of a sine function to T_c (---). The maximum values of W_ℓ, T_c and T_o are assigned on arbitrary ordinate scales.

$r' = 0$ to r. The heat flowing per unit area and unit time equals the product of the thermal conductivity k, the area, and the gradient of the temperature, i.e.,

$$2\pi \int_0^r r' q(r') dr' = -2\pi rk \frac{dT}{dr}.$$

To find the heat flow per unit area from the surface of the UO_2 fuel, we

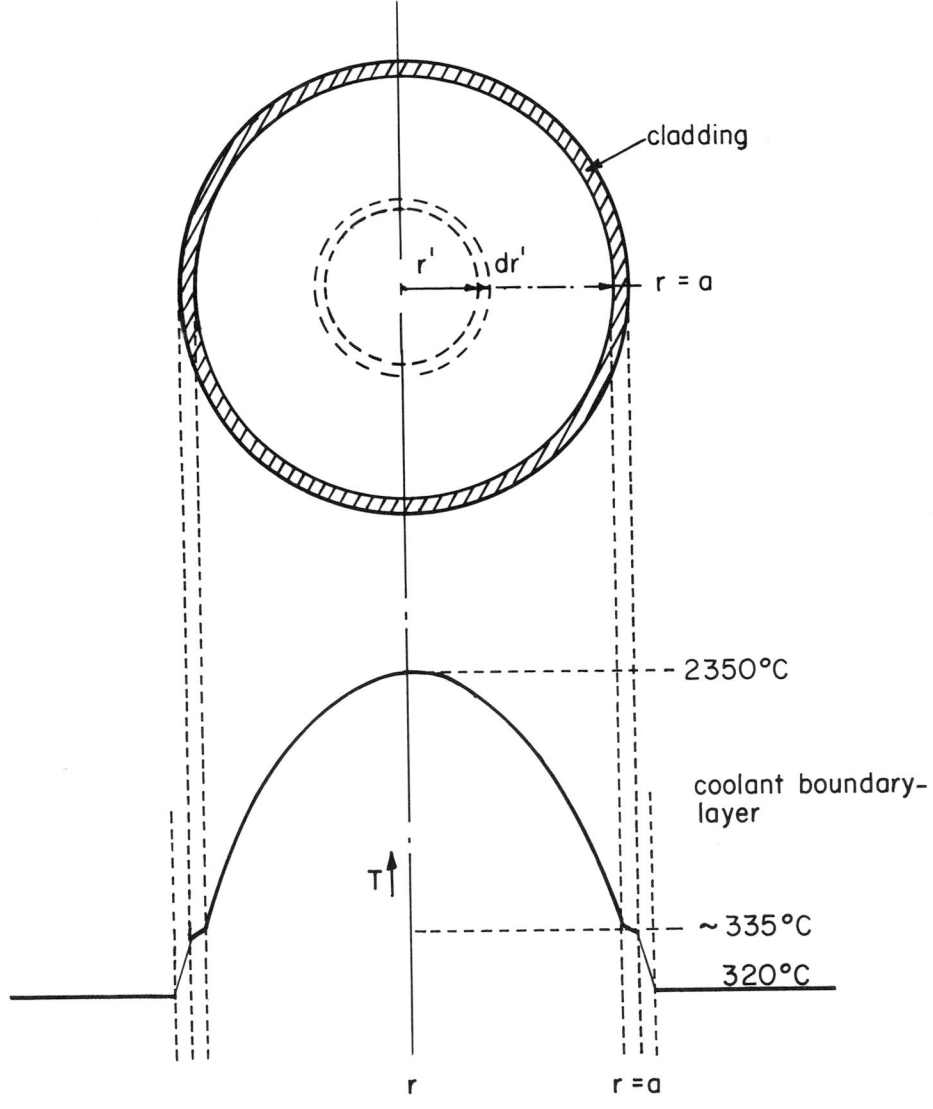

Fig. 21A.5-2 A clad UO_2 fuel element and a schematic plot of temperature
          along a radius projecting from the center of the rod into the
          coolant.

divide by $2\pi r$ and integrate from $r = 0$ to $r = a$ (a = the radius of the UO_2 pellet), viz.

$$\int_0^a \frac{1}{r} \int_0^r r'q(r')dr' \, dr = \int_{T_c}^{T_o} k dT.$$

If heat generation is uniform and the thermal conductivity does not change with temperature, the integrals give a relation between the temperature anywhere in the fuel and the rate of heat generation as follows: $(qr^2/4k) = T_o - T$; at the surface, $(qa^2/4k) = T_o - T_c$. If we multiply both sides of the last relation by π, we find that $\pi q a^2 = W_\ell$ = heat produced per unit length of the fuel element. Setting $W_\ell = \pi a^2 q$, $T_o - T_s = W_\ell/4\pi k$. Thus, the temperature drop from the center to the surface depends only on the power per unit length and is independent of the radius of the element. In the general case, because of strong absorption of neutrons in the fuel (called "self-shielding"), q depends on r and decreases toward the center. In addition, the thermal conductivity of UO_2 varies strongly with temperature. In general, the average power per unit length is

$$W_\ell = 4\pi \int_{T_c}^{T_o} k(T)dT.$$

Much effort has been expended on experimental measurements in reactors of the thermal conductivity integral and the thermal conductivity itself. In practice, the latter quantity is difficult to measure accurately. It varies with composition, porosity of the pellet and burn-up. Its magnitude is between 0.025 and 0.035 w/cm-°C over much of the range of interest. Extensive tests have established that the maximum practical central temperature and power per

unit length are, respectively, about 2,350°C and 650 to 700 w/cm; a reasonable
working maximum lies between 500 and 650 w/cm.

In any practical case, there is a limit on the temperature drop from
the cladding surface to the body of the flowing coolant and, in turn, this temperature drop limits the heat flux per unit area. Thus, we find that there is
a lower limit on the total area of the fuel elements. The fluid boundary layer
next to the fuel element offers the major resistance to heat flow and the drop
in temperature depends on its thickness and thermal conductivity. The effective thickness depends on the velocity of flow, an effective size of the channel,
the roughness of the wall, and the viscosity and density of the coolant. Theoretical studies suggest ways of combining variables to give relationships that are
readily fitted to experimental data. The data may be interpolated or extrapolated to the conditions of flow, channel dimensions and temperature for given
practical designs. The estimation of heat-transfer coefficients from cladding
to coolant is illustrated on pp. 196-199 of Ref. [1a] in Section 21A.1 and pp.
139-141 of Ref. [6] in Section 21A.1.

For a given liquid, velocity and local pressure, an increasing temperature drop will accompany the flow of increasing heat, as is illustrated in
Fig. 21A.5-3, which pertains to naturally-convecting water. When the boiling
point of the liquid at the interface is reached, heat transfer increases rapidly
with ΔT because very small bubbles of vapor form and collapse as they en-
counter cooler liquid. This "nucleate boiling" breaks up the boundary layer
and provides additional heat capacity. However, there is a ΔT at which bubbles tend to coat the surface continuously and give a layer of very poor thermal
conductivity. This "burn out" condition must be avoided in a reactor. Heat
flux must be kept in a safe range, typically well below 3.2×10^6 w/m^2 for an
LWR. In the nucleate boiling range, there is always some danger that a fluctuation will cause a transition to region D on the other side of the "hump" in

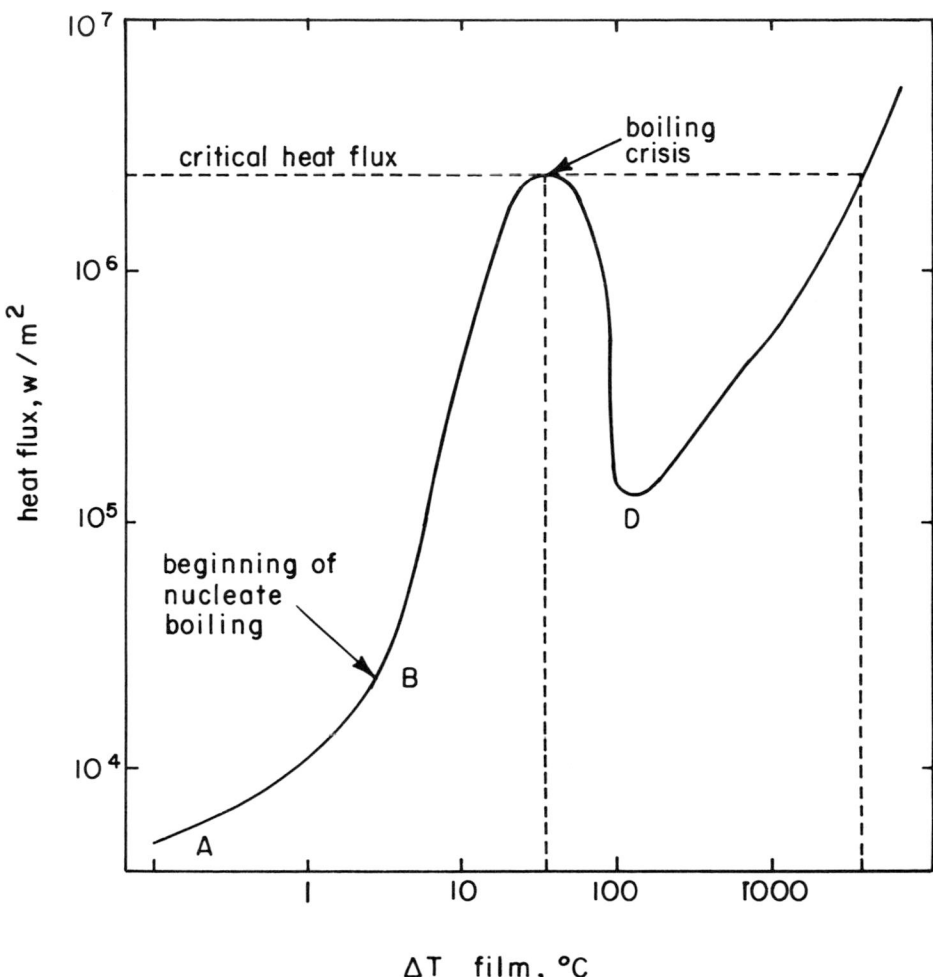

Fig. 21A.5-3 The variation of heat flux with the temperature difference across
a laminar film next to a heated rod in a pool of water.

Fig. 21A.5-3. Extensive studies of this limit provide correlations among local values of heat flux, flow velocity, pressure, temperature, and channel size. These correlations are used in the computer-programmed design calculations to assure safe, maximum performance.

The calculations of heat transfer prescribe the maximum heat production and flux in the hottest channel. Earlier, we indicated how the maximum power along a channel was related to the average for the channel. Similar relations, depending on geometry, the pattern of fuel loading and location of poison elements hold between the hottest channel and the average for the reactor. The enthalpy increase in the hottest channel is about twice the average. These hot channel factors are essential in design and relate to the most explicit connection between the neutronic and heat-transfer portions of the design analysis. In addition to ratios of peak to average fission rate, called the "nuclear hot channel factor," other variations such as tolerance in fuel-pellet size, cladding thickness, errors in adjusting coolant flow and fuel element combine to make the ratio of peak-to-mean heat load about 2.3. Care in design, analysis, manufacturing and operation can reduce these factors and increase the average. Efficiency is promoted by bringing average performance as close as possible to the limiting peak performance.

The approximate maximum limiting heat flux is slightly greater than 200 w/cm^2 (see Fig. 21A.5-3). Although the limiting condition for burn-out occurs downstream of maximum linear power generation, let us use this number at the zone of maximum power as follows: $620(\text{w/cm}) = (2\pi a \text{ cm}) \times 200(\text{w/cm}^2)$ or $2a = 620/(200 \times 3.14) = 0.99 \text{ cm}$. We see that the design choice matches the maximum power per unit length and hence the maximum allowable central temperature with the maximum allowable heat flux and the maximum safe ΔT across the fluid boundary layer.

The ratio of peak-to-average heat flux is 200 (w/cm^2) ÷ 70 (w/cm^2) = 2.86. We may compare this estimate with that obtained from our previous average along a channel multiplied by the average over all channels, 1.57 × 2 = 3.14, and with the value in Table 21A.5-1 which is based on detailed analysis.

Applying the design values of the hot channel factor for heat flux, we obtain a very rough value for the average power per unit length $\overline{W}_\ell$ as follows: $\overline{W}_\ell$ = W/2.80 = 222 w/cm. From this value, an estimate of the approximate total length of fuel rods required to yield the desired power from the fuel is L = 3.34 × 10^9 w/(222 w/cm) = 1.5 × 10^7 cm. The design value is approximately 1.45 × 10^7 cm. Thus, requirements for the transfer of heat determine, within limits, the area, size and total length of the fuel rods.

The length of a single fuel rod is given by the size of the reactor, which is derived from the criticality equation. The height H of the reactor core would be determined by solving the equations given in Section 21A.2 for B^2. Using the value for H given in Table 21A.5-1, the number of rods may be shown to have the value given in this table.

Neutronic calculations show the percentage of U-235 in the fuel, the total mass of uranium, the distribution of coolant flow and the "nuclear portion" of the hot channel factors. The design, for which some values are listed in Table 21A.5-1, requires 3 radial enrichment zones containing 2.25, 2.80 and 3.30% of U-235 in a fresh core. The values change somewhat as refueling occurs.

Another requirement on fuel-element size is that the volume of fuel required be contained. If the radius of the fuel pellets is a_f and their average density in ρ_f (in practice about 90% of the crystalline density of UO$_2$), then $\pi a_f^2 L \rho_f$ = M = the mass of fuel. The requirement on area in terms of heat flux Q is 2πaLQ = total power W. Dividing the second equation by the first, W_s = 2Qa/$a_f^2 \rho_f$, where W_s represents the specific power and is an important design parameter because it shows how well the fuel is being used. The difference between a and a_f is approximately the thickness of the cladding, a value determined by the strength of materials and the methods and expense of fabrication. For zircaloy tubing used in the LWR, this number is about 0.04 cm. Thus, $a/a_f \simeq 1.08$ and $W_s \simeq 2.2\, Q/a_f \rho_f$.

One may make W_s larger, or lessen the requirements on Q by making $\underline{a}_f$ and $\underline{a}$ smaller and increasing the number of fuel elements sufficiently to keep the volume of fuel constant. However, the cost of manufacturing the elements increases directly as the number and inversely as a power of a_f, which is probably greater than unity because the number of pellets, the difficulty of fabricating tubing and the fraction rejected for imperfections increase as the radius and thickness decrease. Therefore, the optimum fuel radius lies close to that required for the maximum permissible values of W_ℓ and Q, as estimated above for the LWR and determined by detailed design in Table 21A.5-1.

The critical mass of fast reactors is much greater than that for the LWR or HTGR. Hence, there is greater incentive to maximize heat flux. At the same time, the peak power per unit length of uranium-plutonium oxide elements is limited to lower values, perhaps 500 to 550 w/cm because plutonium lowers the melting point of the oxide and alters its chemical behavior somewhat. The use of sodium to transfer heat increases the permissible surface-heat flux considerably because the high thermal conductivity of the liquid metal permits the ΔT through the boundary layer to be small. Also, the boiling point is about 837°C, thus permitting a much larger difference in temperature across the boundary layer while reducing the hazard of burn-out. Liquid sodium is more than adequate as an effective heat-transfer medium when uranium dioxide fuel elements are the sources of heat. This property is a major reason why the effectiveness of the LMFBR depends on the development of uranium-plutonium monocarbide elements, which permit W_ℓ to be greater than 1,000 w/cm.

21A. 6 <u>The Analysis of Power Plants</u>

  The detailed analysis of a complete nuclear power plant is a highly developed art. Methods of systems analysis serve to predict and integrate the performance of the reactor, steam generators, turbines, electrical generator, system controls, output switch-gear and the electrical power demanded through the transmission line feeding an electric power network. There are both a very extensive literature and engineering practice.

  El-Wakil[1] describes many of the considerations relating to steam and power generation that we have been unable to mention.

----

[1]Ref. [5] of Section 21A. 1.

CHAPTER 22

NUCLEAR FISSION ENERGY: BREEDER REACTORS

22.1 Introduction

The U.S. budget authorizations requested for the Liquid Metal (Cooled)-Fast Breeder Reactor (LMFBR) program during 1977 are about 555×10^6. This is the largest single energy research and development program in the U.S. Most of the funds are being spent on a demonstration plant for electrical power production to be located on the Clinch River near Oak Ridge, Tennessee, the Clinch River Breeder Reactor (CRBR). Other large portions of the program include a Fast Fuel Testing Facility (FFTF) at Richland, Washington, the development of advanced fuel elements to avoid the thermal limitation of uranium-plutonium oxide fuel elements and the testing of large heat exchangers, pumps and other apparatuses for handling molten sodium. A program to develop a (helium-) Gas-Cooled Fast (breeder-) Reactor (GCFR) supplements the LMFBR program. FBR engineering must be appreciated as a feasible development rather than as an established practice. Probably more advanced programs to develop fast breeder reactors are underway in France, Germany,

England, Russia, and Japan. Phenix, a demonstration plant in France, has been producing 250 Mw_e of electric power since March 1974.

The urgent motivation for these programs is a large increase in total energy production from limited supplies of uranium. The available energy may be increased more than one hundred times above that produced by the thermal neutron reactors described in Chapter 21. Alternative means of extending and conserving fission energy resources receive approximately one tenth of the current effort devoted to the LMFBR. The wisdom of the LMFBR program is being questioned on the grounds that uranium and thorium resources may prove to be much larger than the identified reserves, that fast reactors may be too hazardous to operate and that the large amounts of plutonium flowing in the fuel cycle require impracticably stringent control measures in order to eliminate toxic effects and diversion for the production of nuclear bombs. We shall now discuss design objectives and measures to assure operational safety.

A decision[1] by a circuit court of appeals in June 1973 required the AEC to prepare an environmental statement concerning all of the possible consequences of developing and deploying the LMFBR through the year 2020. This first of a kind environmental study required more than fifty man-years of effort and is recorded in a seven-volume report[2] that includes statements resulting from extensive public reviews. If the stringent requirements for a safely operating LMFBR program can be met, the program will pay back many times the development cost of several billion dollars, largely through reduced demand on resources such as coal and uranium.

---

[1]Judge Wright, "AEC's Fast Breeder Program Requires NEPA Statement," Atomic Energy Law Reports §4103, Commerce Clearing House, Inc., pp. 9268-9282, 1973.

[2]USAEC Proposed Final Environmental Statement: Liquid Metal Fast Breeder Program, WASH-1535, December 1974. Available from NTIS, Springfield, Va. 22151.

The desirability of providing sufficient neutrons to produce as much fissile material (by means of the reactions listed in Section 21.6) as is used up was recognized prior to 1942. During the war, some of the basic data (see Table 21.5-1) were determined. When the research required for the production and use of plutonium for weapons was being successfully completed in late 1944 and early 1945, a few investigators turned to the problem of improving the production of fissile materials and power. Thus, as early as 1945, the following programs were underway in the laboratories that have been successively supported by the USAEC and ERDA: at ANL, Fermi, Zinn and others emphasized the high neutron yield from fast neutron fission and stimulated development of a fast breeder made of metallic uranium cooled with liquid sodium; at LASL, Morrison, Hall and others, for similar reasons, directed attention to the use of plutonium alloys for fast reactor fuels; at ORNL, Wigner, Weinberg and others turned to the study of liquid-fueled thermal-neutron reactors as a means of minimizing requirements and maximizing the efficiency for use of fissile materials through rapid and continuous fuel reprocessing, thus avoiding parasitic neutron losses and nuclide transmutations. The first and last programs have remained with modifications. About 1965, an old theme was reemphasized in the Naval Reactors Program, namely, that of introducing sufficient thorium into an LWR to breed ^{233}U. The possibility of cooling an FBR with helium at 80 to 100 atm has emerged relatively recently as the difficulties of achieving high power densities and low parasitic losses in the LMFBR have become apparent.

22.2 Breeding and Conversion Ratios; Production of Fissile Materials

In any fission reactor, the ratio (R) of fissile nuclides produced to the number destroyed by fission and capture (see Section 21.6) may be expressed as

$$R = \eta\epsilon - 1 - l,$$

where η and ϵ have been defined in Section 21.7 and Appendix 21A, the number 1 represents the neutron or the nuclide required for the chain reaction, and l equals losses by parasitic neutron capture and leakage (which have also been discussed in Appendix 21A).

In design analysis, we calculate η, ϵ, l, and R for any given reactor. It is instructive (see Table 22.2-1) to compare the results extracted from various reports on the neutronic design analyses for two thermal and two fast reactors.

If we think of the fissile material as an investment, then R is related to the increase or decrease of our investment and is the ratio of the new and old investment after one turnover or cycle. The specific inventory is inversely related to the rate of turnover and represents an effective life of the fuel in the reactor.

Considering first the thermal reactors or converters, which convert ^{235}U to either ^{239}Pu or ^{233}U, we note that, if we start with a unit amount of ^{235}U, we would use R units of fertile material during one cycle. During a second cycle, supplemented with feed material, the diminished original investment would convert R^2 units of fertile material. Thus, replacing successive amounts of fertile material consumed by conversion and fission, we obtain the sum

$$R + R^2 + R^3 + \cdots = R/(1 - R).$$

Table 22.2-1 The calculated neutron balances for four reactors, each producing 1,000 Mwe. Also listed are representative data related to the rate and efficiency of consumption of the nuclear fuel. These data have been selected from several reports to the USAEC on the evaluation of nuclear reactors.

	PWR	HTGR[†]	LMFBR	GCFR
η = neutrons produced from fissile nuclides per fissile nuclide destroyed	1.94	2.09	2.24	2.36
α = ratio of gamma emission to fission of fissile nuclides	0.2	0.1	0.28	0.22
ϵ = fast fission effect or augmentation by fertile fission	1.02	1.0	1.14	1.17
$\epsilon\eta$ = neutrons available from all nuclides	1.98	2.09	2.55	2.75
losses per fissile atom consumed				
leakage from core and blanket or reflector	0.04	0.07	0.03	0.04
absorption in { coolant and moderator	0.07	0.03	0.002	0.00
fission products	0.18	0.19	0.02	0.02
structural materials	0.04	-	0.05	0.04
control poisons	0.04	0.07	0.03	0.03
actinide nuclides	0.04	0.11	0.22	0.18
total losses	0.41	0.47	0.32	0.28
R = production of fissile material/destruction	0.57	0.66	1.18	1.44
fraction of energy contained in uranium that could be released	0.015	0.020	0.6-0.7	0.6-0.7
inventory of fissile material in a reactor, kg	2200	1600	3800	4700
specific inventory, kg/Mw$_t$	0.69	0.63	1.6	1.8
ratio of fissile to fertile nuclides in { the core	0.023	0.04	0.15	0.17
the core plus blanket	-	-	0.04	0.04
ratio of electric to thermal power (efficiency)	0.33	0.39	0.42	0.38
metric tons of U_3O_8 required over the life of the reactor (40-year life and 80% service = 32 full power years)	3500[*]	2700[*†]	30	31
y = time to produce the fissile material for one additional reactor, not counting the time for processing	-	-	25	13

[*]This number depends on the fraction of ^{235}U in the tailings produced in the isotope enrichment process, which is here taken to be 0.2%.

[†]A standard design optimized for lowest cost power with contemporary nuclear fuel costs is assumed; the uranium demand may be reduced to about 1600 mt.

Using the numbers in Table 22.2-1 for the LWR and HTGR, the sums are found to be 1.35 and 1.94, respectively. Thus, the fraction of ^{238}U that may be converted and fissioned is $0.0072R/[(1 + \alpha)(1 - R)]$, where the factor $(1 + \alpha)$ accounts for loss of U-235 by neutron capture to form ^{236}U, which does not fission with thermal neutrons. The fractions of the fertile materials that can be used by the two thermal reactors are 0.8 and 1.3%, respectively. Adding the fraction of ^{235}U to each, we find that the fractions of our resources that may be fissioned are 1.52 and 2.02%, respectively.

For the fast breeders, the value of R is greater than unity and our initial investment of fissile plutonium grows. We need to feed only fertile material to the fuel cycle for this reactor. In addition, we may produce the fuel for a second reactor. The time required to produce one loading of fissile material and to double our original investment may then be calculated. The net rate $\dot{M}$ in grams (g) per day (d) of producing fissile material by a reactor producing thermal power P (in Mw_t) may be written as

$$0.95(g/Mw_t d)P(Mw_t)(R - 1)/(1 + \alpha) = \dot{M}(g/d).$$

If M_0 is the initial mass of a loading, the number of days required to produce an amount equal to M_o is $\dot{M}d = M_o$. Thus, $0.95d(g/Mw_t d)P(Mw_t)(R - 1)/(1 + \alpha) = M_o$. Letting $M_o/P \equiv M_s$ = specific inventory, the doubling time in days (t_d) is given by the relation

$$t_d \simeq M_s(1 + \alpha)/(R - 1).$$

The numbers listed at the bottom of Table 22.2-1 are expressed in years. If the U.S. supply of reasonably priced uranium is indeed limited to proved reserves of 1 to 3.5×10^6 tons (see Volume I, Section 1.4A), the case for development of the breeder is compelling. With this supply, the total number of 1,000 Mw$_e$ LWRs that can be operated during their expected life is about 300 to 1,000, corresponding to a total power-generation capacity of 300 to 1,000 Gw$_e$. The lower value is roughly equal to the total installed capacity to be expected by 1990. On the other hand, with FBRs the same uranium resources could support over 30,000 reactor lifetimes and serve for a few hundred years at a reasonable rate of construction and operation. Further increases in the cost of producing fertile material would have little effect on the cost of electric power.

22.3 The Liquid Metal Fast Breeder Reactor (LMFBR)

A. Design Requirements

In order to achieve favorable breeding ratios and specific inventories, general design requirements (which were recognized before 1945) must be met. These are: (a) use of as dense a core as possible holding as many fissile and fertile atoms in the proper ratio, per unit volume, as is physically feasible; (b) minimal slowing down and absorption of neutrons by cooling and structural materials; (c) a stable fuel element after fission of 14% of the fuel material, which is equivalent to an average exposure in the reactor of about 100 Mw$_t$d/kg or about 3 years; (d) retention of fission products within the fuel elements or management of these in a closed system; (e) excellent transfer of heat from the fuel element to the coolant and from the reactor core to external heat exchangers; (f) designing to meet a number of requirements related to safety (such as a negative temperature coefficient while avoiding any increase in core density) and economical construction, maintenance and operation (see Chapters 24 and 21).

The requirements for compactness and high heat transfer directed early attention to the use of metallic uranium-plutonium fuel elements cooled by liquid sodium. Unfortunately, good stability and long life cannot be implemented with metallic fuel elements. Early development work showed that UO_2 might meet the stated requirements but at a considerable decrease in specific power (reciprocal of specific inventory). Sodium technology, the development of fuel elements, and reactor safety have been and continue to be the dominant engineering problems in the LMFBR program.

B. The Choice of Coolant

We may compare the properties and performance of the three reactor coolants listed in Table 22.3-1. Water will serve as a familiar standard for comparison, although it cannot be used in a fast reactor. Sodium is clearly the most effective medium for heat transfer. Most reactors consist of bundles of rods or tubes containing fuel from which heat is removed by the coolant flowing lengthwise. The derived quantity in the eighth row of the table (i.e., the product of density, usable velocity of flow, specific heat, and permissible increase in temperature) shows that, for a specified rate of heat removal, the cross-sectional area and number of channels for coolant flow, and hence the spacing of the rods, will be smallest if sodium is used. Thus, the use of sodium will assure maximum core density or compactness. Its boiling point is $873°C$. In a low pressure system, sodium may be used to transfer heat at the temperature used for steam generation (which is seldom greater than $570°C$). Reference to row 7 in Table 22.3-1 shows that parasitic neutron absorption by sodium may be small.

The disadvantages of using sodium are the result of its chemical reactivity with water and air and of high opacity and readily induced radioactivity.

Table 22.3-1 Properties and heat-transfer performance of reactor coolants. These data have been selected or estimated from standard handbook sources.

Coolant	Water	Helium		Sodium
reactor type	PWR	HTGR	GCFR	LMFBR
average temperature, °C	300	550	450	450
average reactor pressure, atm	150	50	85	5
temperature rise in the reactor, °C	30	420	300	165
density of the coolant as used, g/cm^3	0.70	0.003	0.0058	0.80
specific heat of the coolant, c_p in j/g-°C	5.3	5.2	5.2	1.25
flow velocity of the coolant, V in m/s	4.5	50.	80.	6.
thermal power transported from the reactor per unit of frontal area of the core, $\rho V C_p \Delta T$ in w	50,000	33,000	72,000	100,000
coefficient of heat transfer from the surface of the rod, H in w/cm^2-°C	3	0.25	1.5	10
average temperature drop across the boundary layer in coolant adjacent to fuel rods	20	120	100	20
average heat transfer from a fuel rod per unit surface area, HΔT in w/cm^2	60-70	30	150	200
maximum heat transfer from a fuel rod per unit of surface area as limited by boiling, HΔT in w/cm^2	300	-	-	1,000

If sodium is totally enclosed in steel conduits and vessels, it is relatively inert. It may dissolve small quantities of the constituents of steels in the hottest portions of the cooling system and deposit these in the coolest region. If the dissolved oxygen content of the sodium is less than about 20 ppm, this mass transfer is tolerable. Sodium transports long-lived radioactive elements to external heat exchangers, thus influencing maintenance procedures. Any leaks of water into the sodium system or of sodium to the steam will produce serious consequences, even if the leaks involve limited quantities that are inadequate to increase the pressure (by hydrogen formation or heating). Sodium hydroxide corrodes and embrittles most of the steels that are likely to find application in sodium systems or sodium-water steam generators. In summary, sodium is an excellent reactor coolant when the container walls are entirely free of leaks, stringers of iron oxide or any pathway that will permit contact of sodium with water or air. Consequently, the industrial use of sodium heat-exchange loops will require introduction of carefully selected materials, great care in manufacture and maintenance, and unusually good quality control.

The capture of neutrons by sodium results in considerable radioactivity and the need for shielding around all portions of the primary coolant loop. The nuclear reactions involved are

$$^{23}_{9}Na + n \rightarrow {^{24}_{9}Na}^{*} \overset{\beta}{\rightarrow} {^{24}_{10}Mg} + e, \text{ half life} = 14.9 \text{ h.}$$

To avoid radioactive contamination of the steam-generating system in case of failure, an intermediate non-radioactive loop (including an intermediate sodium to sodium heat exchanger) is interposed between the reactor loop and the steam generators.

C.   General Arrangement of the Components of an LMFBR Steam-
Supply System

Figure 22.3-1 shows how the reactor, the machines for remote handling
of fuel elements, the fuel-storage tank, and the intermediate heat exchangers
may be arranged inside a prestressed, concrete containment structure. This
structure has a steel lining and is filled with argon or nitrogen in order to exclude air, water or any other reactive fluid from the sodium contained in the
reactor vessel and the fuel-storage tank. We note that the motor driving the
pump that circulates the radioactive, primary sodium is accessible for maintenance from the outside of the containment building. Also, the steam generators are well separated from the regions containing radioactive sodium. Maintenance, repair or replacement of equipment within the containment structure
may be accomplished only by remotely controlled operations or by direct contact after complete removal of all sodium and other radioactive substances
from inside the structure.

The reactor vessel, piping, pump and other parts in contact with sodium are made[*] of AISI 316 steel. Extensive study and testing have demonstrated that this material has the desired strength and resistance to corrosion
and is suitable for forging, rolling, drawing, welding and other steps in the
manufacture of pipes and vessels. The exterior surfaces of the system are
well insulated (to minimize heat loss) and electrically heated to avoid solidification of the sodium before start-up or during shutdown (the melting point of
sodium is approximately 98° C).

_____

[*]The American Iron and Steel Institute (AISI) designation of a commercial,
stainless, austenitic steel has the following approximate composition (%): Fe
60-70, Cr 16-18, Ni 10-14, Mo 2-3, Mn 2, C 0.03, Si 1, S 0.03, P 0.045.
Its creep strength at system temperatures is greater than 17×10^7 Pascal
(25,000 psi).

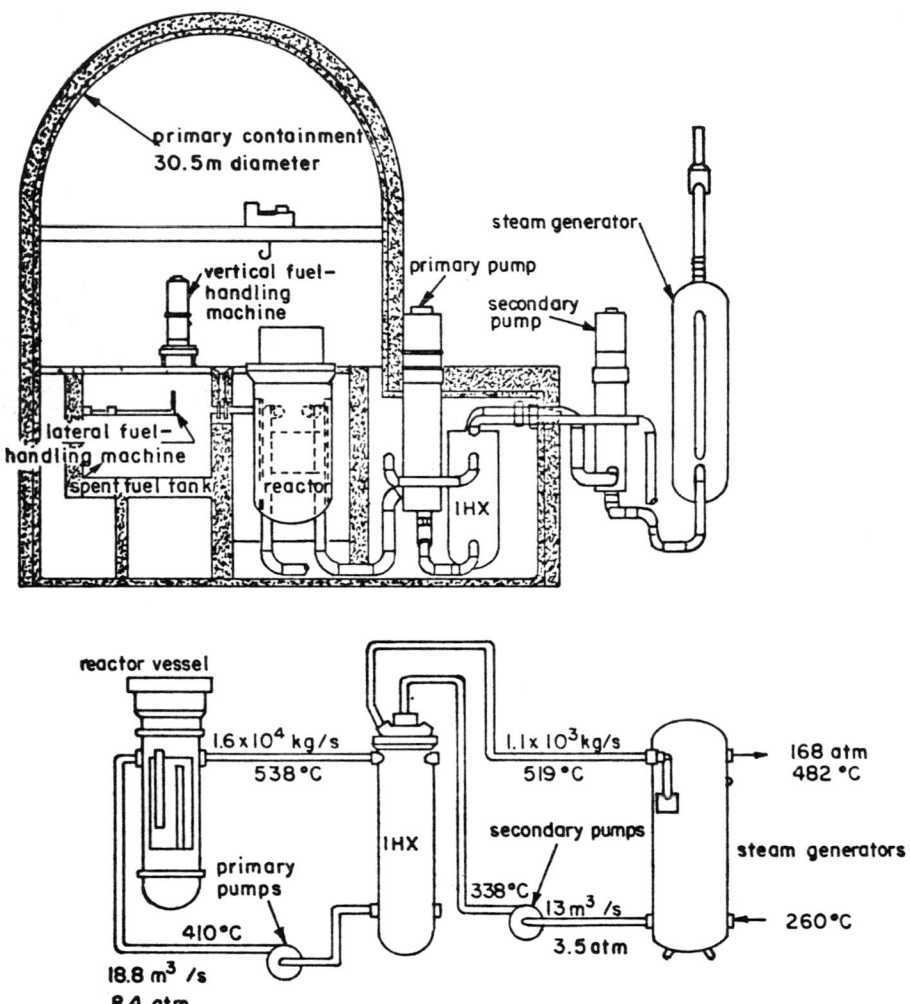

Fig. 22.3-1 A simplified illustration of an LMFBR steam-supply system. The
            schematic view on an elevation (top) shows the relative sizes and
            arrangement of the reactor, the primary and secondary heat-ex-
            change systems and the containment structures. Below are shown
            the temperatures and flows in the loops required for generation of
            approximately 1,000 Mw$_e$. The symbol IHX means intermediate
            heat exchanger. Adapted from USAEC reports.

From the data accompanying the schematic diagram of the heat-exchange loops in the lower portion of Fig. 22.3-1, we see that the excellent heat transfer afforded by the sodium results in a temperature drop of only $56^{\circ}C$ from the reactor outlet to the steam supply to the power turbine. In the steam-power system, reheating of steam, economizing and feed-water heating may be arranged to provide a favorable heat balance and an optimized efficiency as great as 42% for electric power generation.

The arrangements inside the reactor vessel proposed for the CRBR are shown in Fig. 22.3-2. The closely spaced core and blanket elements are centrally located. Receptacles for fuel storage are provided around the reactor. Control rods are driven from above, the drives being located in a sealed cap on the cover of the vessel.

The flow of sodium and pressure drop through the reactor are similar to that of water through a PWR (see Section 21.13), except that the total sodium-system pressure is less than 5 atm. Baffles direct the inlet sodium downward next to the walls of the reactor vessel to a plenum or manifold below the bottom of the core. The bundles of carefully spaced fuel and blanket rods are enclosed in hexagonal steel boxes that are spaced side by side with just enough free volume to allow for expansion and some bowing of the boxes. A duct at the bottom of each box seats in an orifice plate, thus directing the required flow of sodium upward within the box and along the rods. From the top of the core, the heated sodium flows to the intermediate heat exchanger. From the sodium in the secondary loop, heat is transferred to water and steam in the steam generator.

It is necessary to avoid leakage or loss of Na from the core region in the event of a break in the piping. Not only would cooling be lost, but the fission rate would increase unless the decreased neutron absorption were

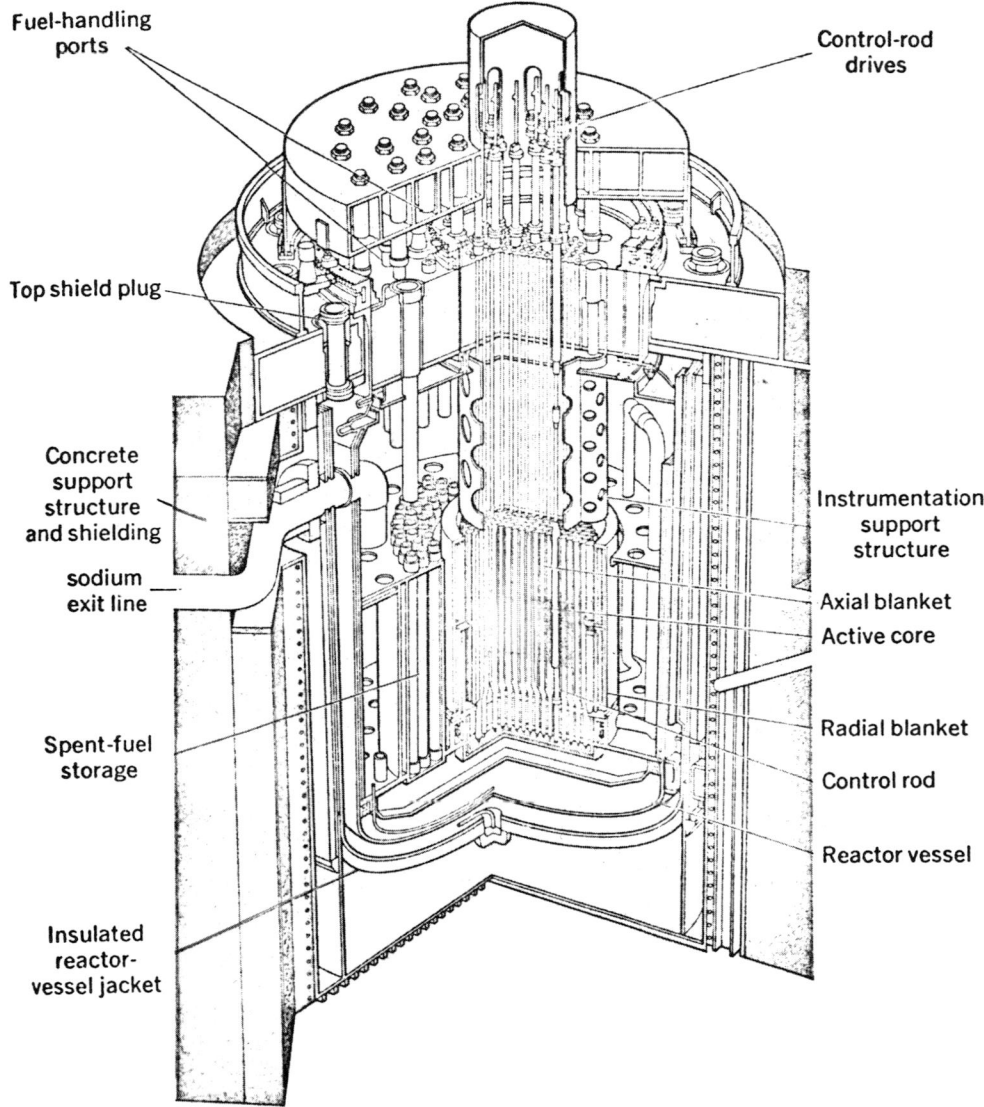

Fuel-handling ports

Control-rod drives

Top shield plug

Concrete support structure and shielding

sodium exit line

Instrumentation support structure

Axial blanket
Active core

Spent-fuel storage

Radial blanket

Control rod

Reactor vessel

Insulated reactor-vessel jacket

Fig. 22.3-2 An isometric cross section showing the core and other essentials inside the reactor vessel of a fast breeder reactor of the liquid-metal-cooled type that will be incorporated in a demonstration power plant to be built near Oak Ridge, Tennessee. Reproduced with permission by the Westinghouse Electric Corporation.

simultaneously compensated for by control mechanisms. Thus, to provide maximum safety and integrity of the reactor vessel, all connections and penetrations are above the level of the core. Natural convection of sodium, either in the loop or within the tank without pumping, will remove the heat of radioactive decay of the fission products in the fuel rods after shutdown of the reactor.

Carrying this idea of avoiding any chance of sodium loss one step further, a pot or tank type design has evolved. The Prototype Fast Reactor (PFR) now ready to operate at Dounreay, Scotland, employs this concept in which the intermediate heat exchanger, pumps and fuel-handling apparatus are placed under the sodium in a large tank surrounding the centrally located reactor. Control rods are operated from above. Fuel elements are moved by remote control from a transfer flask through a port in the cover of the pot into a rotating magazine located at one side of the core. A transfer tube operated from above moves elements from storage to the core or conversely. Although many mechanical operations must be done under sodium, the general arrangement would seem to assure cooling of the core under all circumstances. For repair, a component would be withdrawn from the tank and freed of sodium and radioactivity before removal to a location for maintenance work.

D.   The Fuel Elements, Reactor Core and Blanket

i.   Composition and Dimensions of the Fuel, Core and Blanket

At the present time, the only fuel element with reasonably assured stability for the required life consists of pellets of uranium-plutonium and uranium oxide contained in AISI 316 steel tubes that are 6 to 8 mm in outside diameter, 2 to 3 m long and have a wall thickness between 0. 3 and 0. 5 mm. Steel grids and spiral wire wrapping hold and space the rods in a triangular

lattice with a pitch or spacing of about 7.5 mm. The core containing the fissile plutonium is only about 1 to 1.3 m high and 2 to 3 m in diameter, depending on the operating power desired. It must be surrounded by a blanket of fertile material that is thick enough to absorb most of the neutrons leaking from
the core to give net breeding of fissile material. Thus, the end sections, each
about 1/3 of the length of the fuel rods, are filled with pellets of UO_2. Surrounding the fuel elements are the radial blanket elements, each consisting of
rods 12 to 13 mm in diameter and containing pellets of UO_2. The core and
blanket approximate a cylinder 2.8 to 4.5 m high by 3 to 3.7 m in diameter. At
one end of each fuel rod is an empty space large enough to contain fission gases
at pressures no greater than 65 atm, well within the strength of the tubing;
at least 90% of the gases escape from the pellets at exposures of 100 $Mw_t d/kg$.
The initial composition of the fuel pellets is $U_{1-x}Pu_xO_{2-\delta}$, where $0.1 \leq x \leq$
0.2 and $\delta \simeq 0.04$. The chemical potential of oxygen in the fuel element is extremely important and is adjusted by a final heating in hydrogen at a prescribed
temperature. For the CRBR, 217 rods are placed in each hexagonal box,
which is approximately 12 cm across flats. The reactor comprises 198 to
318 core elements, surrounded by 150 to 234 blanket elements.

The core of the reactor is made somewhat short and wide in order to
increase the leakage of neutrons from the core and cause the neutron multiplication to decrease as the temperature increases. This phenomenon is associated with the negative temperature coefficient of reactivity. In a core having
minimum surface area and neutron leakage, the decrease of neutron absorption
by sodium, steel and fission products with temperature would more than compensate for increased leakage, cause the temperature coefficient to be positive
in spite of the increase of resonance absorption with temperature, and thus
make reactor control more difficult and possibly unsafe.

ii.   The Performance of the Fuel Elements

As is shown in Appendix 21A, the power per unit length of a fuel rod is a property of the fuel material and depends on the permissible central temperature, the required or permissible temperature at the coolant-cladding interface, and the thermal conductivity of the fuel material.   Thus, $W_\ell(T_c, T_m)$ $= 4\pi \int_{T_c}^{T_m} k(T')dT'$ where W_ℓ = power per unit length, $k(T)$ is the thermal conductivity, T_c is the working temperature of the cladding, and T_m is the temperature at the center of the pellet.   For FBR fuel rods, the maximum T_m $= 2,300°C$.   Nominally, T_c can be taken to be $500°C$; W_ℓ has been empirically determined to be about 500 w/cm for these temperatures.   Averaged over the reactor, W_ℓ is approximately 230 w/cm.

Studies to determine the behavior and thus to predict and control the performance of fuel elements in service have proved to be very complex. Recently observed phenomena (associated with the displacement of atoms by fast neutrons and recoiling fission fragments, the accumulation of approximately 20 atom percent of fission products and the large gradients of temperature within the elements) lead to distortion, embrittlement, weakening, and other changes in properties of the elements.   Not only must loss of integrity be avoided, but also the changes in dimensions, shape and strength must be provided for in both engineering design and reactor operation.   The following are a few of the important phenomena.

(1) Materials used for cladding and boxes at 400 to $500°C$ may increase in volume as much as 10% by the end of their desired life, during which each atom is displaced from a crystal lattice site 50 to 100 times by some $2\text{-}4 \times 10^{22}$ fast neutrons/cm^2.   This change in volume is due to the growth of new planes between those of the original crystals from the displaced atoms while 0.3 to 0.5% of the vacant crystal lattice sites agglomerate to form voids permitting

extension of a crystal in its [111] directions. The amount of swelling of an alloy is influenced and may be controlled within limits by adjusting composition and by prior deformation at room temperature (cold work).

(2) The accumulation of small amounts of helium from neutron-induced alpha emission [(n, α)-reactions] in the grain boundaries in steel causes serious loss of ductility of cladding materials.

(3) In the large thermal gradients ($\sim 10^{4}$°C/cm) in oxide fuels, fission products, uranium and plutonium migrate significant distances, depending on the temperatures and rates of atom displacement. A major driving force for migration is the temperature coefficient of the chemical potential times the temperature gradient. The chemical potentials of several constituents of the fuel depend on the amount of oxygen and lead to the requirement on the composition of the oxide mentioned above. For example, if the oxygen-metal (O/M) ratio is 2, plutonium migrates toward the center of the fuel rod.

(4) Above about 1100°C, the oxide recrystallizes and voids and gas bubbles are swept toward the center. The oxide densifies and a central axial void appears.

(5) The more volatile fission products (cesium, iodine, and tellurium) migrate to the cladding. Also, if O/M $\simeq$ 2, intergranular corrosion and weakening of the cladding may occur.

(6) Because of the volume occupied by fission products and the other effects mentioned, the oxide pellet expands and applies some load on the cladding. Because the pellet usually cracks as the result of thermal stresses, the stress in the cladding near the cracks may be excessively concentrated. Fuel rods may be designed to control these effects within tolerable limits but detailed study of the engineering relationships among the variables must continue. It is probable that alloys for cladding will be improved.

Because the use of oxide elements limits the power density and breeding ratio, and thus the rate of production of fissile material, an extensive program to develop uranium-plutonium carbide elements is underway. The thermal conductivity of the carbides is 5 to 7 times that of the oxide. Also, the metal-atom density is greater. These factors combine to reduce the predicted doubling time by factors between 2 and 3. Criteria for a commercial, power-producing LMFBR are likely to be met only as the result of successful conclusion of current research programs.

22.4 Helium-Gas-Cooled Fast Breeder Reactors (GCFR)

A. Reasons for Considering Helium for Cooling an FBR

The data given in Table 22.3-1 show that, at a pressure of 85 atm, helium can transfer heat approximately three fourths as well as sodium although at the cost of higher pumping power, which corresponds to less than 4% of the reactor output. Since the PCRVs can be designed for safe containment at this pressure, cooling with helium is entirely feasible. Reference to the schematic diagrams in Fig. 22.4-1 and comparison with Figs. 22.3-1 and 2 shows that a helium system can be much simpler than a sodium system. We may summarize the reasons for using helium as follows: (a) Adequate heat transfer may be achieved with less probability of thermal shock than corresponds to extremely rapid cooling by sodium immediately following a scram. (b) Very little neutron capture or moderation occur. (c) Because of (b), very little change in fission rate or neutron level (core reactivity) accompanies a drop in the pressure; this behavior should be compared with the large changes taking place if voids develop in sodium during boiling. (d) Helium is chemically inert except for impurities (which may be controlled at low levels) and affords

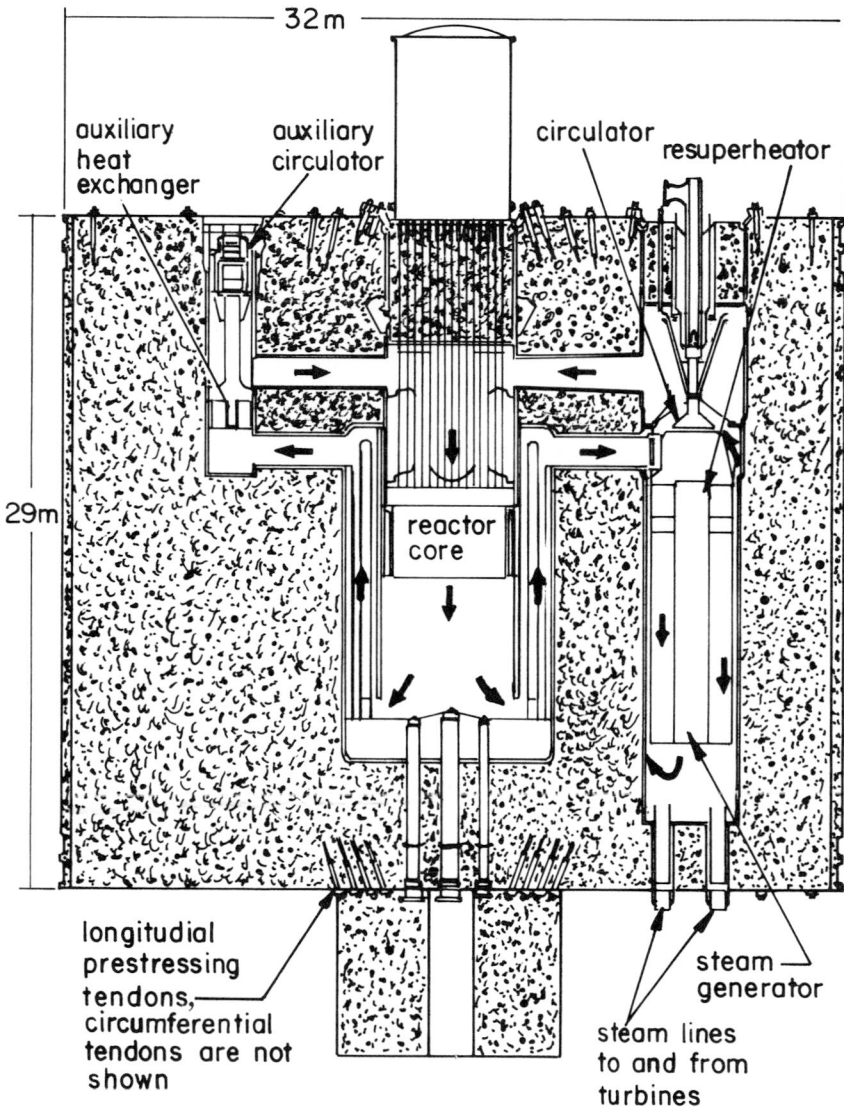

Fig. 22.4-1 A vertical section of a GCFR in a PCRV. It should be noted that
the primary loop is totally enclosed and an intermediate loop is
not required because helium transports little radioactivity and is
completely inert. Two other steam-generating and auxiliary loops
are not visible in this section. Adapted from General Atomic Co.
reports to the AEC.

reasonable freedom for the choice of structural materials. (e) A change of phase will not occur on overheating with He, while Na will boil. (f) Because of the PCRV and containment structure, the helium pressure cannot fall below 2 atm, a pressure at which circulating helium will remove heat from a shutdown reactor. (g) The half-life for accidental loss of helium is no less than 20 sec. (h) Transparency, absence of induced radioactivity and presence of very little entrained radioactivity reduce the time and effort for maintenance of equipment in the helium loop outside of the reactor. (i) Continuous purification and removal of fission products, as they diffuse through designed vents into a fission-gas-recovery system from pressure-equalized fuel elements, maintain relatively low levels of radioactive elements in the cooling system. (j) It has proved difficult to develop fuel elements capable of transferring heat at a rate much higher than that provided by circulating helium. (k) Cores as dense and compact as originally thought desirable are less safe and less easily controlled than those with optimal void space for gas cooling (about 45% void volume). (l) Although limited by the required pumping power and by the temperature of the fuel elements, optimized plant efficiencies between 36 and 38% may be realized. (m) The breeding ratio of 1.4 to 1.5 (see Table 22.2-1) permits the use of a thorium blanket and production of ^{233}U for the HTGR, as well as breeding of ^{239}Pu.

B. General Arrangements

There are many similarities in the conceptual designs of the GCFR and HTGR plants. Figure 22.4-1 shows a vertical section through the PCRV and the reactor of a 1,200 Mw$_e$ GCFR unit. Except for fuel loading and unloading at the bottom of the reactor, the balance of plant is quite similar for the two reactor types. The central core also resembles the LMFBR core closely. The ratio of fissile to fertile material is 0.15 to 0.20 and is thus slightly greater than that for the LMFBR. The core is somewhat less flattened than for the LMFBR because less neutron leakage is required to provide a negative temperature coefficient. The core approximates a right cylinder, about 1.6 m high and 2.6 m in diameter. Fuel elements are 7 to 9 mm in diameter and contain pellets of $U_{1-x}Pu_xO_{2-\delta}$, $0.1 \leq x \leq 0.2$, $\delta \approx 0.04$. The end sections contain the axial blanket of UO_2 pellets and bring the total rod length to about 2 m. Radial blanket elements surround the core and increase the total diameter to about 3 m.

Comparison of Figs. 22.4-1, 22.3-1 and 22.3-2 shows that the GCFR is less complex and less massive than the LMFBR.

C. The Fuel Elements, Reactor Core and Blanket

Most of the requirements and problems associated with LMFBR elements (see Sections 22.3Di and 22.3Dii) apply to GCFR elements. The following essential differences exist:

(1) Fission gases are allowed to escape from the upper, cool end of each element through a tubular manifold that conducts the gases through a charcoal-filled trap to a junction with another manifold within the core-support plate above the reactor core. This is called a pressure-equalizing system (PES). The manifolds of tubular conduits (internal diameters a few mm) conduct the fission gases by means of a controlled flow of helium into a cryogenic gas-adsorption apparatus that is connected to the inlet of the main helium circulators. Thus, the pressure inside the element is one to two atm below the average pressure in the helium-cooling loop. Although complex, the PES affords a way of identifying failed fuel rods and simplifies cooling-system maintenance.

(2) Heat transfer to flowing helium is improved by carefully prepared circumferential ridges on the upper half of the fuel rods. This "roughening" doubles the heat transfer at the cost of trebling the pressure drop along the rods.

(3) The use of high performance fuel elements is essential for the LMFBR but leads to only a small overall improvement for the GCFR. On the other hand, cladding with suitable strength at slightly higher temperature ($800°$C compared to $650°$C) would be highly advantageous and will ultimately be developed.

The axial blanket sections of the fuel rods contain UO_2, as does the LMFBR. The breeding ratio for plutonium, including process losses, is approximately unity. The radial blanket elements may contain either UO_2 or ThO_2. The latter is preferred in order to produce enough ^{233}U to fuel 1 to 3 HTGRs.

22.5 Epilogue

The technological feasibility of fast-neutron-breeder reactors has been demonstrated through measurements of nuclear reaction probabilities. The following studies have been completed: conceptual design; design analyses, reactor and component experiments, tests and fabrication; analysis of safety and studies of benefits, risks and costs. A preliminary evaluation of actual costs, which workers at ERDA hope to reach by 1986, will result from the current reactor-demonstration program, from evaluations of improved fuel elements and fuel cycles, and from learning how to reduce the capital (manufacturing and construction) costs of the power plants. At the present time, the major uncertainties relate to costs and social acceptability.

Progress toward commercial use of the FBR appears to many observers to be more rapid in France, England, Germany, Russia, and Japan than in the U.S. The urgency of having sources of power that minimize dependence on exogenous resources and optimize the use of indigenous ones is relatively greater in Europe and Japan than in the U.S. Of the non-communist nations, France probably has the most comprehensive energy policies and plans.

Metz has published summaries[1] on the status of the foreign programs. Sodium-to-water steam generators installed in the Prototype Fast Reactor (PFR) plant at Dounreay, Scotland, developed leaks which caused extensive down-time. In Russia, similar problems persisted over a period of 3 years. Once sodium hydroxide is formed, corrosion-causing caustic-embrittlement leads to extensive cracking of steel plates and tubes in the generators. In France, the use of special designs and fabrication precautions has led to very large and costly steam generators. The feed materials and fuel-reprocessing costs will be very low compared to LWRs or fossil-fueled plants. Capital costs of the power plants and reprocessing plants will be large.

It is likely that many of the issues and problems associated with extensive use of FBRs have been anticipated, analyzed and debated more thoroughly in the U.S. (see Refs. [1] and [2] in Section 22.1) than in other countries. As Judge Wright stated in his court decision,[2] we must learn to assess all costs, benefits and environmental effects of a major technological program many years in advance of implementation.

---

[1]W. D. Metz, "European Breeders I: France Leads the Way; II: The Nuclear Ports are Not the Problem; III: Fuels and Fuel Cycles are Keys to the Economy," Science 190, 1279-1281 (1975); 191, 368-362, 551-553 (1976).

[2]Ref. [1] in Section 22.1.

CHAPTER 23

CONTROLLED THERMONUCLEAR FUSION

23.1 Introduction to Physical Processes in Nuclear-Fusion

Fusion has been characterized as a potential energy source with a relatively low pollution level (see Section 24.7 for details) and promises to be
virtually inexhaustible, of low cost and available to everyone. Fusion power
in any practical form has not yet been achieved, but it is likely that scientific
feasibility will be demonstrated during the next decade. Engineering implementation on a significant scale may, however, be many decades away. As
pointed out by Post and Ribe,[1] fusion may prove to be a stable solution to the
energy problem politically as well as physically.

It is not possible to treat all of the physical mechanisms involved in
fusion with the same degree of detail. The major branches of physics involved
are electrodynamics, magnetodynamics, nuclear-reaction theory, special
relativity, classical, statistical and quantum mechanics, plasma physics,

---

[1] R. F. Post and R. L. Ribe, "Fusion Reactors as Future Energy Sources,"
Science 136, 397-407 (1974).

many-body theory, thermodynamics, and kinetic and transport theories. Our discussion of the basic topics is analogous to showing only the tip of an iceberg.

The hydrogen isotopes deuterium and tritium are promising fusion fuels; the deuterium nucleus consists of a neutron and a proton; tritium has two neutrons and a proton in its nucleus (see Fig. 23.1-1). Because all nuclear charges are positive, nuclei in the free state repel each other. For fusion of two nuclei to occur, sufficient energy to overcome the relative Coulomb repulsion must be available. When nuclei come sufficiently close, the stronger attractive nuclear force takes over and binding occurs while releasing energy equal to the nuclear binding energy. The relative sizes of the gravitational, electromagnetic and nuclear forces are 1, 2.0×10^{35} and 1.5×10^{41}, respectively.

The fused or composite nucleus is unstable and breaks up into a number of stable decay products (final states). The binding energy of the final state is lower than that in the composite nucleus and energy is therefore released in the form of kinetic energy of the decay products. Some representative candidates for fusion are depicted in Table 23.1-1, with the energy released by the reaction given in Mev (10^6 ev).

The state of matter in which these nuclei must exist for fusion to occur is known as a plasma. To form a plasma, the gas being considered must be heated to such a high temperature that the atoms are stripped of all of their electrons. At these high energies, the probability of two ions (nuclei) colliding with sufficient force to penetrate their relative repulsive Coulomb barriers, enabling the nuclear force to fuse the ions, is small. The concentrations of ions must, therefore, be very high for fusion to occur.

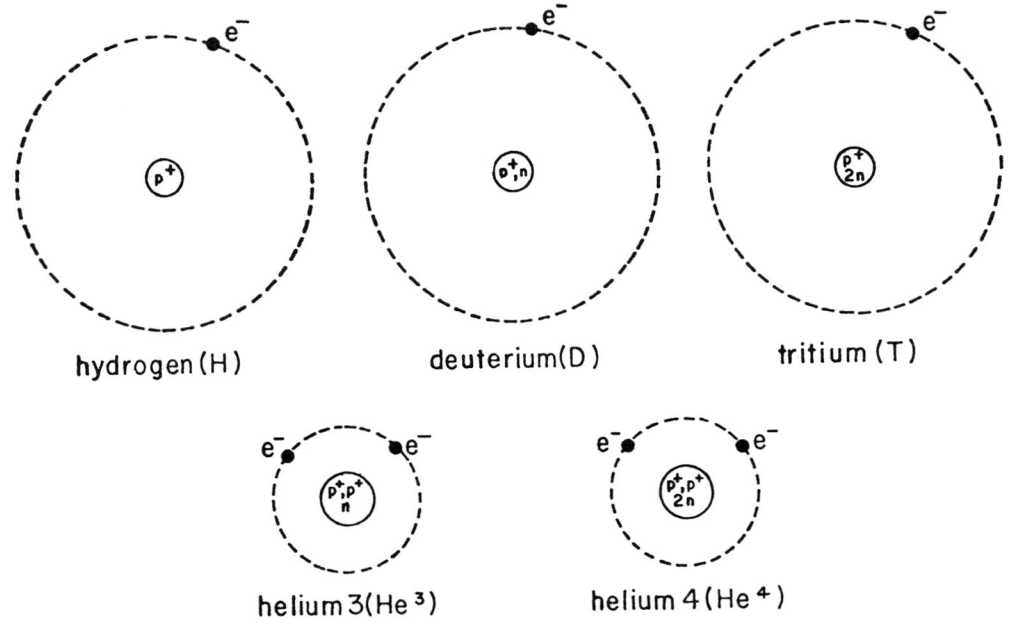

Fig. 23.1-1 Schematic representation of the hydrogen isotopes H^1 (hydrogen), H^2 (deuterium) or D, and H^3 (tritium) or T and of the helium isotopes He^3 (helium 3) and He^4 (helium 4). The symbols have the following meaning: e^- stands for an electron, p for a proton (or ionized hydrogen H^+), n for a neutron, and He^4 for the helium nucleus (α particle). The distance scales have been grossly distortted. The inner circle portrays the positively charged nucleus.

To construct a system in which fusion can be induced one must accomplish the following: (i) create a plasma of the fuel to be ignited, e.g., by completely ionizing a gas of hydrogen isotopes; (ii) contain the plasma in a restricted region of space so that the density remains high; (iii) contain the plasma for a sufficiently long time to allow the fusion process to occur. The processes of heating and containment are at the root of most of the research done on fusion in the past two decades.

Table 23.1-1 Possible fusion reactions in controlled thermonuclear reactors.[*]

	Abbreviation
1st generation fusion reactions (DT); $T \geqslant 4$ kev $D + T \rightarrow \alpha + n + 17.6$ Mev	$T(D,n)He^4$
2nd generation fusion reactions (DD); $T \geqslant 34$ kev $D + D \rightarrow T + p + 4.03$ Mev $D + D \rightarrow He^3 + n + 3.27$ Mev	$D(d,p)T$ $D(d,n)He^3$
3rd generation fusion reaction (DD); $T \geqslant 100$ kev $D + He^3 \rightarrow \alpha + p + 18.3$ Mev	$He^3(d,p)He^4$
4th generation fusion reactions $Li^6 + p \rightarrow He^3 + \alpha + 4.02$ Mev $Li^6 + D \rightarrow Be^7 + n + 3.4$ Mev $Li^6 + D \rightarrow He^3 + n + \alpha + 1.8$ Mev $Li^6 + D \rightarrow Li^7 + p + 5.0$ Mev $Li^6 + D \rightarrow T + p + \alpha + 2.6$ Mev $Li^6 + D \rightarrow 2\alpha + 22.4$ Mev	$Li^6(p,\alpha)He^3$ $Li^6(d,n)Be^7$ $Li^6(d;n,\alpha)He^3$ $Li^6(d,p)Li^7$ $Li^6(d;t,p)He^4$ $Li^6(d,\alpha)He^4$
$B^{11} + H \rightarrow (C^{12})^* \begin{cases} \cdots \alpha \\ \searrow Be^8 \rightarrow 2\alpha \end{cases} + 8.7$ Mev	$B^{11}(p,2\alpha)He^4$

[*]D Stands for D^+, T for T^+, Li^6 for $(Li^6)^{+++}$, etc. [see the footnote to Eq. (23.1-1)].

Consider first the problem of containment. Impurities in the fuel markedly increase energy losses due to radiation and thereby lower the plasma temperature and inhibit fusion. Estimates indicate that even the smallest traces of impurities in the plasma quench the fusion process. To eliminate these impurities and prevent the walls from disintegrating, the plasma must be confined in a "non-material vessel" formed by a strong magnetic field. Under fusion conditions, the plasma behaves like an ordinary gas and exerts an outward pressure $p_N = NkT$, where N is the fuel concentration, k is Boltzmann's constant and T is the plasma temperature. The pressure of the magnetic field p_B must balance p_N to confine the plasma. One method of plasma heating involves injecting a plasma into a cylinder in which there is a weak magnetic field. The field strength is slowly increased, thereby compressing and heating the plasma. Additional methodologies for heating have been studied by using neutral beam injection, wave heating, heating with laser beams, etc. Alternatively, a very short, high-intensity laser pulse may be used to ionize and ignite a solid fuel pellet (see Section 23.7 for details). [2]

A. Energy Balance of Nuclear-Fusion Reactions

Fusion is believed to be the most common form of energy continuously being released in stellar interiors. One cycle of reactions for the formation of helium from hydrogen is called the carbon cycle, where carbon and nitrogen

[2] K. A. Brueckner and S. Jorna, "Laser-driven Fusion," Rev. Mod. Phys. 46, 325-367 (1974); J. L. Emmett, J. Nuckolls and L. Wood, "Fusion Power by Laser Implosion," Scientific American, pp. 24-37, June 1974.

act as catalysts (see Fig. 23.1-2). The overall reaction is[*]

$$4H \rightarrow He^4 + 2e^+ + 24.7 \text{ Mev},$$ (23.1-1)

where e^+ stands for the positvely charged electron (positron). The reactions in Fig. 23.1-2 are too slow to allow their use in controlled thermonuclear fusion. Reactions involving D and T are promising candidates for fusion because of a low ignition temperature. Table 23.1-1 shows possible fuels for future fusion applications.

A nuclear periodic table is shown in Table 23.1-2 for proton numbers less than 9. Stable and unstable nuclei for various numbers of protons and neutrons are listed. It should be noted that the nuclear force is most stable for nearly equal numbers of neutrons and protons in the nucleus for the elements listed. The stable elements are underlined in the table.

The masses of the stable isotopes given in Table 23.1-2 are listed in Table 23.1-3 in atomic mass units (amu). The conversion factors from mass to energy for electrons, protons and neutrons are also given.

Einstein's relation between relativistic mass (m) and energy ($\mathcal{E}$) is

$$\mathcal{E} = mc^2,$$ (23.1-2)

where c is the velocity of light. In a stationary reference frame, m becomes

---

[*] In high-temperature plasmas, the atoms are totally stripped of electrons. Here, as in the literature on fusion, such abbreviations as He for He^{++} are used, where the ++ indicates the loss of two electrons; D stands for D^+ and T for T^+. When ambiguity occurs because of incomplete ionization outside the plasma core or in the presence of high Z impurities, the atomic convention of writing He^{++} explicitly is often followed.

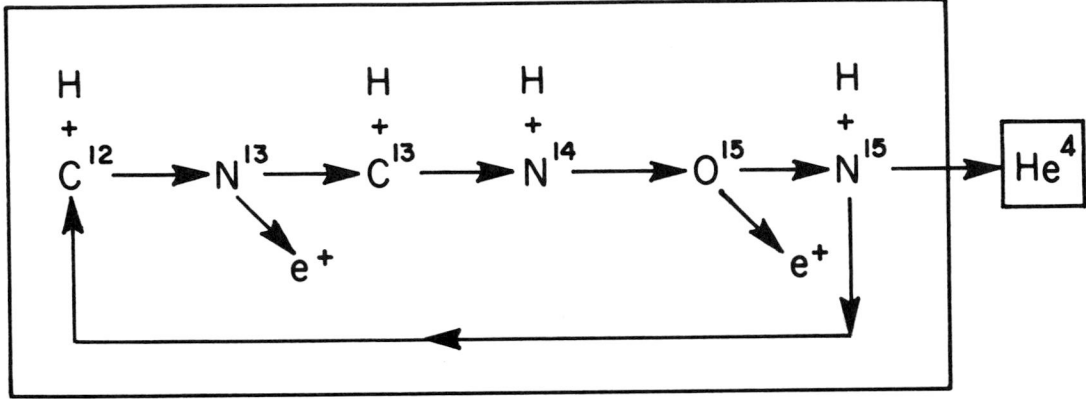

<u>Carbon (C-N) Cycle: $4H \rightarrow He^4 + 2e^+$</u>

$$C^{12} + H \rightarrow N^{13} + \gamma, \qquad\qquad N^{14} + H \rightarrow O^{15} + \gamma,$$

$$N^{13} \rightarrow C^{13} + e^+ + \nu, \qquad\qquad O^{15} \rightarrow N^{15} + e^+ + \nu,$$

$$C^{13} + H \rightarrow N^{14} + \gamma, \qquad\qquad N^{15} + H \rightarrow C^{12} + He^4.$$

Fig. 23.1-2 Helium is generated by the carbon (C-N) catalytic fusion cycle
            inside normal stars (e^+ = positron, γ = gamma ray, ν = neu-
            trino); H stands for the proton, He^4 for the α-particle, etc.

the rest mass m_o. If the particle velocity is v with respect to a second reference frame, then, in this frame,

$$m = m_o[1 - (v^2/c^2)]^{-\frac{1}{2}} \qquad\qquad\qquad (23.1-3)$$

and

$$\mathcal{E} = m_o c^2/[1 - (v^2/c^2)]^{\frac{1}{2}}. \qquad\qquad\qquad (23.1-4)$$

Table 23.1-2 A portion of the periodic table showing elements of interest in fusion application. The numbers refer to lifetimes of nuclei; stable isotopes are underlined (1971 data).

Number of neutrons	Number of protons							
	1	2	3	4	5	6	7	8
0	$\underline{H}(p)$							
1	$\underline{D}(H^2)$	$\underline{He}^3$						
2	$\underline{T}(H^3)$ 12.4y	$\underline{He}^4 (\alpha)$	$\underline{Li}^5$ 10^{-21} sec					
3		He^5 unstable	$\underline{Li}^6$	Be^7 53 days	B^8 0.8 sec	C^9 .13 sec		
4		He^6 0.83 sec	$\underline{Li}^7$	Be^8 10^{-16} sec	B^9 no data	C^{10} 19 sec		
5		He^7 no data	Li^8 0.9 sec	$\underline{Be}^9$	$\underline{B}^{10}$	$\underline{C}^{11}$ 20.4 min	$\underline{N}^{12}$ .013 sec	$\underline{O}^{13}$ 0.009 sec
6		He^8 0.12 sec	Li^9 0.17 sec	Be^{10} 1.9×10^6 y	$\underline{B}^{11}$	$\underline{C}^{12}$	N^{13} 10.1 min	O^{14} 71 sec
7				Be^{11} 13.6 sec	B^{12} 0.02 sec	$\underline{C}^{13}$	$\underline{N}^{14}$	O^{15} 2.0 min
8				Be^{12} 0.01 sec	B^{13} 0.02 sec	C^{14} 5,790y	$\underline{N}^{15}$	$\underline{O}^{16}$
9					B^{14} no data	C^{15} 2.4 sec	N^{16} 7.1 sec	$\underline{O}^{17}$
10					B^{15} no data	C^{16} .74 sec	N^{17} 4.2 sec	$\underline{O}^{18}$
11							N^{18} .63 sec	O^{19} 27.0 sec
12								O^{20} 13.6 sec

Table 23.1-3   Masses of selected particles and fully ionized nuclei. Compiled from standard reference sources.

Particle (or nucleus)	Mass, amu	Nucleus	Mass, amu	Mass-energy conversion factors
e	5.485930×10^{-4}			amu = atomic mass unit (based on at. wt. of C^{12} as 12.00000)
n	1.0086652	$(Be^9)^{4+}$	9.009991	$1 \text{ amu} = 1.660531 \times 10^{-24}$ g
H^+(or p)	1.0072766	$(B^{10})^{5+}$	10.010195	$= 931.4812$ Mev
D^+ or $(H^2)^+$	2.0135536	$(B^{11})^{5+}$	11.0065623	$m_e = 9.109558 \times 10^{-28}$ g
T^+ or $(H^3)^+$	3.0155011	$(C^{12})^{6+}$	11.996708	$= .511004$ Mev
$(He^3)^{++}$	3.014933	$(C^{13})^{6+}$	13.000063	$m_p = 1.672614 \times 10^{-24}$ g
$(He^4)^{++}$ (or α)	4.001506	$(N^{14})^{7+}$	13.999235	$= 938.2592$ Mev
$(Li^6)^{+++}$	6.013479	$(N^{15})^{7+}$	14.996268	$m_n = 1.67492 \times 10^{-24}$ g
$(Li^7)^{+++}$	7.0143581	$(O^{16})^{8+}$	15.990526	$= 939.5527$ Mev
				$1 \text{ Mev} = 10^6 \text{ ev} = 10^3$ kev
				$= 1.60219 \times 10^{-6}$ erg
				$= 3.83 \times 10^{-14}$ cal
				$= 4.45 \times 10^{-20}$ kw-h

One may use the mass-energy relation to determine the energy released in a nuclear reaction. The total energy contained in the mass of the particles before and after a nuclear reaction differs by the amount of energy liberated ($\Delta\mathcal{E}$) in the reaction. For the reaction of nuclei A and B forming nuclei C and D,

$$A + B \rightarrow C + D + \Delta\mathcal{E}, \tag{23.1-5}$$

we have the mass difference or mass defect

$$\Delta m = m_A + m_B - m_C - m_D. \tag{23.1-6}$$

The amount of energy liberated is equivalent to the mass lost in the reaction; it therefore follows from Eqs. (23.1-2) and (23.1-6) that

$$\Delta\mathcal{E} = \Delta(mc^2). \tag{23.1-7}$$

As an example, we consider the following reaction between a deuteron and triton forming an alpha particle and a neutron:

$$D + T \rightarrow \alpha + n + \Delta\mathcal{E}. \tag{23.1-8}$$

Using the amu units in Table 23.1-3 for the particle masses, the mass defect is found to be $\Delta m = 0.01886$ amu and the energy liberated by the reaction is $\Delta\mathcal{E} = 17.5896$ Mev. In cgs units, we obtain the values $\Delta m = 3.13566 \times 10^{-26}$ g and $\Delta\mathcal{E} = 2.81819 \times 10^{-5}$ erg. The reaction products in Eq. (23.1-8) thus share 17.59 Mev in kinetic energy.

As a second example, we consider the reaction involving boron and hydrogen to create three alpha particles by the reaction (see Table 23.1-1)

$$B^{11} + H \rightarrow 3\alpha + \Delta\mathcal{E} \qquad\qquad (23.1-9)$$

while liberating the energy $\Delta\mathcal{E}$ in the reaction. The mass defect and energy released are $\Delta m = 0.009321$ amu and $\Delta\mathcal{E} = 8.68$ Mev, respectively. The $\Delta\mathcal{E}$ is distributed equally and each of the three alpha particles formed according to Eq. (23.1-9) gains 2.894 Mev.

To calculate the distribution of energy among the decay products in the reaction shown in Eq. (23.1-5), we use conservation of momentum and energy. Assuming the net momentum of the initial state to be zero,[*] we have

$$m_A \vec{v}_A + m_B \vec{v}_B = m_C \vec{v}_C + m_D \vec{v}_D = 0, \quad m_A \vec{v}_A = -m_B \vec{v}_B. \quad (23.1-10)$$

Squaring the terms in Eq. (23.1-10), we obtain

$$m_C \mathcal{E}_C = m_D \mathcal{E}_D, \qquad\qquad (23.1-11)$$

where

$$\mathcal{E}_C = \left(\frac{1}{2}\right) m_C v_C^2 \text{ and } \mathcal{E}_D = \left(\frac{1}{2}\right) m_D v_D^2 \qquad (23.1-12)$$

---

[*] The momenta of the decay products are 10^3 times larger than the initial momenta. The smaller initial values are therefore negligible.

are the kinetic energy of the particles C and D, respectively. The total energy available for distribution among the decay products is the mass defect $\Delta \mathcal{E}$ and appears as the kinetic energy

$$\Delta \mathcal{E} = \mathcal{E}_C + \mathcal{E}_D. \qquad\qquad (23.1\text{-}13)$$

Using Eq. (23.1-11) results in

$$\mathcal{E}_C = m_D \Delta \mathcal{E}/(m_C + m_D), \quad \mathcal{E}_D = m_C \Delta \mathcal{E}/(m_C + m_D). \qquad (23.1\text{-}14)$$

The collision between a triton and deuteron forms an alpha particle and a neutron while releasing 17.6 Mev, viz.

$$T + D \rightarrow \alpha + n + 17.6 \text{ Mev}. \qquad\qquad (23.1\text{-}15)$$

Using Eq. (23.1-14), the kinetic energy of the decay products is found to be

$$\left.\begin{array}{l} \mathcal{E}_\alpha = m_n \Delta \mathcal{E}/(m_\alpha + m_n) \approx \Delta \mathcal{E}/5 \approx 3.5 \text{ Mev}, \\[2em] \mathcal{E}_n = m_\alpha \Delta \mathcal{E}/(m_\alpha + m_n) \approx 4\Delta \mathcal{E}/5 \approx 14.1 \text{ Mev}. \end{array}\right\} \qquad (23.1\text{-}16)$$

Thus, the neutron receives four times the kinetic energy of the α-particles.

A particular reaction often leads to a number of distinct final states. The energy liberated will be different for each final state. The classical cross section σ is the effective area of a target experienced by an incoming projectile. The reaction cross section in a quantum system is analogously a measure of

the probable occurrence of reaction.   For a large number of reactions, therefore, the ratio of cross sections for the reaction branches determines the fraction of the final states that corresponds to a given branch.   For example, the
deuteron-deuteron reaction has the two branches

$$D + D \begin{cases} He^3 + n + \Delta \mathcal{E}, \\ T + p + \Delta \mathcal{E}', \end{cases}$$

(see Fig. 23.1-3).   In Table 23.1-4, we list the distribution of energies in Mev
among the fusion fragments for various reactions.

B.   Radiation Losses

In fusion, the critical processes are those which compete with the nuclear reactions for the available energy in the system.   Radiation loss due to
the scattering of electrons by ions is one such competitor.   The radiation is
known as bremsstrahlung and corresponds, at high temperatures, to continuous
X-rays in the kev region.   The X-rays are emitted by accelerating charged
particles, primarily electrons.   The other type of radiation which can be important is cyclotron radiation.   We shall now calculate the power lost from a
plasma of electrons and ions due to these processes.

The total power (W) radiated by an accelerating electron traveling at
nonrelativistic velocities is[3]

$$W = 2e^2 \dot{v}^2 / 3c^3, \quad \dot{v} = dv/dt, \tag{23.1-17}$$

---

[3] J. D. Jackson, <u>Classical Electrodynamics</u>, pp. 505-513, John Wiley & Sons,
New York, 1962.

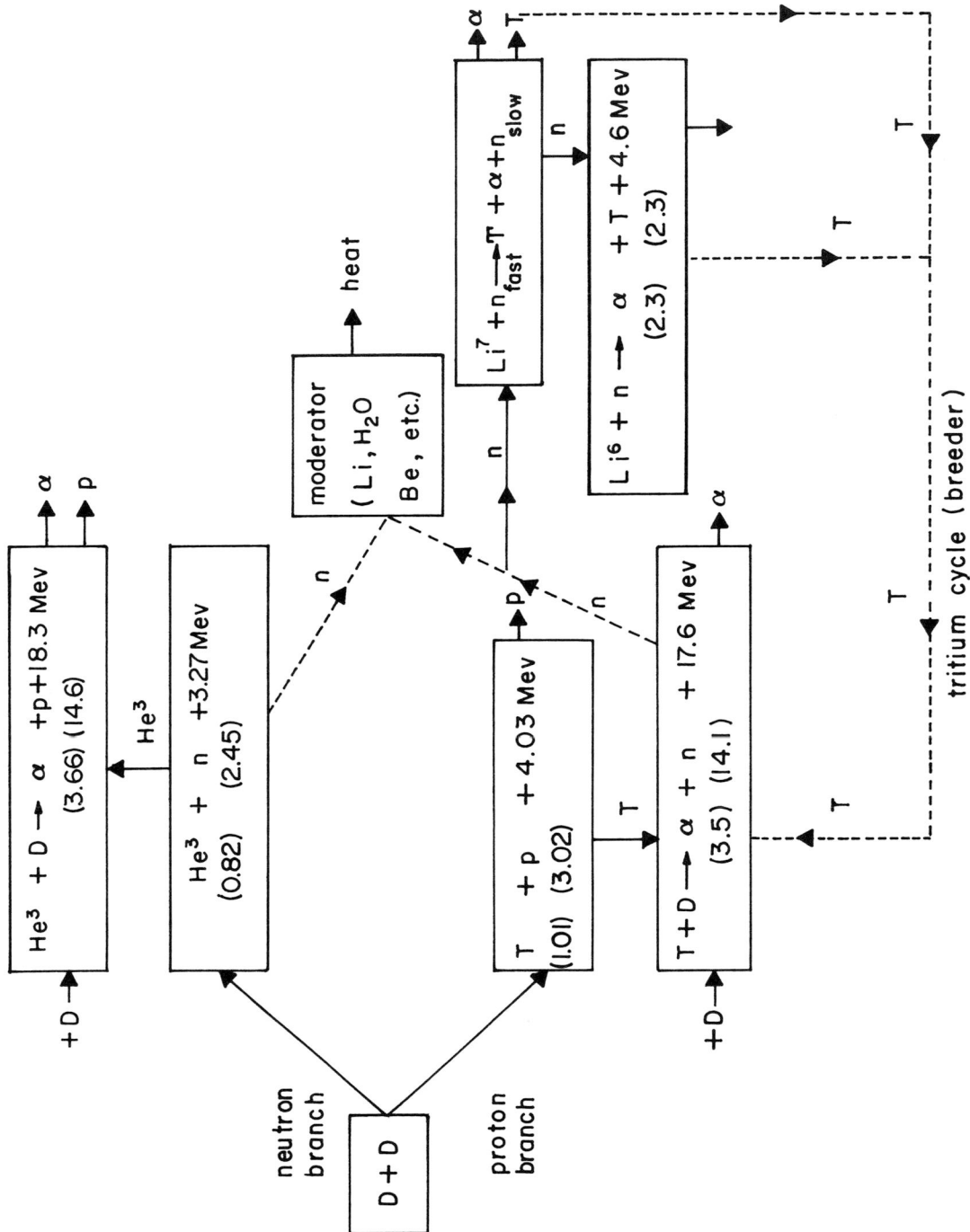

Fig. 23.3-3 Pictorial representation of a possible complete fusion cycle using deuterium and tritium breeder flow diagrams.

Table 23.1-4 Distribution of energy generated in fusion (in Mev).

	Charged Particles	Neutrons	Number of Input Particles
I.	**DT Reaction**		
	α (3.5*)	n (14.1)	1 D, 1 T
	α (2.6) ⎫ processes		1 Li^6
	T (2.0) ⎭ involving Li^6		

The total energy release for heating a plasma and conversion for electricity is 3.5 Mev. The total energy (without Li^6) is 17.6 Mev. The total available energy (with heat released by Li^6) is 22.2 Mev. The energy derived from neutrons is 18.7 Mev.

II.A.   DD Reaction, T < 100 kev (without the He^3 + D reaction)

	Charged Particles	Neutrons	Number of Input Particles
1.	**Proton Branch**		
	p (3.0*) ⎫		2 D
	T (1.0*) ⎭		
	DT reaction α (3.5*)	n (14.1)	1 D
	α (2.6) ⎫ processes		1 Li^6
	T (2.0) ⎭ involving Li^6		
2.	**Neutron Branch**		
	He^3 (0.8*)	n (2.45)	2 D
	α (2.6) ⎫ processes		1 Li^6
	T (2.0) ⎭ involving Li^6		

The total energy released by charged particles for heating a plasma and direct conversion to electricity is 8.3 Mev for 5D or 1.7 Mev/deuteron. The total energy released by neutrons is 16.5 Mev. The additional energy released via Li^6 (heat) is 9.2 Mev. The total available energy is 34.0 Mev or 6.8 Mev/deuteron.

II.B.   DD Reaction, T ≥ 100 kev (with He^3 + D reaction)

	Charged Particles	Neutrons	Number of Input Particles
	α (3.7*) ⎫ (He^3 + D only)		1 D
	p (14.6*) ⎭		

The additional energy released by charged particles for plasma heating is 18.3 Mev.

IIA + IIB,	for plasma heating	26.6 Mev
	for heat	25.7 Mev
	total	52.3 Mev for 6D or
		8.7 Mev/deuteron

---

*The asterisk refers to the energies of charged particles inside the reactor, which are useful in heating the plasma.

where c is the velocity of light and e is the charge of the electron. The acceleration ($\dot{v}$) of the electron is induced by electron interaction with the electric
field of the heavier ions in the plasma. Here $\dot{v}$ may be expressed in terms of
the electric field of the ion. The semi-classical expression for the total energy
radiated away by the electron during scattering is[3]

$$\mathcal{E}^{br} \approx 4 Z_i^2 e^6 / 3 m_e^2 c^3 v b^3 .\tag{23.1-18}$$

The charge of the ion is $Z_i e$, m_e is the mass of the electron, v is the electron
velocity, and b is the distance of closest approach between the electron and ion.

In a plasma, where the gas has been fully ionized, all of the electrons are
free, and there are many "collisions" producing bremsstrahlung. To calculate
the total energy lost due to bremsstrahlung, we sum the contributions over the
scattering events. The number of such events per unit volume of the plasma
per unit time is

$$dN = v \sigma(v) dN_i dN_e ,\tag{23.1-19}$$

where $\sigma(v)$ is the classical scattering cross section, v is the relative velocity
between the electron and ion, and dN_i and dN_e are the numbers of ions and
electrons per unit volume of the plasma.

The total power density lost from the plasma due to bremsstrahlung is
indicated by a superscript br and is

$$W^{br} = \int_0^\infty \mathcal{E}^{br} dN ,\tag{23.1-20}$$

where $\mathcal{E}^{br}$ is given by Eq. (23.1-18). For an electron plasma with a Maxwellian distribution of velocities, Eq. (23.1-20) may be integrated to

$$W^{br} = \left(\frac{2\pi k T_e}{3m_e}\right)^{\frac{1}{2}} \frac{32\pi}{3} \frac{N_e N_i Z_i e^6}{hm_e c^3} g.$$ (23.1-21)

Equation (23.1-21) is the expression for the total power lost from a plasma due to bremsstrahlung if N_e and N_i are the electron and ion concentrations, respectively, and h is Planck's constant. For the semi-classical expression, g = 1. For high electron temperatures, a more accurate quantum mechanical calculation gives $g \simeq 2\sqrt{3}/\pi \simeq 1.11$.

If there is more than one species i of ion in the plasma, we introduce a summation in Eq. (23.1-21). It may then be shown that

$$W^{br} = \left(\frac{2\pi k T_e}{3m_e}\right)^{\frac{1}{2}} \frac{32\pi N_e e^6 g}{3hc^3 m_e} \sum_i Z_i^2 N_i.$$ (23.1-22)

Thus, the power density lost from the plasma due to bremsstrahlung (with concentrations expressed as the number of particles per cm^3) is

$$W^{br} = 1.57 \times 10^{-27} N_e \sum_i N_i Z_i^2 \sqrt{T_e({}^\circ K)} \ erg/cm^3\text{-sec}$$

$$= 5.35 \times 10^{-31} N_e \sum_i N_i Z_i^2 \sqrt{T_e(kev)} \ w/cm^3.$$ (23.1-23)

In a DD reaction (referring to the neutron and proton branches of Fig. 23.3-3), the concentration of electrons equals that of the deuterons, i.e., $N_e = N_D$, $N_i = N_D$, $Z_i = 1$. The power density is then

$$W_{DD}^{br} = 5.35 \times 10^{-31} N_D^2 (cm^{-3}) \sqrt{T_e (kev)} \ w/cm^3.$$

For a deuteron-triton reaction with 50% D and 50% T, we have $N_e = N_D + N_T$ $= 2N_D$, $Z_i = 1$; also $N_e \sum_i N_i Z_i^2 = 2N_D (N_D \times 1 + N_T \times 1) = 4N_D^2 = 4N_D N_T$ so that

$$W_{DT}^{br} = 2.14 \times 10^{-30} N_D N_T (cm^{-3}) \sqrt{T_e (kev)} \ w/cm^3.$$

It should be noted that in a DD or in a 50% D, 50% T plasma,

$$W_{DD}^{br} = W_{DT}^{br} = 5.35 \times 10^{-31} N_e^2 (cm^{-3}) \sqrt{T_e (kev)} \ w/cm^3. \quad (23.1\text{-}24)$$

The confinement of the plasma by a magnetic field will be discussed in Section 23.2. Although the magnetic field confining the plasma does not contribute to the plasma energy, it exerts a force on the charged particles. This force imparts an angular acceleration to the electrons and ions, thereby causing a helical motion along the magnetic field lines. As is indicated in Eq. (23.1-17), this acceleration results in a radiative loss of power.

At low kinetic energies of the electrons, the radiation induced by the angular acceleration is called <u>cyclotron radiation</u> and will be identified by a superscript cyc.[4] The frequency of the emitted radiation ω corresponds to

[4]G. Bekefi, <u>Radiation Processes in Plasmas</u>, pp. 177-213, John Wiley & Sons, New York, N.Y., 1966. Cyclotron radiation is often called synchrotron radiation.

the infrared and microwave regions. The spectrum of emitted radiation is not continuous but rather consists of a number of discrete wavelengths separated on the frequency scale by the gyrofrequency $\omega_c = eB/m_e c$, where B is the intensity of the magnetic field. The intensity of the radiation decreases with increasing frequency.

For electrons having a Maxwellian distribution of velocities, the cyclotron radiation power per unit volume is given by

$$W^{cyc} \approx 4e^2 \omega_c^2 p_e / 3m_e c^3,$$
(23.1-25)

where the angular acceleration is

$$dv/dt = v\omega_c$$
(23.1-26)

and p_e is the kinetic pressure on the electrons. Using the definitions of the cyclotron frequency $\omega_c = eB/m_e c$ and the ideal gas law $p_e = kN_e T_e$, we have

$$W^{cyc} = 4e^4 B^2 kN_e T_e / 3m_e^3 c^5.$$
(23.1-27)

To maintain plasma confinement, the magnetic pressure must equal or exceed the total gas pressure, i.e. (with T_i for the ion temperature)

$$B^2/8\pi \geq p_e + p_i = N_e kT_e + N_i kT_i.$$
(23.1-28)

Substituting for the magnitude of the magnetic field from Eq. (23.1-28) in

Eq. (23.1-27), we obtain (with N_e in cm^{-3} and T_e and T_i in kev)

$$W^{cyc} \approx 32\pi e^4 k^2 N_e^2 T_e^2 [1 + (T_i/T_e)]/3m_e^3 c^5 \beta$$

$$= 2.5 \times 10^{-32} N_e^2 T_e^2 [1 + (T_i/T_e)]/\beta \quad w/cm^3, \qquad (23.1-29)$$

where β is the ratio of the gas pressure to the magnetic pressure. For $T_e = T_i = T$, Eq. (23.1-29) becomes

$$W^{cyc} \approx 5.01 \times 10^{-32} N_e^2 T^2/\beta \ w/cm^3. \qquad (23.1-30)$$

For a DT or DD plasma, the power-density expression for bremsstrahlung radiation (with T in kev) is given by [from Eq. (23.1-24)]

$$W^{br} = 5.35 \times 10^{-31} N_e^2 \sqrt{T} \ w/cm^3. \qquad (23.1-31)$$

The energy lost due to cyclotron radiation in a DT plasma exceeds that due to bremsstrahlung, $W^{cyc} > W^{br}$, when the plasma temperature is[*] (see Fig. 23.1-4) $T \geq 4.85$ kev ($\beta = 1$). The cyclotron radiation will be more of a problem for the DD fuel than for the DT fuel. The bremsstrahlung radiation cannot be prevented from escaping from the plasma. The cyclotron radiation is, however, in the infrared and microwave regions and will probably be partially

---

[*] $5.35 \times 10^{-31} \sqrt{T} = 5.01 \times 10^{-32} T^2$ or $T = (10.69)^{2/3} = 4.85$ kev.

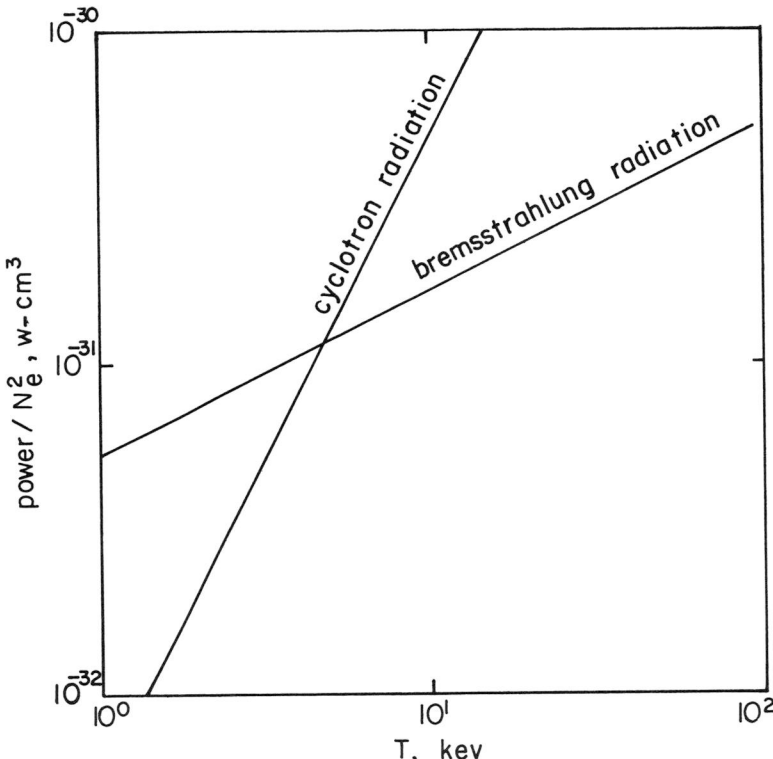

Fig. 23.1-4 The radiation loss from a plasma due to bremsstrahlung and
cyclotron radiation. The cyclotron radiation is in the presence
of the minimum magnetic field for plasma confinement, $\beta = 1$.

reabsorbed by the plasma. Ordinary metals may be used to reflect this radiation back into the reacting system to reduce energy loss.

Thus, for hydrogen plasmas, the cyclotron radiation may become a significant source of energy loss when the temperature exceeds 4.8 kev. This loss
mechanism is not very important for a DT fuel since the ignition temperature

is only 4. 4 kev. However, in a DD fuel the ignition temperature is about 48 kev

so that this effect may become significant.

If the kinetic energy is so high that the electrons have relativistic velo-

cities, the accelerating electrons emit a different type of radiation.[4] This

radiation is expected to be only important in the presence of strong magnetic

fields and for plasmas at very high ignition temperatures but not for DT fusion.

C.   Reaction Cross Sections and the Coulomb Barrier

An initial configuration of nuclei may interact to form a number of dis-

tinct final states by means of a single nuclear reaction. Each final state occurs

with a relative frequency given by its reaction cross section. In a physical

description of these cross sections, we may assume the formation of an inter-

mediate state or compound nucleus in the reaction process. The compound

nucleus decouples the initial and final states of the reaction. The probability

of occurrence of a given final state is given by the joint probability that the

compound nucleus is formed and that it decays to the appropriate final state.

These two events are assumed to be independent so that the joint probability

is given by the product of the probabilities of occurrence of the separate events.

Consider collision of particle $\underline{a}$ with nucleus A (the word particle refers

to a complex structure such as D or an α-particle). There is a finite probabil-

ity that during the collision a compound nucleus (A + $\underline{a}$) is formed. This inter-

mediate state is long lived ($\sim 10^{-21}$ sec) compared to the transit time of the

nucleons in the nucleus. A finite time after its formation, the compound nu-

cleus decays into a number of alternative final states. This process is shown

in Fig. 23. 1-5, where A^{*} depicts an excited state of the nucleus A and $\Delta \mathcal{E}$

refers to the energy released between the initial and final states of the collision.

The subscripts denote the specific reaction branches.

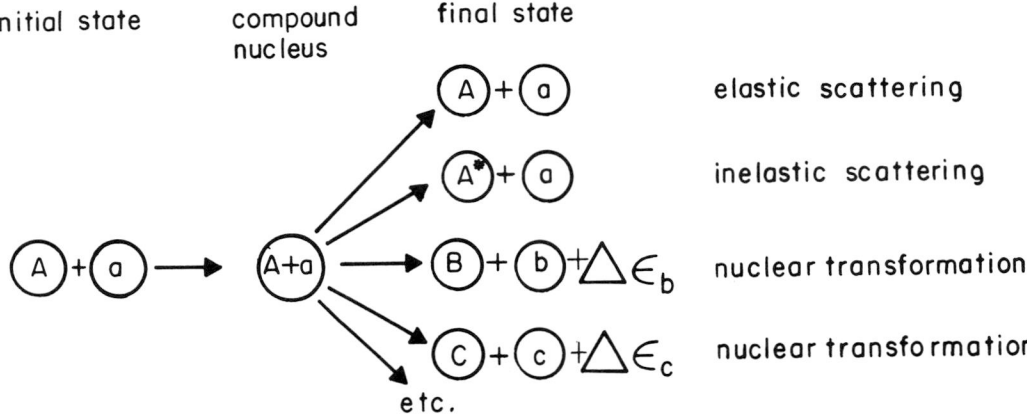

Fig. 23.1-5 The reaction of the nuclei A and a leads to the formation of a
           compound nucleus (A + a). The compound nucleus decays after
           a finite time into a number of distinct final states. An excited
           state of the nucleus A is indicated by A^* and other nuclei are in-
           dicated by B, C, b, c, and a. The energy released in the reaction
           is $\Delta \mathcal{E}$ with a subscript identifying a particular reaction.

The cross section for the reaction producing a compound nucleus in the
intermediate state is

$$\sigma(a, b) = (\text{constant}/\sqrt{\mathcal{E}})P_a . \tag{23.1-32}$$

Here P_a is the probability of penetrating the relative Coulomb barrier and is
given by

$$P_a \approx \eta \exp(-\eta). \tag{23.1-33}$$

Equation (23.1-33) determines[5] the yield for nuclear reaction at low energies. The decay parameter η is given by

$$\eta \equiv \gamma/\sqrt{\mathcal{E}} = [(2\pi)^2 Z_1 Z_2 e^2/h] \sqrt{\mu/2\mathcal{E}} \tag{23.1-34}$$

where μ is the reduced mass of the interacting nuclei $[(1/\mu) = (1/m_a) + (1/m_b)]$ and $\mathcal{E}$ is the energy in the center of mass frame.

Using Eqs. (23.1-33) and (23.1-34) for the penetration factor in Eq. (23.1-32), the total nuclear reaction cross section can be written as[5,6]

$$\sigma = (A/\mathcal{E})\exp(-\gamma/\sqrt{\mathcal{E}}). \tag{23.1-35}$$

The constants A and γ are deduced in practice from fits to experimental data on nuclear scattering; γ usually agrees closely with theoretical values since the Coulomb interaction is well understood. In contrast, the nuclear interactions

[5] L. I. Schiff, Quantum Mechanics, pp. 278-279, McGraw-Hill, Inc., New York, N.Y., 1968.

[6] W. A. Fowler, "Cross Sections for Thermonuclear Reactions," Phys. Rev. 81, 655 (1951); L. Stewart and G. M. Hale, "The T(d,n)He^4 and T(t,2n) Cross Sections at Low Energies," Los Alamos Scientific Laboratory, Report LA-5828-MS, January 1975; H. V. Argo, R. Taschek, H. Agnew, A. Hemmendinger, and W. Leland, "Cross Section of the D(t,n)He^4 Reactions for 80-1200 kev Tritons," Phys. Rev. 87, 612-615 (1952); W. R. Arnold, J. A. Phillips, G. A. Sawyer, E. J. Stovall, Jr., and J. L. Tuck, "Cross Sections for the Reactions D(d,p)T, D(d,n)He^3, T(d,n)He^4, and He^3(d,p)He^4 below 120 kev," Phys. Rev. 93, 483-497 (1954); R. F. Post, "Controlled Fusion Research - An Application of the Physics of High Temperature Plasmas," Rev. Mod. Phys. 28, 338-362 (1956).

which determine the value of the parameter A are more complex.[6] The fits

to A and γ are shown in Table 23.1-5 and a plot of σ as a function of energy is

given in Fig. 23.1-6. More detailed fits to $A = A(\mathcal{E})$ are given in Ref. [6].

The cross section for the nuclear reaction $D + T \rightarrow \alpha + n$, as given

by Eq. (23.1-35), is depicted in Fig. 23.1-6. From Table 23.1-4, one

finds that the cross section for the DT reaction is orders of magnitude larger

than that for the other reactions at deuteron energies below 200 kev. At higher

energies, the DHe^3 reaction has a cross section approaching that of DT.

The power released by a nuclear reaction is indicated by the super-

script nuc. For ions of types i and j, it is given by

$$W_{ij}^{nuc} = N_i N_j \sigma(i, j) v_{ij} \Delta \mathcal{E}_{ij}, \qquad (23.1-36)$$

where N_i and N_j are the concentrations of ions of types i and j, v_{ij} is the re-

lative velocity between the ions i and j, $\sigma(i, j)$ is the reaction cross section,

and $\Delta \mathcal{E}_{ij}$ is the energy released in each reaction. For a plasma with a Maxwel-

lian distribution of ion velocities, the power developed per unit volume of the

plasma is

$$W^{nuc} = \sum_{i, j} N_i N_j \langle \sigma(i, j) v_{ij} \rangle \Delta \mathcal{E}_{ij}, \qquad (23.1-37)$$

where the brackets $\langle \ldots \rangle$ denote an average over the ion velocity distribution

and the sum extends over ion species.

Introducing the reaction-rate coefficient

$$\mathcal{K}_{ij} \equiv \langle \sigma(i, j) v_{ij} \rangle \qquad (23.1-38)$$

Table 23.1-5 Parameters A and γ in the expression $\sigma = \frac{A}{\mathcal{E}} \exp(-\frac{\gamma}{\sqrt{\mathcal{E}}})$ for the cross section of the reaction indicated in the first column; $\mathcal{E}$ = relative energy in collision. The listed data with asterisks are from W. A. Fowler, G. R. Caughton and B. Zimmerman, "Thermonuclear Reaction Rates," Ann. Rev. Astron. Ap. 5, 525-568 (1967), where further corrections can be found. All other values were taken from W. B. Thompson, "Thermonuclear Reactions," Proc. Phys. Soc. 70B, 1-5 (1957), after changing to relative energies.

Reaction process	Standard notation	A, barn-kev	γ, $\sqrt{\text{kev}}$
$D + T \rightarrow n + \alpha$	$T(d,n)He^4$	11,000*	34.4*
$D + D \rightarrow p + T$	$D(d,p)T$	53*	31.4*
$D + D \rightarrow n + He^3$	$D(d,n)He^3$	53*	31.4*
$T + T \rightarrow 2n + \alpha$	$T(t,2n)He^4$	160*	54.3*
$D + He^3 \rightarrow p + \alpha$	$He^3(d,p)He^4$	7,000*	88.8*
$D + Li^7 \rightarrow n + 2\alpha$	$Li^7(d;n,\alpha)He^4$	220	85.4
$D + Li^6 \rightarrow p + Li^7$	$Li^6(d,p)Li^7$	2,040	86.0
$D + Li^6 \rightarrow n + He^3 + \alpha$	$Li^6(d;n,\alpha)He^3$	9,150	99.6
$D + Li^6 \rightarrow n + Be^7$	$Li^6(d,n)Be^7$		
$D + Li^6 \rightarrow 2\alpha$	$Li^6(d,\alpha)He^4$	3,060	104.0
$D + Li^7 \rightarrow n + Be^8$	$Li^7(d,n)Be^8$	35,900	122.0
$p + B^{11} \rightarrow 3\alpha$	$B^{11}(p,2\alpha)He^4$	100,000	150.3

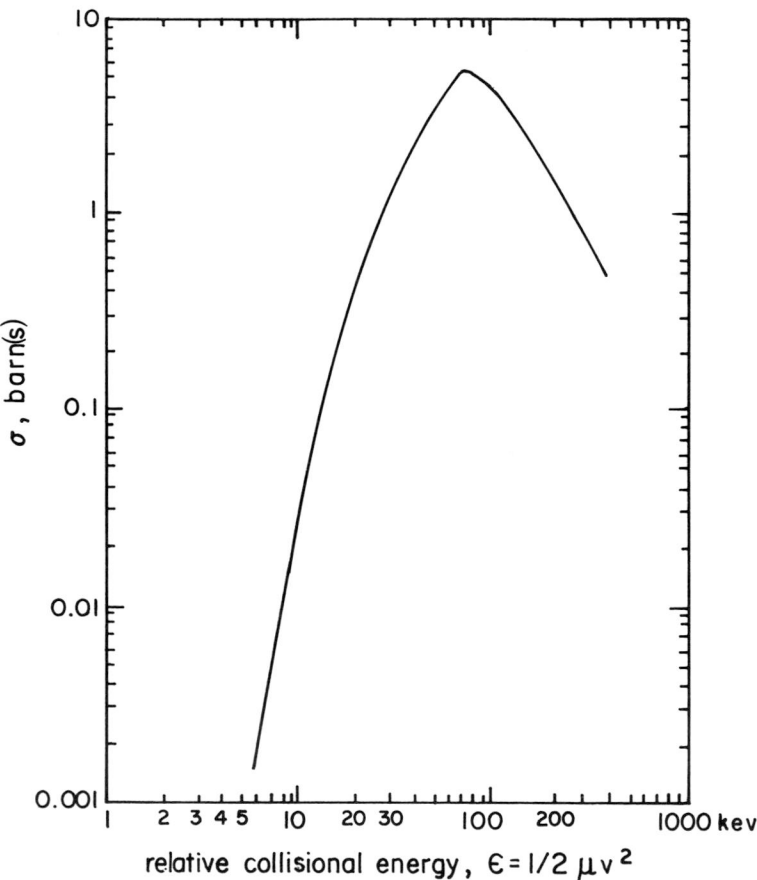

Fig. 23.1-6 Cross sections for the deuteron-triton reaction, $D(t,n)He^4$; 1 barn = 10^{-24} cm^2. Reproduced from L. Stewart and G. M. Hale, "The $T(d,n)He^4$ and $T(t,2n)$ Cross Sections at Low Energies," Los Alamos Scientific Laboratory Report LA-5828-MS, USNDC-CTR-2, Los Alamos, New Mexico, January 1975.

and the energy released per sec by the nuclear reactions,

$$W^{nuc} = d\mathcal{E}^{nuc}/dt, \tag{23.1-39}$$

allows us to write Eq. (23.1-37) in the form of the following chemical kinetic rate equation:

$$d\mathcal{E}^{nuc}/dt = \sum_{i,j} b_{ij} \mathcal{K}_{ij} N_i N_j \Delta\mathcal{E}_{ij} = 1.602 \times 10^{-13} \sum_{i,j} b_{ij} \mathcal{K}_{ij} N_i N_j Q_{ij} (\text{Mev}) \text{w/cm}^3 \tag{23.1-40}$$

where $\mathcal{K}_{ij}$ is in the units of cm^3/sec, N_i and N_j are ion concentrations (e.g., D, T) in number/cm^3. The energy released in a nuclear reaction between the ith and jth ions is $\Delta\mathcal{E}_{ij}$ in erg or $\Delta\mathcal{E}_{ij} \equiv Q_{ij}$ (in Mev) and W^{nuc} is given in w/cm^3, $b_{ij} = 1$ if $i \neq j$, $b_{ij} = 1/2$ if $i = j$.

D. Reaction-Rate Constant

For the reaction A + B, the rate equation for the loss in concentration of either species A or B is

$$dN_A/dt = -\mathcal{K}N_A N_B, \tag{23.1-41}$$

where $\mathcal{K}$ is the reaction-rate constant in the units $(\text{length})^3/\text{sec}$ defined by Eq. (23.1-38). Here N_j is the concentration of ion species j. The product $N_A N_B$ is the number of binary collisions between the ions A and B when A and B are different species.

In a plasma, the concentrations appropriate for fusion are on the order of 10^{14} electrons (or ions) per cm^3. It is impossible to describe this system deterministically. We therefore introduce a distribution function to describe the statistical behavior of the electrons (ions) of the plasma. The one-particle distribution function of electrons $f_e(\vec{v}, \vec{r}, t)$ is defined such that $f_e(\vec{v}, \vec{r}, t)d\vec{v}d\vec{r}$ is the number of electrons at time t in a volume element $d\vec{v}$ ($\equiv dv_x dv_y dv_z$) at the position $\vec{v}$ in velocity space and a corresponding volume element $d\vec{r}$ ($\equiv dxdydz$) located at $\vec{r}$ in physical space.

In general, one may define appropriate averages denoted by brackets $\langle ... \rangle$ for any function which depends on the velocities of the electrons $\vec{v}$ or ions $\vec{V}$ in terms of the distribution functions

$$\langle h(\vec{r}, t) \rangle \equiv \frac{1}{N_e} \int h(\vec{v}) f_e(\vec{v}, \vec{r}, t) d\vec{v}$$

and

$$\langle H(\vec{r}, t) \rangle \equiv \frac{1}{N_i} \int H(\vec{V}, \vec{r}, t) f_i(\vec{V}, \vec{r}, t) d\vec{V}. \tag{23.1-42}$$

The distribution functions f_e and f_i are determined by an equation of evolution, generally a partial differential equation, e.g., Boltzmann's equation in the kinetic theory of gases.

If there is no bulk or average velocity for the electrons and ions, then $\langle \vec{v} \rangle = \langle \vec{V} \rangle = 0$. The differential concentration of ions is

$$dN_i = 4\pi V^2 f_i dv. \tag{23.1-43}$$

Introducing the parameters $V_i = V_i^o \sqrt{W}$ and $V_i^o = (2kT_i/M_i)^{1/2}$ into Eq. (23.1-43) yields

$$dN_i = 2\sqrt{W}\,dW\,\exp(-W)/\sqrt{\pi} \tag{23.1-44}$$

for a Maxwellian distribution of velocities. The average of an arbitrary function of the ion velocity $[h(V)]$ may then be rewritten, in view of Eq. (23.1-44), as

$$\langle h(V) \rangle = \frac{2}{\sqrt{\pi}} \int_o^\infty h(W)\sqrt{W}\,dW\,\exp(-W). \tag{23.1-45}$$

Using Eq. (23.1-45), the expression for the average rate coefficient in Eq. (23.1-38) becomes

$$\Bbbk = \left(\frac{8kT_i}{\pi\mu}\right)^{1/2} \int_o^\infty \sigma(W)W\,(\exp - W)dW, \tag{23.1-46}$$

where μ is the reduced mass and $W = \mathcal{E}/kT_i$ is a dimensionless parameter given in terms of the relative energy $\mathcal{E} = \frac{1}{2}\mu V^2$. Equation (23.1-46) may be integrated by using the reaction cross section of Eq. (23.1-35). In terms of the parameter W, the reaction cross section may be written as

$$\sigma(W) = \frac{(A/kT_i)}{W} \exp\left(-\frac{\gamma/\sqrt{kT_i}}{W}\right) \tag{23.1-47}$$

so that Eq. (23.1-46) becomes

$$\bar{k} = \frac{A}{kT} \sqrt{\frac{8kT}{\pi\mu}} \int_0^\infty \left[\exp - \left(W + \frac{2B}{\sqrt{W}} \right) \right] dW \qquad (23.1\text{-}48)$$

with[*]

$$B = \gamma/2\sqrt{kT} = 1.703 \times 10^3 \, \gamma \, T^{-1/2}. \qquad (23.1\text{-}49)$$

Evaluating the integral in Eq. (23.1-48) approximately about the point of maximum contribution, which occurs for the argument in the exponential at $W = B^{2/3}$, yields the average rate coefficient for the thermonuclear reaction as

$$\bar{k} = \beta T^{-2/3} \exp(-\Lambda T^{-1/3}), \quad \beta = 0.8052 \times 10^{-16} \frac{A}{(\mu/m_H)^{1/2}} \gamma^{1/3},$$

$$\Lambda = 3(\gamma/2)^{2/3}, \qquad (23.1\text{-}50)$$

where β and Λ are constants for a given reaction process and γ and A are obtained from Table 23.1-5. The rate coefficient has a maximum at $T = (\Lambda/2)^3 = (27/8)(\gamma/2)^{2/3}$, i.e., at the temperature of $T = 27\mu\pi^2 Z_1^2 Z_2^2 e^4/16\hbar^2 k$.

---

[*]The integral yields approximately $I(B) \simeq 2(\pi/3)^{1/2} B^{1/3} \exp(-3B^{2/3})$; T is in kev and γ in $(\text{kev})^{\frac{1}{2}}$.

In many instances, it is difficult to fit the best cross sections to the form of Eq. (23.1-35). It is then convenient to perform direct numerical integration, i.e.,

$$\mathcal{K} = \frac{4.95 \times 10^{-17}}{T_i \sqrt{T_i}} \left(\frac{m_H}{\mu}\right)^{1/2} G(T_i), \qquad (23.1-51)$$

where m_H is the mass of proton and

$$G(T_i) = \int_0^\infty \mathcal{E} \sigma(\mathcal{E}) e^{-\mathcal{E}/T_i} d\mathcal{E}$$

with T_i in kev, $\mathcal{E}$ in kev, and σ in barn; for DT, $m_H/\mu = 5/6$ and, for DD, $m_H/\mu = 1$.

In Fig. 23.1-7, the rate coefficient DT is plotted from results of this numerical integration (using Los Alamos cross sections) from Ref. [6]. For the DT reaction, the experimental results of Arnold et al (see Ref. [6]) were integrated and fitted numerically,[2] with the results

$$\mathcal{K}_{DT} = 3.8 \times 10^{-12} T_i^{-2/3} \exp(-19.02 T_i^{-1/3}), \ cm^3/sec \ for \ T_i < 10 \ kev,$$

$$\mathcal{K}_{DT} = 3.41 \times 10^{-14} T_i^{-2/3} \exp(-27.217 T_i^{-2/3} + 3.638 T_i^{-1/3}), \ cm^3/sec \ for \ T_i > 10 \ kev.$$

For the DD reaction, Post (see Ref. [6]) gives

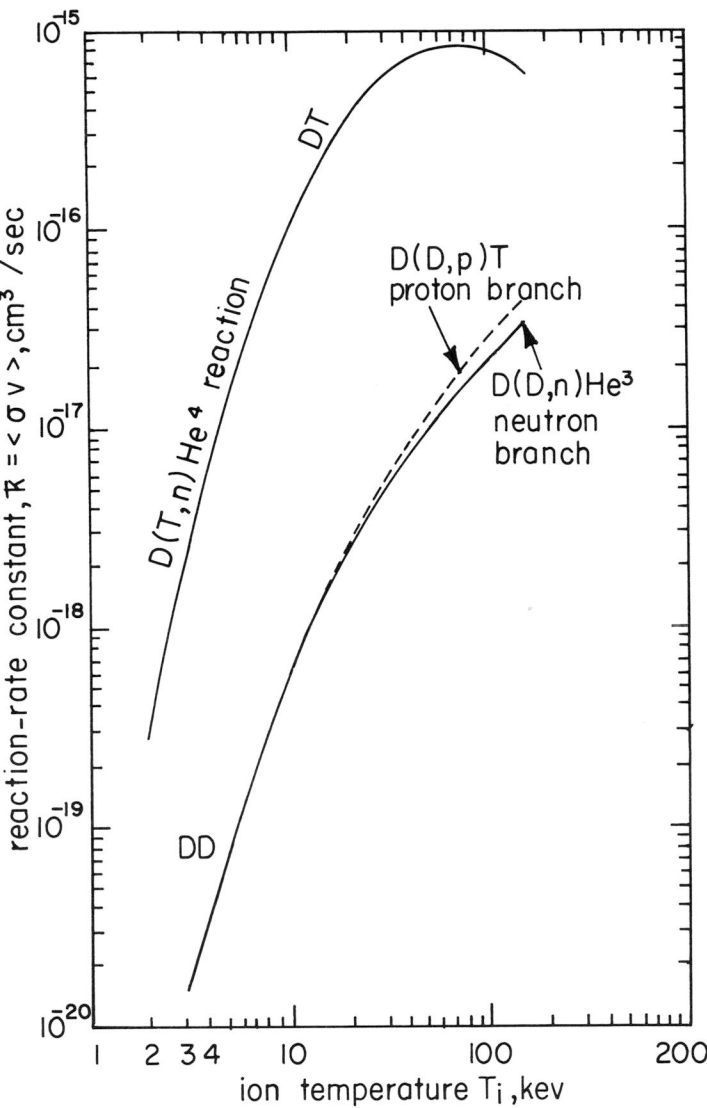

Fig. 23.1-7 Average reaction rate constant k̇ for the deuteron-triton (DT) and
deuteron-deuteron (DD) nuclear reactions (for a Maxwellian distribution of ion velocities) as functions of the ion temperature, T_i.
The data on cross sections were compiled from various sources,
especially Ref. [6], and were integrated numerically.

$$\bar{k}_{DD} = 2.6 \times 10^{-14} T_i^{-2/3} \exp(-18.76 T_i^{-1/3}), \quad cm^3/sec.$$

E. Ignition Temperature and Lawson's Criterion

The criterion for fusion is that the energy made available by nuclear reactions exceeds that lost by radiation and other processes. The power density in w/cm^3 released in a reaction between the ions i and j is given by [see Eq. (23.1-40)]

$$W_{ij}^{nuc} = 1.602 \times 10^{-13} N_i N_j \bar{k}_{ij} Q_{ij}(Mev) b_{ij}, \qquad (23.1-52)$$

where $Q_{ij}(Mev)$ is the average energy released per interaction, $\bar{k}_{ij}$ is the average reaction rate for the interaction of ions i and j, $b_{ij} = 1$ if $i \neq j$, $b_{ij} = 1/2$ if $i = j$.

The radiation power density in w/cm^3 lost due to bremsstrahlung is given by Eq.(23.1-23) for a DT plasma as

$$W^{br} = 5.35 \times 10^{-31} N_e^2 \sqrt{T_e}. \qquad (23.1-53)$$

The condition for fusion, assuming bremsstrahlung to be the dominant loss mechanism for one type of reaction (e.g., only D + T), is

$$W^{nuc} \geq W^{br}. \qquad (23.1-54)$$

Using Eqs. (23.1-52) and (23.1-53) in Eq. (23.1-54), we obtain

$$\frac{k_{ij}(T_i)}{\sqrt{T_e}} \geq \frac{3.34 \times 10^{-18}}{Q_{ij}(\text{Mev})} \frac{N_e^2}{N_i N_j b_{ij}} \qquad (23.1-55)$$

as the condition on the rate coefficients for energy to be released fast enough
to balance and exceed that lost due to bremsstrahlung.

For DD reactions, the average energy released to the charged particles
is 4.85 Mev (see Table 23.1-4) so that the power density in w/cm^3 is given by
(proton and neutron branches only)

$$W_{DD}^{nuc} = 3.9 \times 10^{-13} N_D^2 k_{DD} \ w/cm^3. \qquad (23.1-56)$$

For example, using Fig. 23.1-7 for the reaction coefficient at two plasma
temperatures with $T = 10$ kev, $k_{DD} = 1.210^{-18} \ cm^3/sec$ and $T = 100$ kev,
$k_{DD} = 4.9 \times 10^{-17} \ cm^3/sec$ gives the corresponding power densities as

$$\left.\begin{array}{l} W_{DD}^{nuc} = 4.6 \times 10^{-31} N_D^2 \ w/cm^3, \quad T = 10 \text{ kev;} \\[2em] W_{DD}^{nuc} = 1.9 \times 10^{-29} N_D^2 \ w/cm^3, \quad T = 100 \text{ kev.} \end{array}\right\} \qquad (23.1-57)$$

For a DT reaction, the average energy released to the charged particles
is 3.5 Mev so that, for a reactor with 50% D and 50% T fuel, the power density
is

$$W_{DT}^{nuc} = 5.6 \times 10^{-13} N_D N_T k_{DT} \ w/cm^3. \qquad (23.1-58)$$

Using Fig. 23.1-7 again for the same example at the same temperatures as for the DD reaction, with $\mathcal{k}_{DT}$ (10 kev) = 1.1×10^{-16} cm^3/sec, $\mathcal{k}_{DT}$ (100 kev) = 8.3×10^{-16} cm^3/sec, we obtain (in w/cm^3)

$$
\left.
\begin{aligned}
W_{DT}^{nuc} &= 6.2 \times 10^{-29} \, N_D N_T , \quad T = 10 \text{ kev;} \\[2em]
W_{DT}^{nuc} &= 4.6 \times 10^{-28} \, N_D N_T , \quad T = 100 \text{ kev.}
\end{aligned}
\right\} \tag{23.1-59}
$$

The plasma temperature of interest is the one at which the equality in Eq. (23.1-55) occurs. For example, in the DT reaction, Eq. (23.1-55) reduces to ($N_D = N_T$, $N_D + N_T = N_e$, $Z_i = Z_j = 1$)

$$
\mathcal{k}_{DT}(T_i)/\sqrt{T_e} = 3.8 \times 10^{-18} \text{ cm}^3/\text{sec-(kev)}^{1/2}, \tag{23.1-60}
$$

where the equality sign defines the temperature at which the powers generated and lost are in balance, i.e., the ignition temperature. Equating the ion and electron temperatures in Eq. (23.1-60), $T_i = T_e = T$; taking $\mathcal{k}_{DT}(T)$ from Fig. 23.1-7, we may graph Eq. (23.1-60). The generated power and bremsstrahlung losses for DD and DT fuel are shown in Fig. 23.1-8 as functions of the temperature. The intersection of the power generation and loss curves gives the ideal ignition temperature as the solution to Eq. (23.1-60). For the DT fuel, this ignition temperature is approximately 4.4 kev and for the DD fuel it is approximately 48 kev. In the DT reaction, if we take the energy charged α particles (i.e., 3.5 Mev) only (see Fig. 23.1-3) and if $T_e = T_i$, then we obtain 4.4 kev for the ignition temperature. If we could use some of the neutron energy in the plasma which is dissipated as heat, then the ignition temperature would be between 2.8 and 4.4 kev. For DD reactions, charged particle contribution gives an ignition temperature of 48 kev.

The effect of impurities in the fuel on the power lost by bremsstrahlung is a serious problem in the design of fusion reactors. We can estimate this loss by using Eq. (23.1-23). For example, let us assume that in the DT reaction process, with equal concentrations of deuterons and tritons, we have 0.1% of tungsten impurity ($N_W = 10^{-3} N_D$). If the most abundant ion of tungsten is W^{56+}, we find for the summation over ion species ($N_D = N_T$)

$$N_e \sum_i N_i Z_i^2 = (N_D + N_T + 0.056 N_D)[N_D + N_T + 10^{-3}(56)^2 N_D] = (2.64)4 N_D^2,$$

which gives 2.6 times the power lost by bremsstrahlung in the absence of the tungsten impurity. This shifting of the power-loss curve in Fig. 23.1-8 results in a change in the ignition temperature for the DT fuel. For 0.1% of iron, the ignition temperature shifts from 4.4 to about 8 kev (see Section 23.1J).

The first estimate for the time a plasma must be confined to insure ignition and fusion was made by Lawson[7] and is still used today. In the following, we derive the confinement parameter (the product of time and fuel concentration) as a function of ion temperature.

In the formation of a plasma, a part of the input energy will go to heating the electrons [$(3/2)N_e kT_e$], part to heating the ions [$(3/2)N_i kT_i$] and to ionization, some will escape as X-rays by bremsstrahlung radiation, and some will

---

[7] J. D. Lawson, "Some Criteria for a Power Producing Thermonuclear Reactor," Proc. Phys. Soc., London, B70, 6-12 (1957).

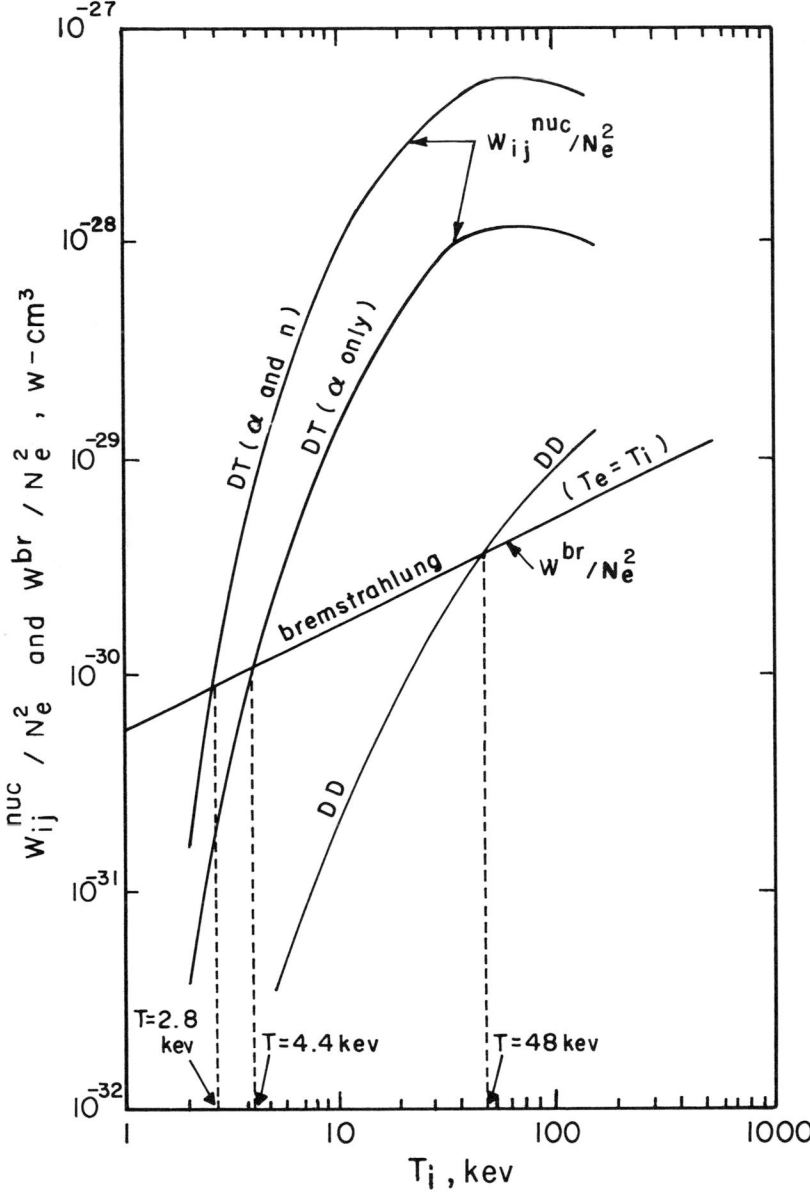

Fig. 23.1-8 The powers generated by DT and DD reactions and by bremsstrahlung radiation are shown as a function of ion temperature, T_i.
The intercept gives an ignition temperature (for break-even) for
each fuel; charged particles contribute to the DD curve only.

be lost in conduction, gas expansion, etc. The thermal heating input to the plasma for the duration of the confinement time τ is

$$\mathcal{E}_{input}^{therm} = \tau W_{input}^{therm} = \frac{3}{2}N_e kT_e + \frac{3}{2}N_i kT_i = \frac{3}{2}N_e k(T_e + T_i)$$

$$= 2.40 \times 10^{-16} N_e (T_e + T_i)_{kev} \; w/cm^3, \qquad (23.1\text{-}61)$$

where we assume a singly-ionized hydrogen plasma with $N_e = N_i$. Then

$$\mathcal{E}_{input} = \mathcal{E}_{input}^{therm} + \mathcal{E}_{rad} + \Delta\mathcal{E} ; \; \mathcal{E}_{rad} = \mathcal{E}_{br} + \mathcal{E}_{cyc}, \qquad (23.1\text{-}62)$$

where $\mathcal{E}^{rad}$ is the energy lost due to bremsstrahlung (and cyclotron radiation in a magnetically confined plasma) while $\Delta\mathcal{E}$ represents the energy loss due to the other processes mentioned above.

When fusion between ion species i and j occurs, the kinetic energy re- leased is $\mathcal{E}_{nuc}$ and the output energy becomes

$$\mathcal{E}_{output} = \mathcal{E}_{input} + \mathcal{E}_{nuc}. \qquad (23.1\text{-}63)$$

The plasma must be confined for a time long enough so that

$$\mathcal{E}^{nuc} > \mathcal{E}_{input}. \qquad (23.1\text{-}64)$$

The useful energy of the system, e.g., that which may be converted to electricity, can be written as an efficiency factor η times $\mathcal{E}_{output}$, which must exceed $\mathcal{E}_{input}$. Thus,

$$\eta \mathcal{E}_{output} \geq \mathcal{E}_{input}, \qquad (23.1\text{-}65)$$

where η is the efficiency of the conversion process.

The condition on the fusion-generated energy in terms of the conversion efficiency η, using Eq. (23.1-65), is

$$\eta(\mathcal{E}_{input} + \mathcal{E}_{nuc}) \geq \mathcal{E}_{input}, \qquad (23.1\text{-}66)$$

$$\mathcal{E}_{nuc} \geq \frac{1-\eta}{\eta} \mathcal{E}_{input} = \frac{1-\eta}{\eta}(\mathcal{E}_{input}^{therm} + \mathcal{E}_{rad}). \qquad (23.1\text{-}67)$$

For $\mathcal{E}_{nuc} = \mathcal{E}_{input}$ ("break-even"), $\eta = 1/2$ and, for $\mathcal{E}_{nuc} = 2\mathcal{E}_{input}$, $\eta = 1/3$. Using $\tau W^{nuc} = \mathcal{E}_{nuc}$, $W^{nuc} = \sum_{i,j} W_{ij}^{nuc}$, $\mathcal{E}_{rad} = W^{rad}\tau$, substituting for the fusion and bremsstrahlung power from Eqs. (23.1-52) and (23.1-53) and for cyclotron radiation power from Eq. (23.1-29), Eq. (23.1-67) becomes

$$1.602 \times 10^{-13} N_i N_j k_{ij} Q_{ij}(Mev) b_{ij} \geq \frac{1-\eta}{\eta} \left[\frac{2.40 \times 10^{-16} N_e (T_e + T_i)}{\tau} \right.$$

$$\left. + 5.35 \times 10^{-31} N_e^2 \sqrt{T_e} + 2.5 \times 10^{-32} N_e^2 T_e^2 (1 + T_i/T_e) \right] \qquad (23.1\text{-}68)$$

or, using $N_i N_j = a_{ij} N_e^2$ where for DT reactions $a = 1/4$ and for DD reactions $a = 1$, we obtain for the density-confinement-time product

$$N_e \tau > q(T_e + T_i) \left[\frac{\eta b_{ij}}{1 - \eta} a_{ij} \hat{k}_{ij}(T_i) Q_{ij} - 3.34 \times 10^{-18} \sqrt{T_e} - 1.56 \times 10^{-19} T_e^2 (1 + T_i/T_e) \right]^{-1},$$

$$q = 0.0015; \tag{23.1-69}$$

here N_e is in cm^{-3}, τ in sec, T_e and T_i in kev, Q_{ij} in Mev; also $\hat{k}_{ij}$ is a function only of T_i and is shown in Fig. 23.1-7. The last term in Eq. (23.1-69) is due to cyclotron radiation. Without cyclotron radiation, $N_e \tau = \infty$ corresponds (break-even) to the condition for the ignition temperature with $\eta = 1/2$.

In Fig. 23.1-9, we have plotted $N_e \tau$ vs. T_i for the DT reaction using an efficiency of $\eta = 1/2$, i.e., the break-even condition, with $T_e = T_i$ and bremsstrahlung but without cyclotron radiation (thus the result is applicable to laser fusion or to a magnetically confined plasma with reflecting walls for reabsorption of cyclotron radiation). For Q_{DT} we used 3.5 Mev, i.e., the efficiency is calculated for the energy gain from charged particles only. We note that $N_e \tau = \infty$ for $T_i = 4.4$ kev, which corresponds to the ignition temperature [see Eq. (23.1-55)]. Thus, we obtain the commonly used Lawson criterion of $N_e \tau > 10^{14}$ cm^{-3} sec for practical achievement of fusion reaction. For the magnetic confinement of a plasma if the electron density is 10^{14} cm^{-3}, the confinement times must be longer than 1 sec. If, on the other hand, the plasma is confined for 10^{-9} sec (e.g., in fusion resulting from the heating of a DT pellet by high intensity lasers), N_e must exceed 10^{23} cm^{-3}. For a magnetically confined DT plasma, the pressure exerted by the magnetic field $p_B = B^2/8\pi$

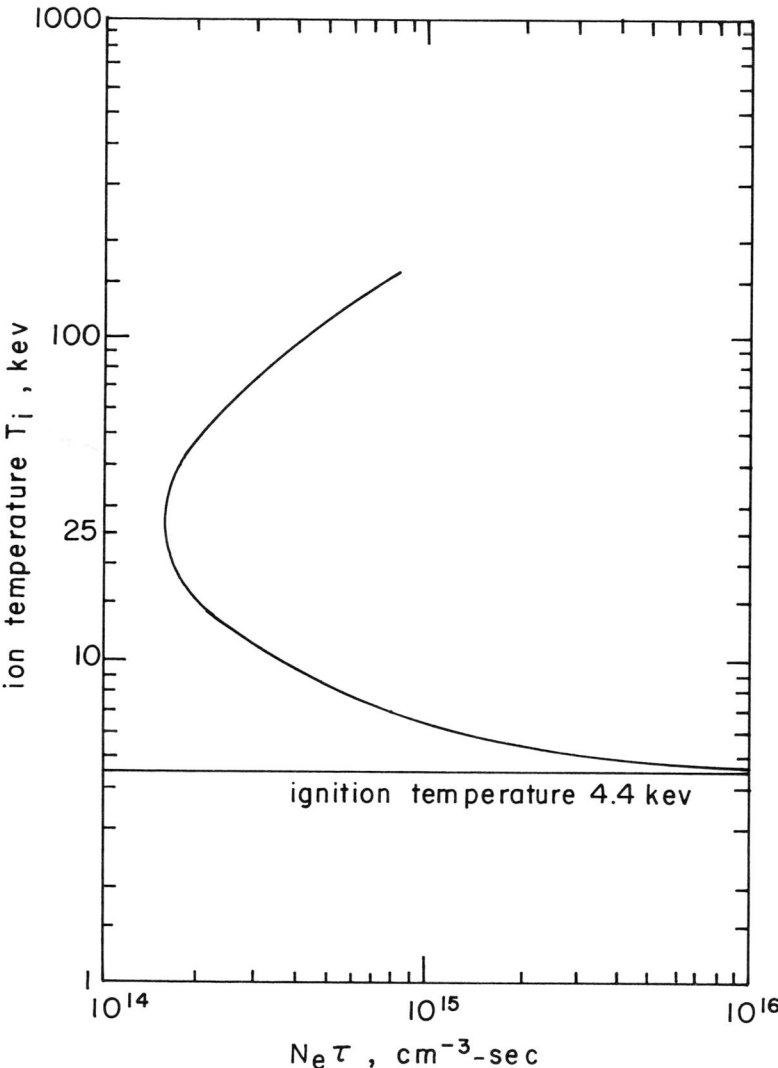

Fig. 23.1-9 The density-confinement time product $N_e\tau$ for DT fusion for
break-even (generated energy = energy input, i.e., $\eta = 1/2$) as a
function of ion temperature, T_i. Only radiative losses by bremsstrahlung are included and $T_e = T_i$. For DT fusion, $T > 4.4$ kev,
$N_e\tau > 1.5 \times 10^{14}$ cm^{-3} sec; $Q_{DT} = 3.5$ Mev.

balances the gas kinetic pressure $p = (N_D + N_T)kT_i + N_e kT_e = N_e k(T_i + T_e)$ of charged particles so that

$$B \geq 2.01 \times 10^{-4} \sqrt{N_e(T_e + T_i)}, \text{ G.} \qquad (23.1\text{-}70)$$

For example, for $T_e = T_i = 25$ kev, $N_e = 1.23 \times 10^{14}$ cm^{-3}, the required magnetic field is $B \geq 1.6 \times 10^4$ G = 16 kG.

In Fig. 23.1-10, the efficiency used is $\eta = 1/3$ (energy doubling) for $T_e = T_i$, $T_e = 10 T_i$ and $T_e = 0.1 T_i$. In addition to the curves for $Q_{DT} = 3.5$ Mev (charged α-particles), curves are plotted for $Q_{DT} = 17.6$ Mev, which (see Table 23.1-4) allows not only for energy carried by charged particles but also for that carried by the fast (14.1 Mev) neutrons. If we are concerned only with electricity from charged particles, then $Q_{DT} = 3.5$ Mev and, for $T_i = T_e$ and $\eta = 1/3$, $(N\tau)_{min} = 3 \times 10^{14}$ cm^{-3}-sec at $T_i = 25$ kev.

Figure 23.1-11 indicates the progress to date in approaching the Lawson criterion in the magnetically-confined plasmas of various machines. In laser fusion, $N\tau \simeq 5 \times 10^{13}$ cm^{-3}-sec at $T \sim 0.7$ kev. Recently (in early 1976), a high magnetic field in an Alcator machine at MIT achieved $N = 6.5 \times 10^{14}$ cm^{-3} with $\tau \geq 20$ msec, i.e., $N\tau \geq 1.3 \times 10^{13}$ cm^{-3}-sec with $T \sim 1$ kev and B = 75 kG. The Tokamak-10 machine in the USSR has recently reported a similar value of $N\tau$, i.e., $N\tau \sim 10^{13}$ cm^{-3}-sec.

F.   A Model Nuclear Reaction

In this section, we consider an isolated nuclear reaction in order to construct rate equations for the reaction processes. The reaction we consider is

$$T + D \xrightarrow{k} \alpha + n \qquad (23.1\text{-}71)$$

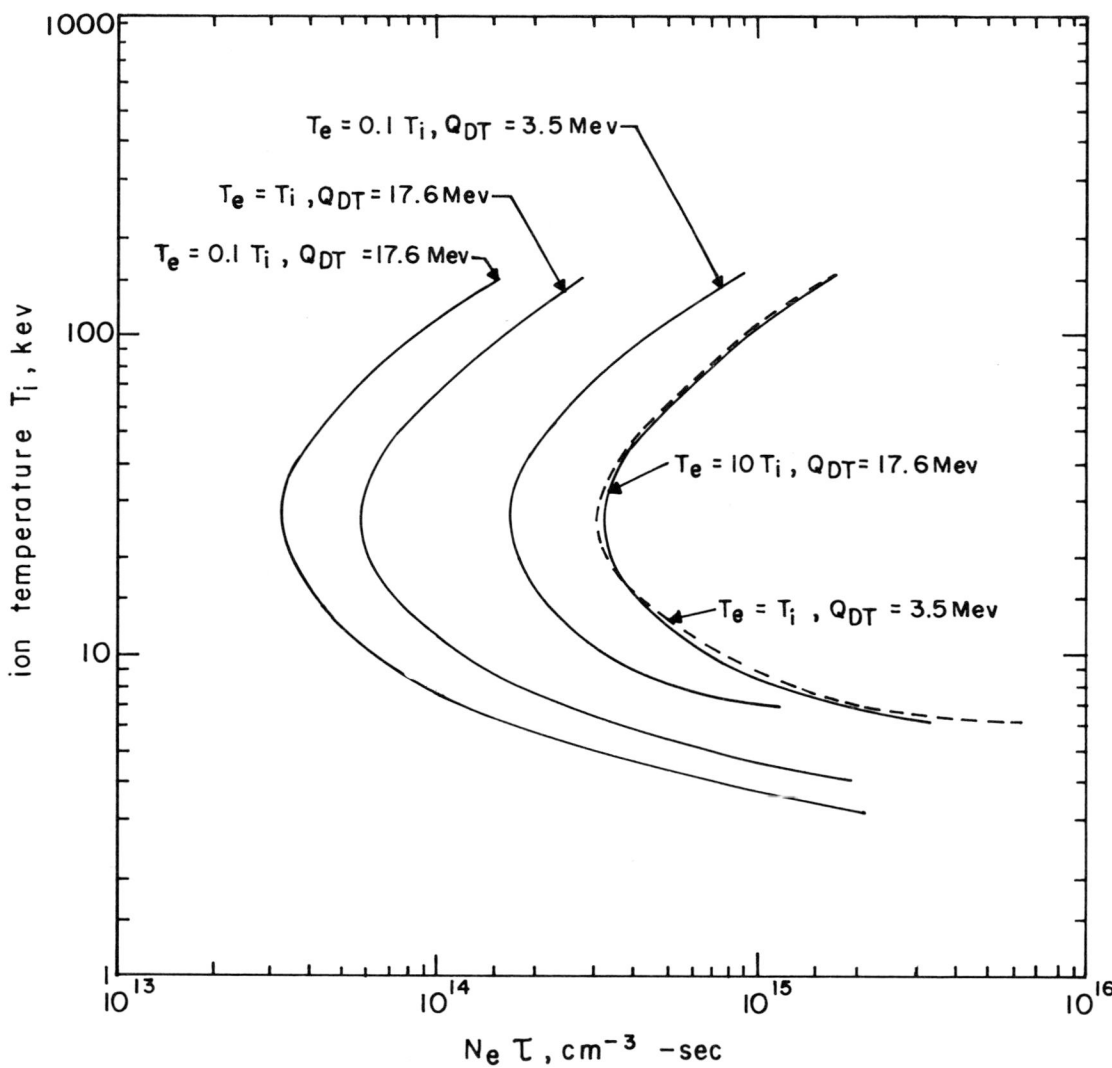

Fig. 23.1-10 Density-confinement-time product $N_e\tau$ for DT fusion, when the generated energy = twice the energy input ($\eta = 1/3$), as a function of the ion temperature, T_i for different ratios of T_e/T_i and different Q_{DT}. Only radiative losses by bremsstrahlung are included. The lowest curve holds also for $T_e = T_i$, $Q_{DT} = 17.6$ Mev and $\eta = 1/2$ (break-even point).

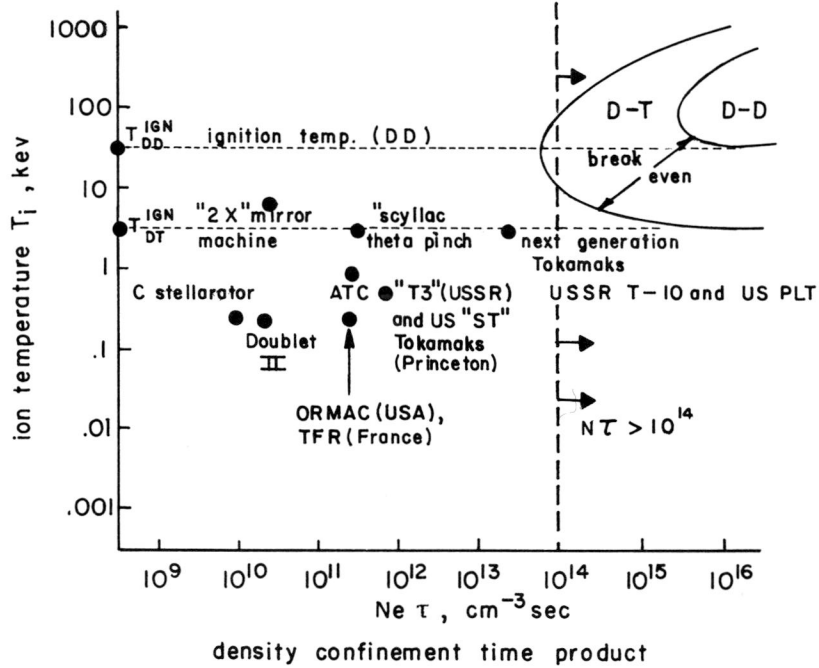

Fig. 23.1-11 The approach to break-even plasma conditions for thermonuclear
fusion.   Compiled from A. L. Hammond et al, Energy and the
Future, p. 83, AAAS, Washington, D. C., 1973, and from Sta-
tus and Objectives of Tokamak Systems for Fusion Research,
AEC, WASH-1295, UC-20, U.S. Government Printing Office,
Washington, D. C., 1974.

and occurs by the collision of a triton and deuteron.   The collision engenders
nuclear reaction between T and D and yields the decay products α and n.   The
nuclear reaction proceeds at a rate $\mathcal{K}$.

The number of decay products is determined by the number of T and D
collisions.   The number of collisions is given by the product of the concentrations $N_D N_T$.   The rate at which the decay products on the right are produced
is, therefore, given by the reaction rate $\mathcal{K}$ times the number of collisions, i.e.,

$$dN_\alpha/dt = dN_n/dt = \mathcal{K}N_D N_T. \tag{23.1-72}$$

Similarly, the rate at which the deuterium and tritium are lost from the system is (in the absence of diffusion loss)

$$dN_D/dt = dN_T/dt = -\mathcal{K}N_D N_T. \tag{23.1-73}$$

The rates at which D and T are lost from the system are the same so that the preceding equation may be integrated to yield $N_D - N_T = N_D^0 - N_T^0$, where N_j^0 is the initial concentration of ion species j. We may introduce the concentration of α particles and neutrons and thus obtain $N_D = N_D^0 - N_\alpha = N_D^0 - N_n$, $N_T = N_T^0 - N_\alpha = N_T^0 - N_n$. The rate equations may then be rewritten as

$$dN_\alpha/dt = \mathcal{k}(N_D^0 - N_\alpha)(N_T^0 - N_\alpha) \tag{23.1-74}$$

and, for equal initial concentrations of D and T, we find

$$dN_\alpha/(N_D^0 - N_\alpha)^2 = \mathcal{K}dt. \tag{23.1-75}$$

Integration now yields

$$N_\alpha = N_D^0 \left[1 + \left(N_D^0 \int_0^t \mathcal{k}dt \right)^{-1} \right]^{-1}, \tag{23.1-76}$$

where the rate coefficient (k) is in general a function of ion temperature and therefore of time.

We assume that an average rate coefficient $\bar{k}$ can be chosen to be independent of time. Then Eq. (23.1-75) integrates to $N_\alpha = N_D^0 \tau/(1 + \tau)$ in terms of the parameter $\tau = \bar{k} N_D^0 t$. There is an identical expression for N_n.

Again assuming the initial concentrations of D and T to be equal, we have for the rate of deuterium loss from Eq. (23.1-73)

$$dN_D/dt = - k N_D^2,$$

which integrates to

$$N_D = N_D^0 \Big/ \Big(1 + N_D^0 \int_0^t k\, dt\Big);$$

the same expression is obtained for tritons. Again using an average rate constant ($\bar{k}$) independent of time,

$$N_D = N_D^0 \Big/ (1 + \tau). \qquad (23.1-77)$$

The deuteron and triton fuel is therefore used up at a rate that depends on the inverse of $(1 + \tau)$. The neutron and alpha particle concentrations increase linearly with time for early times and approach the constant value N_D^0 asymptotically.

The growth in concentration of the decay products is related to the energy liberated by the DT reaction. If an energy $\Delta\mathcal{E}$ is liberated in each DT reaction then, since there are $kN_D N_T$ reactions per sec, the rate at which energy is liberated is

$$d\mathcal{E}/dt = kN_D N_T \Delta\mathcal{E}. \tag{23.1-78}$$

For an average rate coefficient $(\bar{k})$, we have the concentrations for N_D and N_T as functions of time so that Eq. (23.1-78) becomes

$$d\mathcal{E}/dt = N_D^0 \Delta\mathcal{E}/(1 + \tau)^2 \tag{23.1-79}$$

in terms of the parameter τ. Equation (23.1-79) immediately integrates to $\mathcal{E} = N_D^0 \Delta\mathcal{E}\,\tau/(1 + \tau)$, which is just $N_\alpha \Delta\mathcal{E}$. If the released energies are written as $\Delta\mathcal{E}_\alpha$ and $\Delta\mathcal{E}_n$ for the kinetic energies of the α-particle and neutron [as given by the mass-defect relations, Eq. (23.1-7)] and $\Delta\mathcal{E} = \Delta\mathcal{E}_\alpha + \Delta\mathcal{E}_n$, then the total energy released in the reaction is

$$\mathcal{E} = N_\alpha \Delta\mathcal{E}_\alpha + N_n \Delta\mathcal{E}_n, \tag{23.1-80}$$

where N_α and N_n are the functions of time given earlier.

The simple result in Eq. (23.1-80) follows from the simplifying assumption that the rate constant is independent of time and $N_D^0 = N_T^0$. In the following section, we will consider more complex reactions and relax the restrictions made above.

It should be noted that, although $\mathcal{E}$ is the total energy released in the DT reaction, only $N_\alpha \Delta\mathcal{E}_\alpha$ is available for fusion. The fast neutrons exit from the plasma and deposit their energy elsewhere. It is possible, however, that this energy may be returned to the plasma in a usable form (see Fig. 23.1-3).

G. A Description of the Kinetics of Fusion Reactions

The energy released by the reactions

$$D + D \xrightarrow{\mathit{k}_{DD}^{(p)}} T + p + \Delta\mathcal{E}_{T,p}, \quad D + D \xrightarrow{\mathit{k}_{DD}^{(n)}} He^3 + n + \Delta\mathcal{E}_{He^3,n},$$

$$D + T \xrightarrow{\mathit{k}_{DT}} \alpha + n + \Delta\mathcal{E}_{\alpha,n}, \tag{23.1-81}$$

and the kinetics of the reactions will now be discussed. The DD reaction yields protons and neutrons at equal rates (k_{DD}), i.e., $\mathit{k}_{DD}^{(p)} = \mathit{k}_{DD}^{(n)}$, $\mathit{k}_{DD} = 2\mathit{k}_{DD}^{(n)} = 2\mathit{k}_{DD}^{(p)}$. The number of decay products is determined by the number of DD and DT collisions. The number of collisions times the average rate coefficient yields the rate at which the energy of a particular decay product is generated.

The number of binary collisions between dissimilar species such as D and T is given by the product of species concentration $N_D N_T$. The number of binary collisions in the DD reactions is $N_D(N_D - 1)/2 \simeq N_D^2/2$. These collisions lead to generation terms of decay products such as those modeled in Section 23.1F. There are, however, a number of particle-loss mechanisms.

Neutrons are assumed to escape from the plasma without interaction since they are neutral. The charged particles have a finite confinement time

(τ), as discussed in Section 21.3E. To obtain a distribution of ion velocities, we assume that the particle loss by ejection from the plasma follows an exponential decay. Finally, introducing S_D and S_T as sources of the deuteron and triton fuels, we can write the kinetic equations for production and loss of particles as[8]

$$dN_p/dt = k_{DD} N_D^2/4 - N_p/\tau, \quad dN_{He^3}/dt = k_{DD} N_D^2/2 - N_{He^3}/\tau,$$

$$dN_\alpha/dt = k_{DT} N_D N_T - N_\alpha/\tau, \quad dN_D/dt = S_D - 2k_{DD} N_D^2 - k_{DT} N_D N_T - N_D/\tau,$$

$$dN_T/dt = S_T + k_{DD} N_D^2/4 - k_{DT} N_D N_T - N_T/\tau. \tag{23.1-82}$$

The rate at which energy is transferred to different ion species is described by the following system of equations:

$$d\mathcal{E}_p/dt = k_{DD} N_D^2 \Delta\mathcal{E}_p/4, \quad d\mathcal{E}_{He^3}/dt = k_{DD} N_D^2 \Delta\mathcal{E}_{He^3}/4, \quad d\mathcal{E}_\alpha/dt = k_{DT} N_D N_T \Delta\mathcal{E}_\alpha,$$

$$d\mathcal{E}_n/dt = k_{DD} N_D^2 \Delta\mathcal{E}_n^{DD}/4 + k_{DT} N_D N_T \Delta\mathcal{E}_n^{DT},$$

$$d(\mathcal{E}_D + \mathcal{E}_T)/dt = -\tfrac{1}{2} k_{DD} N_D^2 [\Delta\mathcal{E}_p + \Delta\mathcal{E}_{He^3} + \Delta\mathcal{E}_n^{DD}] - k_{DT} N_D N_T [\Delta\mathcal{E}_\alpha + \Delta\mathcal{E}_n^{DT}]$$

$$= -[k_{DD} N_D^2 \Delta\mathcal{E}_{DD} + k_{DT} N_D N_T \Delta\mathcal{E}_{DT}]. \tag{23.1-83}$$

[8]C. Powell and O. J. Hahn, "Energy Balance Instabilities in Fusion Plasma," Nuclear Fusion 12, 667-672 (1972).

The superscripts on $\Delta \mathcal{E}_n$ distinguish the contribution made by the DD and DT reactions.

The maximum rate at which the fuel can be used up by the reactions can be estimated by employing the balance between the outward pressure of the plasma and the containment pressure of the magnetic field, $(N_D + N_T)kT = B^2/8\pi$. For equal concentrations of deuterons and tritons, we obtain $kTN_D = B^2/16\pi$ and the rate of energy release becomes

$$\frac{d}{dt}(\mathcal{E}_D + \mathcal{E}_T) = -\left(\frac{B^2}{16\pi k}\right)^2 \left\{ \frac{\mathcal{K}_{DD}(T)}{T^2} \Delta\mathcal{E}_{DD} + \frac{\mathcal{K}_{DT}(T)}{T^2} \Delta\mathcal{E}_{DT} \right\} \qquad (23.1-84)$$

in view of the last of the relations in Eq. (23.1-83). The maximum rate of energy release occurs when the derivative with respect to T of the right-hand side of Eq. (23.1-84) vanishes.

Using the form of the average rate coefficient from Eq. (23.1-50), we have

$$\mathcal{K}_{ij} = \beta_{ij} T^{-2/3} \exp(-\Lambda_{ij}/T^{1/3}), \quad \beta_{ij} = 0.8052 \times 10^{-16} A_{ij} \gamma_{ij}^{1/3} (m_H/\mu_{ij})^{1/2},$$

$$\Lambda_{ij} = 3(\gamma_{ij}/2)^{2/3}, \qquad (23.1-85)$$

with A_{ij} and γ_{ij} being the empirical parameters listed in Table 23.1-5 for the various nuclear reactions $i + j$ while μ_{ij} is the reduced mass for particles i and j. The function to be maximized is

$$f(T) = \frac{1}{T^{8/3}}\left\{\exp\left[-3\left(\frac{\gamma_{DD}}{2\sqrt{T}}\right)^{2/3}\right] + \frac{\beta_{DT}\Delta\mathcal{E}_{DT}}{\beta_{DD}\Delta\mathcal{E}_{DD}}\exp\left[-3\left(\frac{\gamma_{DT}}{2\sqrt{T}}\right)^{2/3}\right]\right\}. \quad (23.1-86)$$

Setting $df(T)/dT = 0$, we obtain $T\mathcal{E}_{max} \approx (3/8)^3(\gamma_{DD}/2)^2$ since γ_{DD} and γ_{DT} are within 8% of each other in magnitude. The calculated T at $\mathcal{E}_{max}$ is approximately 95 kev, which is very much higher than the operational temperature presently anticipated in fusion reactors.[8]

Equation (23.1-83) describes the energy released to the decay products in the reaction (23.1-81). The energy-balance process in the plasma is the critical relation for fusion to occur however. Considering both the energy generation and loss mechanisms, we have for the plasma-power density

$$d\mathcal{E}_{plasma}/dt = W^F - W^c - W^n - W^{br}. \quad (23.1-87)$$

The plasma-energy density is

$$\mathcal{E}_{plasma} = \frac{3}{2}NkT, \quad N = N_D + N_T = N_e.$$

The fusion power density minus the power density of the lost neutrons is, from Eq. (23.1-83),

$$W^F - W^n = k_{DD}N_D^2(\Delta\mathcal{E}_p + \Delta\mathcal{E}_{He^3}) + k_{DT}N_D N_T \Delta\mathcal{E}_\alpha; \quad (23.1-88)$$

the charged-particle power density is, from Eq. (23.1-61),

$$W^c = \frac{3}{2} NkT/\tau = 2.40 \times 10^{-16} NT_{kev}/\tau \; w/cm^3, \qquad (23.1-89)$$

where τ is the particle-confinement time and the bremsstrahlung power density W^{br} is given by Eq. (23.1-53).

To describe the dynamics of the plasma, we must solve the system comprising the energy equation (23.1-83) and the nonlinear, coupled particle-conservation equations (23.1-82). Numerical solutions of these equations may be found in Ref. [9].

The plasma-confinement condition, in terms of the balance between the magnetic pressure and the plasma pressure, has been used to determine the overall state of the reacting fuel. Similarly, the rate at which the nuclear reactions proceed has been described by an average reaction-rate coefficient $k_{ij}(T)$, which is assumed to be only a function of the ion temperature. The local deviations in the plasma from equilibrium may, however, be quite violent when the reaction products are considered. How these fast, high energy, charged particles give up their energy to the ambient plasma subsequent to reaching thermal equilibrium is an important problem. Of particular importance is the modification of reaction rates in such high density fuels as the fuel pellets used in laser-driven fusion.[9, 10]

We consider a fuel with initially equal concentrations of deuterons and tritons. As the reactions shown in Eq. (23.1-81) proceed, increasing fractions of the ions in the plasma are the reaction products p, He^3 and α-particles,

[9] J. Nuckolls, L. Wood, A. Thiessen, and G. Zimmermann, "Laser Compression of Matter to Super-High Densities: Thermonuclear (CTR) Applications," Nature 239, 139-142 (1972).

[10] J. S. Clarke, H. N. Fisher and F. J. Mason, "Laser Driven Implosion of Spherical DT Targets," Phys. Rev. Let. 30, 89-92 (1973).

assuming the neutrons have escaped.   These products will lose energy and slow down in the DT medium by Coulomb scattering with the ambient ions.   In a dense medium such as a fuel pellet, the high energy of the reaction products make nuclear scattering a viable mechanism for energy loss.[11]

The reaction products have high energy and will presumably interact with the DT fuel particles of the plasma in a high density fuel.   This interaction slows down the reactants and an α-particle in a DT medium will collide with D and T particles as it slows; these will, in turn, be slowed down by elastic collisions and will react along the way; they ultimately reach thermal equilibrium.   An argument to estimate[12] the enhanced probability of a reaction occurring during the slowing of an α-particle in an infinite DT medium goes as follows.   If P is the probability of occurrence of reaction by a D or T particle knocked-on by an α-particle, then, on the average, there are P α-particles produced by such a reaction.   These newly produced α-particles, in turn, slow down by collision, each with a probability P of activating another reaction.   This reaction chain leads to the geometric series

$$P + P^2 + P^3 + \cdots = P/(1 - P), \hspace{2cm} (23.1-90)$$

in which the factor $(1 - P)^{-1}$ gives the fractional increase in the probable number of reactions due to the slowing of α-particles in the DT medium.   Therefore, the reaction rates $\mathcal{K}_{DT}$ and $\mathcal{K}_{DD}$ should be increased for the reactions in the DT pellets.

---

[11] J. J. Devanly, M. L. Stein, "Plasma Energy Deposition for Nuclear Elastic Scattering," Nuc. Sci. and Eng. 46, 323-333 (1971).

[12] K. A. Brueckner and H. Brysk, "Fast Charged Particle Reactions in a Plasma," J. Plasma Phys. 10, 141-147 (1973).

The reaction probability as a function of electron temperature (taken equal to the ion temperature) is given in Ref. [12] for both Coulomb and nuclear scattering of particles in a DT fuel. The nuclear scattering force is found to be at least as important a mechanism as Coulomb scattering for precipitating nuclear reactions.[12]

H. Tritium Breeding Reactions

Of the total energy released in the DT reaction, as given in Eq. (23.1-83), only $N_\alpha \Delta \mathcal{E}_\alpha$ is available for fusion. The fast neutrons exit from the plasma and deposit their energy elsewhere. It is possible, however, that this energy may be returned to the plasma in a usable form. Consider the breeder reaction depicted in Fig. 23.1-12. The fast, high-energy neutrons are shown as escaping from the plasma into a moderating blanket (e.g., H_2O), which encases the plasma region. The neutrons give up energy by inelastic collisions and heat the water.

Slow neutrons leave the moderator sheath and enter a region of solid made from a lithium compound (Li^6F). The reaction

$$Li^6 + n \rightarrow T + \alpha + 4.785 \text{ Mev} \tag{23.1-91}$$

"breeds" tritons at 2.7 Mev and α-particles at 2.0 Mev. Fast neutrons can also produce tritons by the following reaction:

$$n(\mathcal{E} > 2.467 \text{ Mev}) + Li^7 \rightarrow T + \alpha + n_{slow} + (\mathcal{E} - 2.467) \text{ Mev}. \tag{23.1-92}$$

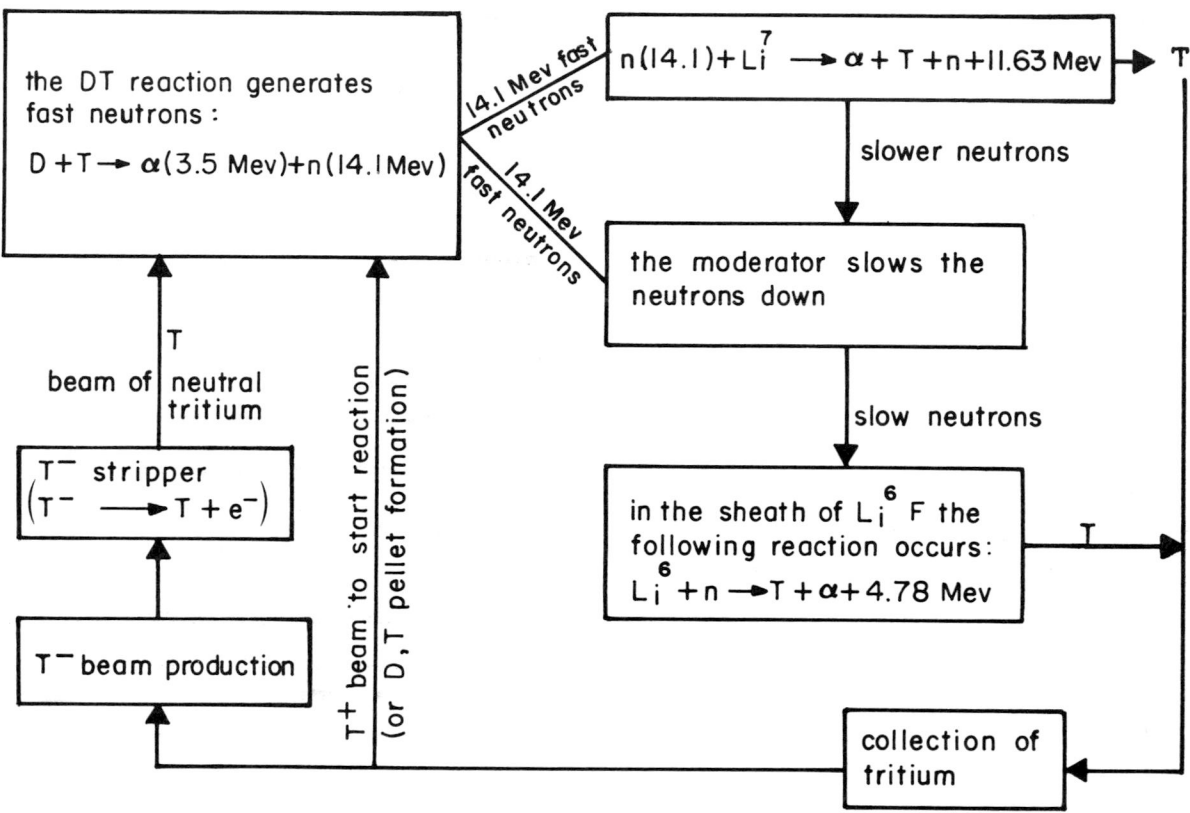

Fig. 23.1-12 A conceptual representation of a tritium breeder reactor is
             shown.

        The reactions involving the hydrogen isotopes are shown in Fig. 23.1-3;
included is a branch for the tritium-breeder reaction. The distribution of the
liberated energy over the decay products is shown in parantheses. The total
energy released by charged particles for heating the plasma and direct conversion to electricity is 3.5 Mev per deuteron (see Fig. 23.1-3) in the absence of

the Li^6 breeder cycle.   The total energy carried by neutrons which is available for heat is 14.1 Mev.   The additional energy released to Li^6 is 4.6 Mev, 2.3 Mev by α-particles and 2.3 Mev by neutrons.   The total energy available is 22.2 Mev with heat released by Li^{6}F (see also Table 23.1-5).

For plasma temperatures above 10 kev, the He^{3}D reaction rate becomes larger than the DD rate so that additional 18.3 Mev are released to the charged particles.   The energy available from all processes to heat the plasma is 26.6 Mev; the energy from neutrons going into heat is 25.7 Mev.   The total energy released for six deuterons is therefore 8.7 Mev per deuteron.

I.   Neutral Particle Injection

For a magnetically confined plasma, the density and temperature can be increased by injection of a neutral beam of deuterium (or tritium).   Workers at many laboratories in the U.S. and U.S.S.R. are currently engaged in the design of these negative ion sources.[13]   These sources use a cesium-coated surface which is bombarded by deuterium and generates a beam of negative ions of deuterium (D$^-$).   An accelerator will then be used to accelerate D$^-$ to the needed energies ranging from 20 to about 200 kev.   This energetic negative ion beam will be stripped of the loosely bound outer electron, which has a binding energy of 0.7 ev as compared with 13.6 ev for the inner electron. Many methods have been suggested for the stripping mechanisms: passage

---

[13]K. Prelec and Th. Sluyters, "Formation of Negative Hydrogen Ions in Direct Extraction Sources," Rev. Sci. Instr. 44, 1451-1463 (1973); Y. I. Belchenko, G. I. Dimov and V. G. Dudnikov, "Emission of Intense Negative-Ion Fluxes From a Surface Bombarded by Fast Particles from a Discharge," Izv. Akademii Nauk SSSR, Seriya Fizicheskaya 37, 2573-2577 (1973); Y. I. Belchenko, G. I. Dimov and V. G. Dudnikov, Nuclear Fusion 14, 113-118 (1974).

through a cell of D_2 gas, a plasma cell, stripping by a laser beam, Lorentz stripping by using magnetic fields, and the use of electric fields. Finally, the neutral beam of energetic D is injected across a magnetic field into the plasma, where the temperature and density will be raised simultaneously.

J. Effect of Impurities on Radiative Losses

Impurities have been found in experiments on CTR in magnetically confined plasmas. These impurities are produced when ions, neutrals and neutrons bombard and remove wall material (sputtering) by evaporation and blistering (explosion of gas bubbles underneath the wall surface) at locally overheated areas,[14] and also by the use of imperfect vacuum systems. The low-Z impurities (C, N, O) and high-Z metallic impurities (Cr, Fe, Ni, Mo, W) arise from the liner, limiter, etc. The high-Z impurities have a tendency to concentrate in the plasma center and are responsible for a significant energy loss by radiation. Unless the formation of impurities is carefully prevented, they will cause problems in the practical achievement of fusion energy in a magnetically-confined plasma. This conclusion will be emphasized by quantitative data.

i. The Concentration of High-Z Impurity Ions

Increasing amounts of energy are required to strip successive bound electrons from partially ionized atoms. For carbon, the successive ionization energies are approximately 11, 24, 48, 64, 392, and 492 ev. The ionization energies for nitrogen are 14, 30, 47, 77, 98, 552, and 667 ev and for oxygen 14, 35, 55, 77, 114, 138, 739, and 871 ev. In hot plasmas at kev temperatures,

---

[14] T. Kammask, Fusion Reactor Technology, pp. 351-404, Ann Arbor Science Publishers, Inc., Ann Arbor, Mich., 1976.

the low-Z impurities (C, N, O) are expected to be completely stripped of elec-trons. For high-Z impurities, some of the ions will, however, not be com-pletely stripped even at the hot plasma center. The ionization energy for stripping the last (24th) electron from Cr is 7.9 kev; for the last (26th) elec-tron of Fe, it is 9.3 kev; for the last (28th) electron from Ni, 10.8 kev; for the last (42nd) electron from Mo, 24.6 kev; and for the last (72nd) electron of W, 81 kev. The successive ionization energies for W are 8, 18, 24, 35, 48, 61, 114, ...,19680, 79377, and 81009 ev.

In laser-induced fusion, one needs to consider local quasi-thermody-namic equilibrium of the plasma with temperatures T_e and T_i.[15] In a low-density, magnetically-confined plasma, energy loss by radiation must be esti-mated by considering appropriate rate processes. The concentration of each ion species must be obtained from rate processes (averaged cross sections) and not from equilibrium constants, as is done when the plasma is in thermo-dynamic equilibrium. In accurate calculations of the radiation loss for a plas-ma in a steady state, multitudes of excitation and de-excitation cross sections (collisional radiative calculations) are required. Steady-state calculations for a plasma with iron impurities have been made.[16, 17] The results shown in Fig. 23.1-13 suggest that, for $N_e = 10^{14}$ cm^{-3}, Fe^{25+} will peak near 10 kev and the iron will be completely ionized for $T > 10$ kev. Partially stripped, complex high-Z ions (Mo, W) are expected to persist to higher temperatures. An estimate has been made for molybdenum.[16, 17] In general, the concentra-tions of various ions $[N_i'^{(n)}]$ of a given element follow a distribution of the type

---

[15]See p. 357 in Ref. [2]

[16]R. C. Vik and I. Katz, "Effect of Impurities on Tokamak-like Plasmas," Bull. Am. Phys. Soc. 19, 893 (1974).

[17]R. C. Vik and I. Katz, courtesy of Systems, Science and Software, La Jolla, Calif.

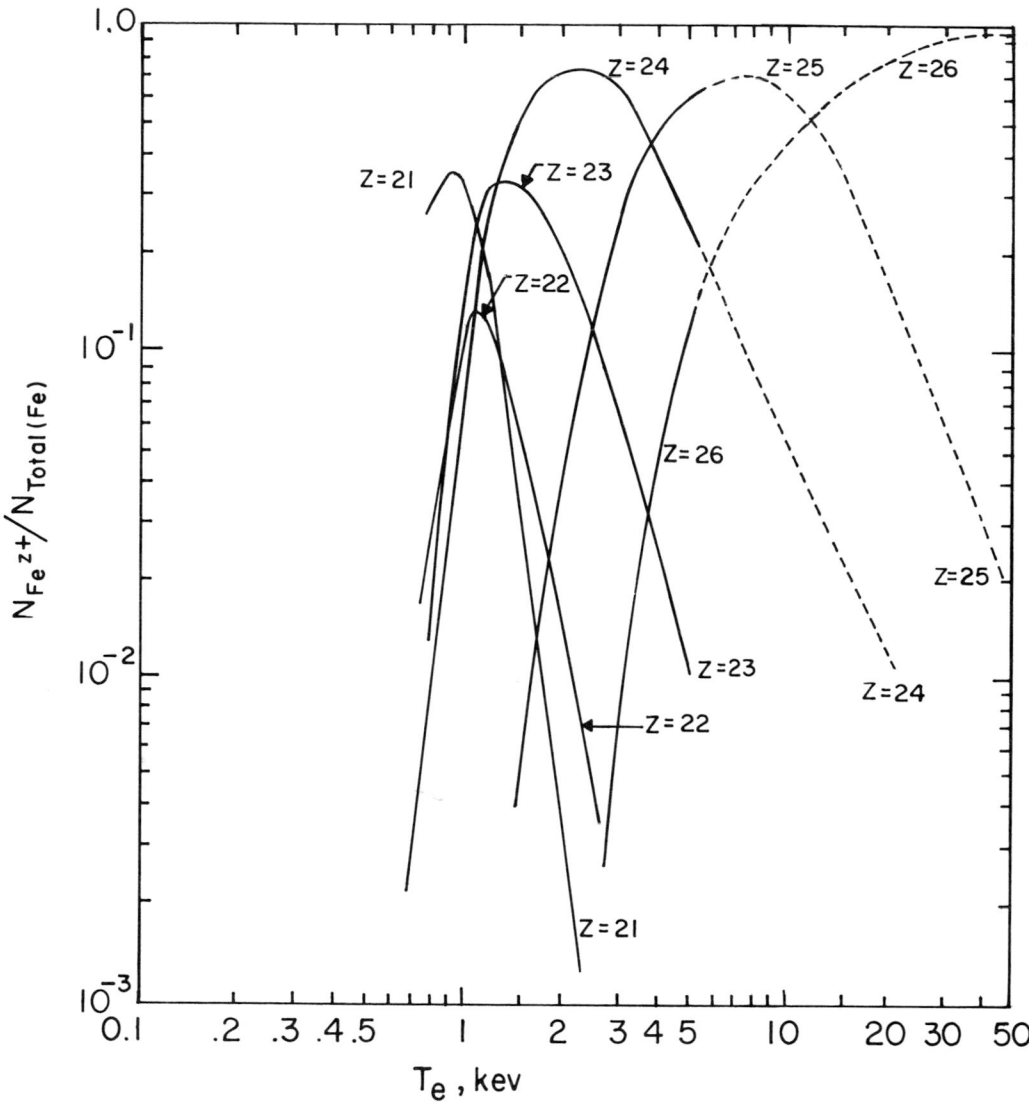

Fig. 23.1-13 Steady-state, collisional radiative results for the distribution of iron ions as a function of temperature. The curve Z = 25 shows the fractional concentration of one-electron (hydrogenic) iron ions (Fe^{25+}), the curve labelled Z = 24 shows the fraction of He-like ions (Fe^{24+}), etc.; $N_e = 10^{14}$ cm^{-3}. Reproduced from R. C. Vik and I. Katz, courtesy of Systems, Science and Software, La Jolla, Calif. The dotted lines are from simple graphical extrapolation.

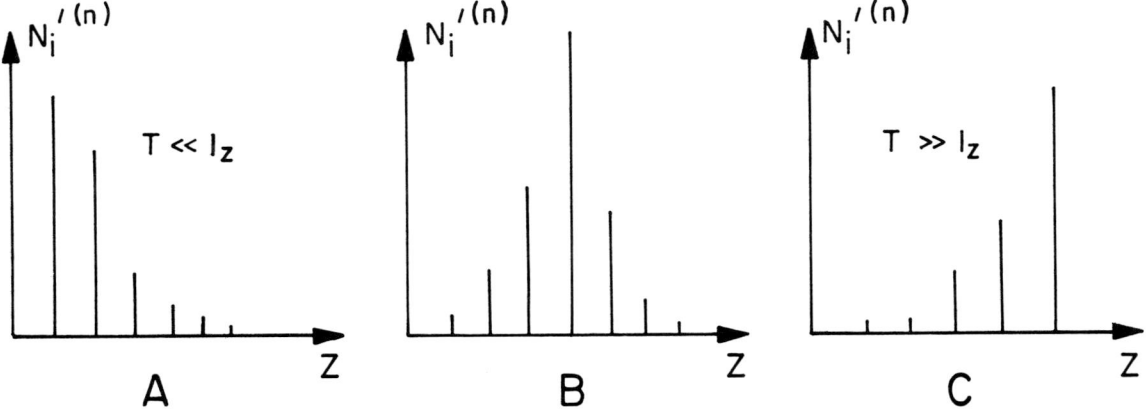

Fig. 23.1-14 Distribution of concentrations of various positive ions in a plasma as a function of ionic charge Z; I_Z is the ionization energy
required to ionize Z - 1 times ionized ion into Z times ionized
ion. In A, $T \ll I_Z$ and the distribution applies at very low temperatures; in B, the plasma temperature is high but high-Z ions
are far from being ionized; in C, the most abundant ion is the
completely stripped ion.

shown in Fig. 23.1-14.

For one impurity element with fractional concentration $N_i' = \alpha(N_D + N_T)$,
where N_i' is the number of impurity atoms per cm^3, $N_i' = \sum_{n=0} N_i'^{(n)}$, where
$N_i'^{(n)}$ is the number of n times ionized impurity ions per cm^3. Let the
fraction of n-times ionized ions be X_n (so $N_i^{(n)} = X_n N_i'$, $\sum_n X_n = 1$] and
let Z_n be the residual charge of the atom; then in a DT plasma

$$N_e = N_D + N_T + \sum_n N_i'^{(n)} Z_n = (N_D + N_T)[1 + \alpha \sum_n X_n Z_n]$$

and

$$\sum_n N_i^{\prime(n)} Z_n^p = \alpha(N_D + N_T) \sum_n X_n Z_n^p,$$

where p is an integer.

ii. Bremsstrahlung Radiation from Impurities

From Eq. (23.1-23), with T_e in kev, N_D and N_T in cm^{-3},

$$W^{br} = 5.35 \times 10^{-31} \sqrt{T_e} (N_D + N_T)^2 (1 + \alpha \sum_n X_n Z_n)(1 + \alpha \sum_n X_n Z_n^2) \text{ in } w/cm^3.$$

$$(23.1-93)$$

For iron with temperatures above 10 kev (Z = 26), the iron impurity will contribute significantly to bremsstrahlung for $\alpha > (676)^{-1} = 0.15\%$. If $\alpha = 0.15\%$, the bremsstrahlung radiative energy loss will be twice as large as without the iron impurity. Normally, the bremsstrahlung radiative energy loss from impurities for T < 10 kev is smaller than the sum of the line and recombination radiation.

iii. Line Radiation from Impurities

Bound electrons, on impurity ions (Fe^{25+}) in a plasma, will be distributed among the various excited states and will radiate whenever electrons drop to lower energy states. The radiation from partially ionized impurities appears in a line spectrum in the far ultraviolet and X-ray regions of the spectrum. Accurate calculations of radiative power loss from the plasma are rather complicated. However, using a fit to oxygen and carbon calculations for T > 0.1 kev, it is found to be approximately (for the ions of one impurity)[18,19,20]

$$W_{(line)}^{rad} \simeq 2.2 \times 10^{-34}(N_D + N_T)^2 \alpha(1 + \alpha \sum_n X_n Z_n) \sum_n X_n Z_n^6 / T_e \sqrt{T_e} \text{ in w/cm}^3.$$

(23.1-94)

iv.   Recombination Radiation from Impurities

Another radiative energy loss is associated with radiation from free-bound collisions between electrons and partially ionized, high-Z impurities. In this process, a Z times ionized ion captures a free electron to become a (Z - 1) times ionized ion.   The power in recombination radiation has also been approximated by making a fit to oxygen and carbon calculation[18,19,20] (for T > 0.1 kev).   For the ions of one impurity, the result is

$$W_{(recomb)}^{rad} \simeq 1.8 \times 10^{-32}(N_D + N_T)^2 \alpha(1 + \alpha \sum_n X_n Z_n) \sum_n X_n Z_n^4 / \sqrt{T_e} \text{ in w/cm}^3.$$

(23.1-95)

[18] A. P. Vasilev, G. G. Dolgov-Savelev and V. I. Kogan, "Radiation by Impurities in Rarified Hot Plasma," Nuclear Fusion, Suppl. 2, 655-663 (1962).

[19] V. I. Kogan in Physics of Plasma and Problems of Controlled Thermonuclear Reactions, Vol. 3, in Russian, Atomizdat, Moscow, 1958; Y. I. Galushkin and V. I. Kogan, "Radiation Losses in a Dense High-Temperature Hydrogen Plasma containing Impurities," Nuclear Fusion 11, 597-604 (1971).

[20] L. A. Artsimovich, Controlled Thermonuclear Reactions (translated from the Russian), p. 54, Gordon & Breach Science Publishers, Inc., New York, 1964.

In Fig. 23.1-15, we have plotted results of detailed collisional, radiative power-loss calculations by Vik and Katz[16, 17] for iron and approximate calculations for molybdenum. On the same figure, we have plotted results for iron derived from the approximate relations [Eqs. (23.1-94) and (23.1-95)] using the Zs from Fig. 23.1-14. We note from Fig. 23.1-16 that the ignition temperature of a DT plasma changes from 4.4 to about 8 kev for a 0.1% Fe impurity. In Fig. 23.1-17, the effects of other impurities are shown.

Several techniques are currently being studied to reduce the concentrations of impurities. In one technique, a divertor (magnetic limiter) is used which may prolong the build-up time of impurities. A solid limiter placed inside Tokamaks prevents the plasma from coming in direct contact with the walls but is a major source of impurities. If the limiter is replaced by a divertor, this problem may be partially avoided.[21]

Other suggestions for removing impurities include the following: use of a gas blanket, plasma rotation and selective acceleration of impurities.[21] The presence and influence of impurities represent an as yet unsolved problem in achieving feasibility of fusion by magnetic confinement.

K. Transport Coefficients in Plasmas

We now describe some of the more important aspects of transport

---

[21]R. Behrisch and B. B. Kadomtsev, "Impurities" in Nuclear Fusion, Special Supplement, 1974; see also Fusion Reactor Design Problems, pp. 451-454, International Atomic Energy Agency, Vienna, 1974.

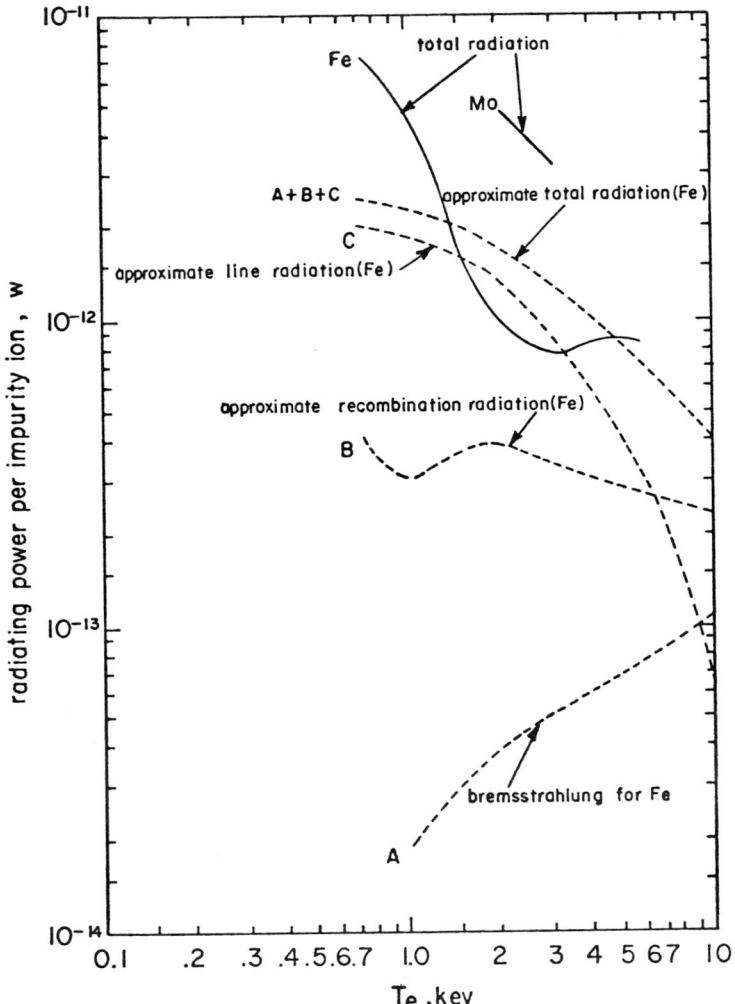

Fig. 23.1-15 Plots showing the radiated power in w per impurity ion as a function of temperature in kev for iron (Fe) and molybdenum (Mo) impurity ions; $N_e = 10^{14}$ cm^{-3}. The solid curves are based on steady-state, collisional radiative calculation (from R. C. Vik and I. Katz, Systems, Science and Software, La Jolla, Calif.). A: radiation from bremsstrahlung for Fe; B: approximate recombination radiation for Fe; C: approximate line radiation for Fe. In A, B and C, the Z_is were obtained from Fig. 23.1-13 for Fe.

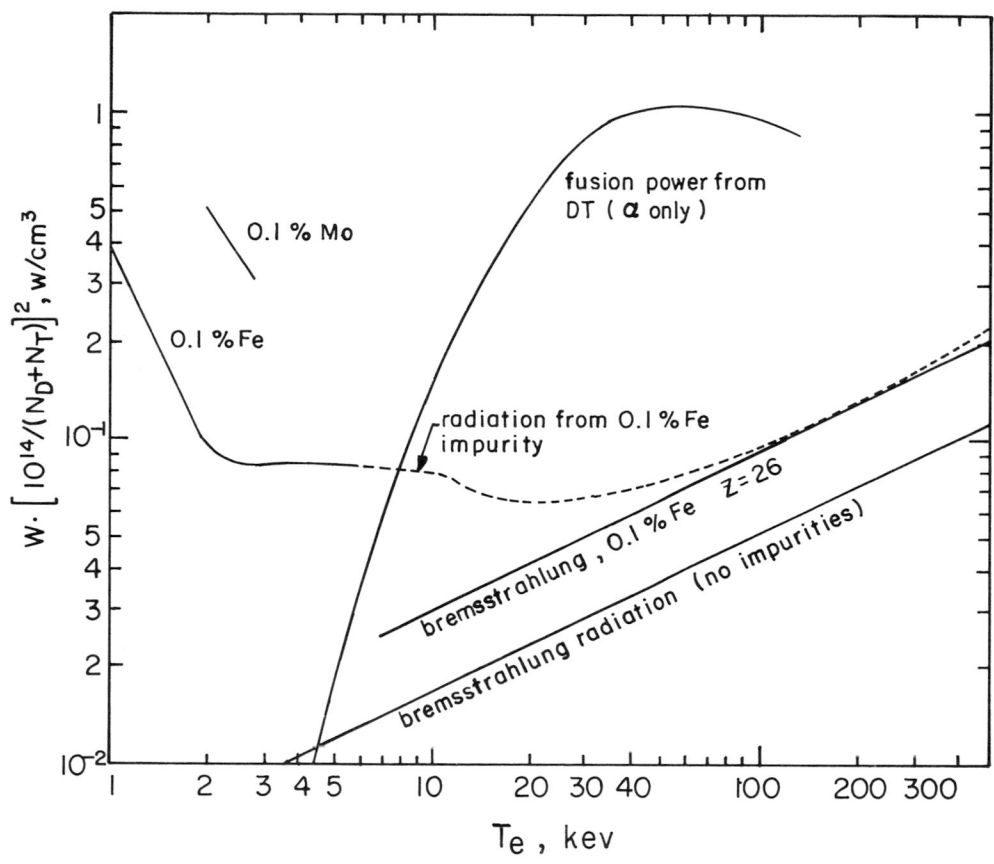

Fig. 23.1-16 Radiant power in w/cm³ is shown as a function of the electron temperature in kev for 0.1% Fe and 0.1% Mo. Also plotted are the corresponding curves for (a) bremsstrahlung radiation without impurities and (b) the fusion power released for the DT reaction (with the formation of α-particles only).

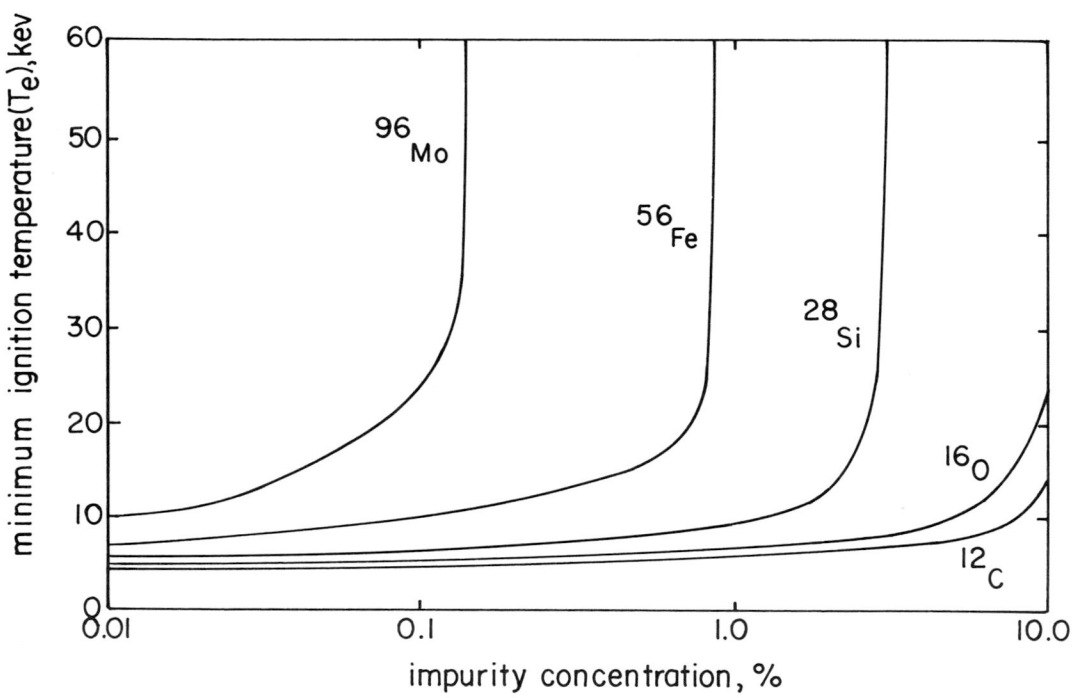

Fig. 23.1-17 Approximate effect of various impurities on ignition tempera-
ture (from W. M. Stacey et al, "Tokamak Experimental Power
Reactor Studies," p. I-17, Argonne National Laboratory,
Argonne, Ill., Report ANL/CTR-75-2 for ERDA, June 1975).

processes in plasmas. For details, the reader is referred to Refs. [22], [23] and [24].

The particle current densities for electrons and ions in a plasma are defined as

$$\vec{j}_e = - eN_e \langle \vec{v}_e \rangle, \quad \vec{j}_i = ZeN_i \langle \vec{v}_i \rangle,$$

where $\langle \vec{v}_e \rangle$ and $\langle \vec{v}_i \rangle$ are the average (e.g., over a Maxwellian velocity distribution) electron and ion velocities, respectively, at a given spatial location. The total particle current density in a plasma is

$$\vec{j} = \vec{j}_i + \vec{j}_e = e[Z_i N_i \langle \vec{v}_i \rangle - N_e \langle \vec{v}_e \rangle]. \qquad (23.1\text{-}96)$$

For a hydrogen plasma (e.g., DD or DT plasma), $Z_i = 1$, $N_e = N_i$ and this current density is

$$\vec{j} = N_e e[\langle \vec{v}_i \rangle - \langle \vec{v}_e \rangle]. \qquad (23.1\text{-}97)$$

In the absence of temperature, density or pressure gradients, the current density is proportional to the electric field (Ohm's law). Thus, for electrons,

[22] L. Spitzer, Jr., Physics of Fully Ionized Gases, pp. 65-87, Interscience Publishers, Inc., New York, 1956.

[23] A. R. Hochstim and G. A. Massel, "Calculation of Transport Coefficients In Ionized Gases," in Kinetic Processes in Gases and Plasmas, pp. 141-255, edited by A. R. Hochstim, Academic Press, New York, 1969.

[24] I. P. Shkarofsky, T. W. Johnston and M. P. Bachynski, The Particle Kinetics of Plasma, pp. 315-400, Addison-Wesley Publishing Co., Reading, Mass., 1966.

$$\vec{j}_e = \sigma \vec{E}, \qquad\qquad\qquad (23.1\text{-}98)$$

where σ is electron conductivity. In more general cases when only temperature gradients are absent but electron-density gradients exist,

$$\vec{j}_e = \sigma \vec{E} + e D_e \nabla N_e, \qquad\qquad (23.1\text{-}99)$$

where D_e is the electron-current diffusion coefficient. In the presence of magnetic fields, one needs to consider the current density in different directions
as well as the electron conductivities and diffusion coefficients in different directions. The conductivities and diffusion coefficients are then components of
a tensor.

The direct current (d.c.) electron conductivity parallel to the applied
magnetic field is given by

$$\sigma_\| = N_e e^2 / m_e \bar{\nu} g_\sigma^o, \quad g_\sigma^o = 0.5064, \qquad (23.1\text{-}100)$$

where $\bar{\nu} = \bar{\nu}_{ei} + \bar{\nu}_{en}$ is the collision frequency of electrons with ions and neutrals if any are present (e.g., in the laser heating of a pellet) and g_σ^o is a numerical constant. The collision frequency for fully ionized plasmas is given
by

$$\bar{\nu}_{ei} = 4(2\pi k T_e / m_e)^{\frac{1}{2}} (e^2/kT_e)^2 (\ell n \Lambda) \sum_i N_i Z_i^2 /3 = 3.63 \, \bar{Z} N_e (\ell n \Lambda)/T_e^{3/2} \text{ in sec}^{-1}.$$

$$(23.1\text{-}101)$$

For a hydrogen plasma ($Z = 1$), the Coulomb logarithm $\ln\Lambda$ is about 20 for fusion. The electric conductivity perpendicular to the magnetic field is given
by[22, 23]

$$\sigma_\perp = \frac{N_e e^2 g_\sigma}{m_e \bar{v}} [g_\sigma^2 + (\omega_c \sqrt{\bar{v}})^2 h_\sigma^2]^{-1}, \qquad (23.1\text{-}102)$$

where $\omega_c = eB/m_e c$ is the electron Larmor frequency and g_σ and h_σ are numerical coefficients tabulated in Ref. [23]; here $g_\sigma = 0.5064$, $h_\sigma = 1.1989$ for zero
magnetic field and $g_\sigma = 1$, $h_\sigma = 1$ for an infinitely strong magnetic field.

The particle diffusion coefficients are related to particle conductivity
by the Einstein relation

$$D_e = kT_e \sigma / N_e e^2. \qquad (23.1\text{-}103)$$

Thus D_e parallel and perpendicular to the magnetic field can be obtained from
the above expression by using $\sigma_{||}$ or $\sigma_\perp$. In magnetically confined plasmas, the
diffusion of charged particles to the wall is an important problem because both
particle loss and sputtering of impurities occur at the wall. We see that, for
strong magnetic fields,

$$(D_e)_\perp \simeq kT_e \bar{v}/m_e \omega_c^2 \sim B^{-2}. \qquad (23.1\text{-}104)$$

However, experiments in controlled thermonuclear fusion research (CTR)
show an anomalously fast escape of plasma across the magnetic field, often

indicating not a B^{-2} but a B^{-1} dependence of the diffusion coefficient, in qualitative agreement with the formula postulated by Bohm[23]

$$D_B = ckT_e/16eB. \tag{23.1-105}$$

The problem of anomalous diffusion is not yet solved satisfactorily.[23]

In addition to the problem of charged-particle transport, there is also
the problem of energy transport in different directions. The heat flux for electrons is given by

$$\vec{g}_e = -\alpha\vec{j}_e - K_{eff}\nabla T_e, \tag{23.1-106}$$

where the effective electron thermal conductivity K_{eff} in the absence of a magnetic field or parallel to the magnetic field is given by[23]

$$K^{(e)}_{eff\,\|} = 5N_e k^2 T_e/m_e\bar{\nu}(g^o_K)_{eff}. \tag{23.1-107}$$

For a fully ionized plasma, in the absence of magnetic fields, $(g^o_K)_{eff} = 1.562$
and, in the presence of magnetic fields, $(g^o_K)_{eff} = 0.358$. For a fully ionized
plasma in a strong magnetic field, the ion-ion collisions are more important
than the electron-ion and electron-electron collisions for the thermal conductivity transverse to the magnetic field. The thermal conductivity for ions perpendicular to the magnetic field is[23]

$$K^{(i)} = \frac{5}{2} \frac{N_i k^2 T_i}{M_i \overline{\nu}_{ii}} \frac{g_{Ki}}{h_{Ki}^2} \left(\frac{\nu_{ii}}{\omega_c} \right)^2 , \tag{23.1-108}$$

where $g_{Ki} = 0.56568$, $h_{Ki} = 1.000$ and

$$\overline{\nu}_{ii} = Z^2 \left(\frac{T_e}{T_i} \right)^{3/2} \left(\frac{m_e}{M_i} \right)^{1/2} \overline{\nu}_{ei}, \quad \omega_c^{(i)} = ZeB/M_i c. \tag{23.1-109}$$

L.   Motion of Charged Particles in a Magnetic Field

At present, there is considerable effort to obtain fusion by magnetic confinement of plasmas. Since plasmas can be confined in a vessel by proper design of magnetic fields, this "magnetic bottle" is the focus of considerable effort on CTR in the U.S., U.S.S.R. and in other countries. In the preceding sections, we have considered physical processes common to both magnetically confined plasmas and to laser heating of pellets. In the remaining sections, we introduce fundamental notions for the motion of electrons and ions in simple (constant in space and time) magnetic and electric fields. In the following sections, these concepts will be applied to the problems inherent in magnetic confinement.

Lines of force are generally introduced to visualize the electric and magnetic fields. The density of lines indicates the field strength and directed arrows for an electric field indicate the direction in which a positive charge would be accelerated by the force. For an electron, the charge is $q = -e$ and for an ion $q = Z_i e$, where Z_i is a positive number and $e > 0$. The Lorentz force on an electron with mass m_e in externally applied $\vec{E}$ and $\vec{B}$ fields is

$$m_e d\vec{v}_e/dt = -e[\vec{E} + (\vec{v}_e \times \vec{B}/c)] \tag{23.1-110}$$

and, for an ion with mass m_i,

$$m_i d\vec{v}_i/dt = Z_i e[\vec{E} + (\vec{v}_i \times \vec{B}/c)]. \tag{23.1-111}$$

The magnetic field does not contribute to the energy of the system even though it exerts a force on the charges. To verify this, we write Eqs. (23.1-110) and (23.1-111) in Cartesian coordinates (with m representing either m_e or m_i) as follows:

$$mdv_x/dt = q(cE_x + v_y B_z - v_z B_y)/c, \quad mdv_y/dt = q(cE_y + v_z B_x - v_x B_z)/c,$$

$$mdv_z/dt = q(cE_z + v_x B_y - v_y B_x)/c. \tag{23.1-112}$$

Multiplying each expression in Eq. (23.1-112) by the appropriate velocity component and adding, we obtain

$$\frac{d}{dt}\left(\frac{1}{2} m \vec{v} \cdot \vec{v}\right) = q\vec{E} \cdot \vec{v}. \tag{23.1-113}$$

Introducing a scalar electrical potential (Φ), we write

$$\vec{E} = -\nabla\Phi. \tag{23.1-114}$$

Using the definition of the total derivative, we have

$$\frac{d\Phi}{dt} = v_x \frac{\partial \Phi}{\partial x} + v_y \frac{\partial \Phi}{\partial y} + v_z \frac{\partial \Phi}{\partial z} \equiv \vec{v} \cdot \nabla \Phi = - \vec{v} \cdot \vec{E}. \qquad (23.1\text{-}115)$$

Substituting Eq. (23.1-115) in Eq. (23.1-113) yields

$$\frac{d}{dt} [\frac{1}{2}m \vec{v} \cdot \vec{v} + q\Phi] = 0 \qquad (23.1\text{-}116)$$

which integrates to

$$\frac{1}{2}mv^2 + q\Phi = \text{constant} = U_o, \qquad (23.1\text{-}117)$$

where U_o is a constant of integration and is the initial total energy in the system. It should be noted that Eq. (23.1-117) is independent of the magnetic field, i.e., $\vec{B}$ does not contribute to the energy balance.

The general properties of the motion of charged particles in an electromagnetic field will be illustrated by considering a number of restricted examples.

i.   Constant Magnetic Field in the Absence of the Electric Field

Consider a particle of charge Ze and mass m in an externally applied, constant $\vec{B}$ field without an electric field present ($\vec{E} = 0$). The component equations of motion are

$$dv_x/dt = \omega_z v_y - \omega_y v_z, \quad dv_y/dt = \omega_x v_z - \omega_z v_x, \quad dv_z/dt = \omega_y v_x - \omega_x v_y, \quad (23.1\text{-}118)$$

where we have introduced for convenience the vector

$$\vec{w} = Ze\vec{B}/mc. \tag{23.1-119}$$

If we take the z-axis to be aligned with the magnetic field, Eq. (23.1-118) reduces to

$$dv_x/dt = w_z v_y, \, dv_y/dt = -w_z v_x, \, dv_z/dt = 0. \tag{23.1-120}$$

We define $w_z = \pm w_c$, where w_c is the cyclotron frequency (or gyromagnetic frequency). The quantity $w_c = |(Ze/mc)B|$ is always a positive number. The + sign (for $w_z > 0$) is used for ions and the - sign ($w_z < 0$) for electrons and negative ions. Integrating Eq. (23.1-120) yields

$$v_x = v_x^o + w_z y, \, v_y = v_y^o - w_z x, \, v_z = v_z^o, \tag{23.1-121}$$

with the initial velocities v_x^o, v_y^o and v_z^o. From Eq. (23.1-121), we obtain $v_\perp^2 = [w_z y + v_x^o]^2 + [w_z x - v_y^o]^2$ where $v_\perp^2 \equiv v_x^2 + v_y^2$ is the component of the velocity perpendicular to the magnetic field $\vec{B}$. Using Eq. (23.1-117) and the condition that $\vec{E} = 0$, the above relation simplifies to

$$v_\perp^2 = w_z^2 \left[\left(y + \frac{v_x^o}{w_z} \right)^2 + \left(x - \frac{v_y^o}{w_z} \right)^2 \right] = (v_x^o)^2 + (v_y^o)^2. \tag{23.1-122}$$

Introducing the notion of a Larmor radius (radius of gyration) by the expression

$$R_L = v_\perp / \omega_c \tag{23.1-123}$$

or, equivalently, by

$$R_L^2 = [x - (v_y^o/\omega_z)]^2 + [y + (v_x^o/\omega_z)]^2, \tag{23.1-124}$$

the motion of the charged particle may be described as a circle in the (x, y)
plane. Integrating the last expression in Eq. (23.1-121), however, yields a
linear displacement along the direction of the magnetic field for the center of
the circle (drift of guiding center), i.e.,

$$z = v_z^o t + z_o \tag{23.1-125}$$

so that the motion of the particle in three dimensions is that of a right helix
as shown in Fig. 23.1-18.

Let us introduce polar coordinates (R_L, θ) as follows:

$$y + v_x^o/\omega_z = R_L \sin \theta, \quad x - v_y^o/\omega_z = R_L \cos \theta;$$

differentiating with respect to time, we obtain

$$v_y = R_L(\cos \theta)\dot{\theta}, \quad \dot{\theta} \equiv d\theta/dt, \quad v_x = -R_L(\sin \theta)\dot{\theta}, \quad v_\perp^2 = v_x^2 + v_y^2 = R_L^2 \dot{\theta}^2,$$

or

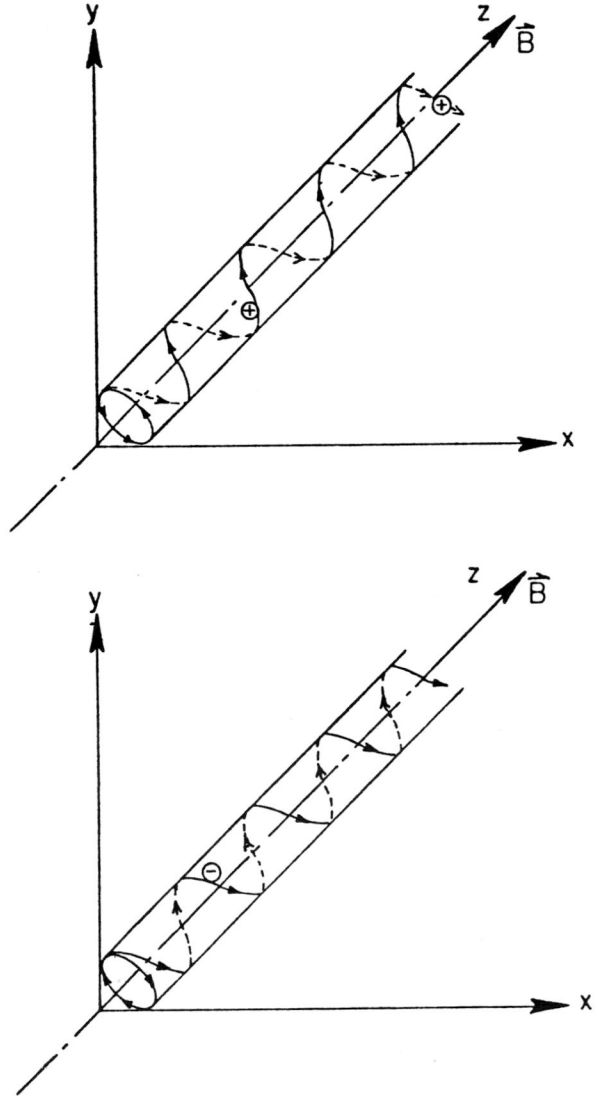

Fig. 23.1-18   Electrons spiral in the sense of a right-handed screw about the
magnetic field; the x-y projection shows clockwise motion (lower
diagram).   Positive ions spiral in the sense of a left-handed
screw about the magnetic field; the x-y projection shows anti-
clockwise motion (upper diagram).   The directions of motion
for electrons and ions may be in opposite directions, depending
on the value of the initial velocities.

$$\theta = \pm v_\perp / R_L = \pm \omega_c, \qquad\qquad (23.1\text{-}126)$$

i.e., the angular frequency is the gyrofrequency (Larmor) frequency. For
ions, the negative sign applies and

$$\omega_c^{(i)} = \left| ZeB/M_i c \right|, \quad R_L^{(i)} = v_\perp^{(i)} / \omega_c^{(i)}. \qquad\qquad (23.1\text{-}127)$$

For electrons, the positive sign applies and

$$\omega_c = \left| eB/m_e c \right|, \quad R_L^{(e)} = v_\perp^{(e)} / \omega_c. \qquad\qquad (23.1\text{-}128)$$

The law of conservation of angular momentum is

$$mv_\perp R_L = \text{constant.} \qquad\qquad (23.1\text{-}129)$$

Substituting for R_L from Eq. (23.1-123),

$$mv_\perp R_L = (2mc/Ze)\mu = (m^2 c v_\perp^2 / ZeB) = \text{constant,} \qquad\qquad (23.1\text{-}130)$$

where the magnetic moment μ is defined by

$$\mu = (mv_\perp^2 / 2B). \qquad\qquad (23.1\text{-}131)$$

Thus we see that the magnetic moment and $v_\perp^2 / B$ are constants.

When the magnetic field changes slowly in space over distances large compared with the Larmor radius,

$$\Delta B/B \ll \Delta x/R_L, \qquad (23.1\text{-}132)$$

or else changes with time on a scale that is long compared with the rotation of a charged particle around the magnetic lines of force (ω_c^{-1}),

$$\Delta B/B \ll \Delta t/\omega_c^{-1}, \qquad (23.1\text{-}133)$$

then $v_\perp^2/B$ (or μ) is only approximately constant. In this case, $v_\perp^2/B$ (or μ) is called an adiabatic invariant, where the term adiabatic refers to slow changes in B in either space or time.

   ii.   Mutually Perpendicular, Constant $\vec{B}$ and $\vec{E}$

We consider the case of nonzero applied $\vec{E}$ and $\vec{B}$ fields, constant in magnitude and direction, with $\vec{E}$ perpendicular to $\vec{B}$. For $\vec{E} = \vec{E}(E_x, 0, 0)$ and $\vec{B} = \vec{B}(0, 0, B_z)$, the equations of motion for a particle of mass m and charge Ze are

$$dv_x/dt = (Ze/m)E_x + \omega_z v_y, \quad dv_y/dt = -\omega_z v_x, \quad dv_z/dt = 0. \qquad (23.1\text{-}134)$$

The motion of the particle in the z-direction is given by $v_z = v_z^o$, $z = z_o$. We choose the initial condition $v_z^o = 0$.

Multiplying the first expression in Eq. (23.1-134) by v_x and the second by v_y and adding yields the conservation of energy expression [Eq. (23.1-113)], viz.

$$\frac{1}{2}m\frac{d}{dt}(v_x^2 + v_y^2) = Zev_xE_x.$$ (23.1-135)

Equation (23.1-135) integrates to

$$\frac{1}{2}m(v_x^2 + v_y^2) - ZexE_x = \text{constant} = U_o,$$ (23.1-136)

where U_o is the constant total energy of the system. We integrate the v_x and v_y expressions in Eq. (23.1-134) to obtain

$$v_x = v_x^o + (Ze/m)E_x t + \omega_z y, \quad v_y = v_y^o - \omega_z x.$$ (23.1-137)

By squaring and adding the expressions in Eq. (23.1-137) and using Eq. (23.1-136), we obtain

$$\frac{2}{m}(U_o + ZexE_x) = \omega_z^2\left[\left(x - \frac{v_y^o}{\omega_z}\right)^2 + \left(y + \frac{ZeE_x}{m\omega_z}t + \frac{v_x^o}{\omega_z}\right)^2\right].$$ (23.1-138)

The terms in Eq. (23.1-138) may be regrouped to yield the following expression for the particle motion:

$$R_L^2 = (x - a)^2 + [y - b(t)]^2,$$ (23.1-139)

where

$$a = \frac{1}{\omega_z}\left(v_y^o + \frac{ZeE_x}{m\omega_z}\right), \quad b(t) = -\frac{1}{m\omega_z}(v_x^o + ZeE_x t),$$

$$R_L^2 = \frac{2}{m\omega_z^2}\left[U_o + \frac{ZeE_x}{m\omega_z}v_y^o + \frac{1}{2m}\left(\frac{ZeE_x}{\omega_z}\right)^2\right]. \tag{23.1-140}$$

Equation (23.1-139) is a circle in the horizontal plane of radius R_L and is centered at the points $x = a$ and $y = b(t)$. The ordinate of the center of the circle is, however, a linear function of time so that the circle "drifts" along the y-axis, as is shown in Fig. 23.1-18. The drift velocity is given by v_D $= ZeE_x/m\omega_z = \pm cE_x/B_z$, with the positive sign applying to positive ions and the negative sign to electrons.

There are a number of possible motions for the particle to execute in this field configuration. The motions depend on the ranges of the parameters given in Eq. (23.1-140). Differentiating Eq. (23.1-134) again, $\ddot{v}_x = -\omega_z^2 v_x$, $\ddot{v}_y = -\omega_z^2(v_y + v_D)$ where $\ddot{v}_x \equiv d^2 v_x/dt^2$, etc.; the solution is

$$v_x = A_1 \sin \omega_z t + A_2 \cos \omega_z t, \quad v_y = C_1 \sin \omega_z t + C_2 \cos \omega_z t - v_D.$$

At $t = 0$, let $v_x^o = A_2 = 0$ and $v_y^o = C_2 - v_D$; then

$$v_x = A_1 \sin \omega_z t, \quad v_y = C_1 \sin \omega_z t + (v_y^o + v_D) \cos \omega_c t - v_D. \tag{23.1-141}$$

Let

$$x = R_L(1 - \cos \theta), \quad \theta = \theta(t), \quad y = - v_D t + R_L \sin \theta; \qquad (23.1\text{-}142)$$

then

$$v_x = R_L \dot{\theta} \sin \theta \quad \text{and} \quad v_y = - v_D + R_L \dot{\theta} \cos \theta. \qquad (23.1\text{-}143)$$

Comparing the two expression for v_x and v_y, we obtain $A_1 = R_L \dot{\theta}$, $\theta = \omega_z t$, $A_1 = R_L \omega_z$, $C_1 = 0$, and $v_y^o + v_D = R_L \dot{\theta} = R_L \omega_z$. Thus,

$$\theta = \omega_z t, \qquad (23.1\text{-}144)$$

$$x = R_L(1 - \cos \theta), \qquad (23.1\text{-}145)$$

$$y = - (v_D / \omega_z)\theta + R_L \sin \theta, \qquad (23.1\text{-}146)$$

$$R_L = (v_y^o + v_D)/\omega_z, \qquad (23.1\text{-}147)$$

$$v_D = \pm cE_x / B_z. \qquad (23.1\text{-}148)$$

Equations (23.1-145) and (23.1-146) are the expressions for a cycloid (for $v_y^o = 0$, we have a normal cycloid).

We notice that the electric and magnetic fields always appear in the same combination in Eqs. (23.1-145) to (23.1-148), i.e., in the ratio $ZeE_x / m\omega_z = \pm cE_x / B_z$ which defines the drift velocity v_D. This notion can be generalized by writing the total velocity of the charged particle as

$$\vec{v} = \vec{v}_o + \vec{v}_D,$$
(23.1-149)

where $\vec{v}_o$ is the velocity in the absence of fields and $\vec{v}_D$ is the drift velocity induced by the $\vec{E}$ and $\vec{B}$ fields. Substituting Eq. (23.1-149) into the Lorentz force
equation, with the drift velocity defined by

$$\vec{v}_D \equiv c(\vec{E} \times \vec{B}/B^2),$$
(23.1-150)

yields

$$\frac{d\vec{v}_o}{dt} = \frac{Ze}{m}\left[\vec{E} + \frac{\vec{v}_o \times \vec{B}}{c} + \frac{(\vec{E} \times \vec{B}) \times \vec{B}}{B^2}\right].$$
(23.1-151)

Using appropriate vector identities,

$$(\vec{E} \times \vec{B}) \times \vec{B} = \vec{B} \times (\vec{B} \times \vec{E}) = \vec{B}(\vec{B} \cdot \vec{E}) - B^2\vec{E};$$
(23.1-152)

$\vec{B} \cdot \vec{E} = BE \cos \theta_{BE} = 0$ since $\vec{B}$ is perpendicular to $\vec{E}$ ($\theta_{BE} = \pi/2$) and we obtain from Eq. (23.1-151)

$$d\vec{v}_o/dt = \frac{Ze}{m}(\vec{v}_o \times \vec{B}/c).$$
(23.1-153)

Equation (23.1-153) is the same expression as would be obtained if $\vec{E} = 0$. Also,

$\vec{v}_o$ is the velocity component along the helical motion around the direction of the magnetic field and $\vec{v}_D$ is the $\vec{E} \times \vec{B}$ drift.

It is clear from Eq. (23.1-153) that $\vec{v}_o$ is the source of the helical motion and $\vec{v}_D$ is the source of drift or translation (see Fig. 23.1-19). In general, we have

$$\vec{v}_D = c(\vec{E} \times \vec{B}/B^2) = c(E/B) \sin \theta_{EB} \qquad (23.1\text{-}154)$$

or

$$v_D(\text{m/sec}) = \frac{3 \times 10^{10} \ E(\text{stat volt/cm})}{B(G)} = \frac{10^8 \ E(\text{volt/cm})}{B(G)}. \qquad (23.1\text{-}155)$$

We see that electrons and ions rotate in opposite directions. However, $\vec{v}_D$ is independent of charge so that the electrons and ions have the same drift velocity.

iii. $\vec{E}$ and $\vec{B}$ Constant and Parallel

As the third example, we consider the case of constant $\vec{E}$ and $\vec{B}$ fields, both aligned with the z-axis. The equations of motion may be written as

$$dv_x/dt = \omega_z v_y, \quad dv_y/dt = -\omega_z v_x,$$

$$dv_z/dt = ZeE_z. \qquad (23.1\text{-}156)$$

The last equation integrates immediately to

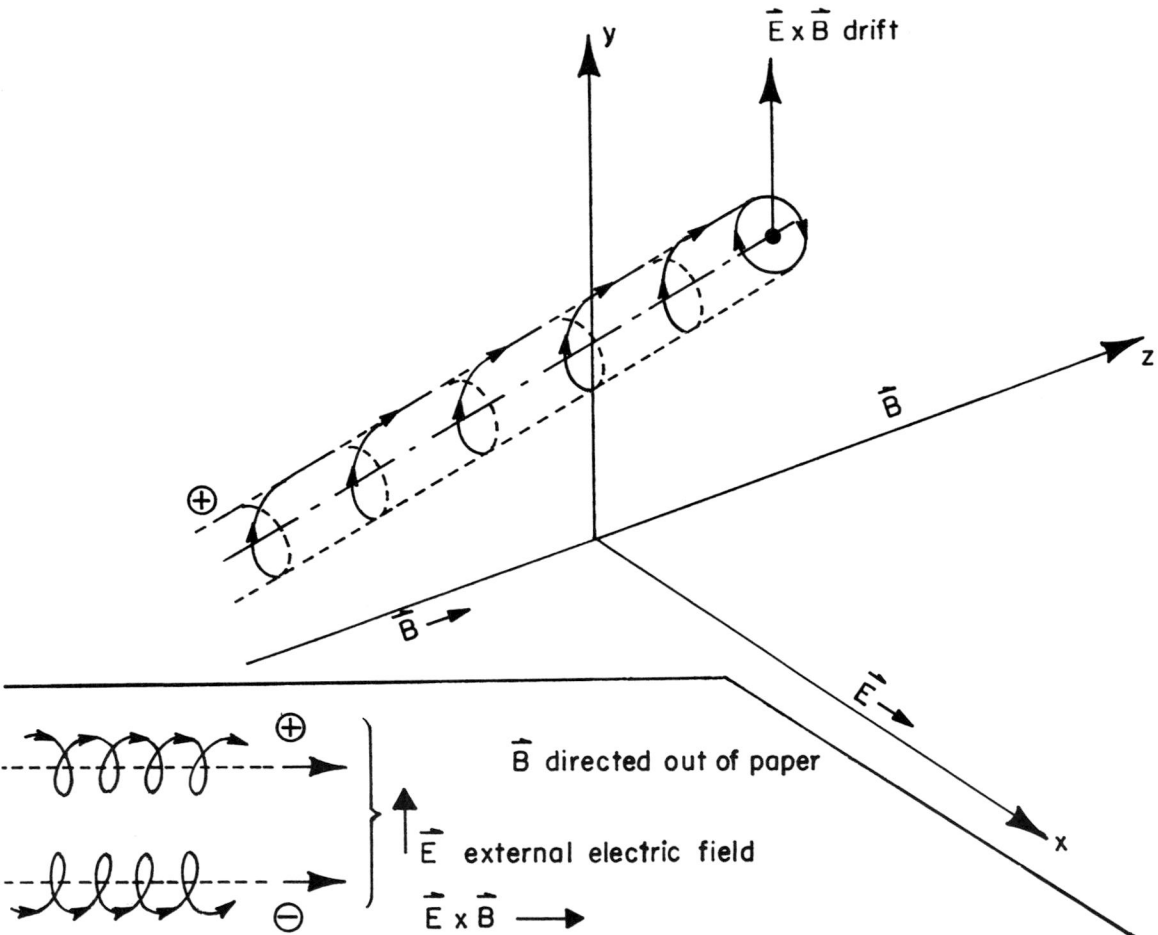

Fig. 23.1-19 Mutually perpendicular constant electric and magnetic fields showing $\vec{E} \times \vec{B}$ drift. In the lower figure, $v_y^o = 0$.

$$v_z = (ZeE_z/m)t + v_z^o, \quad z = (ZeE_z/2m)t^2 + v_z^o t + z_o, \qquad (23.1\text{-}157)$$

and the remaining two equations may be integrated to yield

$$v_x = \omega_z y + v_x^o; \; v_y = -\omega_z x + v_y^o; \; v_\perp^2 = v_x^2 + v_y^2 = \omega_z^2 \{[y + (v_x^o/\omega_z)]^2 + [x - (y_y^o/\omega_z)]^2\}.$$

$$(23.1\text{-}158)$$

Since $v_\perp^2/B$ is constant* [see Eq. (23.1-130)], we obtain the equation of a circle in the (x, y)-plane with radius

$$R_L = v_\perp/\omega_c.$$

$$(23.1\text{-}159)$$

Now let $X = x - (v_y^o/\omega_z) = \rho \cos \theta$, $Y = y + (v_x^o/\omega_z) = \rho \sin \theta$; then

$$X^2 + Y^2 = R_L^2$$

$$(23.1\text{-}160)$$

and, differentiating with respect to time, we obtain

$$V_x = \dot{x} = \dot{X} = \dot{\rho} \cos \theta - \rho \dot{\theta} \sin \theta,$$

$$(23.1\text{-}161)$$

$$V_y = \dot{y} = \dot{Y} = \dot{\rho} \sin \theta + \rho \dot{\theta} \cos \theta.$$

$$(23.1\text{-}162)$$

Combining V_x and V_y, we have

$$V_\perp^2 = V_x^2 + V_y^2 = \dot{\rho}^2 + \rho^2 \dot{\theta}^2,$$

$$(23.1\text{-}163)$$

$$X^2 + Y^2 = (V_\perp/\omega_z)^2,$$

$$(23.1\text{-}164)$$

---

*Since $(dv_x/dt)v_x = (\omega_z v_y)v_x$ and $(dv_y/dt)v_y = (-\omega_z v_x)v_y$, adding both of these yields $(1/2)d(v_x^2 + v_y^2)/dt = 0$, $v_x^2 + v_y^2 = $ constant.

with

$$X^2 + Y^2 = R_L^2 = \rho^2, \quad R_L = \rho = \text{constant}, \tag{23.1-165}$$

and $\rho^2 \omega_z^2 = \dot{\rho}^2 + \rho^2 \dot{\theta}^2 = \rho^2 \dot{\theta}^2$. Finally, we have

$$\dot{\theta} = \omega_z, \tag{23.1-166}$$

i.e., the angular velocity is equal to the Larmor frequency. For ions and electrons, respectively,

$$(d\theta/dt)_i = \omega_c^{(i)} \quad \text{and} \quad (d\theta/dt)_e = -\omega_c. \tag{23.1-167}$$

The Larmor radii for ions and electrons are, respectively,

$$R_L^{(i)} = v_\perp^{(i)} / \omega_c^{(i)}, \quad R_L^{(e)} = v_\perp^{(e)} / \omega_c. \tag{23.1-168}$$

23.2 Magnetic Confinement and Thermonuclear Fusion

The next five sections (23.2-23.6) are concerned with the magnetic confinement approach to the thermonuclear fusion problem. This research has a 25 year history and has required the development of two new branches of physics, magnetohydrodynamics and plasma physics, neither of which is familiar to most students of physics and engineering, hence a good deal of the material in the next four sections is introductory and concerned with basic concepts and vocabulary. The reader who is interested only in the present experimental situation and the engineering speculations arising from it   may proceed directly to Section 23.6 and consider the intervening sections as an extended glossary.

In Section 23.1E, the conditions that must be satisfied if thermonuclear energy is to be harnessed were discussed. Briefly, they require that the fuel be in a state not too far from thermal equilibrium at a temperature of 5 kev and that the fuel be confined for a time τ related to the particle density by $n\tau \simeq 4 \times 10^{14}$ sec cm^{-3}.

One way of satisfying this last criterion is to complete the reaction in a very short time, or to assemble a large mass of fuel and heat it very suddenly, for, unless contained by external forces, the fuel will expand at the sound or thermal speed, which for deuterium at 5 kev is approximately $v_\theta^i = 7 \times 10^7$ cm/sec. The reaction time is then given, approximately, by $\tau = L/v_\theta^i = (1/7) \times 10^{-7} L$, L being the size of the fuel element; the $n\tau$ criterion becomes $nL \simeq 2.8 \times 10^{-22}$ cm^{-2} or, noting that the mass of a deuteron is 3.4×10^{-24} g, $\rho L \simeq 0.1$ g/cm^2, the energy that must be provided to the fuel is $3NkT = 3nkT \times (4\pi/3)L^3 \simeq (4\pi kT/m) \times (0.1)^3/\rho^2 = 3.6\rho^{-2}$ Mj.

If the density is of order unity, the pressure at 5 kev is $p \simeq 10^7$ atm and the energy must be released in a violent explosion. There is dramatic evidence that devices of this kind have been made to work on the scale of the hydrogen bomb, the input energy being supplied by a nuclear explosion. With the development of high-powered lasers, which appear capable of delivering very large amounts of energy in extremely short times, it is possible to consider micro-explosions of miniature deuterium-tritium pellets as sources of controlled thermonuclear power. Investigations of this scheme form the subject of Section 23.7. Here, we will consider devices for which we attempt to satisfy the Lawson criterion by using forces to contain the plasma.

There have been many suggestions made of methods to contain a hot plasma but the most promising seems to be confinement of the fuel in a strong magnetic field.[1] As we have seen, at reacting temperatures, the fuel must certainly be completely ionized and thus is composed of charged particles. However, as was shown in Section 23.1J, a magnetic field has a profound influence on the motion of charged particles; instead of moving across magnetic field lines, charged particles travel about the field lines in circles of radius $r_L = v_\perp/\omega_c$ cm, ω_c being the gyrofrequency in $\sec^{-1}$,

$$\omega_c^e \simeq 1.76 \times 10^{10} B \ \sec^{-1}, \quad \omega_c^i \simeq 0.96 \times 10^7 B \ \sec^{-1}, \qquad (23.2\text{-}1)$$

where B is measured in kG, and on substituting for the perpendicular component of the velocity $v_\perp$, the thermal speed, we find that

$$r_L^e \simeq 0.01 \ T^{\frac{1}{2}} B^{-1} \ \text{cm}, \quad r_L^i \simeq 4.6 \ T^{\frac{1}{2}} B^{-1} \ \text{cm}, \qquad (23.2\text{-}2)$$

---

[1] L. A. Artsimovich, Controlled Thermonuclear Reactions, Gordon and Breach, New York, N.Y., 1972.

where T is the temperature in kev. For a field of 50 kG and a temperature of

5 kev, $r_{L-} \simeq 0.00045$ cm, $r_{L+} \simeq 0.02$ cm.

    To get an idea of the effectiveness of a magnetic field in confining a

plasma, let us first observe that a particle can move across the magnetic

field only when it experiences a collision, and on collision it moves a distance

$\delta x = r_L$. Moreover, as is easily shown, when two like particles collide, the

displacement experienced by one is just the negative of that experienced by the

other. The effect of collisions then is to produce random motions of the par-

ticles across the field lines. If the plasma is uniform, this produces no ma-

croscopic effect; however, if there is a density gradient, a net current will

flow.[2]

    Consider (see Fig. 23.2-1) an imaginary surface through the plasma

at the point x. Particles will cross that surface from left to right at a rate

per unit area given by the product of the jump distance δx, the collision fre-

quency ν, and the density to the left of S, and we will make no great error if

we call this $n[x - (\delta x/2)]$ and $f_+ = \nu \delta x\, n[x - (\delta x/2)]$. Particles will cross the

surface from right to left at the rate $f_- = \nu \delta x n[x + (\delta x/2)]$. The net flux of

particles will be $f = f_+ - f_- = \nu \delta x \{n[x - (\delta x/2)] - n[x + (\delta x/2)]\} = - \nu \delta x^2 (\partial n/\partial x)$.

Moreover, unless the slope of $n(x)$ is uniform, there will be a loss of particles;

for, if we consider two surfaces, one at x and one at $x + \Delta x$, the flux in through

the side x will be $- \nu \delta x^2 (\partial n/\partial x)$ and that out through $x + \Delta x$ will be $- \nu \delta x^2$

$\times [\partial n(x + \Delta x)/\partial x]$. The difference between these is $\Delta x \nu \delta x^2 (\partial^2 n/\partial x^2)$, assuming

that $\nu \delta x^2$ is constant. But the difference between the flux into a surface and

that out gives the rate at which particles are accumulating between the surfaces,

---

[2]S. Chandrasekhar, Plasma Physics, Chapter 7, The University of Chicago
   Press, Chicago, Ill., 1960.

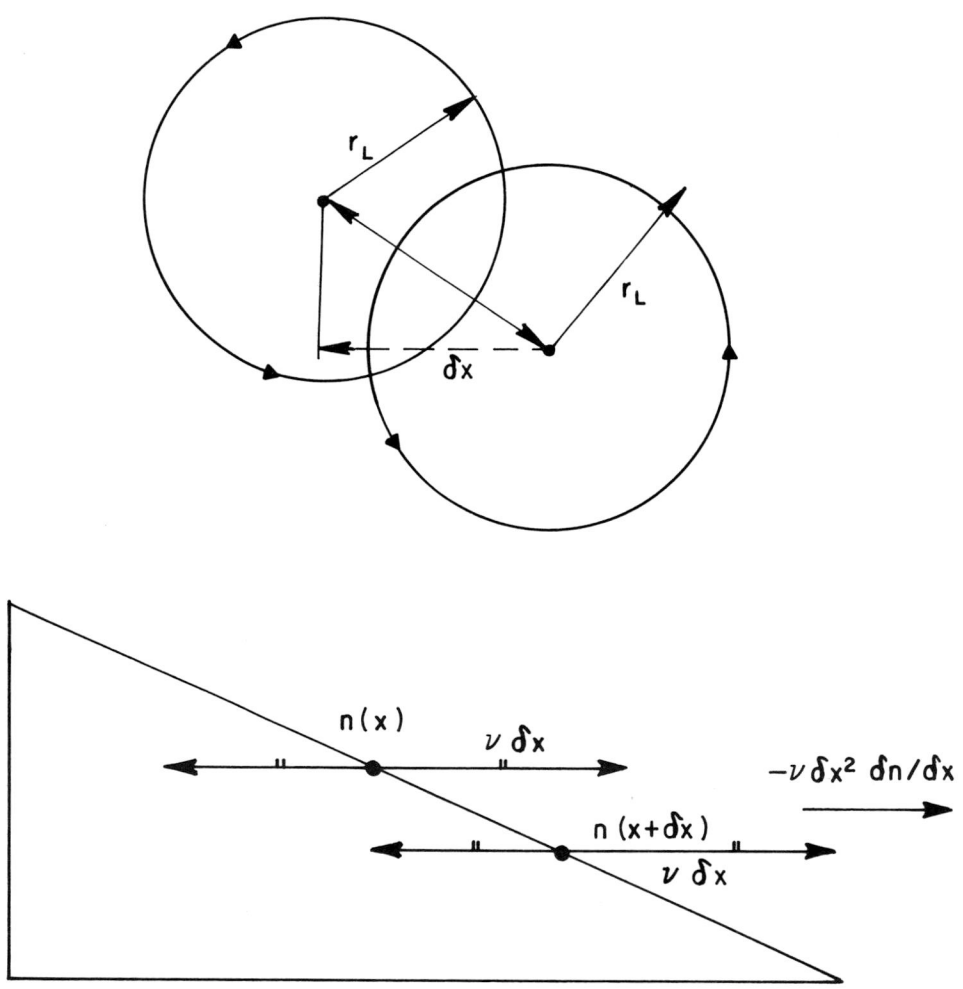

Fig. 23.2-1 The diffusion process in a magnetic field. On collision, the direction of particle motion is changed and the center of the circular orbit shifted by δx, with an average magnitude r_L (the gyroradius). Although shifts to the right or the left are equally frequent, a density gradient gives rise to a current.

$(\partial n/\partial t)\Delta x = \nu \delta x^2 (\partial^2 n/\partial x^2)\Delta x$, and by canceling the Δx, we obtain the diffusion equation, $\partial n/\partial t = D\partial^2 n/\partial x^2$, where the diffusion constant $D = \nu \delta x^2$ cm^2 sec^{-1}.

For particles diffusing through a magnetic field, the diffusion coefficient is the same for ions and electrons and, using the value of the collision frequency given in Section 23.1K,

$$D_c = (m/M)n \langle \sigma v_{\theta-} \rangle r_{L+}^2 = 1.3 \times 10^{-10} nT^{-\frac{1}{2}}B^{-2} \text{ cm}^2/\text{sec}, \qquad (23.2-3)$$

where n is in cm^{-3}, B in kG and T in kev. To solve the diffusion equation is very difficult, especially since D is a function of position, so we will be content with a scaling argument, writing $(\partial^2 n/\partial x^2) = n/L^2$, where L is a length characteristic of the equilibrium and $(\partial n/\partial t) = n/\tau$, where τ is the characteristic life time against diffusion loss. Then $\tau = L^2/D \simeq L^2/(r_L^2 \nu) = \tau_{coll} L^2/r_L^2$, where τ_{coll} is the collision period (ν^{-1}) and the Lawson criterion becomes $n\tau = nL^2/D > 4 \times 10^{14}$ cm^{-3}-sec. Using the value of D given above, $n\tau = 8 \times 10^9 L^2 T^{\frac{1}{2}} B^2$ sec-cm^{-3} and, for B = 50 kG, T = 5 kev, the Lawson criterion is satisfied for values of L as low as 2.5 cm. Since the radius of the fuel element in any practical reactor would be several m, the classical diffusion time, $\tau_{clas} = \tau_{coll} L^2/r_L^2$ sec, amply satisfies the Lawson criterion; it is this fact that makes magnetic confinement so attractive (see Fig. 23.2-2).

In spite of this, and in spite of twenty years of theoretical speculation and laboratory experiment, the goal of a magnetically confined thermonuclear plasma satisfying the Lawson criterion has not been met (at the time of this writing). However, several experimental groups have come within an order of magnitude of satisfying both Lawson's criterion and the temperature

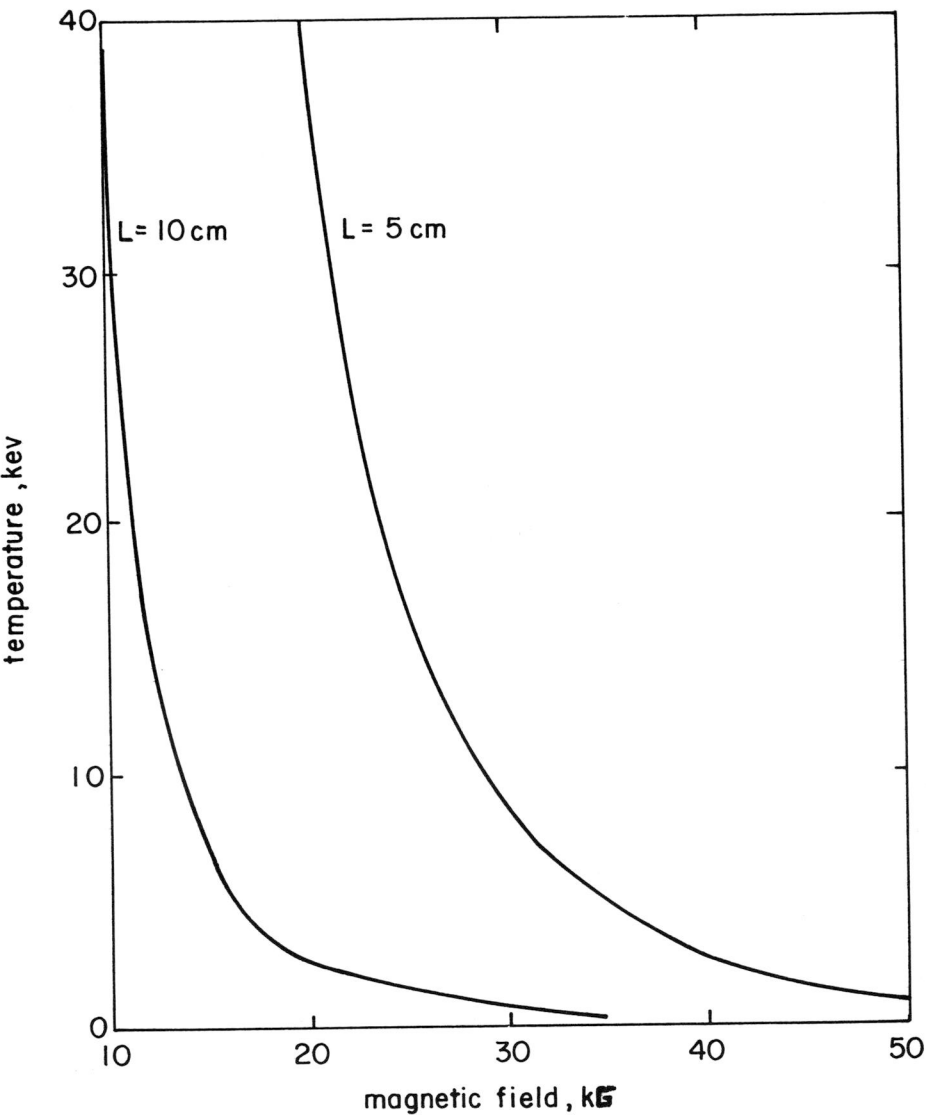

Fig. 23.2-2 Classical diffusion and magnetic confinement. Contours showing constant values of $n\tau = 5 \times 10^{14}$ sec-cm^{-3} for two values of the density scale length L, as functions of temperature and magnetic field strength, assuming classical diffusion.

requirement, $n\tau = 4 \times 10^{13}$, $T \simeq 1$ kev, [3] and since research is proceeding on a reasonably large scale, the subject is worth discussing in a volume of this kind.

As is often the case, the attempt to extend engineering practice into a completely new region has required an extension of the underlying physics, and again, as is usually the case, understanding of the operation of a device. or the nature of an experiment requires a grasp of the theory which underlies it; so we turn next to an exposition of the theoretical basis for experiments on magnetic confinement. We will then consider the more successful experiments that have been performed in attempts to confine a plasma and will end with a brief discussion of engineering speculations about possible fusion reactors.

When a plasma is confined in a magnetic field, the force that prevents the escape of any particle is, as was shown in Section 23.1J, the Lorentz force $\vec{F} = e\vec{v} \times \vec{B}/c$.

To discover the macroscopic consequences of this, we can sum over all the particles, positive and negative in a unit volume, thus obtaining the macroscopic force/unit volume acting on the plasma, $\vec{F} = \Sigma e_i \vec{v}_i \times \vec{B}/c = \vec{j} \times \vec{B}$, where $\vec{j}$ is the electric current density. Since $\vec{j}$ in a steady state is related to the spatial variation of the magnetic field by Ampère's law, $4\pi\vec{j} = \vec{\nabla} \times \vec{B}$, the existence of a force requires that the magnetic field be nonuniform. When the plasma is confined, the plasma density and plasma pressure are also nonuniform and the force density causing the plasma to expand is the pressure gradient, $\vec{\nabla}p$. As we shall show, the pressure for some kinds of confined plasma need not be the simple hydrodynamic scalar pressure, but for plasmas confined for the classical diffusion time, which as we have seen is $(L^2/r_L^2) \times$ the collision time, the pressure will indeed be a scalar, and the equilibrium

_____

[3] Fusion Forefront <u>8</u>, No. 3, December 1975, ERDA CTR Division, Washington, D.C.

between the hydrodynamic forces causing expansion and the electromagnetic
confining force may be written as

$$- \vec{\nabla} p + \vec{j} \times \vec{B} = 0.$$

(23.2-4)

To get an idea of the magnitudes of the magnetic fields required, we
consider a one-dimensional case in which p and B vary only in the x-direction,
while B points along 0Z. The magnitude of j is then given by Ampère's law,
$j = (4\pi)^{-1}(\partial B/\partial x)$, and the equilibrium becomes $\partial p/\partial x + (\partial B^2/\partial x)/8\pi = 0$ or
$[\partial(p + B^2/8\pi)/\partial x] = 0$.

Suppose that in the center of the plasma B goes to zero, while at the
edge p goes to zero, then at the edge $B^2 = 8\pi p$. If p is 1,000 atm, which is
probably an upper limit, $B \simeq 160,000$ G. In this kind of equilibrium, the magnetic energy and the energy in the pressure are equal. However, for stability
reasons, it appears often necessary to consider plasmas for which the magnetic energy density is much greater than the thermal, i.e., for which the
parameter $\beta = 4\pi p B^{-2} \ll 1$. For a pressure of 50 atm, B = 35 kG at $\beta = 1$,
while at $\beta = 0.1$, B = 110 kG. This is a high field, but one that can be reached,
even with superconducting magnets (see Fig. 23.2-3).

23.3 The Geometry of Magnetic Confinement

A. Magnetohydrodynamic Confinement

We turn now to a consideration of the geometrical constraints on the
magnetic field needed for confinement[1] and begin with the case of a scalar

_____

[1] W. B. Thompson, An Introduction to Plasma Physics, Chapter 4, Pergamon
Press, Oxford, U.K., 1962.

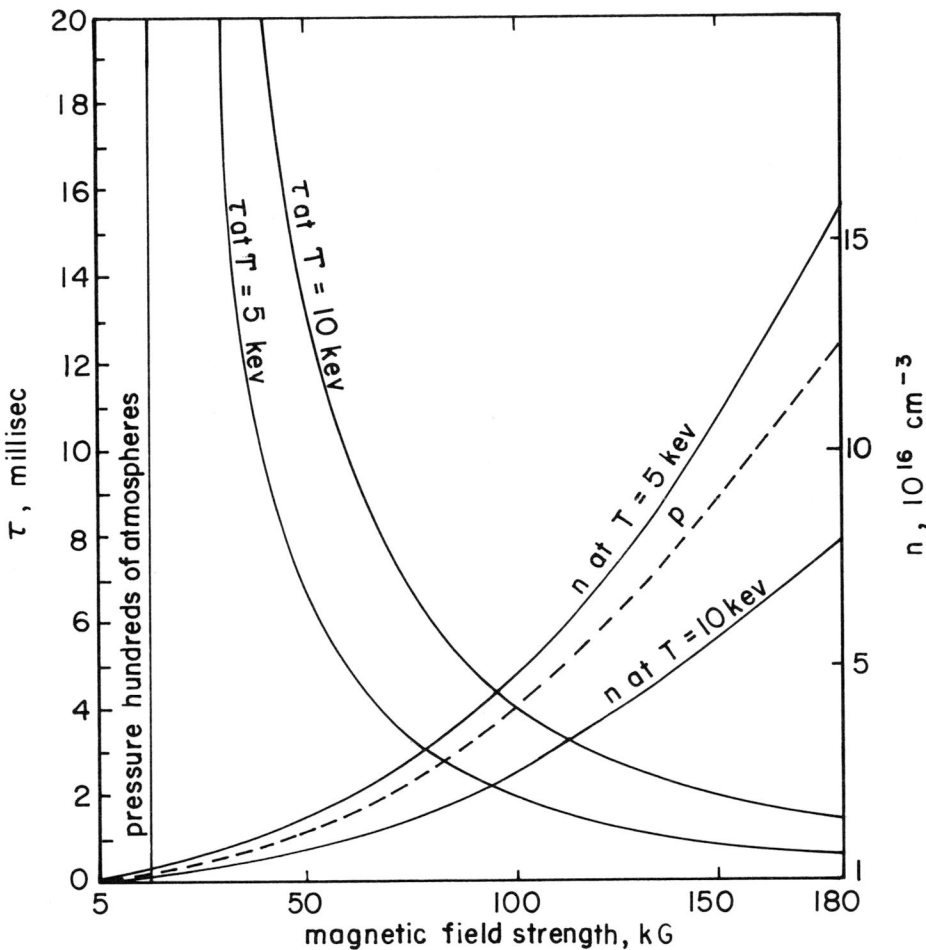

Fig. 23.2-3 Magnetic confinement showing the pressure p (atm), the density n (cm^{-3}) for two different temperatures, and the confinement time needed to satisfy Lawson's criterion as functions of the magnetic field.

pressure for which the equilibrium is described by Eq. (23.2-4), $\vec{\nabla}p = \vec{j} \times \vec{B}$. The vector on the left is orthogonal to the surfaces on which p is constant, while the vector on the right is orthogonal to both $\vec{j}$ and $\vec{B}$; hence, both the current and the magnetic field must lie on surfaces of constant pressure. However, if the plasma is to be confined, the surfaces of constant p must be both closed and nested, the magnetic surfaces (see Fig. 23.3-1). In a stationary system, moreover, both the vectors $\vec{j}$ and $\vec{B}$ are solenoidal, that is the number of lines entering any simply-connected closed curve lying on the surface must be equal to the number leaving it. This has the important consequence, intuitively obvious, and moreover demonstrable, that the surfaces containing $\vec{B}$ and $\vec{j}$, the magnetic surfaces, cannot themselves be simply connected, and the simplest containing geometry is toroidal (doughnut shaped!).[2] As is easily shown, plasma cannot be contained in a torus in which the magnetic field goes only around the doughnut the long way, $B = B_\varphi$ (a purely toroidal field); for confinement, we need in addition a field, B_θ, encircling the doughnut the short way (a poloidal field) so that the magnetic field is twisted into a helix. Now, if we imagine cutting a slice across the doughnut, then the magnetic surfaces will form closed and nested curves on this slice and, in the simplest case, those curves will have a limit point, the magnetic axis. Any given magnetic line will now be represented by a point on the curve, and after going once around the large circle, the field line will strike the slice at some other point on the magnetic surface curve. The angle between two lines drawn on the slice from the magnetic axis to the two points where a single magnetic field line hits the slice separated by a single rotation about the major circle is called the rotational transform, ι. If ι is the same for all magnetic

---

[2]L. Spitzer, "The Stellerator Concept," Physics of Fluids 1, 253-264 (1958).

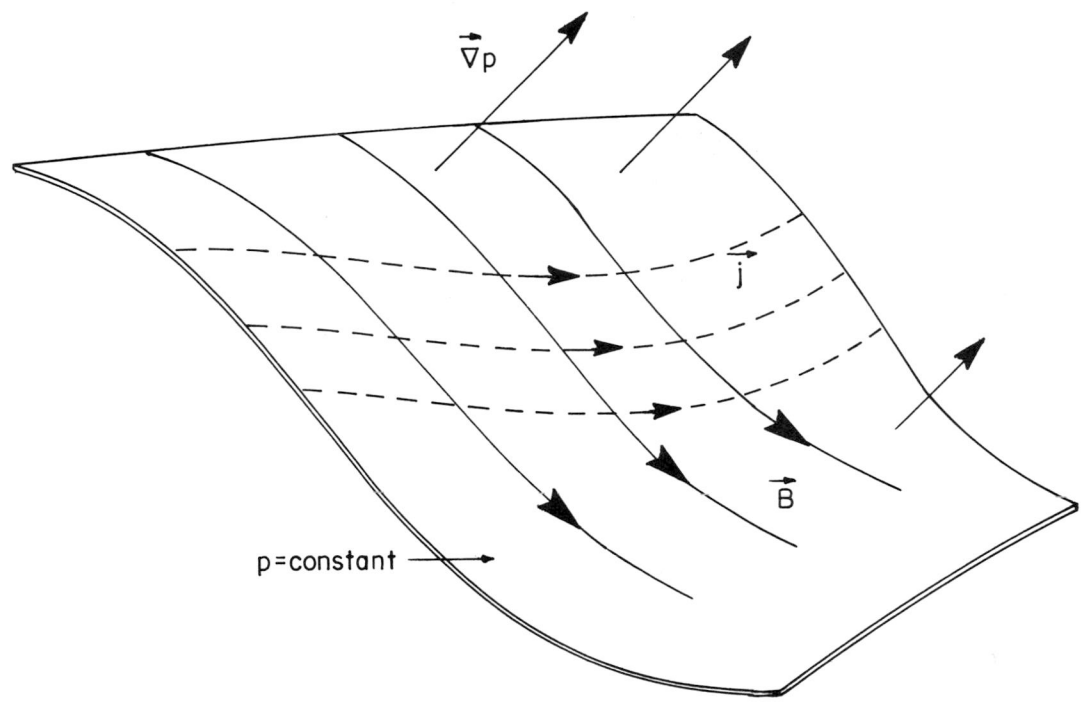

Fig. 23.3-1 A magnetic surface: a surface of constant pressure
             containing the current and magnetic field and ortho-
             gonal to $\vec{\nabla}p$. For magnetohydrodynamic confinement,
             these surfaces must be closed and nested.

surfaces, the magnetic field is shear-free, otherwise there is magnetic shear in the system. The quantity $q = 2\pi/\iota$ is the inverse winding number of the field line, the number of big circles that must be completed by a field line before it encircles the magnetic axis. In Tokamaks, q is known as the safety factor because of its bearing on stability.

There has been some interest in devices which are not closed, but are cylinders containing an axial field and an azimuthal current.[3] These are the θ-pinch machines which can confine a plasma against end losses for a time determined by the length and the thermal speed (see Fig. 23.3-2):

$$\tau \simeq L/2v_\theta \simeq 2.4 \times 10^{-6} T^{-\frac{1}{2}} L \text{ sec} \qquad (23.3-1)$$

where L is the length in m. Devices of this kind have produced plasmas with densities of order 10^{16} cm^{-3} and temperatures of about 1 kev, the plasma being heated by radial shock waves, resulting from an impulsive rise in magnetic field. A reactor might be produced in this way if the confinement time exceeded $\sim 10^{-2}$ sec, i.e., if the length exceeded 1 km. Experimental devices of length 8 m have given confinement times of about 10 μsec and appear rather well behaved, but no one plans to build a giant. There is more interest in a simple toroidal device, the Tokamak,[4,5] which is a circular torus, usually of circular cross section in which $B_\varphi \sim R^{-1}$, where R is the major radius of the torus and $B_\theta \sim \frac{1}{q}(r/R)B_\varphi$, q being the safety factor. The magnetic surfaces

[3]A. Kolb, "Magnetic Compression of Plasmas," Rev. Mod. Phys. 32, 748-756 (1960).

[4]Chapter 6 in Ref. [1] of Section 23.2.

[5]B. Coppi and J. Rem, "The Tokamak Approach in Fusion Research," Scientific American 227, 65-75 (1971).

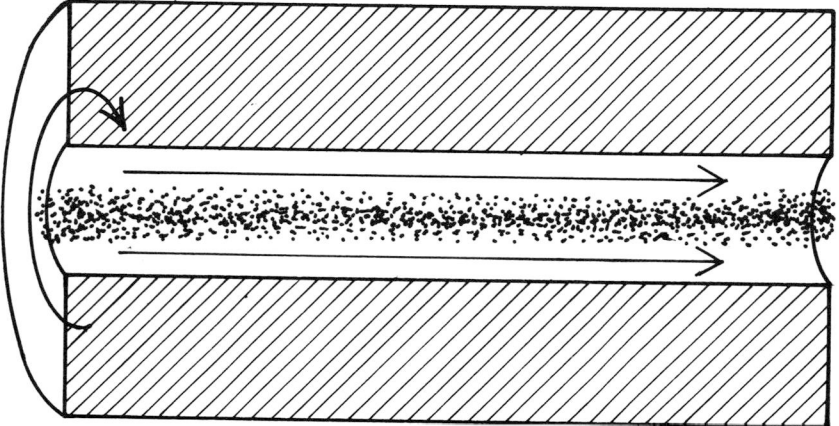

Fig. 23.3-2 In a θ-pinch, a rapidly rising current around the
cylindrical conductor produces an axial magnetic
field which compresses the plasma. The same
diagram represents the operation of the straight
Z-pinch but in this case the arrows must be rein-
terpreted; the current flowing along the axis produces an azimuthal magnetic field. Reproduced
with permission from W. C. Gough and B. J.
Eastland, "The Prospects of Fusion Power,"
Scientific American 224, 50 (1971).

can then be characterized by $\Psi = \int B_\theta R dr$ where r is measured from the magnetic axis, $B_\theta = 0$, and Ψ is the flux encircling that magnetic axis. For stability reasons, we require that q > 1. In the simplest Tokamak, B_φ is a vacuum field and the confining current flows in the φ-direction; thus confinement is entirely by the small component of the magnetic field, so that Tokamaks are necessarily low β devices. Moreover, Ohmic heating is difficult since the current density is limited by the magnitude of B_φ and the resistivity falls as $T^{-3/2}$ so that the energy deposited is proportional to $B^2 T^{-3/2}$ and the heating must decrease at high temperature. Nonetheless, the geometric simplicity of Tokamaks makes it possible both to fabricate and understand them, and they appear to be the most stable toroidal device so far invented. In experimental apparatus, temperatures of about 1 kev, with densities approaching 10^{14} cm^{-3} and confinement times of about 0.01 sec have been obtained.[6] Furthermore, the plasma behaves in a way that is reasonably close to theoretical predictions, hence Tokamaks are now the white hope of M.H.D. confinement research projects (see Fig. 23.3-3).

If the poloidal (B_θ) field is made much greater so that $(B_\theta/B_\varphi) \gtrsim 1$, the Tokamak is transformed into the toroidal Z-pinch, a favorite device in early experiments.[7] These systems must be stabilized by magnetic shear, and early attempts to construct them were defeated by resistive instabilities produced by the high current densities required. Although these devices have produced extremely hot plasmas, disruptive instabilities have kept confinement

---

[6] Ref. [3] of Section 23.2.

[7] Ref. [2] of Section 23.2, Chapter 6.

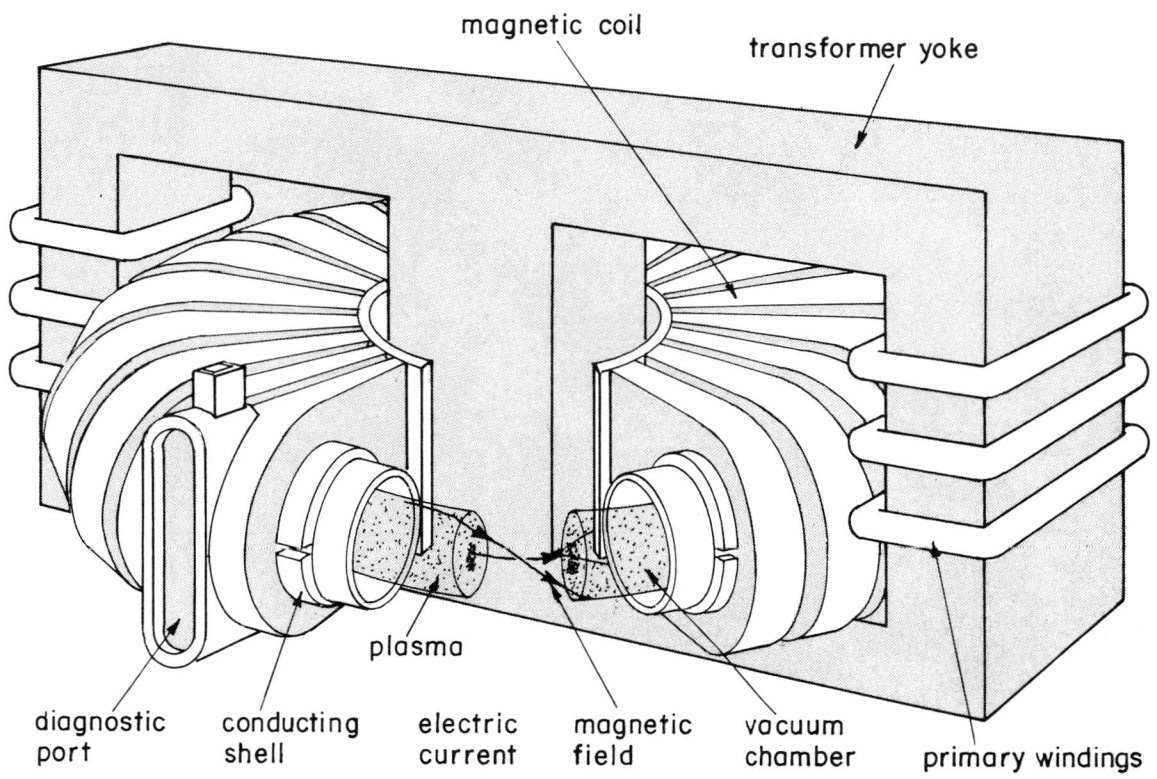

Fig. 23.3-3 In the Tokamak the conducting shell is cut in both directions. After the toroidal field has been established, a condenser bank is switched into the primary winding and a large current is induced in the plasma secondary. The geometry of a toroidal Z-pinch is exactly the same, although the ratio B_θ/B_φ is greatly increased. Reproduced with permission from B. Coppi and J. Rem, "The Tokamak Approach to Fusion," Scientific American $\underline{227}$, 65 (1972).

times short.[8] If the stability problem can be solved, the high β values possible in the Z-pinch make it an attractive system.

In the confinement systems we have described so far, the magnetic surfaces were determined by currents flowing in the plasma. Again, for stability reasons, it is sometimes desirable to have these currents flowing in rigid conductors. In the simplest of such devices, the unpinch, the magnetic axis of a pinch is replaced by a solid conductor carrying a toroidal current and generating the magnetic surfaces, the plasma being confined to a hollow torus encircling this conductor. To avoid the need for supports, the conductor may be levitated by magnetic fields, whereupon the device is called a levitron (see Fig. 23.3-4).[9]

There is enhanced stability if the single conducting core is replaced by a number carrying parallel currents as in the multipole. These systems do not seem promising as potential reactors, but have proved useful in demonstrating stable confinement (see Fig. 23.3-5).[10]

Another important device is the stellerator,[2] in which the flux surfaces are produced by external windings. This is not possible with axial symmetry but, if the external windings are helical, closed flux surfaces can be produced. Consider a cylindrical case for simplicity; the helical fields are then

---

[8]D. C. Robinson, J. E. Crow, C. W. Gowels, G. F. Malesso, A. A. Newton, A. J. L. Varhage, and H. A. B. Bodin in Fifth European Conference on Controlled Fusion and Plasma Physics, Association EURATOM-CEA, Grenoble, France, 2, 47-57 (1972).

[9]Ref. [1] in Section 23.2, Chapter 6.

[10]T. Ohkawa, J. R. Gilleland, T. Tamano, T. Takeda, and D. K. Bhadra, "Collisional Diffusion in the d.c. Octopole," Physical Review Letters 27, 1179-1181 (1971).

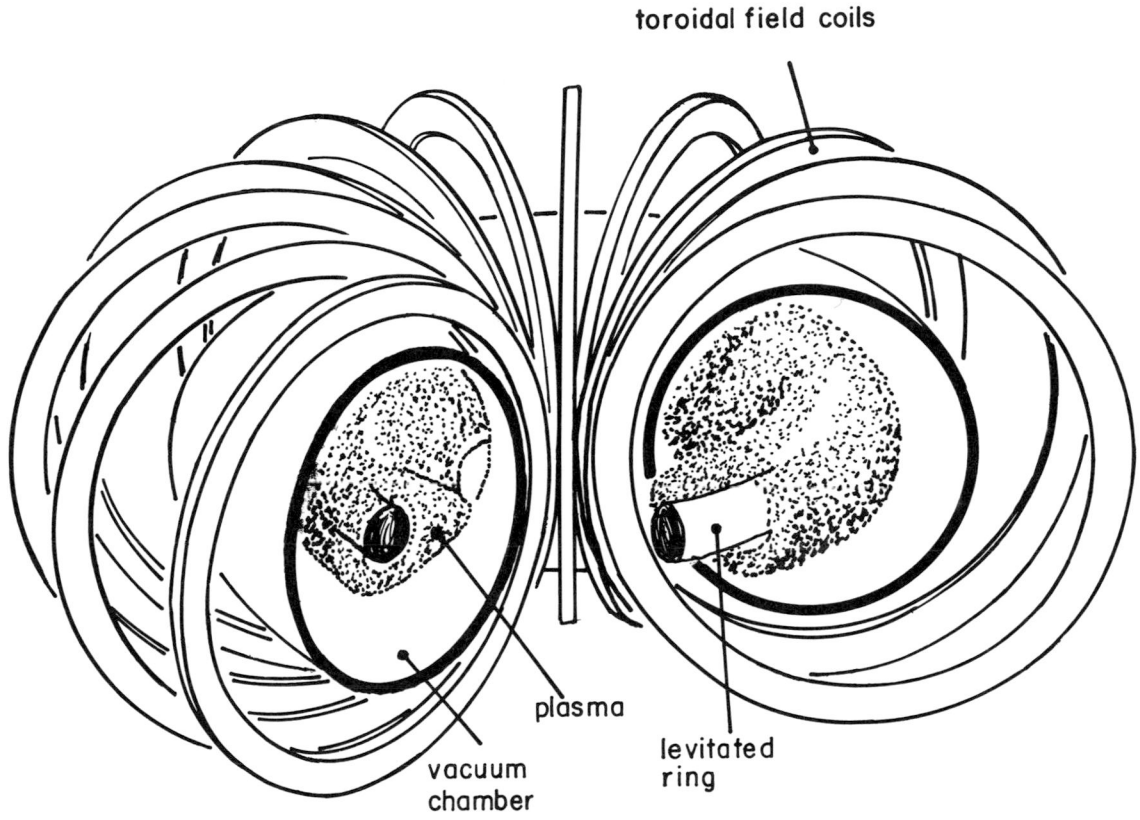

toroidal field coils

plasma

vacuum
chamber

levitated
ring

Fig. 23.3-4 Operating principle of the levitron: for enhanced
stability, a large fraction of the current is carried
by the rigid levitated ring, which is itself supported
by magnetic fields. Reproduced by courtesy of the
Director from the Culham Laboratory Annual Report
for 1972, United Kingdom Atomic Energy Authority,
London, 1972.

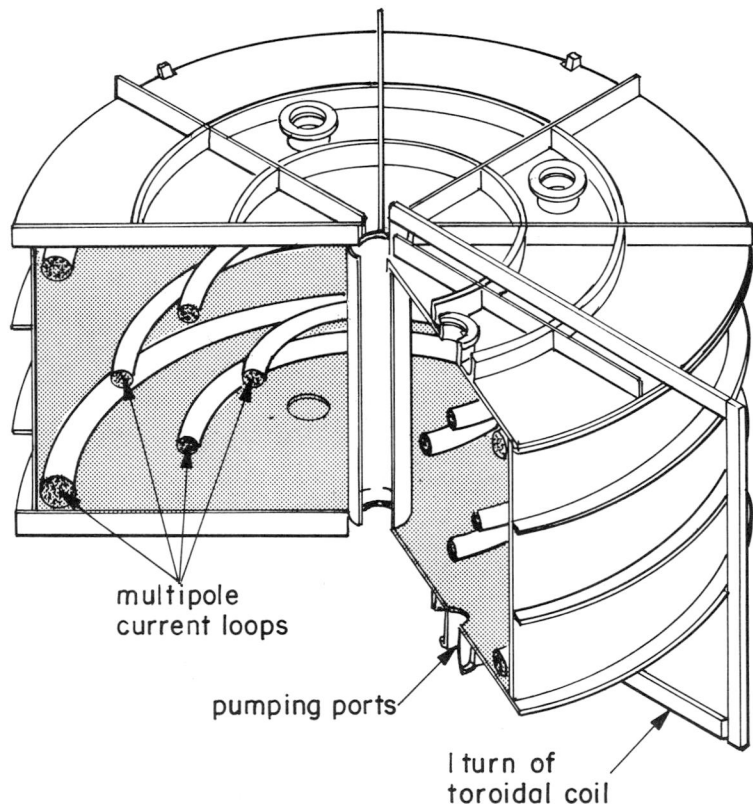

multipole
current loops

pumping ports

I turn of
toroidal coil

Fig. 23.3-5 Operating principle of the multipole: currents carried
in the internal rings produce a stable magnetic well.
Plasma losses are enhanced by coil supports. Repro-
duced by permission of the author from the Gulf General
Atomic CTR Summary Report for 1971, Gulf General
Atomic Corp., San Diego, Ca., 1971.

$$B_Z = B_0[1 + \varepsilon r^{\ell} \cos \ell(\theta - \alpha z)],$$

$$B_r = B_0 \ell \, \varepsilon r^{\ell-1} \sin \ell(\theta - \alpha z),$$ (23.3-2)

$$B_\theta = B_0 \ell \, \varepsilon r^{\ell-1} \cos \ell(\theta - \alpha z),$$

and the flux surfaces $\Psi = \frac{1}{2} r^2 + \varepsilon r^{\ell} \cos \ell(\theta - \alpha z)$, which are helicoids, closed and nested even in the absence of plasma. The addition of small amounts of plasma produces small corrections to the magnetic fields. Stellerators are difficult to make, but successful low β confining experiments have been performed in them (see Fig. 23.3-6).[11]

B. Collisionless or Mirror Confinement

There is a second class of confining geometries which escapes the limitations described above, the magnetic mirrors.[12] Instead of confining particles for a time of order $\tau_{coll}(L/r_L)^2$, their confinement is only for a time of order τ_{coll}, which at temperatures of 5 kev, the threshold temperature, yields a confinement time too small to satisfy Lawson's criterion. The Rutherford scattering cross section, however, is

[11] D. J. Lees, W. Millar, R. A. E. Bolton, G. Cattanei, and P. L. Pilante, "Confinement and R. F. Heating in the Proto-Cleo Stellerator" in Ref. [8], pp. 135-146.

[12] R. F. Post, "Summary of UCRL Pyrotron Program," Second United Nations Conference on the Peaceful Uses of Atomic Energy, 32, 245-265 (1958).

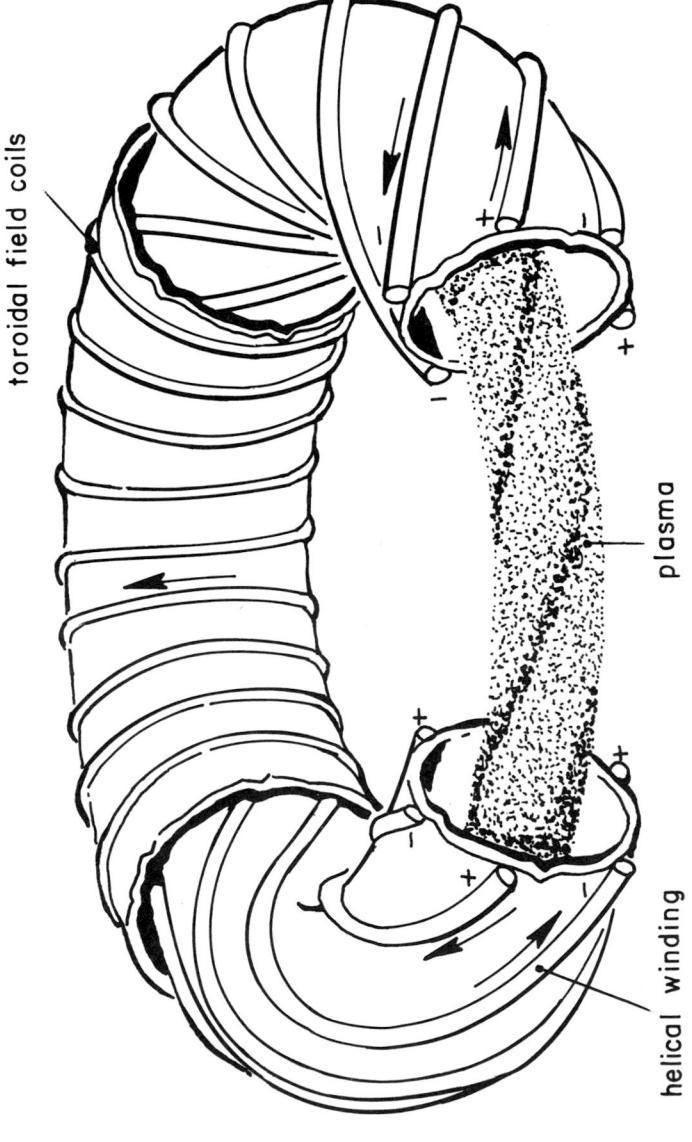

Fig. 23.3-6 Operating principle of the stellerator: the helical windings induce a rotational transform in the magnetic field converting the purely toroidal field into a confining geometry. Reproduced by permission from the <u>Culham Laboratory Annual Report for 1972</u>, U.K.A.E.A., London.

$$\sigma_R \simeq 3 \times 10^{-18} \, E^{-2} \tag{23.3-3}$$

and, at $E = 200$ kev, $\sigma_R \approx 7.5 \times 10^{-23}$. At this energy, the DT cross section has the value $\sigma_{DT} \simeq 3 \times 10^{-24}$, and the probability of a particle making a nuclear reaction before being scattered out of the system is $\sim 1/20$. The nuclear energy released, however, is 17 Mev; hence the gain in energy over the thermal energy of the particle is ~ 40 and the ratio of nuclear energy released to energy supplied to make up the losses is about $R \simeq 2$. Therefore, at high temperature, Lawson's criterion is satisfied. Thus, if energy can be circulated through the system with high efficiency, a magnetic mirror device, in spite of its short containment time, could act as a fusion reactor.

The nature of collisionless confinement must be found by examining the motion of particles in magnetic fields. As we have seen in Section 23.1L, when a particle moves in a uniform field, motion along the magnetic field is undisturbed but, in response to the Lorentz force, particles encircle the magnetic field at a frequency $\omega_c = eB/(mc)$, so that the circles have a radius $r_L = v_\perp/\omega_c$. Moreover, it is possible to show that for slowly varying magnetic fields the quantity

$$\mu = \tfrac{1}{2} m v_\perp^2 B^{-1} \tag{23.3-4}$$

is a constant of motion, an adiabatic invariant.

Suppose now the field is nonuniform; then the motion along the magnetic field is also constrained for, since the energy $E = mv^2/2$ is a second constant of motion, the parallel component of the velocity becomes a function of position,

$$\tfrac{1}{2}mv_{\parallel}^2 = E - \tfrac{1}{2}mv_{\perp}^2 = E - \mu B(x). \tag{23.3-5}$$

Suppose, on a field line, B has a minimum B_m at some point x_m, increasing to some maximum value B_M. The parallel velocity will then vanish at any point x_0 where $E - \mu B(x_0) = 0$ and the particle will be reflected. Clearly, not all particles will be so reflected, for if μ is zero, $v_{\parallel}$ does not depend on position. We can get at the condition for escape by writing μ in terms of B_m and $\tfrac{1}{2}mv_{\perp}^2(0)$ at B_m, the value of $\tfrac{1}{2}mv_{\perp}^2$ at $B = B_m$. Then $\tfrac{1}{2}mv_{\parallel}^2 = E - \tfrac{1}{2}mv_{\perp}^2(0)$ $\times B(x)/B_m$ and particles will be reflected provided

$$E < \tfrac{1}{2}mv_{\perp}^2(0)B_M/B_m,$$

$$1 < \tfrac{1}{2}[mv_{\perp}^2(0)/E]B_M/B_m = (\sin \theta_0)^2 R_M, \tag{23.3-6}$$

where θ_0 is the angle between the particle trajectory and the magnetic field at the field minimum and R_M, defined as B_M/B_m, is the mirror ratio. This is the basis of magnetic mirror confinement. It differs from toroidal or magnetohydrodynamic confinement in two ways: the magnetic field geometry is not so constrained as in the MHD case since field lines can leave the plasma, but the particles are not in thermal equilibrium (see Fig. 23.3-7).

The earliest mirror traps had axial symmetry, the magnetic field being produced by a pair of Helmholtz coils. These mirrors could be filled with plasma in one of several ways and showed evidence for containment, although for times much shorter than collision times. The anomalous loss was associated with instabilities in the confined plasma and to avoid these it has proved necessary to go to more complex geometries.

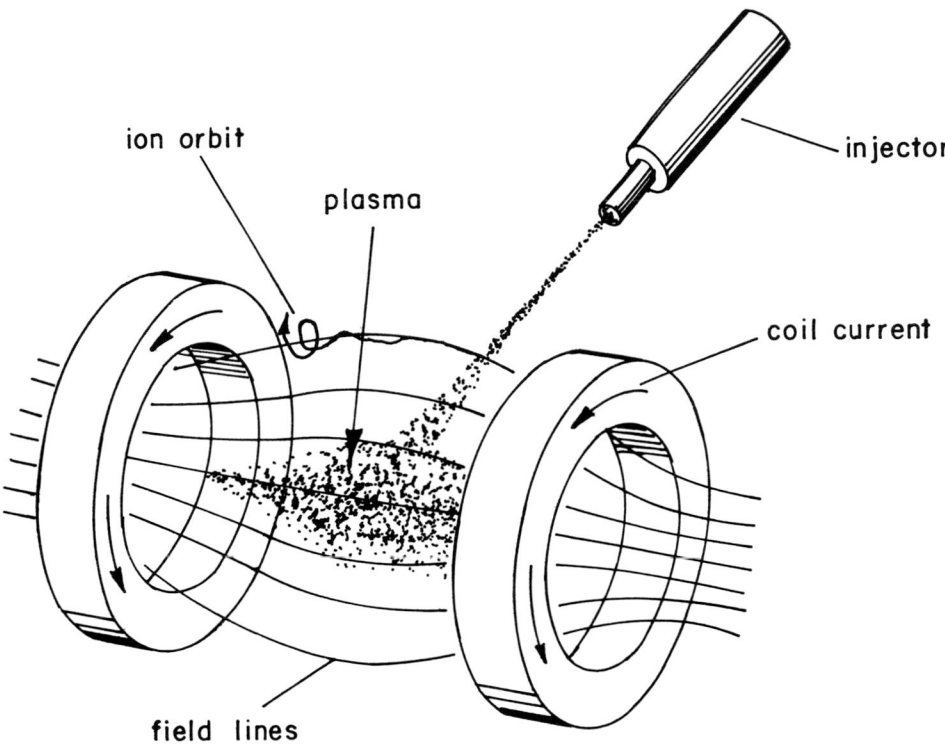

Fig. 23.3-7 Operating principle of mirror confinement. In
              this diagram, a neutral particle injector is shown,
              which is not an essential feature of mirror con-
              finement. Reproduced with permission by the
              author from F. L. Ribe, "Fusion Reactor Systems,"
              Reviews of Modern Physics 47, 7 (1975).

To study more complex geometries, we must look further into the motion of the particles.[14, 15] In a mirror trap, the motion of the particle along the field line, as well as its motion about the field line, is periodic, with a period given by

$$\tau_B = 2 \int_{x_1}^{x_2} v_{||}^{-1} d\ell = 2 \int_{x_1}^{x_2} d\ell [2(E - \mu B)/m]^{-\frac{1}{2}}, \qquad (23.3-7)$$

x_1, x_2 being the turning points where $E - \mu B = 0$. The associated frequency is the "bounce frequency"

$$\omega_B = 2\pi/\tau_B. \qquad (23.3-8)$$

If the field geometry changes slowly as measured on the τ_B scale, then there is an additional adiabatic invariant, J, associated with the periodic longitudinal motion,

$$J = \oint v_{||} d\ell = \oint [2(E - \mu B)/m]^{\frac{1}{2}} d\ell. \qquad (23.3-9)$$

---

[14]Ref. [1], Chapter 7.

[15]M. D. Kruskal, "Elementary Orbit and Drift Theory" in Plasma Physics, pp. 67-90, International Atomic Energy Agency, Vienna, Austria, 1965.

Frequently there may be an electrostatic potential Φ also present, whereupon

$$J = \oint [2(E - \mu B - e\Phi)/m]^{\frac{1}{2}} d\ell.$$

Suppose now that, in some way, the particle is slowly displaced from the field line upon which it travelled by an amount $\xi(\ell)$. This slow variation does not alter J, nor μ, but since B and Φ are functions of position, they will be altered and since, in general, work will be done in displacing the particle, E will be changed. Moreover, the length of the new field line will not be the same as the original one although, to lowest order, the turning points stay fixed. Then

$$0 = \delta J = \frac{1}{m} \oint d\ell \frac{(\delta E - \mu \delta B - e\delta\Phi)}{[2(E - \mu B - e\Phi)/m]^{\frac{1}{2}}} + m[\frac{2}{m}(E - \mu B - e\Phi)]^{\frac{1}{2}} d\ell'/d\ell$$

and the variation in energy becomes

$$\delta E \times \oint \frac{d\ell}{[2(E - \mu B - e\Phi)/m]^{\frac{1}{2}}} = \oint d\ell \frac{[\mu \delta B + e\delta\Phi - 2(E - \mu B - e\Phi)d\ell'/d\ell]}{[2(E - \mu B - e\Phi)/m]^{\frac{1}{2}}}.$$

The integral on the left is just the bounce period, while on the right $d\ell/[2(E - \mu B - e\Phi)/m]^{\frac{1}{2}} = d\ell/v_{\parallel} = dt$; hence δE_J is given by a time average,

$$\delta E_J = \langle \mu \delta B + e\delta\Phi - 2(E - \mu B - e\Phi)d\ell'/d\ell \rangle. \tag{23.3-10}$$

Now, on a displacement, $\vec{\xi}$, with fixed fields B, Φ, $\delta B = \vec{\xi} \cdot \vec{\nabla} B$, $\delta \Phi = \vec{\xi} \cdot \vec{\nabla} \Phi$, $d\ell'/d\ell = -\vec{\xi} \cdot \vec{n}/R$, where $\vec{n}$ is the inward drawn normal to the field lines and R their radius of curvature. δE_J will be zero if $\vec{\xi}$ satisfies the condition $\vec{\xi} \cdot [\frac{1}{2}v_\perp^2 (\vec{\nabla} B)/B + v_\parallel^2 \vec{n}/R + \frac{e}{m} \vec{\nabla} \Phi] = 0$ or if $\vec{\xi}$ is in the direction

$$\vec{b} \times [\frac{1}{2}v_\perp^2 (\vec{\nabla} B)/B + v_\parallel^2 \vec{n}/R + \frac{e}{m} \vec{\nabla} \Phi] = 0. \tag{23.3-11}$$

Thus, motions of the particle will preserve E, J, μ provided they lie in the surface specified by this vector and the magnetic field line, the drift surfaces. Note that not only can such motion occur, but it does, for, in uniform electric and magnetic fields, particles undergo the electric drift

$$\vec{V}_E = e(\vec{E} \times \vec{b}/B) = (e/m)(\vec{b} \times \vec{\nabla} \Phi)/\Omega,$$

$\vec{b}$ being the unit vector along $\vec{B}$; hence, in a nonuniform magnetic field, the drift velocity becomes

$$\vec{v}_D = \vec{b} \times [\frac{1}{2}v_\perp^2 \vec{\nabla} \log B + v_\parallel^2 \vec{n}/R]/\Omega + \vec{E} \times \vec{b}c/B. \tag{23.3-12}$$

We can now discuss possible magnetic equilibria, at least in some special simple cases.[16] Suppose that there is no electric field and that the pressure is extremely small so that $0 = \vec{B} \times \vec{j} = (4\pi)^{-1} \vec{B} \times (\vec{\nabla} \times \vec{B})$. Writing $\vec{B} = B\vec{b}$, where $\vec{b}$ is a unit vector, $0 = \vec{b} \times (\vec{\nabla} \times B\vec{b}) = (\vec{\nabla} \cdot \vec{B})\vec{b} - [\vec{\nabla}(B\vec{b})] \cdot \vec{b}$ and, since $\vec{\nabla} \cdot \vec{B} = 0$, $0 = -\vec{\nabla}_\perp B - \vec{b} \cdot \vec{\nabla} \vec{b} B$; but $(\vec{b} \cdot \vec{\nabla})\vec{b} = -\vec{n}/R$ and therefore

---

[16] J. B. Taylor, "Plasma Confinement in Magnetic Wells," pp. 449-480 in Ref. [15].

$\vec{\nabla} \log B = \vec{n}/R$.   Thus,   if the pressure is low, the drift surface becomes

$$\vec{b} \times (v_{\parallel}^2 + \tfrac{1}{2}v_{\perp}^2)\vec{n}/R \qquad\qquad (23.3\text{-}13)$$

and depends only on field geometry. For the simple mirror field produced by Helmholtz coils, $\vec{n}$ is directed away from the axis of symmetry (along r) while $\vec{b}$ lies in the v, z plane; hence the drift surfaces encircle the axis and are closed within the plasma.

More complicated geometries can also be stable if the drift surfaces close. Observe that, if the magnetic field geometry is given, E is a function of J and μ on any given field line. Thus, if it is required of a distribution function f that it be a function of all constants of motion, $f(E,\mu,J)$, we find that f must be constant on the drift surfaces specified by these three variables; hence, if the drift surfaces are closed and remain within the system, for some range of J and μ, particles will be trapped (see Fig. 23.3-8). The problem of determining mirror equilibria now reduces to that of finding magnetic fields such that the countours in the α, β plane (α, β being coordinates perpendicular to the magnetic field) determined by requiring that E, J, and μ be simultaneously constant should be closed and lie within the containing vessel. Solving this is a fairly formidable task, especially if an electrostatic potential appears, which must be included self consistently and has used up a good deal of computer time; however, the advantage of these special equilibria is that stability conditions are easily applied. Indeed, the plasma is stable for most important modes if $\partial f/\partial E < 0$. To find what this means for the spatial confinement of plasma, note that the spatial variation of f is $\partial f/\partial \alpha = (\partial E/\partial \alpha)(\partial f/\partial E)$, and if $\partial f/\partial E < 0$ and $\partial E/\partial \alpha > 0$, the density of particle having

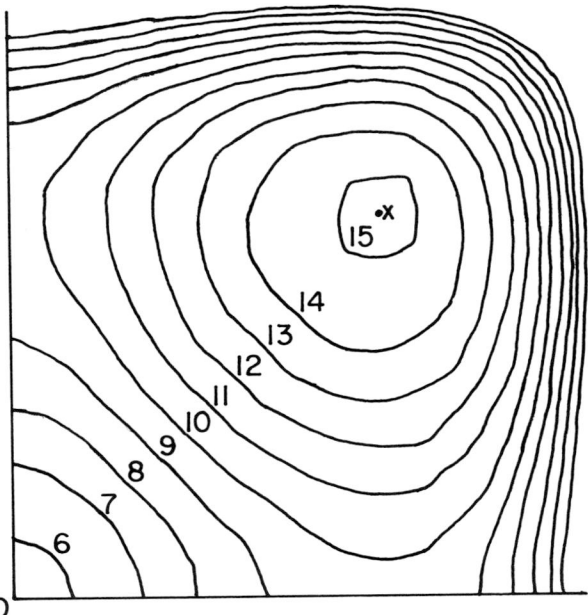

Fig. 23.3-8   Computed surfaces of constant J, μ, E on the
α, β plane. The contours are labelled by the
values of E in arbitrary units, showing the production of a magnetic well. Reproduced from
Ref. [16] of Section 23.3 by permission of the
author.

given E, μ, and J decreases outward. However, at low β, $\delta E = \langle (\frac{1}{2}mv_\perp^2 + mv_\parallel^2)$ $\times \vec{n} \cdot \vec{\xi}/R \rangle$ and this will be positive for outward displacements only if the time average of $\vec{n}$ along the particle orbits lies outside of the plasma.

23.4 The Stability of Magnetically Confined Plasmas

We have now described some of the ways in which a plasma may be placed in equilibrium. This by itself, however, is not enough to ensure that the plasma remains in equilibrium; to persist, the equilibrium must be stable, while, if it is unstable, it may destroy itself in times of the order of a few microseconds.

As an illustration, consider a bar pivoted at one end; this has two equilibrium positions, one with the bar pointing up and one with the bar pointing down, but only the latter persists for any length of time. The underlying dynamical reason for this result lies in the configurational dependence of the energy. If the position of the bar is characterized by the angle θ between the downward-drawn normal and the bar, the potential energy can be written as $V(\theta) = -Mg \times \ell \cos \theta$, where M is the mass and ℓ the distance from the pivot to the center of mass of the bar. Now $V' = Mg\ell \sin \theta$ vanishes at the two positions of equilibrium $(\theta = 0, \theta = \pi)$ but $V'' = Mg\ell \cos \theta$ is positive at $\theta = 0$ and negative at $\theta = \pi$; thus, if the pendulum is displaced by $\delta\theta$, it experiences a restoring stabilizing force, $-V'\delta\theta = -Mg\ell\,\delta\theta$ at $\theta = 0$, and a destabilizing force, $-V''\delta\theta = Mg\ell\,\delta\theta$ at $\theta = \pi$ (see Fig. 23.4-1).

For a plasma to be stable, a similar condition must be satisfied; any change in the plasma conditions must result in an increase in the potential energy of the configuration.

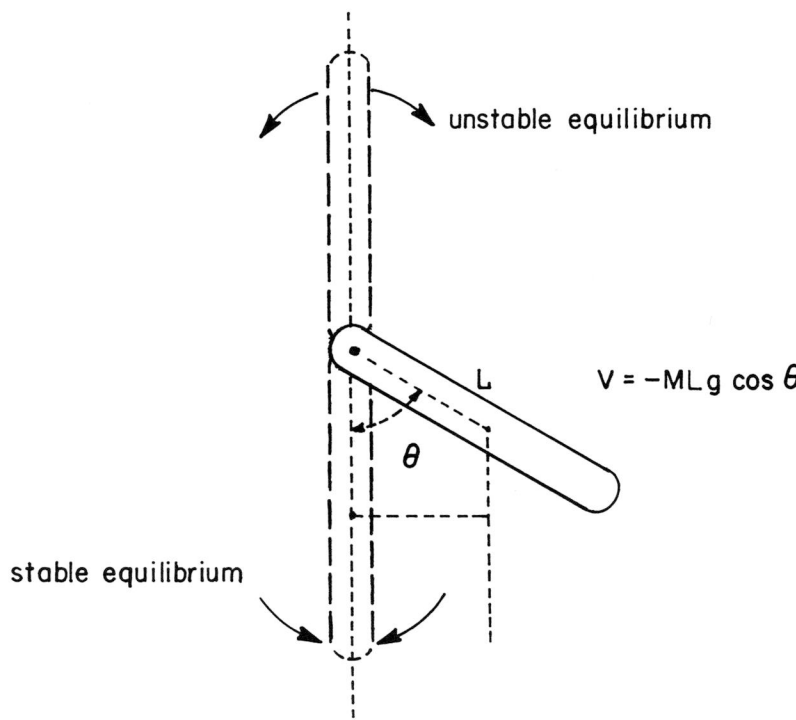

unstable equilibrium

$V = -MLg \cos \theta$

L

θ

stable equilibrium

Fig. 23.4-1 The principle of stability. The potential energy
$V(\theta) = -MLg \cos \theta$ and the torque, $\partial V/\partial \theta$, vanish
when the pendulum is either hanging down or
balanced upright, but only the former, the potential minimum, is stable.

A. Sources of Destabilizing Energy

We consider now the sources of energy available to a plasma. In evaluating these, we can compare the energy of the plasma configuration with that of some configuration known to be stable. If we imagine the plasma as confined by an impervious wall, it is clear that a stable equilibrium is obtained when the plasma is in thermal equilibrium, i.e., uniform, isothermal, and carrying no current. If by some displacement $\vec{\xi}(\vec{x})$ the plasma can move toward this state [$\vec{\xi}(\vec{x})$ being a vector displacement experienced by the element of the plasma at $\vec{x}$], energy can be released and become available to cause the displacement.

(a) Magnetic Energy

The first source of energy lies in the magnetic field which has energy content $E_B = (1/8\pi)\int B^2 d\tau$, the volume integral extending over the whole system. It is well known, although perhaps not immediately obvious, that a vacuum field is a state of minimum energy so, as we might anticipate, a minimum energy state appears if the plasma supports no current. If the plasma is to be confined, $\vec{B}$ cannot be a vacuum field since currents must flow to balance the plasma pressure gradient. If $\vec{\delta B}$ represents the contribution of these non-vacuum fields assumed much less than the vacuum field B_0, however, the pressure balance becomes $\vec{\nabla}p \simeq (\vec{\nabla} \times \vec{\delta B}) \times \vec{B}/4\pi$ and $\delta B/B \simeq 4\pi p/B_0^2 \simeq \beta$; for low enough β, the fractional energy in the non-vacuum part of the field is small. For large β, however, magnetic instabilities are important.

In Tokomak devices, magnetic energy can drive instabilities if q is too small, and it is the free magnetic energy released by "kinking" the plasma that limits q.

We can get an estimate of the time scales of magnetic instabilities by the following argument. If magnetic energy is to be tapped, the energy released by a displacement $\xi \simeq \xi B^2/L$, where L is some scale length of the system. The disturbed force, which is the gradient of this displacement, is then $\sim \xi B^2/L^2$ and the equation of motion becomes $\rho_0 \ddot{\xi} \simeq (B^2/L^2)\xi$. Therefore, the time scale becomes $\tau_B \simeq L/C_A$, where C_A is the Alfvén speed $C_A^2 = B^2/4\pi\rho$. In low β systems, the available energy is only of order βB^2; hence, C_A must be replaced by $(p/B^2)C_A^2 \simeq C_S^2$, C_S being the speed of sound, $C_S^2 \simeq p/\rho$; also, the magnetic time scale is replaced by the hydrodynamic scale, $\tau_H = C_S/L$. At the temperature and density of a fusion plasma, $C_S \simeq 3 \times 10^7$ cm/sec, and the characteristic time, τ_H, is of order a few microseconds. Although this is a rapid process, its characteristic frequency is well below the ion cyclotron frequency since $1/\omega_c^i \tau \simeq C_S/\omega_c^i L \simeq r_L^i/L$ and r_L^i/L must be small.

(b) Thermal Energy

It may be that the magnetic energy is already at a minimum, but that some displacement allows the plasma to expand by a small amount δV, in which case work, $p\delta V$, can be done at the expense of the thermal energy, $\int p/(\gamma - 1) \times d\tau$. The same argument as was used in the magnetic case again shows that characteristic frequencies are $\omega_H \simeq C_S/L$.

(c) Energy in the Distribution Function

If we fix the macroscopic parameters of a distribution function $f(x, v)$, we know that the minimum energy is reached when the system gets close to the thermal equilibrium distribution f_0 and, in relaxing to this, the plasma can release energy $\delta E \simeq \int E(f - f_0)d^3v\, d\tau$.

Relaxation to thermal equilibrium, however, will take a time determined by the collision frequency; but there may be much more rapid instabilities, which take the distribution not to thermal equilibrium but to some distribution f_s known to be stable. The available energy is then $\delta E \simeq \int E(f - f_s)d^3v d\tau$. In many cases, it is sufficient that f_s be a monotone decreasing function of the energy.[1] Frequently, the energy that is released comes from a small part of the particle distribution and is not large enough to disrupt the plasma. Such microinstabilities do, however, have an effect on the transport properties of the plasma such as the cross-field diffusion and energy transport and hence may determine containment times.

B.   Magnetohydrodynamic Instabilities

We consider now the effect of free energy of the first two classes, the macroscopic degrees of freedom, on plasma stability.[2] It is not enough to have energy available, there must be a possible motion by which the lower energy state can be reached and this motion is subject to certain constraints.

(a)   Ohm's Law

We consider first the relation between plasma displacement and the perturbation in the magnetic field. For arbitrary disturbances, this is a diffi-cult task but, for a large scale, low-frequency disturbance, one can make use of the generalized form of Ohm's law, which itself can be obtained by multiply-ing the separate fluid equations for electrons and ions by -e/m and e/M, res-pectively, and adding. If the electron drift velocity and the macroscopic accel-eration are both small, as is frequently the case,

---

[1]C. S. Gardner, "Bound on the Energy Available from a Plasma," The Physics of Fluids 6, 839-840 (1963).

[2]Chapter 6 in Ref. [1] of Section 23. 3.

$$\widetilde{\eta}\vec{j} + \frac{mc}{ne^2}\frac{\partial\vec{j}}{\partial t} + \frac{1}{ne}(\vec{\nabla}p^e - \vec{j} \times \vec{B}) = \vec{E} + \frac{\vec{V}}{c} \times \vec{B}, \qquad (23.4\text{-}1)$$

where p^e is the electron pressure and $\vec{V}$ the center of mass motion of the combined fluids. If the conductivity were perfect (the idealized case), the terms on the left would vanish. The first of these is the ohmic resistance ($\widetilde{\eta}$ being the resistivity, which is given in terms of the collision frequency by $\widetilde{\eta} = \nu mc/e^2$). The second term represents the effect of electron inertia and is negligible at low frequencies. The third term represents the Hall effect in the plasma and may be written as $\vec{\nabla}p^i/ne$, p^i being the ion pressure.

To use Ohm's law as a method of determining the magnetic field, we employ Faraday's law of induction,

$$\vec{\nabla} \times \vec{E} = -(1/c)(\partial\vec{B}/\partial t), \qquad (23.4\text{-}2)$$

to eliminate $\vec{E}$, whereupon the right-hand side becomes

$$-[\frac{\partial\vec{B}}{\partial t} + (\vec{V} \cdot \vec{\nabla})\vec{B} + \vec{B}(\vec{\nabla} \cdot \vec{V}) - (\vec{B} \cdot \vec{\nabla})\vec{V}]. \qquad (23.4\text{-}3)$$

The left-hand side becomes

$$\vec{\nabla} \times (\widetilde{\eta}c\,\vec{j}) + \vec{\nabla} \times \frac{mc^2}{me^2}\frac{\partial\vec{j}}{\partial t} + \vec{\nabla} \times \frac{e}{ne}\vec{\nabla}p^i,$$

where $\widetilde{\eta}c = \dfrac{\nu mc^2}{ne^2} = \eta$ is the electromagnetic conductivity. To get orders of the various terms, let us replace derivatives by length scales so that $\vec{\nabla} \times \vec{j}$ $\simeq \nabla^2 \vec{B} \sim B/L^2$. We may then approximate Eq. (23.4-1) as $\frac{\partial B}{\partial t} (1 + \ell_0^2/L^2)$ $+ VB/L + \eta B/L^2 + v_\theta \cdot \frac{mv_\theta c}{L^2 eB} \cdot B \simeq 0$. The quantity ℓ_0 here is the collisionless screening length $\ell_0^2 = (ne^2/mc^2)^{-1} = (3.6 \times 10^{11})/n$, which for a fusion plasma is $\ll 0.1$ cm; for $L > 10$ cm, the ℓ_0^2/L^2 factor becomes negligible. The remaining terms are of order $1 : \eta/VL : (v_\theta/V)(v_L/L)$. The number $\mathcal{L} = VL/\eta$ is called the Lundquist number and plays the same role in plasma dynamics as does the Reynolds number in the flow of fluids. For fusion plasmas, the high temperature makes the resistivity small, $\eta \simeq 10^{-3} T^{-3/2}$ ohm-cm, which, in rationalized units, becomes $\eta = 10^4 T^{-3/2}$ cm^2 sec, where T is in kv. If V scales as v_θ^i, the Lundquist number becomes $10^3 \times T^2$ LM, where M is the Mach number, V/v_θ; hence, for motions on the hydrodynamic scale, $\mathcal{L}$ is very large. The last term is of order $(1/M)(r_L/LM)$ and, unless M is small and of order r_L/L, this too is negligible. When these conditions are satisfied, the time development of the magnetic field is determined by the vanishing of the expression given in Eq. (23.4-3) or, equivalently, by the vanishing of E^*,

$$\vec{E}^* = \vec{E} + \vec{V} \times \vec{B}/c = 0. \qquad (23.4-4)$$

This expression, however, is the force acting on an electric charge moving with the local velocity of the fluid, i.e., the electric field in the fluid. If we consider now the magnetic flux, Φ, threading a closed loop sharing the motion of the fluid, we find from Faraday's law

$$d\Phi/dt = d/dt \int \vec{B} \cdot \vec{n} \, ds = c \oint \vec{E}^* \cdot d\vec{\ell};$$

hence motions on the hydrodynamic scale, the fastest growing instabilities, are subject to the constraint that in their motions the magnetic flux must be conserved.[3]

(b) Interchange Instabilities

The conservation of flux enables us to develop a criterion for the stability of the plasma against certain types of motions. At low β, energy can be released only by the expansion of the plasma but, since flux must be conserved, this can only be affected by the interchange of the fluid on equivalent lines of force. Now suppose two such flux tubes labelled 1, 2 with pressure p_1, p_2 and volume V_1, V_2 are interchanged. Then the work done by the fluid initially in 1 is approximately $p_1(V_2 - V_1)$ and that done by the fluid in 2 is $p_2(V_1 - V_2)$ so that the net energy release is $\delta W = - (p_2 - p_1)(V_2 - V_1) = - \delta p\, \delta V$. The volume of a length of flux tube is clearly $V = A \int d\ell$, the flux is $\Phi = BA$, and $V = \int \Phi\, d\ell/B$, where the integral extends along the flux tube (see Fig. 23.4-2).

If the equilibrium is characterized by a closed and nested set of flux surfaces Ψ, $\delta p = \delta\Psi(\partial p/\partial\Psi)$ and $\delta V = \delta\Psi(\partial^2 V/\partial\Psi^2)$; therefore, the plasma is stable provided

$$\frac{\partial p}{\partial\Psi}\, \frac{\partial}{\partial\Psi}\left(\int \frac{d\ell}{B}\right) > 0. \qquad\qquad (23.4\text{-}5)$$

If the equilibrium is characterized also by the total volume enclosed by a flux surface Ψ, then $V = V(\Psi)$ and the criterion can clearly be written as

---

[3]I. B. Bernstein, E. A. Frieman, M. D. Kruskal, and R. M. Kulsrad, "An Energy Principle for Hydromagnetic Stability Problems," Proceedings of the Royal Society A 244, 17-40 (1958).

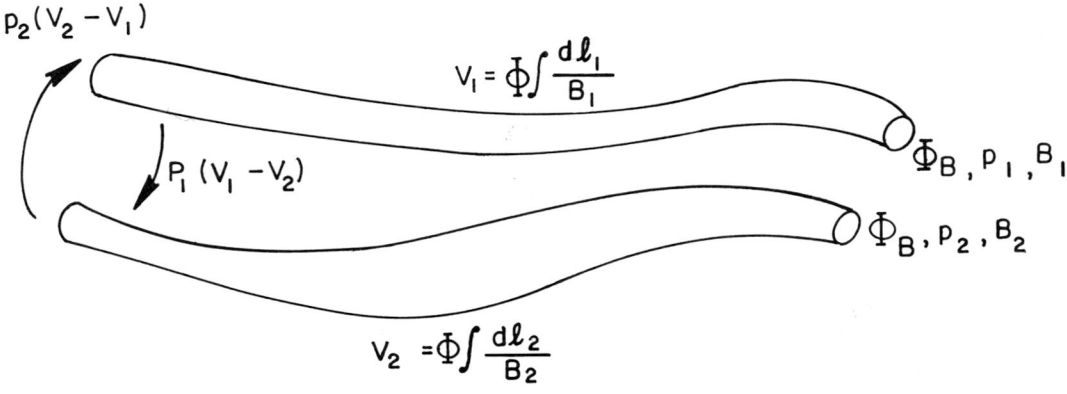

Fig. 23.4-2   Interchange modes.   The volumes shown are defined by surfaces enclosing equal fluxes Φ_B.   On
interchange, the work done is approximately
$(p_2 - p_1)(v_2 - v_1)$.

$$\frac{dp}{d\Psi} \frac{d^2V}{d\Psi^2} > 0. \tag{23.4-6}$$

For most simple containment geometries, the simple pinch or simple mirror, this criterion is violated, since the center of curvature of the magnetic field lies inside the plasma. An important simple geometry for which the interchange criterion is satisfied is the magnetic cusp, in which the field is again produced by two Helmholtz coils but now the currents are opposed, so that, instead of joining on the center line, the field lines are forced out and the center of curvature lies outside the plasma. This, however, is not strictly a containing geometry since there are lines around the center and at the ends where plasma can leak out. The magnitude of these leaks is such that only for a very large system could a cusp form a reactor (see Fig. 23.4-3).[4]

A geometry which combined the stability of the cusp with the adiabatic confinement of the mirror was invented by Joffe[5] in 1960 and gave the first evidence of plasma confinement for times much greater than the hydrodynamic time. In this device, the mirror coils providing the main confining field were supplemented with four conductors parallel to the main field carrying currents in alternating directions, to produce a secondary field with a cusp-like configuration.

In the earliest experiments, the confined plasma volume was small and, although the temperature was modest, the magnetohydrodynamic time was a

---

[4] I. Spalding, "Cusp Containment," Advances in Plasma Physics 4, 79-120 (1971).

[5] M. S. Joffe, "Mirror Traps" in Ref. [15] of Section 23.2, pp. 421-448.

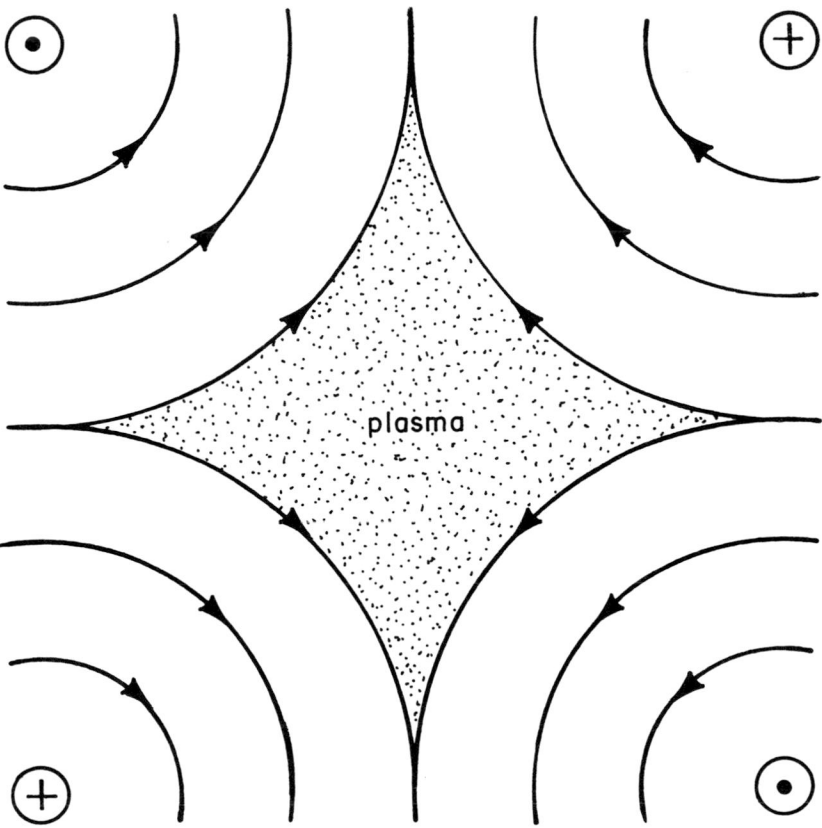

Fig. 23.4-3   Principle of cusp geometry: the physical geometry is produced by rotating the figure about
the horizontal axis. Particle loss occurs through
the point cusp and the ring cusp so formed.

few microseconds, while the plasma was confined for a few milliseconds, about $1,000 \times \tau_H$. Since that time, magnetic wells, as stabilized mirrors are called, have become very popular.

(c) Shear Stabilization

Even if a confined plasma is unstable on the interchange criterion, it may still be stable if it is impossible to interchange the field lines. Since we can stretch, compress, or bend but not break field lines, only topologically equivalent field lines can be interchanged. In a toroidal system, the ratio of the number of toroidal turns about the axis of symmetry to the number of poloidal encirclements of the magnetic axis is a topological property and only for a uniform rotational transform ι = constant can interchanges be performed. Otherwise, work must be done against the magnetic field and the stability criterion becomes

$$\frac{dp}{d\Psi} \frac{d^2 V}{d\Psi^2} + B^2 \left(\frac{d\iota}{d\Psi} \right)^2 \frac{dV}{d\Psi} > 0. \tag{23.4-7}$$

(d) Rayleigh-Taylor Instability

A very simple instability is that of a layer of fluid held up against gravity by a uniform magnetic field. The energy released by an interchange is then clearly

$$\delta W = (\partial \rho / \partial z) \times g \tag{23.4-8}$$

and, if g acts downward, ρ must decrease upward for stability. In a discussion of confined plasmas, this is seldom important although, if the plasma is accelerated, as in a magnetically shocked θ pinch, the negative of the acceleration plays the role of g. On the other hand, particles in motion along a line of force experience a centripetal acceleration of order $v_{\parallel}^2/R$, where R is the radius of curvature of a field line. When summed over the particles, the equivalent gravitation acquires a magnitude of $(p/\rho R)\vec{n}$ acting away from the center of curvature (see Fig. 23.4-4).

(e) Ballooning Modes

We can model a nonuniform magnetic field by considering a plasma contained by a uniform field and representing the curvature by an effective gravity. Suppose, as will be the case in many systems, that the curvature changes sign as one moves along the field line. We could represent this by a gravity that varies periodically in the x-direction, $g = (0, 0 - g \cos kx)$, and the energy released by a perturbation ξ in the z direction becomes $\delta W = -\int[B^2(\partial\xi/\partial x)^2 - g \cos kx(\partial\rho/\partial z)\xi^2]d\tau$, the first term representing the magnetic energy added by stretching the magnetic field lines. If ξ is uniform in x, this vanishes but so does the last term. If, on the other hand, ξ is periodic in x, $\xi = \xi_0 \cos(kx/2)$, then $\xi^2 = \frac{1}{2}\xi_0^2(1 + \cos kx)$ and $\delta W \simeq -\int[(1/4)B^2\xi_0^2k^2\sin^2\frac{1}{2}(kx) - (1/4)(\partial\rho/\partial z)g\xi_0^2(1 + \cos kx)\cos kx]\,dx \simeq (1/8)[g(\partial\rho/\partial z) - B^2k^2]\xi_0^2$. Regions in which $\partial\rho/\partial z$ and g are antiparallel are destabilizing and those in which they are parallel are stabilizing, but a growing mode will stretch field lines. If the distance between stabilizing and destabilizing regions (the connection length L) is small enough, the increased magnetic energy produced by stretching field lines prevents the motion from growing.

If the "gravitation" is now expressed in terms of the radius of curvature R, $g \simeq p/(R\rho)$ and we introduce the density scale length $\ell^{-1} = (1/\rho) \times (\partial\rho/\partial z)$, we may write a stability criterion as $1/L^2 > B/R\ell$ and the plasma

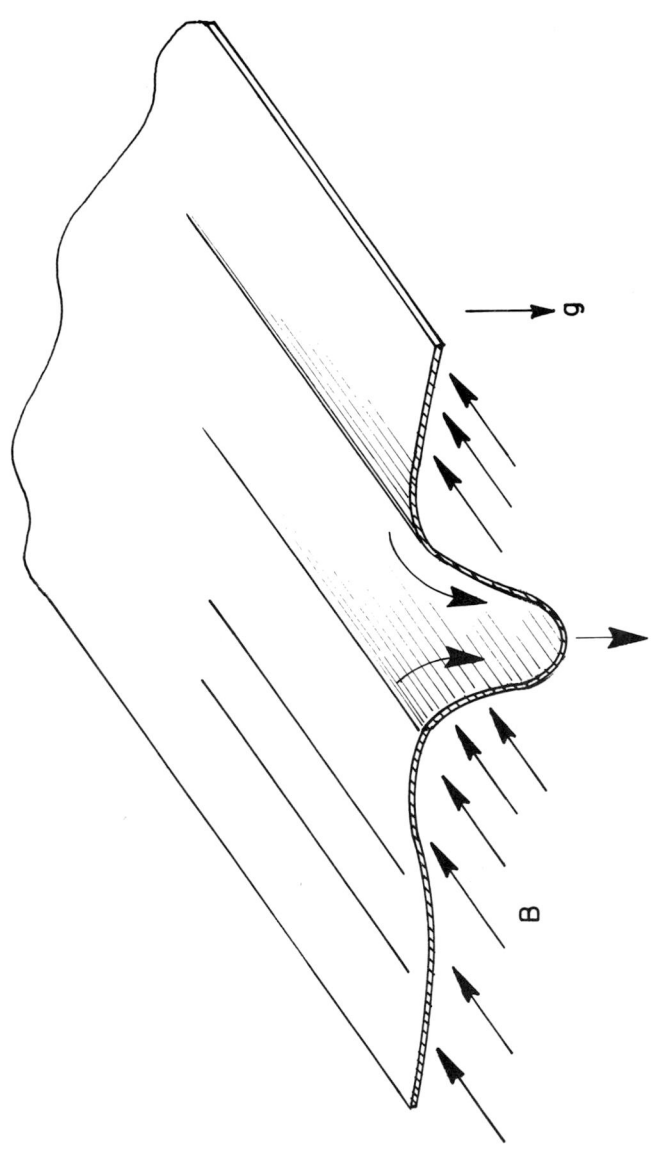

Fig. 23.4-4 The Rayleigh–Taylor instability: a plasma is supported against gravity by a uniform magnetic field. A symmetric downward ripple is shown and releases the energy $-\xi_z g(\partial\rho/\partial z)$.

will be stable against ballooning modes if

$$\beta < R\ell/L^2. \tag{23.4-9}$$

In a Tokamak with $q \gg 1$, the radius of curvature is close to R and the center of curvature lies near the axis of symmetry, so that the inside of the torus has good curvature while the outside has bad (destabilizing) curvature. The connection length must be measured along the field lines and L $\simeq qR$, while the scale length is approximately the minor radius r, thus the maximum β is $\beta_M \approx R r_0/(q^2 R^2) \simeq (1/q^2) r_0/R$. It is sometimes more revealing to ignore the non-confining toroidal field in computing β, replacing it by $\beta_\theta = 4\pi p/B_\theta^2$, whereupon the limiting value is (see Fig. 23.4-5)

$$\beta_\theta \lesssim R/r_0. \tag{23.4-10}$$

(f) Resistive Modes

Effects of finite electrical resistance can be neglected only if the Lundquist number VL/η is large but, in analyzing plasma stability, all possible motions must be considered, the values of V and L being determined by the scale of the motion. The analysis of these effects is complicated by the fact that the energy is no longer a constant and the magnetic flux is no longer conserved so that one must discuss the details of the plasma motion.[6] Resistive effects are particularly important where the undisturbed current density is high.

---

[6] H. P. Furth, J. Killeen and M. N. Rosenbluth, "Finite Resistive Instabilities of a Sheet Pinch," Physics of Fluids 6, 459-484 (1963).

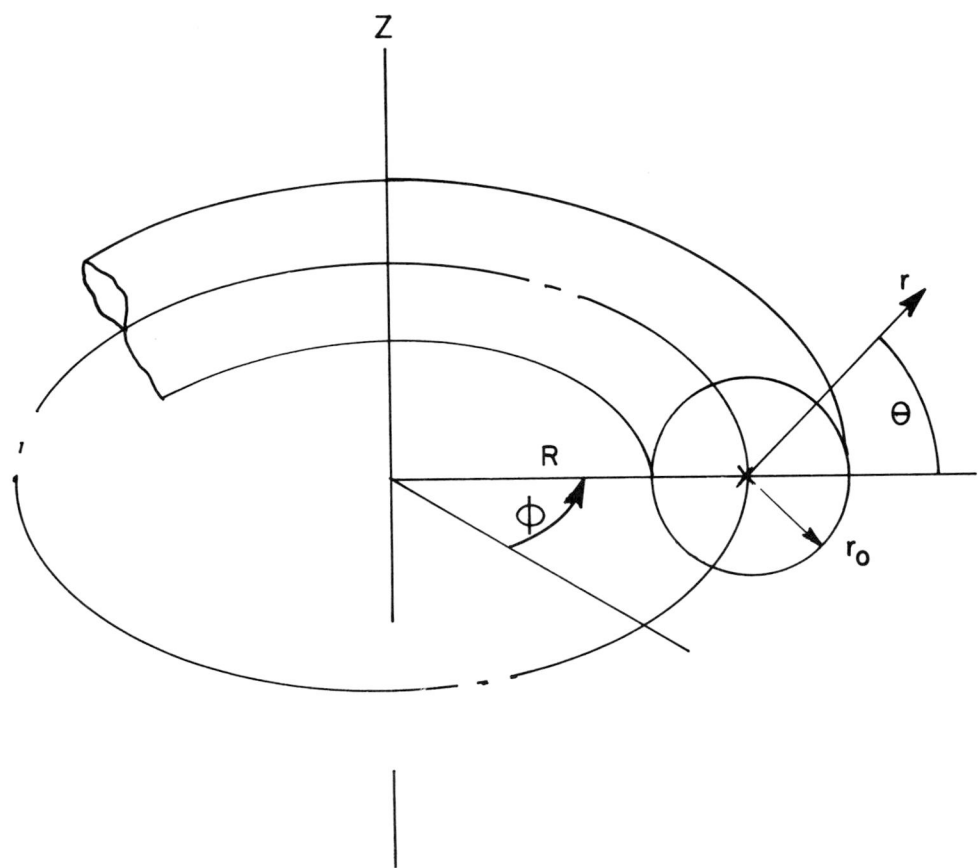

Fig. 23.4-5 Coordinates used in describing a toroidal system such
as Tokamak with circular cross section; the azimuthal
angle is φ, the radius of the magnetic axis is R, the
distance in the plane of constant φ to the magnetic axis
is r, and the poloidal angle around the magnetic axis
is θ.

A sheet current of high density, for example, responds to exactly the same force that is responsible for the "pinch effect", the collapse of a Z pinch and, by the resistive cutting and rejoining of field lines, shreds into a system of current filaments. The growth rates of these disturbances is determined, in part, by the magnetohydrodynamic time scale $\tau_H = L/v_\theta$ and, in part, by the resistive time scale $\tau_\eta \simeq L^2/\eta$ and by formulae of the form $\tau = \tau_\eta^r \tau_H^{(1-r)}$, where r depends on the form of the instability but is typically $\sim 1/3$. The appearance of instabilities of this type in a device called the "unpinch" in which a large current flows down a rigid conductor surrounded by a plasma, a configuration of good curvature everywhere, discouraged the development of shear-stabilized toroidal Z pinches, which had shown some theoretical promise.

C. Instabilities in a Collisionless Plasma

The growth rate of the MHD instabilities is much larger than the collision frequency so that, not only in a magnetic mirror, but in toroidally confined plasma, the use of a locally determined scalar pressure is unjustified and some theory which considers the free motion of the particles is required. We may develop this theory by introducing the particle distribution function $f(v, x, t)$ for each type of particle. The distribution function satisfies the Boltzmann equation but with the collision term omitted, viz.

$$\frac{\partial f}{\partial t} + \vec{v} \cdot \vec{\nabla} f + \frac{e}{m} (\vec{E} + \frac{\vec{v}}{c} \times \vec{B}) \cdot \frac{\partial f}{\partial \vec{v}} = 0. \qquad (23.4-11)$$

(a) MHD Instabilities

For frequencies of the hydrodynamic scale and for motions on a scale large compared to the Larmor radius, this equation for a disturbance δf about equilibrium f_0 may be shown to take the form[7, 8]

$$\frac{\partial}{\partial t}\, \delta f + v_{\parallel}\, \frac{\partial}{\partial x_{\parallel}}\, \delta f = -\, \dot{\vec{\xi}} \cdot \left[\vec{\nabla} f(E, \mu) + (\mu \vec{\nabla} B + m v_{\parallel}^2 \frac{\vec{n}}{R}) \frac{\partial f}{\partial E} \right]. \qquad (23.4\text{-}12)$$

In this equation, it is understood that f is a function of E, μ and that $v_{\parallel}$, the component of velocity along the field lines, must be calculated for fixed E, μ while $\dot{\vec{\xi}}$ is the electric drift

$$\dot{\vec{\xi}} = \vec{V}_E = (\vec{E} \times \vec{B}/B^2)c. \qquad (23.4\text{-}13)$$

If the frequencies are less than the bounce frequency, or the circulating frequencies for particles going around the system, the solution becomes

$$\delta f = -\, \langle \vec{\xi} \cdot \vec{\nabla} f \rangle - \delta E_J\, \frac{\partial f}{\partial E}\, .$$

From this relation, the disturbed pressure can be calculated and depends not on local quantities but on the average of $\langle \xi \rangle$ along a magnetic field line. Moreover, as one might expect, the pressure is no longer a scalar but, if written

[7]Ref. [1] of Section 23.3.

[8]S. Chandrasekhar, A. N. Kaufman and K. M. Watson, "The Properties of a Diffuse Ionized Gas in a Magnetic Field," Am. Phys. 2, 433-470 (1957); ibid 5, 2-25 (1958).

$$\delta\vec{\underline{p}} = \int d^3v\,\delta f(\vec{c} - \vec{V}_E)(\vec{c} - \vec{V}_E),$$ (23.4-14)

becomes a diagonal tensor. Using this relation, one can calculate the work done against the perturbed pressure and hence the energy released by a displacement $\vec{\xi}$. Two rather elegant results follow. It has been shown that δW, the energy released, takes the following form on each flux tube and for each μ:

$$\delta W_{E,\mu} = \delta E_J \langle \xi \cdot \nabla f \rangle + \delta E_J^2 (\partial f/\delta E).$$ (23.4-15)

If the initial distribution is isotropic, then it has been shown that this is always less than the energy that would be released if the MHD pressure were used since the anisotropic distribution produced still has excess non-thermal energy; thus, MHD underestimates the stability of actual plasmas.[9, 10]

In a Tokamak, for instance, the curvature is good on the inside of the torus and bad on the outside. The poloidal field contribution adds a little bad curvature and the MHD stability of a Tokamak depends on the difficulty of fitting an interchange disturbance into the system. If allowance is made for the slowing up of particles in the mirror fields, the time average weights the inner stable side, forming a magnetic well which is stable.

If the distribution function is a function of E, μ and J alone,[11] then

[9]N. M. Rosenbluth and N. Rostoker, "Theoretical Structure of Plasma Equations," Physics of Fluids 2, 23-39 (1959).

[10]M. D. Kruskal and C. Oberman, "On the Stability of a Plasma in Static Equilibrium," Physics of Fluids 1, 275-280 (1958).

[11]Ref. [16] of Section 23.3.

the first term disappears and δW becomes negative if $\partial f / \partial E$ is negative, which may hold in a magnetic well.

(b) Microinstabilities

We turn now to those instabilities that derive their energy from the form of the distribution function[12] and, to gain some insight into the underlying process, we consider first the problem of the propagation of electrostatic waves through a spatially uniform unmagnetized plasma.[13] To do this, we imagine that a potential of the form $\Phi = \Phi_0 \exp\left[\epsilon t + i(\omega t + \vec{k} \cdot \vec{x})\right]$ was switched onto the plasma at $t = -\infty$ and we examine the results near $t = 0$. We then consider the limit as $\epsilon \to 0$. This procedure ensures that the solution is casual, that cause precedes effect, without introducing the transients of a sudden switch on. We find that the B.E. is satisfied by $\delta f = \dfrac{(e/m)\vec{k} \cdot (\partial f / \partial \vec{v})}{(\omega + \vec{k} \cdot \vec{v}) - i\epsilon} \Phi$, and the charge density

$$q = \lim_{\epsilon \to 0} \quad \sum_{species} \quad (e_i^2/m_i)\int d^3v \; \frac{\vec{k} \cdot \dfrac{\partial f_0}{\partial \vec{v}}}{(\omega + \vec{k} \cdot \vec{v}) - i\epsilon} \; \Phi.$$

Taking the limit rather carefully shows that

$$q = [\Sigma(e_i^2/m_i)\int d^3v \; \vec{k} \cdot (\partial f_0/\partial \vec{v})][P\frac{1}{\omega + \vec{k} \cdot \vec{v}} + i\pi\delta(\omega + \vec{k} \cdot \vec{v})]\Phi, \qquad (23.4\text{-}16)$$

---

[12]M. N. Rosenbluth, "Microinstabilities," pp. 485-513 in Ref. [15] of Section 23. 3.

[13]L. D. Landau, "On the Oscillations of an Electric Gas," Journal of Physics of the U. S. S. R., 10, 25-34 (1946).

where $P\dfrac{1}{x}$ indicates the principal part of $1/x$ and δ is the Dirac function $\int f(x)\delta(x)dx = f(0)$. Poisson's equation takes the form

$$k^2 \Phi = 4\pi q = \Sigma(4\pi n e_i^2/m_i)\int_L d^3v \; \frac{\vec{k}\,(\partial\hat{f}/\partial\vec{v})}{(\omega + \vec{k}\cdot\vec{v})}\,\Phi, \qquad (23.4\text{-}17)$$

where the subscript L (for Landau) represents the special value given above for the integral and $\hat{f}$ has been normalized to 1.

This result has the form of Poisson's equation in matter, $k^2 \epsilon \Phi = 0$, with the frequency- and wave-number-dependent dielectric

$$\epsilon(\omega, \vec{k}) = 1 - \Sigma(4\pi n e^2/mk^2)\int_L d^3v \; \frac{\vec{k}(\partial\hat{f}/\partial\vec{v})}{(\omega + \vec{k}\cdot\vec{v})}. \qquad (23.4\text{-}18)$$

If $\epsilon = 0$, there exists a non-trivial solution and a potential wave can propagate. If $\omega/k \ll v_\theta^i$,

$$\epsilon = 1 + k_D^2/k^2, \qquad (23.4\text{-}19)$$

where $k_D = (4\pi n e^2/kT)^{\frac{1}{2}}$ is the inverse Debye length and no waves can propagate. For $\omega/k \gg v_\theta^e$,

$$\epsilon \rightarrow 1 - (4\pi n e^2/m\omega^2) = 1 - \omega_p^2/\omega^2, \qquad (23.4\text{-}20)$$

and a wave can propagate at the plasma frequency ω_p, $\omega_p^2 = 4\pi n e^2/m$.

Since ϵ is complex, ω will not in general be real but will have the form $\omega + i\gamma$. If we assume that both γ and $\mathrm{Im}\,\epsilon$ are small, we obtain

$$R_i \epsilon(\omega, k) = 0, \quad \gamma = -\mathrm{Im}\,\epsilon/(\partial\epsilon/\epsilon\omega). \tag{23.4-21}$$

This is the effect known as Landau damping and it depends on the value of

$$\int d^3v \, \vec{k} \cdot (\partial f/\partial\vec{v})\,\delta(\omega + \vec{k} \cdot \vec{v}), \tag{23.4-22}$$

that is, on the value of the distribution function for particles travelling with the wave. Physically, the origin of the damping is as follows: only particles travelling near the speed of the wave can exchange energy with it; those having velocities slightly less than the wave, are pulled forward by it, while those having velocities slightly greater are pulled back (the principle of phase stability). Hence, the slower particles gain energy from the wave and the fast ones give energy to it; the damping then depends on the difference between those slightly slower and those slightly faster (see Fig. 23.4-6).

If $\partial\epsilon/\partial\omega$ is negative, then the wave itself has negative energy since the electrical energy in a wave is[14]

$$E = (1/8\pi)\partial[\omega\epsilon\langle\vec{E}^2\rangle]/\partial\omega$$

[14] L. D. Landau and E. M. Lifschitz, Electrodynamics of Continuous Media, translated by J. R. Sykes and J. S. Bell, p. 253, Pergamon Press, Oxford, 1962.

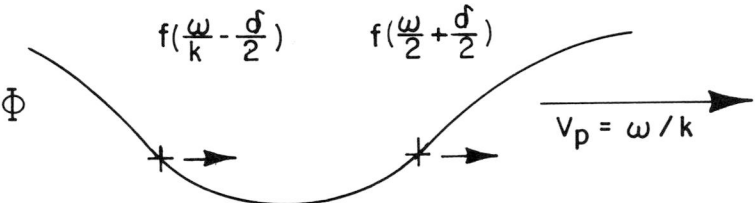

Fig. 23.4-6 Landau damping. In a reference frame moving
           with the phase velocity (ω/k) of the wave, the
           particles see a static potential Φ. Particles
           having velocities $v_- = (\omega/k) - \delta$ (slightly less
           than the wave speed) are accelerated, while
           those having velocity $v_+ = (\omega/k) + \delta$ (slightly
           greater than the wave speed) are decelerated.
           The energy extracted from the electric field
           is thus proportional to $- (\partial f/\partial v)$ evaluated at
           (ω/k).

and, for longitudinal waves, $\epsilon = 0$! Hence the wave grows, paradoxically, as it loses energy.

In thermal equilibrium, $\gamma \geq 0$, and $\partial\epsilon/\partial\omega \geq 0$; hence, all waves are damped but, if the distribution is non-Maxwellian, waves can grow.[12] For instance, if a beam of electrons passes through the plasma, $\partial f/\partial v$ will be positive for velocities just a little less than the beam speed, γ will be negative, and a wave of the correct speed will grow as $\exp(\gamma t)$.

If the electrons are much hotter than the ions, a wave can propagate with a phase velocity $v_p = [kT^e/M^i]^{\frac{1}{2}}$, the ion acoustic mode, and this can become unstable if the electron drift $v^e > v_p$; hence, at high current densities, a plasma becomes unstable to ion-acoustic modes.

What the Landau effect does is to select from the entire distribution a particular class, the resonant particles, which determine the growth or damping of the wave. For example, in a Tokamak, where the toroidal magnetic field varies as $B_\varphi = B^o(1 - \epsilon\cos\theta)$, particles can be trapped in the region near $\theta = 0$. If a wave resonates with these particles, which sit in a region of bad curvature, then a local trapped particle instability can appear.[15] There are many such trapped instabilities in Tokamaks, the most important being associated with low-frequency drift waves. These are waves peculiar[16] to nonuniform plasmas that propagate along the drift surfaces with the magnetic drift velocity $v_D \simeq (p_\| + p_\perp)\,\vec{b}\times\vec{n}/(R\omega_c^i)$.

---

[15] B. B. Kadomtsev and O. P. Pogutse, "Trapped Particles in Toroidal Magnetic Systems," Nuclear Fusion 11, 67-91 (1971).

[16] N. Krall, "Drift Waves," Advances in Plasma Physics 1, 153-199 (1968).

In magnetic mirrors, the distribution is necessarily non-Maxwellian since f must vanish inside the loss cone, i.e., for $v_\perp^2 < \tan^2 \theta_0 v_{||}^2$, where $\sin^2 \theta_0 = B_{min}/B_{max}$.

Now consider a wave propagating across the magnetic field in the y direction[17] with a wave length much shorter than the ion Larmor radius and a phase velocity V. The Landau term then becomes

$$\pounds \simeq \int (\partial f/\partial v_y) \delta(v_y - V) d^3v = \pounds \int (v_y/v_\perp)(\partial f/\partial v_\perp) \delta(v_y - V) d^3v$$

$$= \int \sin \varphi \, (\partial f/\partial v_\perp) \delta(v_\perp \sin \varphi - V) dv_{||} v_\perp dv_\perp d\varphi$$

$$= \int_{-\infty}^{\infty} dv_{||} \int_{V}^{\infty} dv_\perp (\partial f/\partial v_\perp)(v_\perp^2 - V^2)^{-\frac{1}{2}}$$

Because of the loss cone, f must increase for small $v_\perp$ and any $v_{||}$; if V is small, this positive region determines the sign of the integrals and hence the plasma develops the loss-cone instability. Loss cone modes generally do not propagate strictly perpendicular to the magnetic field but travel along it slowly, growing as they travel. At the ends of the plasma, they are effectively absorbed and can be controlled if the mirror is short enough. In addition to this simple mode, there are even more slowly growing drift loss cone modes that appear to determine loss rates from mirrors (see Fig. 23.4-7).

---

[17]R. F. Post and M. N. Rosenbluth, "Electrostatic Instabilities in Finite Mirror Confined Plasmas," Physics of Fluids 9, 730-749 (1966).

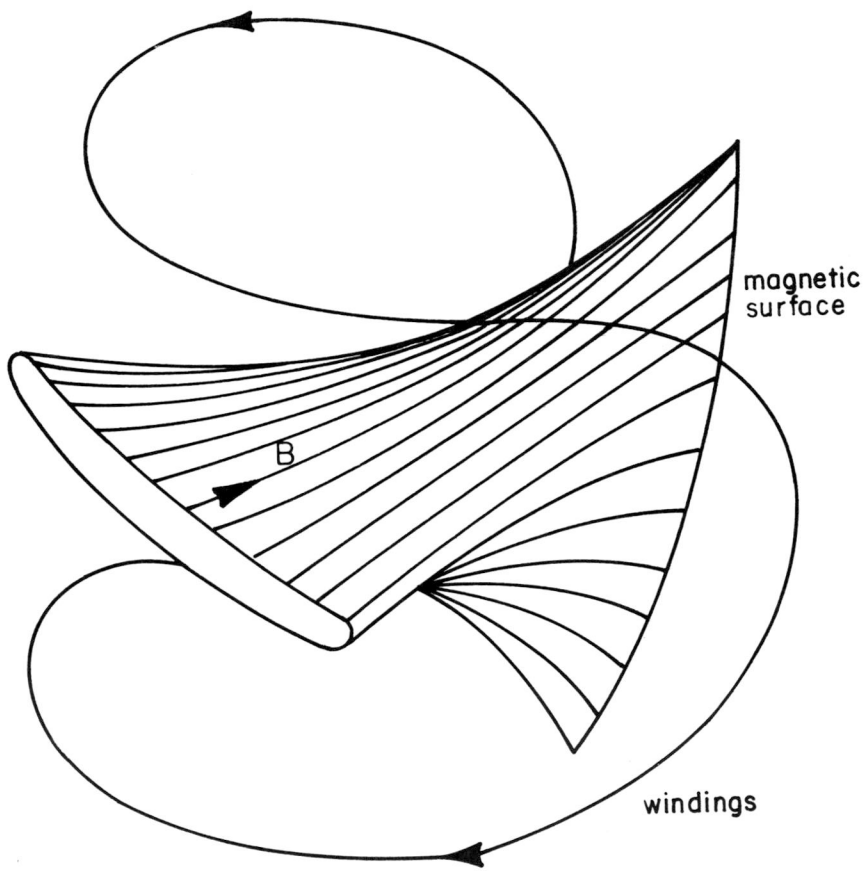

Fig. 23.4-7   Principle of the baseball geometry: the field
              coils produce a magnetic well preventing gross
              instabilities, while the short length inhibits loss
              cone modes.

In most plasmas, the distribution function is not too far from a stable one and the energy available for microinstabilities is not usually enough to disrupt the plasma. Instead, the fluctuating electric-fields lead to enhanced particle scattering, to increased electrical resistance, and to enhanced cross-field diffusion.

23.5 Transport Processes in Confined Plasmas

We turn now to those processes which are responsible for the transportation of matter across the confining magnetic field. As we have seen, a principal process here is particle collisions; this is responsible for the classical cross-field diffusion in a toroidally confined system and for the losses from magnetic mirrors.[1] Our first task must be to understand the effect of collisions.

A. Kinematics of the Collision Process

We begin by considering the collision between two particles, one having mass m_1 and velocity v_1 and the other having mass m_2 and velocity v_2. From the laws of conservation of energy and momentum, we may construct two constants, the center of mass velocity, $\vec{V} = (m_1\vec{v}_1 + m_2\vec{v}_2)/(m_1 + m_2)$ and the relative speed $g = |\vec{v}_1 - \vec{v}_2|$, the last quantity being constant only for elastic collisions. On collision, only one process can happen: the direction of the vector $\vec{g} = \vec{v}_1 - \vec{v}_2$ can change.

Solving for $\vec{v}_1$, $\vec{v}_2$, the initial and $\vec{v}_1'$, $\vec{v}_2'$ the final velocities yields
$\vec{v}_1 = \vec{V} = [m_2/(m_1 + m_2)]\vec{g}$, $\vec{v}_2 = \vec{V} - [m_1/(m_1 + m_2)]\vec{g}$ and $\Delta v_1 = [m_2/(m_1 + m_2)]\Delta\vec{g}$, $\Delta\vec{v}_2 = -[m_1/(m_1 + m_2)]\Delta\vec{g}$.

---

[1] Ref. [2] of Section 23.2.

If we introduce θ, the angle between the initial and final values of $\vec{g}$, and observe that the component of $\Delta\vec{g}$ along $\vec{g}$ is $- g(1 - \cos\theta)$ and that transverse to $\vec{g}$ is $\vec{g}\sin\theta$, with two components $g\sin\theta\cos\varphi$, $g\sin\theta\sin\varphi$, we find that the average change in momentum of particles on collision is

$$\langle\Delta\vec{p}_1\rangle = [m_1 m_2/(m_1 + m_2)]\langle\Delta\vec{g}\rangle = - [m_1 m_2/(m_1 + m_2)]\langle 1 - \cos\theta\rangle(\vec{v}_1 - \vec{v}_2) =$$

$- \langle\Delta\vec{p}_2\rangle$ and the average change in energy is

$$\langle\Delta E_1\rangle = \langle\tfrac{1}{2}m_1\{\vec{V} + [m_2/(m_1 + m_2)](\vec{g} + \Delta\vec{g})\}^2$$

$$- \tfrac{1}{2}m_1\{\vec{V} + [m_2/(m_1 + m_2)]\vec{g}\}^2\rangle$$

$$= \langle m_1 m_2/(m_1 + m_2)\vec{V} \cdot \Delta\vec{g}\rangle$$

$$= - 2[m_1 m_2/(m_1 + m_2)][(E_1 - E_2) + (m_1 - m_2)\vec{v}_1 \cdot \vec{v}_2]\langle 1 - \cos\theta\rangle$$

$$= - \langle\Delta E_2\rangle,$$

where the angular brackets represent an average over all possible values of θ.

For collisions between like particles, $m_1 = m_2$,

$$\Delta\vec{p} = - \langle 1 - \cos\theta\rangle(\vec{p}_1 - \vec{p}_2), \quad \Delta E_1 = \langle 1 - \cos\theta\rangle \times (E_1 - E_2). \qquad (23.5\text{-}1)$$

For collisions between electrons and ions,

$$\Delta \vec{p}^{\,i} = \langle 1 - \cos\,\theta \rangle [\vec{p}^{\,e} - (m/M)\vec{p}^{\,i}),$$

$$\Delta E^{i} = (m/M)\langle 1 - \cos\,\theta \rangle (E^{i} - E^{e} - M\vec{v}^{\,i} \cdot \vec{v}^{\,e}).$$

(23.5-2)

If collisions occur between symmetric distributions, the last term here disappears.

We now need the average of $\langle 1 - \cos\,\theta \rangle$ and to get this observe that collisions are characterized by the Rutherford cross section,[2] $\sigma_R = e^{4}[\sec^{4}(\theta/2)]/(m_r g^{2})^{2}$

$$\langle 1 - \cos\,\theta \rangle = \tfrac{1}{2} \int_{\theta_{min}}^{\pi} (1 - \cos\,\theta) \sec^{4}(\theta/2) \sin\,\theta\,d\theta$$

$$= 4 \int_{\theta_{min}}^{\pi/2} \cos\,(\theta/2) \sin^{-1}(\theta/2) d(\theta/2)$$

(23.5-3)

$$= 4 \log \sin\,(\theta_{m}/2).$$

This relation diverges as θ_{m} approaches zero, but very small scattering represents an interaction between particles that are far apart; if their closest approach is b, $\sin\,\theta \simeq (\pi/m_r g^{2})(e^{2}/b)$. However, the effect of the plasma is, roughly, to screen off the plasma interaction at the Debye length

---

[2] Chapter 8 (Section 3) in Ref. [1] of Section 23.3.

λ_D; $\lambda_D^2 = kT/4\pi n e^2$ and $\sin(\theta_{min}/2) \simeq \theta_{min}/2 \simeq \pi e^2/(2m_r v^2 \lambda_D)$. If we replace $m_r v^2$ here by kT and note that $4\pi n e^2/kT = \lambda_D^{-2}$, this becomes $\sin(\theta_{min}/2) \simeq 1/(8n\lambda_D^3)$. The number of particles in a Debye shielding sphere, $n\lambda_D^3$, is usually very large in a fusion plasma, e.g., $n = 10^{14}$ cm^{-3}, $T = 5$ kev, $n\lambda_D^3 \simeq 1.5 \times 10^7$ and

$$- \log \sin(\theta_m/2) \simeq \log(8n\lambda_D^3) = \log \Lambda \simeq 20. \qquad (23.5\text{-}4)$$

The collision rate is given by

$$\nu \simeq \sigma g \langle 1 - \cos \theta \rangle \simeq 16\pi[e^4/(m_r g^2)^2]\log \Lambda. \qquad (23.5\text{-}5)$$

For collisions between electrons or between ions and electrons, the value of g is determined by the electron speed, while for ion-ion collisions, it is the ion speed that matters. If g is replaced by the thermal speed,

$$\nu_e = 10^{-9}\, n \log \Lambda\, T^{-3/2}\, \sec^{-1},$$

$$\left.\vphantom{\begin{array}{c}1\\1\\1\end{array}}\right\} \quad (23.5\text{-}6)$$

$$\nu_i = 2.3 \times 10^{-11}\, n \log \Lambda\, T^{-3/2}\, \sec^{-1}.$$

(a) Mirror Losses

As an application of this analysis, let us consider the rate at which particles are lost from a mirror.[3] Since this process is dominated by

_____

[3]Ref. [12] of Section 23.3.

frequent small-angle collisions, it is a two-dimensional diffusion process and depends only logarithmically on the mirror ratio, being determined only by the collision frequency. Now, since $\nu_e \gg \nu_i$, electrons are scattered out of the mirror a good deal more rapidly than ions; as a result, the mirror acquires a positive charge and the resulting electric field holds back the electrons so that the energy loss is determined by the ion collision frequency. Therefore, the confinement time is approximately $\tau \simeq 2.6 \times 10^{10} \, T^{3/2}/n$. The product

$$n\tau \simeq 2.6 \times 10^{10} \, T^{3/2} \qquad (23.5-7)$$

and, at T = 200 kev, $n\tau \simeq 7.3 \times 10^{13}$, which at these high temperatures satisfies Lawson's criterion for DT. However, only if highly efficient means of circulating the energy can be developed is the magnetic mirror a probable reactor.

(b) Beam Heating

An important method of heating a plasma is to inject into it a beam of energetic ions. Of course, the very magnetic fields that confine the plasma will exclude an ion beam. This obstacle can be overcome if, instead of injecting a beam of ions, we inject a beam of fast neutrals. As the neutrals go into the plasma, the hot electrons will ionize them, cross sections for this being of order 10^{-18} cm^2 at 5 kev so that, in a plasma at n = 10^{14} cm^{-3}, the distance that the beam travels before being ionized is about 1 m. The charge-exchange cross section is also very large ($\sim 10^{-17}$ cm^2) so that this too contributes to the ionization of the beam. Elastic scattering, however, proceeds only slowly.

Suppose now that the beam has a velocity that greatly exceeds the velocity of both ions and electrons.[4] Then, the energy transfer rates are

$$dE_e/dt \simeq 8\pi \log \Lambda \, e^4 (m/M)[m^2 V_B^3]^{-1} (E_B - E_e),$$

$$dE_i/dt \simeq 8\pi \log \Lambda \, e^4 (M^2 V_B^3)^{-1} (E_B - E_i),$$

$$(23.5-8)$$

and, therefore,

$$dE_e/dt : dE_i/dt :: M : m;$$

hence the energy goes almost entirely to electrons (the plasma analogue of the ionization loss of a fast beam going through cold matter).

If, however, the electrons are already hot, then

$$dE_e/dt = 8\pi \log \Lambda \, e^4 (m/M)(m^2 v_e^3)^{-1} (E_B - E_e) \qquad (23.5-9)$$

and $dE_e/dt : dE_i/dt :: (Mmv_e^3)^{-1} : (M^2 V_B^3)^{-1}$; but $(M/m)V_B^3/v_e^3 \simeq m/M \times (E_B/E_e)^{3/2}$. Therefore, if

$$E_e > (m/M)^{1/3} E_B \simeq 0.08 \, E_B, \qquad (23.5-10)$$

---

[4]J. M. Dawson, H. D. Furth and F. H. Tenney, "Production of Thermonuclear Power by Non-Maxwellian Ions in a Closed Magnetic Field Configuration," Physical Review Letters 26, 1156-1160 (1971).

energy loss to electrons is negligible. If the beam energy is moved to 300 kev,
the large angle Coulomb cross section is only about twice the reaction cross
section and a considerable fraction of the beam will undergo nuclear reactions
before slowing down. Each reaction gives a gain of ~60 over the incident
energy. The rest of the beam, of course, heats the plasma ions, and this
method of heating and increasing the energy release may significantly reduce
the value of $n\tau$ needed to construct a break-even reactor.

As a result of calculations of this kind, a great deal of effort has been
put into the development of intense neutral beams. One of the most successful
of recent experiments, the mirror XIIB, was heated by 360A equivalent of
20 kev[5] neutral hydrogen and a design for a break-even reactor,[6] the
Tokamak fusion test reactor, will be heated by neutral beams.

B.   Diffusion in Toroidal Systems

(a) Classical Diffusion

By considering how a random walk could occur, we derived an expres-
sion for the cross-field diffusion coefficient in a magnetic field, viz. $D_{clas} \simeq$
$\nu r_L^2 = (m/M)\nu_e r_L^{i\,2} \simeq 6 \times 10^{-14} nT^{-\frac{1}{2}} B^{-2}$, where T is in kev, B in kG and n in
cm^{-3}, and have shown that this classical diffusion gives ample containment
for a fusion plasma.

There is an alternative and instructive way of deducing this result,
based on the general expression for the particle drifts. Since any acceleration

---

[5]Ref. [3] of Section 23.2.

[6]"Status and Objectives of Tokamak Systems for Fusion Research," U.S.
Atomic Energy Commission, WASH-1295, Washington, D.C., 1974.

produces a magnetic drift, any force F produces a drift $\vec{v}_D = \vec{F} \times \vec{b}/(M\,\omega_c)$, which is equal and opposite for electrons and ions. If a plasma carries a current $\vec{j}$, then there is such a force acting because of the momentum transfer by collision. On the ions, it is $\vec{F} = \nu_e m(\vec{v}_e - \vec{v}_i)$ and, on the electrons, the negative of this. The corresponding flux of ions is

$$n\vec{v}_D = \nu_e nm(\vec{v}_e - \vec{v}_i) \times \vec{B}/(eB^2)$$

$$= \nu_e (mc^2/e^2B^2)ne(\vec{v}_e - \vec{v}_i) \times \vec{B}/c$$

$$= -\nu_e (mc^2/e^2B^2)\vec{j} \times \vec{B} = (n\vec{v}_D)_e. \qquad (23.5\text{-}11)$$

Now $\vec{j} \times \vec{B} = \vec{\nabla}p$ and, if the temperature is uniform, $n\vec{v}_D = -\nu_e (mc^2/e^2B^2)$
$\times (kT)\vec{\nabla}n = \nu_e (m/M)(M^2c^2/e^2B^2)(v_\theta^2)\vec{\nabla}n = -D_{clas}\vec{\nabla}n.$
    In this derivation, however, we have taken no account of the nonuniformity in the magnetic field but, in a fusion plasma, the mean free path can be very long,

$$\lambda = 2 \times 10^{18}\, T^2(n \log \Lambda)^{-1} \simeq 10^{17}\, T^2 n^{-1}. \qquad (23.5\text{-}12)$$

At $n = 10^{14}$ cm^{-3}, $T = 5$ kev, $\lambda \simeq 250$ m and, instead of considering the particle motions as determined by collisions, we should think of the particle motion as almost free, in which case the drifts will play a role.[7, 8]

(b) Pfirsch-Schlüter Diffusion

An immediate consequence of the drifts is an addition to classical diffusion. In a purely toroidal magnetic field, there is a vertical drift of magnitude $(\frac{1}{2}v_\perp^2 + v_\|^2)/(\omega_c R) = 2p/(m\omega_c R)$. This is opposite in sign for ions and electrons and hence leads to a charge separation and the appearance of a vertical electric field, and a further $\vec{E} \times \vec{b}/B$ drift which forces the plasma out. To overcome this in the Tokamak, a small poloidal field is added so that a neutralizing current can flow along the field line. If there are collisions, however, a residual electric field is required to drive this current, and a small cross-field drift results. If we consider only the effect of the toroidal drift, the current flowing across an element of flux surface (r = constant) is $(2pcr/BR) \times (\sin\theta)d\theta$ and the rate at which charge is deposited between the surface at r and r + dr becomes $(2c/BR)(\partial p/\partial r)(\sin\theta)d\theta$. The current that must flow along the field to carry off this increase in charge is $j_\| = (2qr/BR)(\partial p/\partial r)$.

The resistance now requires a parallel electric field of order $E_\| = \eta j_\|$. Since this is due to a charge separation and hence can be derived from a potential, $E_\perp \simeq q(R/r)E_\| \simeq 2\eta(q^2/B)\partial p/\partial r$. The associated outward flux is

$$nv_{ps} = nE_\perp c/B = 2nc\eta q^2 B^{-2}\partial p/\partial r, \qquad (23.5-13)$$

and, on introducing the value of the resistivity $\eta = (2m_e c/ne^2)$, this yields,

---

[7] R. D. Hazeltine, "Review of Neo-Classical Transport Theory," Advances in Plasma Physics 6, 273-309 (1976).

[8] A. A. Galeev and R. Z. Sagdeev, "Neo-Classical Theory of Diffusion," Advances in Plasma Physics 6, 311-420 (1976).

for the isothermal case,

$$nv = 2q^2(m/M)\nu r_L^2(\partial n/\partial r), = D_{ps}\,\partial n/\partial r,$$

where

$$D_{ps} \simeq 2q^2 D_{clas} \tag{23.5-14}$$

is the Pfirsch-Schlüter diffusion coefficient.

This result, which is valid at large q, already shows an enhanced diffusion in toroidal systems.

(c) Neoclassical Diffusion

The drift velocities cause yet a further enhancement of diffusion in toroidal systems, this time produced by those few particles that are trapped in mirror fields. A Tokamak system, for example, has a toroidal vacuum field $B_\varphi = I(R + r \cos \theta)^{-1} \simeq B_\varphi^O[1 - (r/R) \cos \theta] \simeq B_\varphi^O(1 - \epsilon \cos \theta)$, where ϵ is the inverse aspect ratio r/R, assumed small. As the particle moves around the field lines, it is carried from the outside to the inside of the torus and, if at the outside, $(v_\parallel^2/v_\perp^2) < 2\epsilon$, the particle will be reflected. For these trapped particles, the drift differs rather widely from the magnetic surface. The easiest way of seeing this is to observe that, in an axially symmetric system, the quantity $p_\varphi = mRv_\varphi + (e/c)RA_\varphi$ is a constant of motion. If a particle is reflected at a mirror point, then $v_\varphi \simeq v_\parallel$ is changed from $v_\parallel$ to $-v_\parallel$ and a corresponding change in position δr is required to keep p_φ constant, i.e., $e/c \times$ $\times \delta r \cdot \partial A_\varphi/\partial r = 2mv_\parallel$. But $\partial A_\varphi/\partial r \simeq B_\theta$ and, therefore,

$$\delta r \simeq (2mc/eB_\theta)v_\parallel \simeq (B/B_\theta)v_\parallel/\omega_c$$

$$\simeq (B/B_\theta)\sqrt{\epsilon}\, v_\theta/\omega_c \simeq (q/\epsilon)\sqrt{\epsilon}\, r_L$$

$$\simeq (q/\sqrt{\epsilon})r_L \gg r_L. \tag{23.5-15}$$

From the shape of the drift surfaces projected on the cross section of the torus, these trapped particle orbits are called banana orbits (see Fig. 23.5-1).

When collisions occur, particles are knocked out into the loss cone but are replaced at the same rate. Since scattering into an angle $> \theta_0$, $\theta_0 \simeq (v_\parallel/v)$ $= \sqrt{\epsilon}$, will knock a particle out, the effective collision frequency is $v_{eff} \simeq$ $\langle 2\pi e^4 g/(mrg^2)^2 \rangle \times \int_{\theta_0}^{\pi/2}[\sec^4(\theta/2)](\sin\theta)d\theta \simeq v_0\theta_0^{-2} \simeq v_0\epsilon^{-1}$. Finally, the fraction of particles trapped are those going almost normal to the magnetic field in a band of width θ_0 and this is $\sim \theta_0 \simeq \sqrt{\epsilon}$.

Now, the diffusion may be calculated by considering a random walk of step length δr and frequency $v_{eff} = v\epsilon^{-1}$ involving $\sqrt{\epsilon}$ of the particles. It may be written

$$D_b \simeq n_{eff}v_{eff}(\delta r)^2 = \sqrt{\epsilon}\, n\,\epsilon^{-1}v(q^2/\epsilon)v_L^2$$

$$= \epsilon^{-3/2}q^2 D_{clas}, \tag{23.5-16}$$

D_b being the "banana" diffusion coefficient. For large q and small ϵ, this is much larger than the classical diffusion, scaling with B_θ instead of the much larger B. If we ask how the diffusion scales with collision frequency at fixed

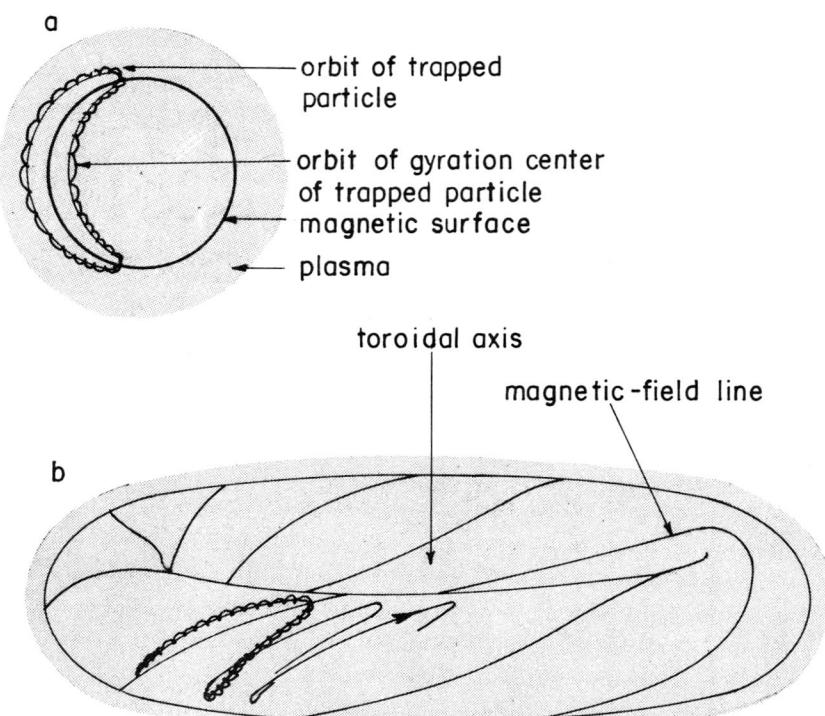

Fig. 23.5-1   Particle orbits in a Tokamak. Reproduced from
B. Coppi and J. Rem, "The Tokamak Approach to
Fusion," Scientific American 277, 65 (1972), by
permission of the authors and publishers.

density and temperature, we find that, for high collision frequency ($\omega_c > \nu >$
$> \omega_B = v_\theta/qR$), the diffusion scales as B^{-2} but when the collision frequency
becomes less than the bounce frequency, $\nu < \omega_b = v_\theta/qR$, then the diffusion
decreases more slowly, scaling as B_θ^2. Between these lies a region in which
the diffusion is only slightly dependent on B. These are the classical, banana,
and plateau regions (see Fig. 23.5-2).

An elaborate and complex long-mean-free-path kinetic theory has been
developed, [7, 8] the neoclassical theory, and the transport coefficients, electrical conductivity, thermal conductivity, diffusion and a host of other quantities have been calculated.

Experimental verification of these processes was provided in the multipole device. [9] In this device, life times could be measured, as well as the
density distribution and the magnetic fields. When the basic poloidal confining
field in this device was supplemented by a large toroidal field and the collision
frequency varied, the transition from classical to neoclassical diffusion could
be observed. Since the field geometry was quite complex, a good deal of computation was needed to compare theory and experiment (see Fig. 23.5-3).

Further confirmation has been provided by particle life times in stellerator devices, although here complications are introduced in the banana region by the lack of axial symmetry (see Figs. 23.5-4 and 23.5-5). [10]

Attempts have been made to describe theoretically the history of the
plasma in a Tokamak using the neoclassical transport theory. For hot plasmas,
the results seem to be an improvement on classical theory but there remains
a substantial disagreement with experiment--diffusion coefficients are as much
as an order of magnitude too small. This result may reflect the presence of

---

[9]Ref. [10] of Section 23.3.
[10]Ref. [12] of Section 23.3.

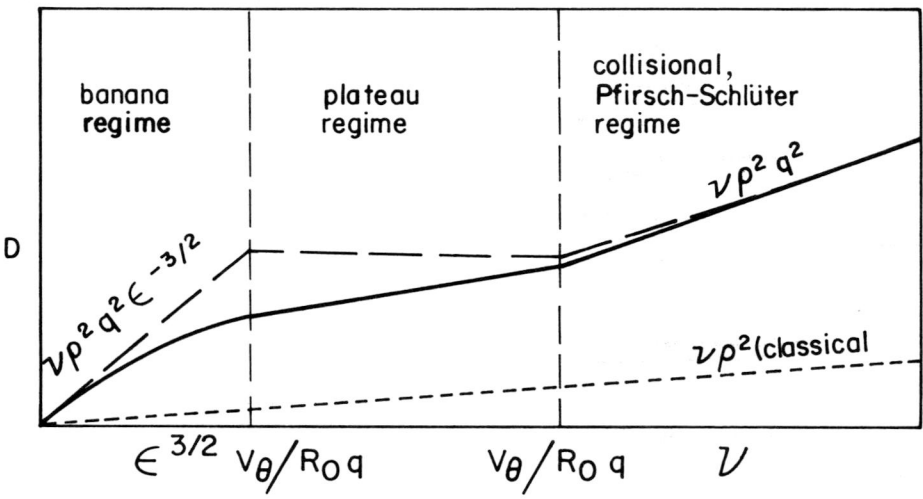

Fig. 23.5-2 Neoclassical diffusion in a torus, showing the
             Pfirsch-Schlüter, plateau and banana regimes.
             Reproduced with permission from Ref. [6] of
             Section 23.5.

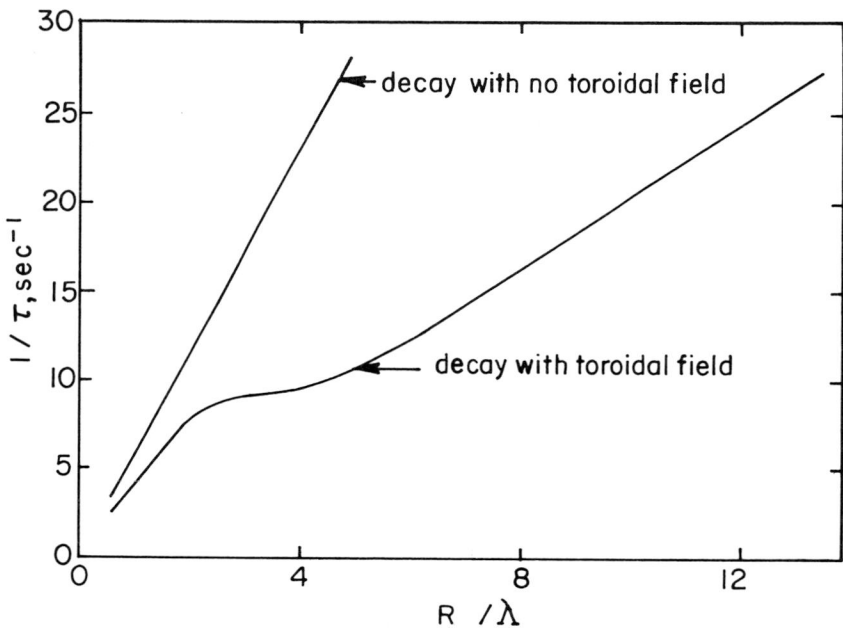

Fig. 23.5-3   Neoclassical diffusion in a multipole with super-
              posed toroidal field: a schematic of the observed
              confinement times.   The reciprocal of the con-
              finement time $1/\tau$ is shown as a function of the
              ratio of scale height to mean-free path, R/λ.
              Reproduced from Ref. [10] of Section 23.3, with
              permission of the author.

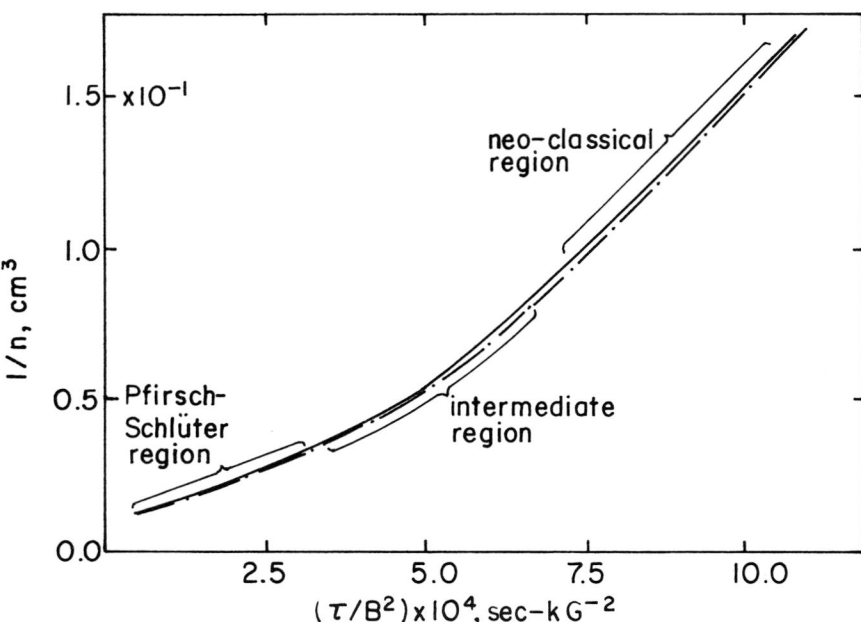

Fig. 23.5-4 Neoclassical diffusion in a multipole: the reciprocal
of the density 1/n is plotted as a function of the quantity τ/B^2, where τ is the confinement time and B the
magnetic field.   The unbroken line refers to results
obtained in a field of 101 G and the dotted line to those
obtained in a field of 427 G.   Reproduced with permission from Ref. [10] of Section 23.3.

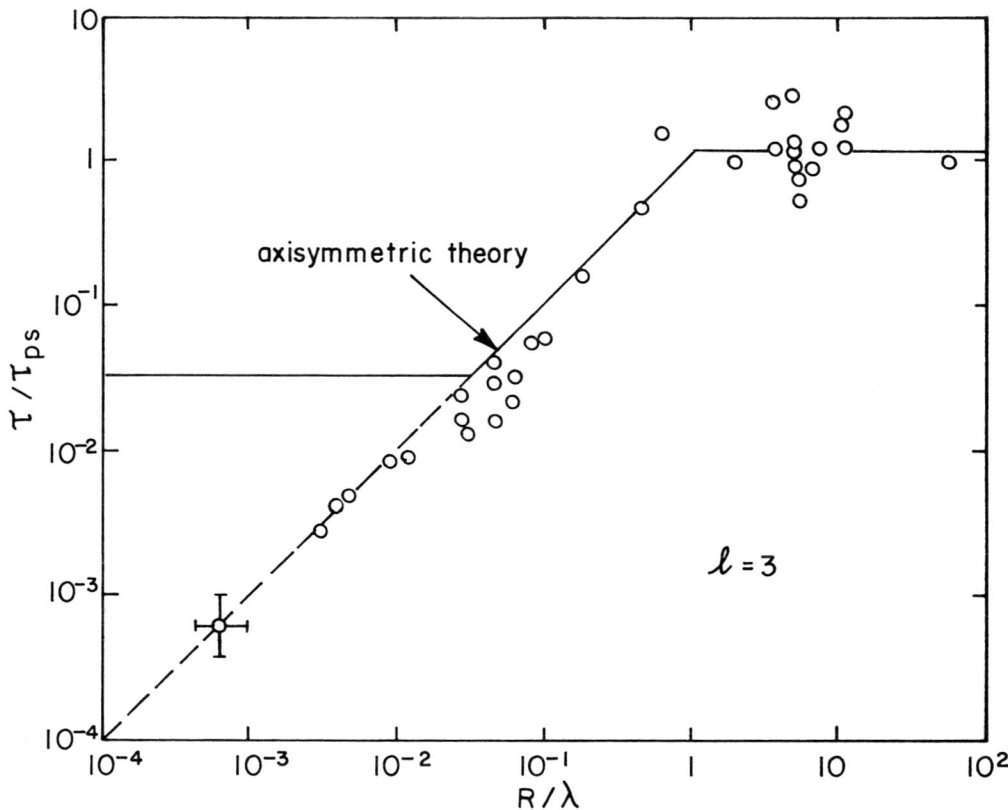

Fig. 23.5-5 Neoclassical transport in a stellerator. The confinement time τ normalized to the Pfirsch-Schlüter value is shown as a function of the ratio of scale height to mean free path. Reproduced with permission from the <u>Culham Laboratory Annual Report for 1973</u>, U.K.A.E.A., London.

high Z impurities in the plasma, for these will rapidly be stripped, and since the Rutherford cross section varies as Z^2, could have a large effect--2% of charge Z = 10 impurities would dominate scattering. Direct measurements of impurity concentration have been attempted, however, and at present it seems that some other explanation must be sought. The anomalies may represent additional diffusion induced by microinstabilities and the consequent random fluctuating fields: turbulent or pseudo-classical diffusion. [6]

(d) Pseudo-Classical Diffusion

To see how fluctuating fields can lead to diffusion, consider a plasma in a uniform magnetic field acted upon by a randomly fluctuating electric field $\vec{E}$ at frequencies much below the cyclotron frequency. Here $\vec{E}$ will induce a drift velocity $\vec{v}_E = \vec{E} \times \vec{B} \, c/B^2$ and a variation in the density n, which satisfies the equation of continuity $\partial n/\partial t + \vec{\nabla} \cdot n\vec{v} = 0$.

Since E and hence the velocity $\vec{v}$ is a random variable, the density n may be split into a slowly varying part n_0 and a rapidly fluctuating part δn and the equation of continuity in turn splits into its average and fluctuating parts,

$$\frac{\partial}{\partial t} n_0 + \langle \vec{\nabla} \cdot \vec{v} \delta n \rangle = 0,$$

$$\frac{\partial}{\partial t} \delta n + \vec{\nabla} \cdot \vec{v} n_0 \simeq 0,$$

where we have neglected a non-linear term $\vec{\nabla} \cdot (\vec{v} \delta n)$ in the fluctuating equation.

If $\vec{E}$ is derivable from a potential, $\vec{\nabla} \cdot \vec{v}$ vanishes and, on the fast time scale, $\delta n = \int_{-\infty}^{t} dt' \vec{v}(t') \cdot \vec{\nabla} n_0$, and

$$\partial n_0 / \partial t + \vec{\nabla} \cdot \langle \vec{v}(t) \int_{-\infty}^{t} dt' \vec{v}(t') \rangle \cdot \vec{\nabla} n_0 = 0.$$

If E is a stationary random variable and the system is isotropic, this relation may be written $\partial n_0 / \partial t = D \nabla^2 n_0 = 0$ with

$$D = (c^2/B^2) \langle \int_{-\infty}^{t} dt' \vec{E}_\perp(t) \cdot \vec{E}_\perp(t') \rangle = (c^2/B^2) \langle E_\perp^2 \rangle \tau, \qquad (23.5-17)$$

where τ as defined above is the correlation time for the fluctuating field. To calculate this type of diffusion, we need a method for calculating the field auto-correlation function. This depends on the processes going on in the plasma. In thermal equilibrium, the calculation is straightforward but, in unstable plasmas, the field spectrum $\langle E^2 \rangle$ depends not only on the instabilities developed but also on the saturation process. In some cases, if the instability is a propagating wave, growing in some regions and damped in others, there may be an equilibrium amplitude that can be calculated fairly simply. In other cases, the non-linear saturation must be calculated, a major preoccupation of current theory. In principle, the fluctuating field could be measured independently of diffusion but such measurements have not yet been made.

23.6 Current Status of Magnetic Confinement

A. Recent Experimental Developments

(a) An Outline of History

Magnetic confinement of a neutralized electron beam was suggested by

Bennett[1] in 1934, and was invoked by Tonks[2] to explain the current limitation in mercury arc rectifiers. Early experiments directed toward magnetic
confinement were performed in 1950 by Thonemann[3] and Ware[4] in England
and, shortly thereafter, workers at the national nuclear research laboratories
in Great Britain, the USSR and the USA began classified studies of the possibility of the controlled release of thermonuclear energy. The first large-scale
disclosure of the results of this research occurred at the 2nd International
Conference on the Peaceful Uses of Atomic Energy held in 1958. By that time,
the major paths of investigation had been defined. Large and small scale Z-
pinches, both straight and toroidal and pulsed on microsecond or millisecond
time scales, had been constructed, and these had produced temperatures of
100 ev, well below fusion level, but orders of magnitude greater than had for-
merly been produced in the laboratory.[5, 6] They also produced measurable
fluxes of neutrons but the confinement time was limited by instabilities.

[1]W. A. Bennett, "Magnetically Self Focussing Streams," Physical Review
   45, 890-897 (1933).

[2]L. Tonks, "Theory of Magnetic Effects in the Plasma of an Arc," Physical
   Review 56, 360-373 (1939).

[3]P. C. Thonemann and W. T. Cowhig, "The Role of the Self Magnetic-Field
   in High Current Gas Discharges," Proceedings of the Physical Society B64,
   345-354 (1951).

[4]A. W. Cousins and A. A. Ware, "Pinch Effect Oscillations in a High Current Ring Discharge," Proceedings of the Physical Society B64, 159-166
   (1951).

[5]J. L. Tuck, "Controlled Thermonuclear Research at Los Alamos," in Ref.
   [12] of Section 23.3, pp. 3-25.

[6]E. P. Butt, R. Carruthers, J. T. D. Mitchell, R. S. Pease, P. C. Thonemann, M. A. Bird, J. Blears, and E. P. Hartill, "The Design and Performance of ZETA," Ref. [12] of Section 23.3, pp. 42-64.

Higher temperatures, 100-1000 ev but shorter confinement times, a few micro-seconds, were produced in dynamic θ-pinches.[7] Several magnetic traps of the stellerator type had been constructed[8] and confined single particles well, but plasma life was short and temperatures low. Simple magnetic mirrors[9] had been constructed, filled by fast neutral beams or by plasma guns, devices which produce a blob of plasma by means of a violent electrical discharge. In at least one of these systems, the plasma was further heated by compression. Neutral beams were produced from kilovolt proton beams by passing them through charge-exchange chambers. The beams were ionized in magnetic traps and produced kilovolt plasmas, but densities were extremely low.

By this time, it was appreciated that instabilities formed the major barrier in the way of realizing magnetic confinement,[10] the magnetohydro-dynamic theory had been well developed and partially verified, while a start had been made on the collisionless theory.[11] Preliminary experiments had been made on the stable cusp geometry, but cusp losses proved embarrassingly high.

During the next few years, attempts were made to test the theory of stability by the unpinch experiments in which plasma is held away from an internal conductor, a magnetohydrodynamically stable configuration.[12]

---

[7] Ref. [3] of Section 23.3.

[8] Ref. [2] of Section 23.3.

[9] Ref. [12] of Section 23.3.

[10] Ref. [3] of Section 23.4.

[11] Ref. [10] of Section 23.4.

[12] K. Aitken, R. Bickerton, R. Hardcastle, J. Jukes, P. Reynolds, and I. Spalding, "Pinch Stability: Theory and Experiment," Nuclear Fusion, Supplement 3, 979-989 (1962).

Unpinches, however, proved to be as unstable as pinches and drew attention to resistive instabilities. The discovery of these modes discouraged further research on the shear stabilized Z-pinch for about 12 years. It has been resumed only recently.

On the other hand, Joffe in Moscow invented a geometry[13] which combined the adiabatic containment of the magnetic mirror with the stability of the cusp, a geometry we have already described. In this device, he managed for the first time to confine a hot plasma for a time that greatly exceeded the hydrodynamic time, showing freedom from gross instabilities.

A further development in mirror research was the identification of loss-cone modes[14] at Livermore and the anomalous diffusion so produced, which required that mirror devices be kept short. This somewhat reduced the possibilities of a mirror reactor, a discouragement that was balanced by the discovery of high field superconductors.

Experiments in the straight θ-pinch at Los Alamos showed a highly symmetrical configuration during early times, while measurements at Culham[15] indicated a density profile determined by classical diffusion during this early phase, anomalies appearing after a time of order L/v_θ. Classical diffusion[16] was demonstrated in the low-density, low-temperature multipole at General Atomic and, as we have observed, later experiments identified neoclassical diffusion (see Figs. 23.6-1 and 23.6-2).

---

[13] Ref. [5] of Section 23.4.

[14] Ref. [172 of Section 23.4.

[15] H. A. B. Bodin, J. McCartan, A. A. Newton and G. H. Wolf, "Diffusion and Stability of High-β Plasma in an 8-metre Theta Pinch," Proceedings of the Third International Conference on Plasma Physics and Controlled Nuclear Fusion Research 2, 533-553 (1968).

[16] Ref. [10] in Section 23.3.

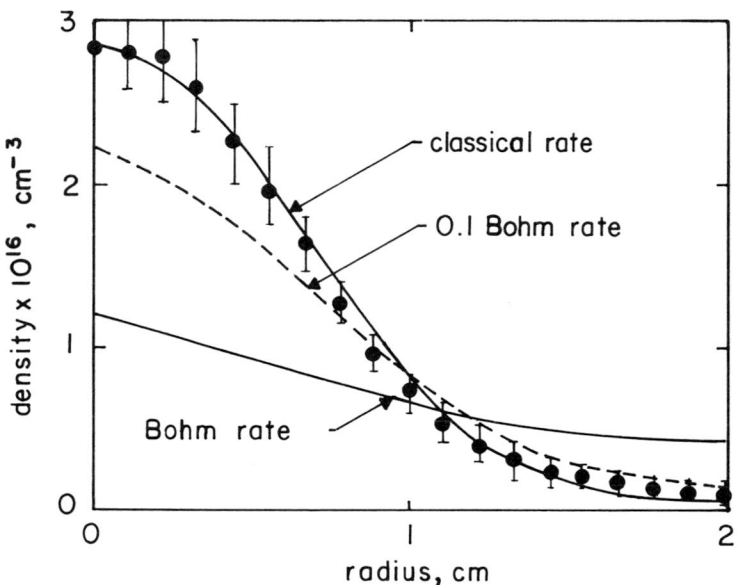

Fig. 23.6-1 Classical diffusion in a θ-pinch showing the theoretical and observed density profiles. Reproduced with permission from Ref. [15].

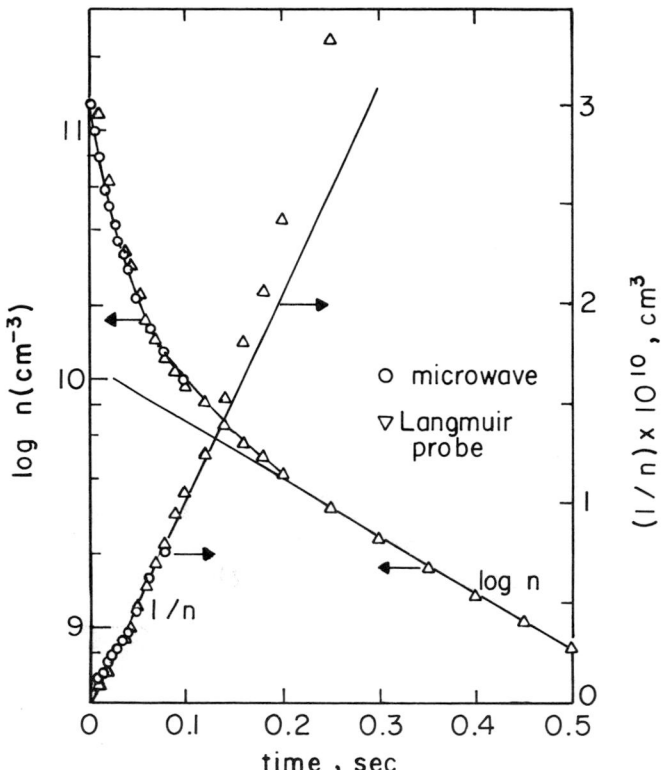

Fig. 23.6-2 Classical diffusion in a low-beta multipole. Two
curves are shown yielding the density n as a function
of time. One curve showing the reciprocal of n is
a straight line when classical collisional diffusion
dominates, while the second logarithmic curve,
log n, at low densities shows the effect of losses
to the supports and losses by support-induced
convection. Reproduced with permission from
Ref. [16].

In 1970, successful containment experiments were performed on the Tokamak system in Moscow.[17] This resembles a Z-pinch with the current kept so low that a magnetic field line must encircle the axis of symmetry many times before encircling the magnetic axis, thus inhibiting the large scale interchange modes. High temperature, well confined plasmas were produced, and from these results proceeds a major part of present day research effort, the Tokamak programs.

B. Plasma Experimentation

(a) Containment

The plasma experimenter must face many difficulties.[18] A confining geometry must be produced, which will hold electrons for 10^6 transit times; even a small error in vacuum magnetic field geometry will prevent this, except in systems of extremely high symmetry. The device must then be filled with plasma, which must be produced and heated. Finally, since plasma densities can seldom exceed $n \simeq 10^{14}$ cm^{-3}, which corresponds at normal temperature to a pressure of $< 10^{-5}$ atm, extreme care must be taken to ensure the purity of the plasma; and vacuum systems must be of the highest quality, with base pressures of 10^{-8} to 10^{-9} atm.

To confine low-β plasmas at modest pressures, large magnetic fields are required and the forces on the conductors producing these fields may be measured in tons. Providing the electric current to power these coils is frequently a formidable engineering task and, for pulsed high currents, may require the transfer of intertial energy from flywheel to generator. To produce

---

[17]L. A. Artsimovich, "Tokamak Devices," Nuclear Fusion 12, 215-251 (1972).

[18]For an excellent review, see Ref. [2] of Section 23.2.

the plasma current, a sudden release of a large quantity of energy is required and plasma laboratories usually include large condenser banks for rapid access energy storage.

(b) Plasma Production

The plasma is frequently produced in situ by a preionization radio-frequency discharge produced in the vacuum vessel before the main current condensor bank is fired. In d. c. mirrors, the plasma is usually supplied from outside by a plasma gun. In one popular type, a titanium electrode is saturated with absorbed hydrogen or deuterium. This now forms one side of a spark gap and, when a high voltage is applied, hydrogen is evolved rapidly from the titanium; plasma is then allowed to drift along a magnetic field where the light hydrogen component outdistances any metal that might be present. The light hydrogen is allowed to enter the magnetic bottle by suddenly reducing the mirror field to open a gate which is shut again as the contaminants arrive. An alternative procedure is to inject powerful beams of neutrals into a low-density, low-temperature, preexisting plasma, which serves to ionize the incoming neutrals. This procedure calls for sophisticated ion beams and charge-exchange devices but has the advantage that the plasma can be produced at kilovolt temperatures.

(c) Plasma Heating

Aside from beam heating, plasmas are most directly heated by the energy used in maintaining a d. c. current, the rate of adding heat per unit volume being $dE/dt = \eta j^2 \simeq 1.6 \times 10^{-3} \, T^{-3/2} \, j^2$ w/cm^3. The rate of loss of heat is usually $\propto T$; hence, if for some reason j is limited, for example, by constraints on the magnetic geometry, then Ohmic heating drops with increasing temperature and it appears impossible to reach temperatures much above ~ 1 kev by this method.

An alternative method of using Ohmic heating is to apply an alternating field to the plasma when, if the frequency is properly chosen, energy may be absorbed from the wave by a resonant process similar to Landau damping. The immediate effect of this is to distort the distribution function which then relaxes collisionally. It may be that the r.f. field or the steady current produces microinstabilities, in which case the consequent fluctuations increase the effective resistivity and induce turbulent heating. If the added heating exceeds the added diffusive losses, this can be effective.

A plasma may have its temperature increased by experiencing external work: by adiabatic compression. The constant entropy in a collisional gas is replaced for collisionless gases by the constant adiabatic invariants, J, and μ. If the compression speed exceeds the velocity of the fast magneto-acoustic mode, $v > (c_S^2 + c_A^2)^{\frac{1}{2}}$, where c_S and c_A are the sonic and Alfvén speeds, then a shock can be produced in the plasma and energy added in that way. This seems an effective way of heating collisional high-β plasmas.

(d) Diagnostics

Once a plasma has been produced, confined, and heated, there remains the problem of determining what has happened. One can measure the overall voltage and current flowing in a toroidal plasma, which gives some of its average properties, including the extremely important energy balance. One can measure the flux and spectrum of particles and radiation leaving the plasma. In magnetic mirrors, the plasma escaping through the loss cone can be directly analyzed, as can that leaving the end of a cusp or a linear θ-pinch. For toroidal systems, no such direct access is possible; the plasma can be sampled only at the edge. At low densities, the flux of fast neutrals produced by charge exchange on the background gas gives information about the ion-energy distribution, but these disappear as the background is ionized. Line radiation from

incompletely ionized impurities gives information about the electron temperature, and its Doppler broadening gives a clue as to the ion temperature. In
a fully ionized plasma, bremsstrahlung gives some evidence for the electron
temperature, especially when examined in the soft X-ray region; but is rather
sensitive to distortions of the electron distribution. Cyclotron radiation is a
little more useful and has recently been studied as a diagnostic tool. It may
give the electron temperature distribution but does not propagate too readily.
If instabilities are present, the plasma may emit characteristic radio frequencies, and this has proved useful in identifying electron cyclotron instabilities. At low β, however, the most important instabilities are potential and do
not couple to the vacuum. Their presence can sometimes be deduced from
fluctuations in surface probes.

In cool plasmas, it is sometimes possible to use probes.[19] By
studying the voltage-current characteristics of a small electrode or of two adjacent small electrodes placed in the plasma, it is possible to deduce the electron density and temperature, while a coil will pick up variations in the magnetic flux. However, such intruders act as a serious source of impurity in a
hot, diffuse plasma.

It is possible to get information about a plasma by observing how particles or radiation are scattered in passing through it. Electron beams are
only useful if propagated along magnetic field lines but fast and heavy ions can
be projected across the magnetic field. Since magnetic fields are strong and
length scales large, the ion beam must be quite energetic. Beams of radio
frequency radiation form another useful probe. Since these are reflected from
surfaces where the density is such that $\omega = \omega_\rho = 5.6 \times 10^4\, n^{\frac{1}{2}}$, these surfaces

[19]G. Francis, "Experiments on Plasma," pp. 273-285 in Ref. [15] of Section 23.3.

can be located. Millimeter waves, for example, are reflected from surfaces having densities of 10^{14} cm^{-3}. Optical radiation is another useful probe.[20, 21] It is easily shown that the incoherent scattered spectrum measures the electron correlation function, $\langle \delta n^2(\omega, k) \rangle$, and one of the successes of plasma kinetic theory has been a reasonably straightforward and experimentally verified connection between this and the plasma parameters. At high wave-number shifts (large scattering angles), the Doppler-broadened normal Thomson scattering gives the electron temperature. At small scattering angles, the spectrum is dominated by the Debye shielding spheres and yields the ion temperature while there is a Raman line at a frequency shift ω_ρ which yields the electron density. If microinstabilities are present, they produce characteristic signals on the scattered profile. Unfortunately, the Thomson cross section is about 10^{-24} cm^2, the fraction of the incident light scattered is extremely small, and the use of this tool requires sophisticated laser technology.

C. Current Status

Today research is being carried out by large groups in the U.S.S.R., in Germany, France, England, and Japan, as well as by smaller groups in Holland, Italy, Sweden, Switzerland, and Denmark. The West European groups have recently combined their efforts under the sponsorship of Euratom and are embarking on large-scale, joint projects. In the United States, the major

---

[20] J. P. Dougherty and D. T. Farley, "A Theory of Incoherent Scattering of Radio Waves by Plasmas," Proceedings of the Royal Society A259, 79-99 (1960).

[21] W. B. Thompson, "The Transport Equation for a Plasma," pp. 225-229, in Ref. [15] of Section 23.3.

research centers are at the National laboratories:[22] Livermore, Los Alamos
and Oak Ridge, with the largest effort at the Princeton Plasma Laboratory, all
these groups being sponsored by the C. T. R. division of ERDA from a total
budget of about $\$1 \times 10^8$ (1975). There is a further large, ERDA financed effort
at General Atomic, a research organization shared by the Gulf and Shell Oil
Companies, while smaller but still substantial programs are supported at
M. I. T. and the Universities of Wisconsin and Texas.

The main lines of research are on three confinement schemes, the θ-
pinch, the magnetic mirror and the Tokamak. The θ-pinch research is centered on Los Alamos, the mirror program on Livermore, while there are
Tokamak systems at Princeton, Oak Ridge, Texas, and M. I. T. Wisconsin has
a modest stellerator.

(a) θ-Pinches

The Syllac θ-pinch studies derive from the demonstration that a straight
θ-pinch had short time stability; but, since a straight system would need to be
several km long to satisfy Lawson's criterion, a considerable experimental
effort has been made to bend a θ-pinch into a circle. Of course, a toroidal θ-
pinch is neither stable nor in equilibrium. To correct this, added windings
are needed, producing a poloidal component and a helical component to the
magnetic field. The plasma is produced in situ by a prepulse and heated by a
strong shock as the main field, provided by a large capacitor bank, is fired.
This yields ion temperatures of about 1 kev and densities of $\sim 10^6 \text{cm}^{-3}$ (in a
system of 8 m radius, a bore of 14 cm, and an azimuthal field of 50 kG, which
rises in 5 μsec, powered by a capacitor of $\sim$ 10 Mj at 60 kv). At present, the

---

[22]"Fusion Power by Magnetic Confinement," WASH-1290, USAEC, Washington, D. C., 1974.

confinement time has reached 30 μsec but this requires, in addition to the added field coils, some dynamic feed-back stabilization. [23]

(b) Magnetic Mirrors

The mirror program at Livermore is divided between two experimental devices, 2XIIB and Baseball, both stable magnetic wells. Baseball is named from the shape of the field coils, like the seam on a baseball. The magnet is superconducting, has a diameter of 1.2 m and will produce maximum field of 20 kG. The plasma is produced by injecting neutral beams and has been contained at n $\sim 4 \times 10^9$ cm^{-3} and 1 kev for 0.7 sec (see Fig. 23.6-3). [23]

The other device, 2XIIB, was originally designed to study a gun-produced plasma; a burst of plasma travels for several m along a guide field of 2 kG. Near one end of the chamber, there is a magnetic mirror so that the plasma is reflected. It is then trapped in a rapidly rising gate field and confined and compressed in a field of ~ 2.0 kG, produced in the stable Yin-Yang configuration (see Figs. 23.6-4 and 23.6-5). [23] This target plasma, at a density $\simeq 10^{13}$ cm^{-3} and a temperature of 3 kev, was then bombarded by up to 360 A equivalent of neutral atoms at an energy of 20 kev, and the plasma density and temperature increased to 3×10^{13} m and 13 kev, where it was confined for 2-4 millisec so that n$\tau = 1.6 \times 10^{11}$ at T = 13 kev. In the magnetic mirrors, the major loss-cone modes are stabilized, in part, by short length (1 m) and, in part, by the injection of a cooler plasma component that partially fills the loss cone and provides electron Landau damping. There remains a residual propagating instability representing the effect of the loss cone on the electron-drift wave. The wave, however, propagates from unstable to absorbing zones and saturates at a modest level, but still makes a significant contribution to

---

[23]Ref. [3] of Section 23.2.

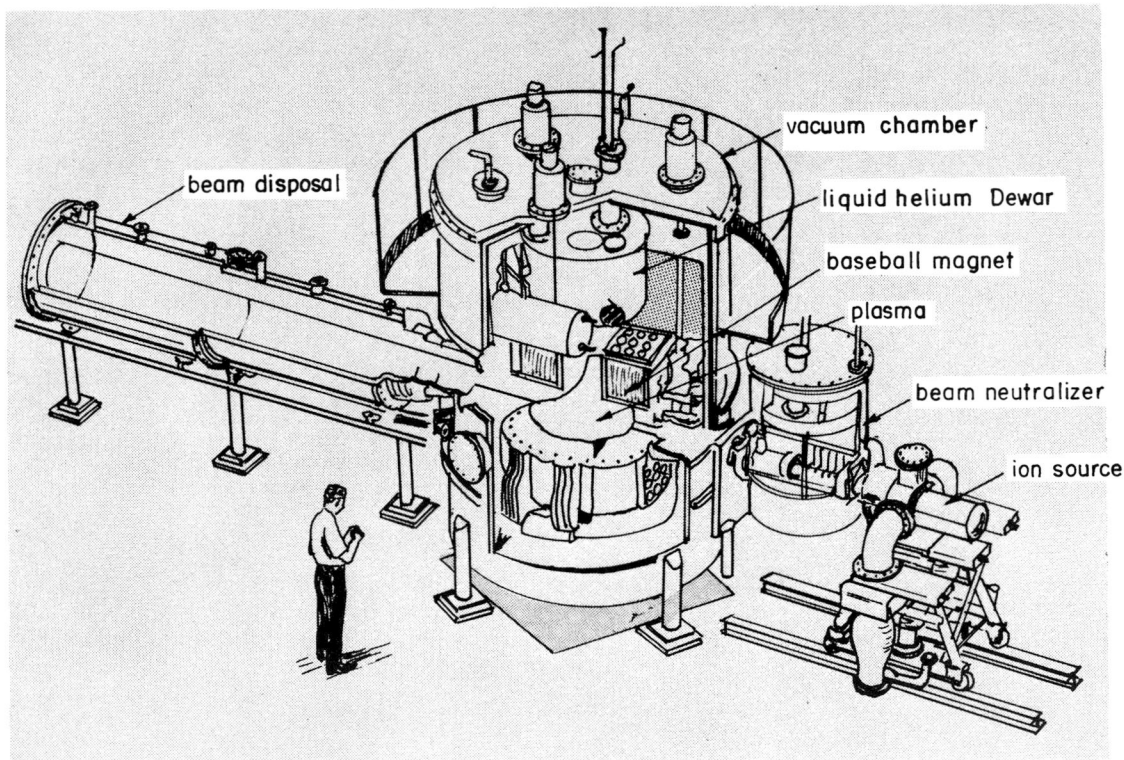

Fig. 23.6-3 Baseball geometry, showing the field coils, vacuum
chamber and beam injector. Reproduced with permission from the Livermore Annual CTR Report
for 1972, UCRL-50002-72, University of California,
Lawrence Livermore Laboratory, Livermore, Ca.,
1971.

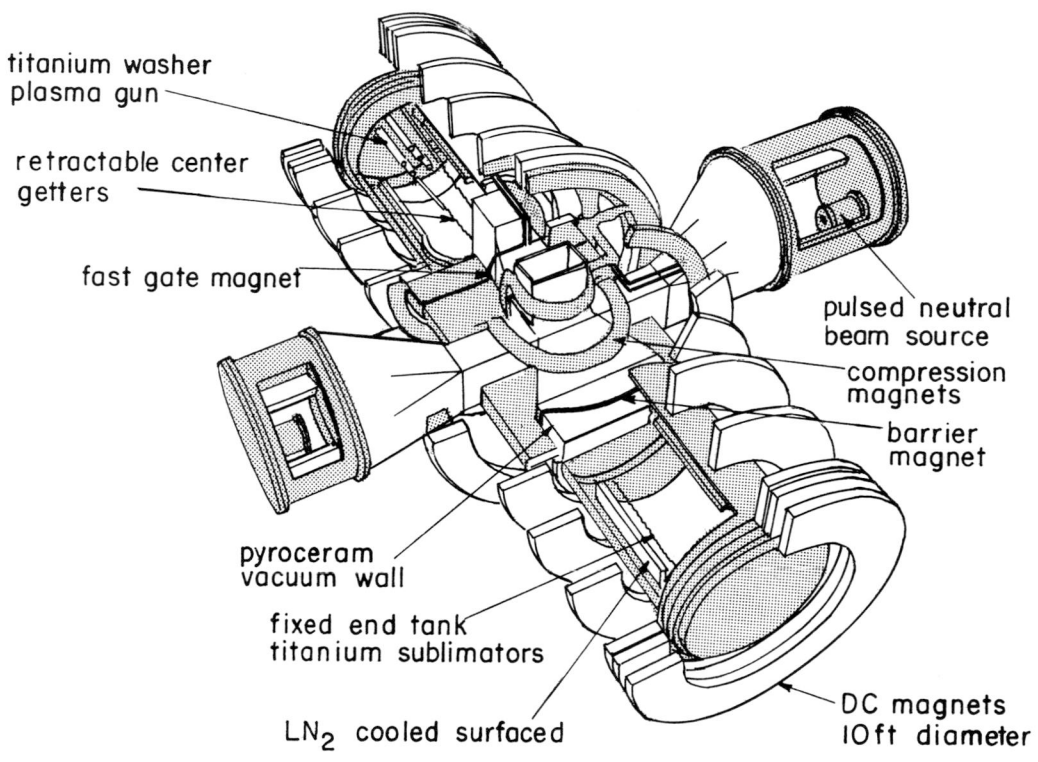

titanium washer
plasma gun

retractable center
getters

fast gate magnet

pulsed neutral
beam source

compression
magnets

barrier
magnet

pyroceram
vacuum wall

fixed end tank
titanium sublimators

LN$_2$ cooled surfaced

DC magnets
10ft diameter

Fig. 23.6-4 Artist's sketch of 2XIIB, showing the vacuum system,
position of the plasma gun, the guide field coils, the
confining and gate magnets, and the neutral beam injectors. Reproduced, with permission, from Fusion
Power by Magnetic Confinement, ERDA 11, Energy
Research and Development Agency, Washington, D.C.,
1972.

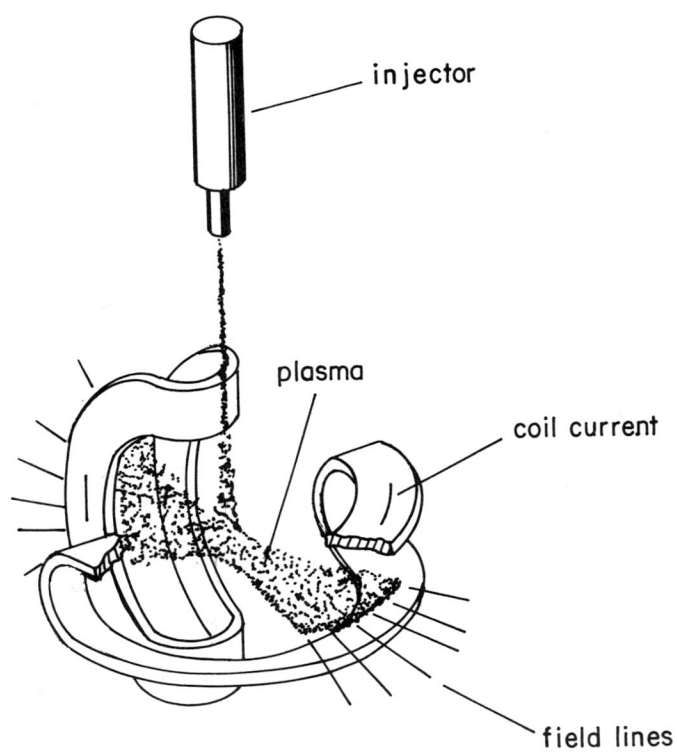

Fig. 23.6-5 The Yin-Yang field configuration. Reproduced
with permission from Ref. [27].

ion scattering and end loss. It has proved possible to analyze the propagation and growth of these waves, and to calculate the anomalous contribution to scattering in a way that agrees with observation, so that mirrors seem to be reasonably well understood (see Fig. 23.6-6)!

(c) Tokamaks

There are 21 large Tokamaks (see Fig. 23.6-7) in operation or being commissioned at present (December 1975).[24] Most of these are of the type we have discussed: axisymmetric tori with circular cross sections with modest aspect ratios $3 \leq R/r \leq 10$, with approximately 1 m major radius. Several, however, have non-circular cross section, elliptical, D-shaped, or the double D, doublet shape, since, with these non-circular cross sections, the effective aspect ratio can be reduced and, keeping the same stability, the total β can be increased (see Fig. 23.6-8).

The Princeton Symmetric Tokamak[24] is typical; this has a major radius of $R = 109$ cm, a minor radius of $r = 13$ cm, and a toroidal field of $B_\theta = 50$ kG. The toroidal plasma acts as the secondary of a step-down transformer and, by discharging a condenser bank through the primary, a current of 130 kA is produced in the plasma. At a plasma density of $n = 4 \times 10^{13}$ cm^{-3}, the Ohmic heating then drives the electron temperature to ~ 2.5 kev while the ion temperature rises to 600 ev. With a safety factor $q = 5$, the containment time is ~ 10 msec, yielding $n\tau \simeq 4 \times 10^{11}$.

The Ohmic heating in this device is a little high and there is some evidence of local magnetohydrodynamic instability near the magnetic axis, which has only a modest effect on transport but prevents peaking of the current channel. If the value of q is decreased, kink modes feeding on the magnetic energy

---

[24]Ref. [6] of Section 23.5.

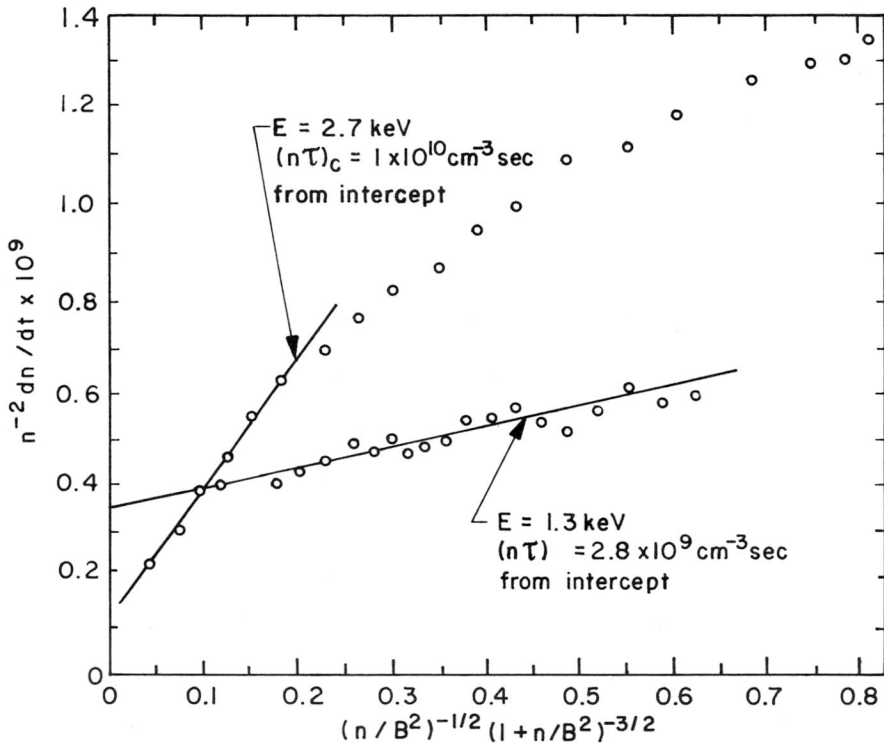

Fig. 23.6-6   Confinement in a magnetic mirror. The horizontal
              lines represent the predictions of the quasi-linear
              turbulence theory. Reproduced with permission
              from the Livermore Laboratory Annual CTR Report
              for 1973, UCRL-50002-73, University of California,
              Lawrence Livermore Laboratory, Livermore, Ca.,
              1973.

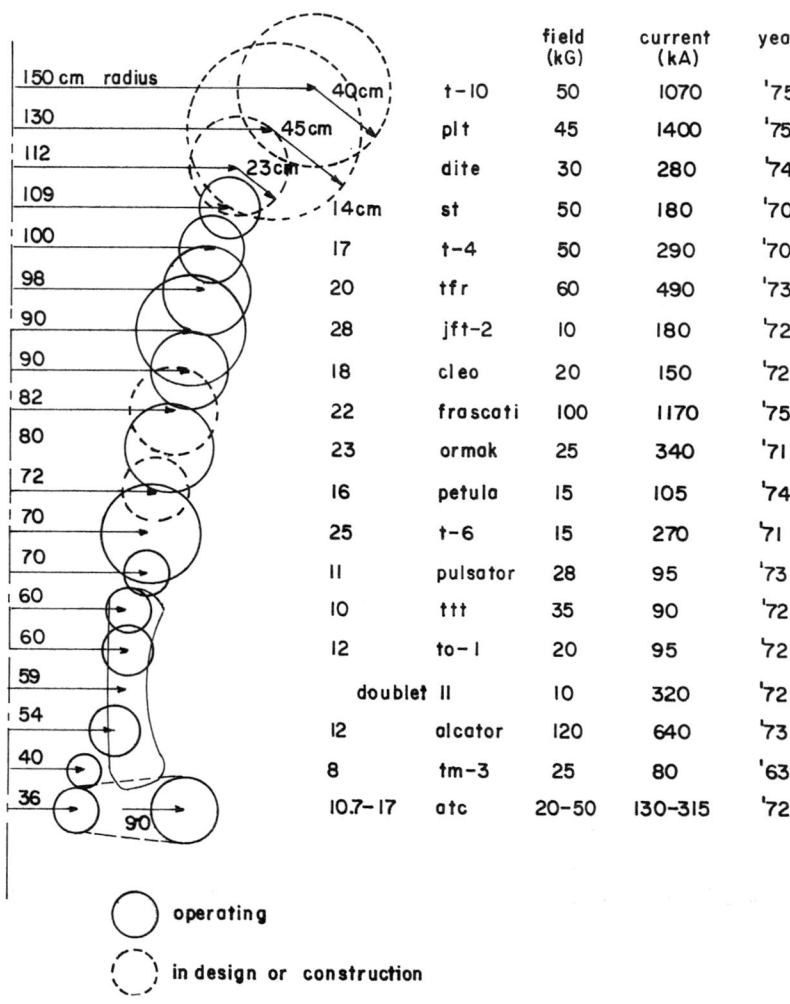

			field (kG)	current (kA)	year
150 cm radius	40cm	t-10	50	1070	'75
130	45cm	plt	45	1400	'75
112	23cm	dite	30	280	'74
109	14cm	st	50	180	'70
100	17	t-4	50	290	'70
98	20	tfr	60	490	'73
90	28	jft-2	10	180	'72
90	18	cleo	20	150	'72
82	22	frascati	100	1170	'75
80	23	ormak	25	340	'71
72	16	petula	15	105	'74
70	25	t-6	15	270	'71
70	11	pulsator	28	95	'73
60	10	ttt	35	90	'72
60	12	to-1	20	95	'72
59		doublet II	10	320	'72
54	12	alcator	120	640	'73
40	8	tm-3	25	80	'63
36	10.7-17	atc	20-50	130-315	'72
90					

⭕ operating

⭕ (dashed) in design or construction

Fig. 23.6-7 A survey of Tokamak parameters. Reproduced with permission from Ref. [24].

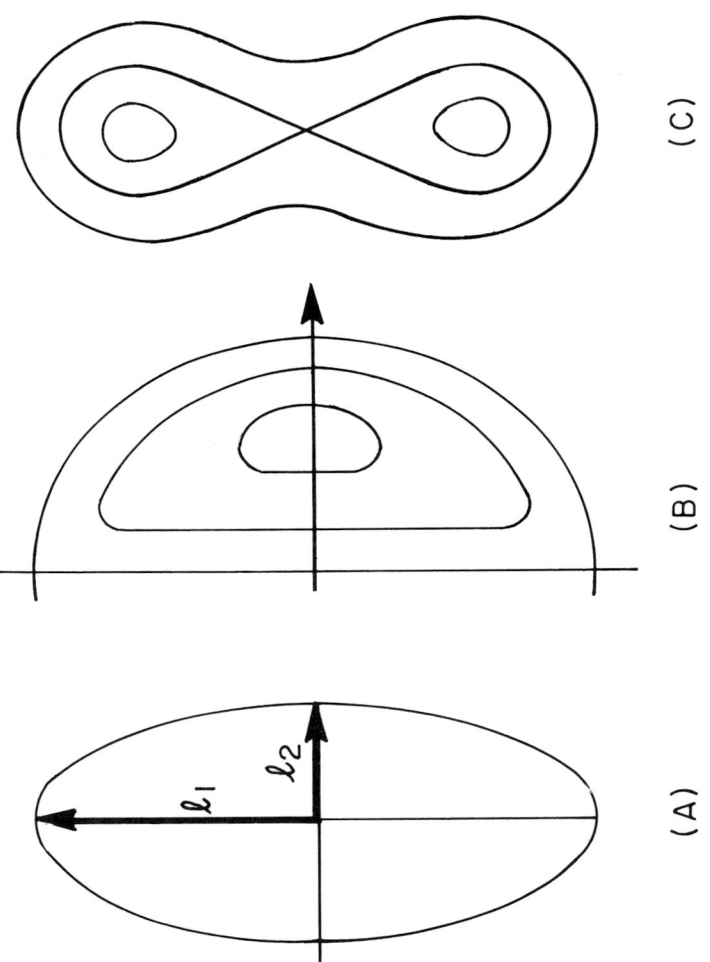

Fig. 23.6-8 Non-circular cross sections used in Tokamak designs. Reproduced with permission from Ref. [24].

in B_θ begin to appear, while near surfaces, on which q is rational, local interchange modes and resistive modes appear. Some of these are disruptive and
produce large spikes in the e.m.f. around the torus which, in stable operation,
is at most only a few volts.

A second interesting experiment is the Adiabatic Toroidal Compression
experiment, also at Princeton. In this, the vacuum chamber containing the
plasma is elongated horizontally and so shaped that the plasma can be compressed by increasing the magnetic field. As the toroidal field was increased
from 15 to 35 kG, the plasma radius was decreased by a factor of 2, the plasma
density was increased from 1.5×10^{13} to 8×10^{13} and the ion temperature from
350 to 750 ev. In this apparatus, the metallic liner that plays a role in preventing large-scale, slowly-growing magnetic modes by supporting eddy currents that inhibit plasma motion was missing. To stabilize these magnetic
modes, an external coil was fed currents that compensated motion of the plasma detected by magnetic pickup loops (feedback stabilization).

The most effective confinement so far has been achieved in the Alcator
device at M.I.T.[23] This is a rather small apparatus with R = 54 cm, r = 12
cm, but is designed to operate at high fields of up to 120 kG. It has been operated at 82 kG with plasma currents between 20 and 22 kA and plasma densities
from $n = 3 \times 10^{12}$ cm^{-3} to 7×10^{14} cm^{-3}. At low density and high current,
the electron drift velocity exceeds the ion thermal velocity and, since the electron temperature is also much higher than the ion temperature, microinstabilities appear which enhance the effective resistivity and give energy to the ions.
In a heating period of 650 msec, the ions reach a temperature of 1.2 kev. At
this point, the density is increased to 6×10^{14} cm^{-3}, the ion temperature
drops to 1 kev, but the instabilities are quenched and the confinement time
rises to $\tau \simeq 1.2 \times 10^{13}$, which is within an order of magnitude of the ignition

threshold (see Fig. 23.6-9). It was encouraging and surprising that, perhaps because of the suppression of instabilities, confinement increased with density.

(d) Projected Experiments

The Tokamak results have been so encouraging that current designs are being extrapolated to build "scientific demonstration" experiments, i.e., machines producing plasmas at the required density and temperature for ignition, producing significant thermonuclear energy (see Fig. 23.6-10). Because of the disagreements we have noted between the calculated value of the diffusion coefficients and the measured density profiles and energy containment times, the design of demonstration experiments is hazardous, and several intermediate experiments are planned: T-10 in the U.S.S.R., the Joint European Tokamak by Euratom, and P.L.T. in the U.S.A.

P.L.T., which is being commissioned at present in Princeton,[23] has a major radius $R \simeq 150$ cm, a minor radius $r = 45$ cm and is designed to carry a plasma current of 1,400,000 A. Using auxiliary heating to supplement the Ohmic losses, it is hoped to reach values of $T_e \simeq 6$ kev and $n\tau \simeq 6 \times 10^{13}$ in this machine (see Fig. 23.6-11).

In the meantime, detailed designs are being prepared for the next stage "scientific demonstration" experiments; in the U.S.A., the Toroidal Fusion Test reactor[24] and, in the U.S.S.R., T-20. In T-20,[25] it is hoped to reach a value of $T_i > 7$ kev and $n\tau > 10^{14}$ sec cm^{-3}, which would satisfy the Lawson criterion for a modest efficiency in circulating energy. This apparatus will have a major radius $R = 5$ m, a minor radius $r = 2$m, and a toroidal field of 35 kG. The plasma current is expected to be 6×10^6 A, the current pulse

---

[25]"Experimental Thermonuclear Installation T-20," National Committee for Utilization of Atomic Energy, U.S.S.R., Moscow, 1975; translation in ERDA-TR 58, Washington, D.C., 1975.

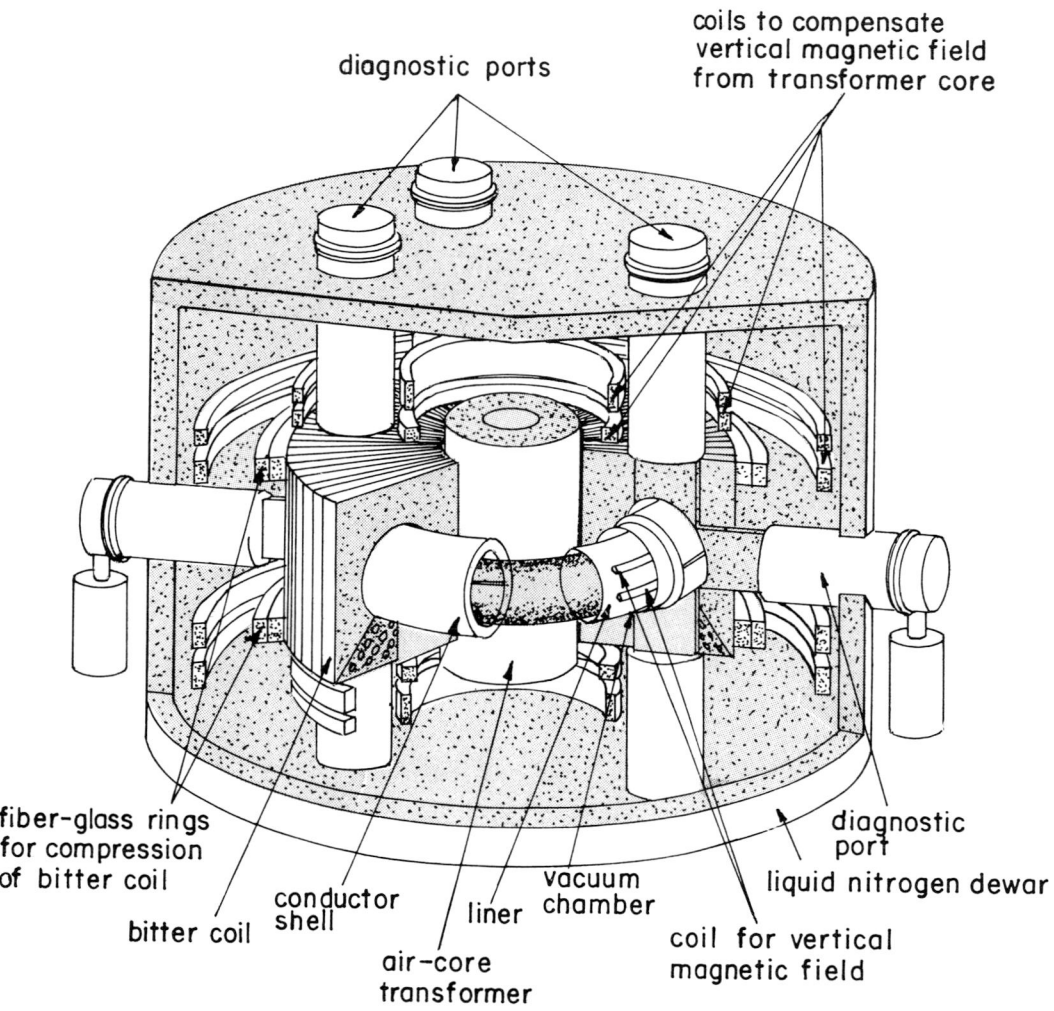

Fig. 23.6-9   Schematic diagram of the Alcator, reproduced with
permission from B. Coppi and J. Rem, "The Tokamak
Approach to Fusion," Scientific American 227, 65 (1972).

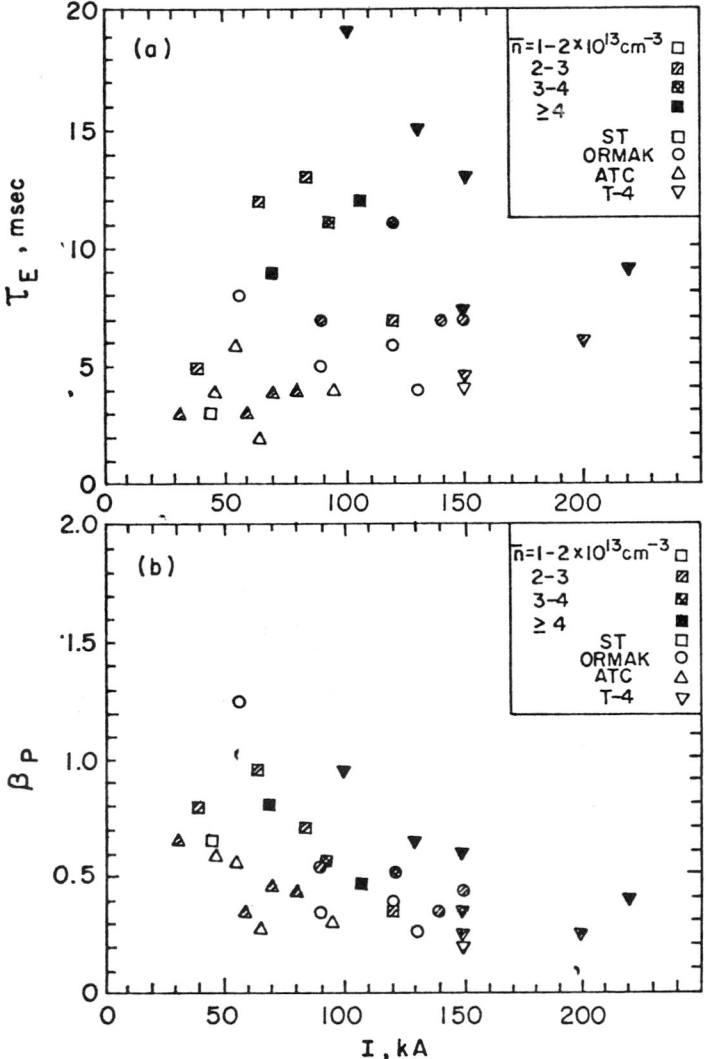

Fig. 23.6-10 Results of Tokamak experiments, showing the energy confine-
ment time τ_E, and the β-poloidal as functions of current. Den-
sity ranges are indicated by shading and devices by shape. ST:
Princeton symmetric Tokamak; ORMACK: Oak Ridge Tokamak;
ATC: adiabatic toroidal compression experiment, Princeton;
T-4: Tokamak 4, Moscow. Reproduced with permission from
Ref. [24].

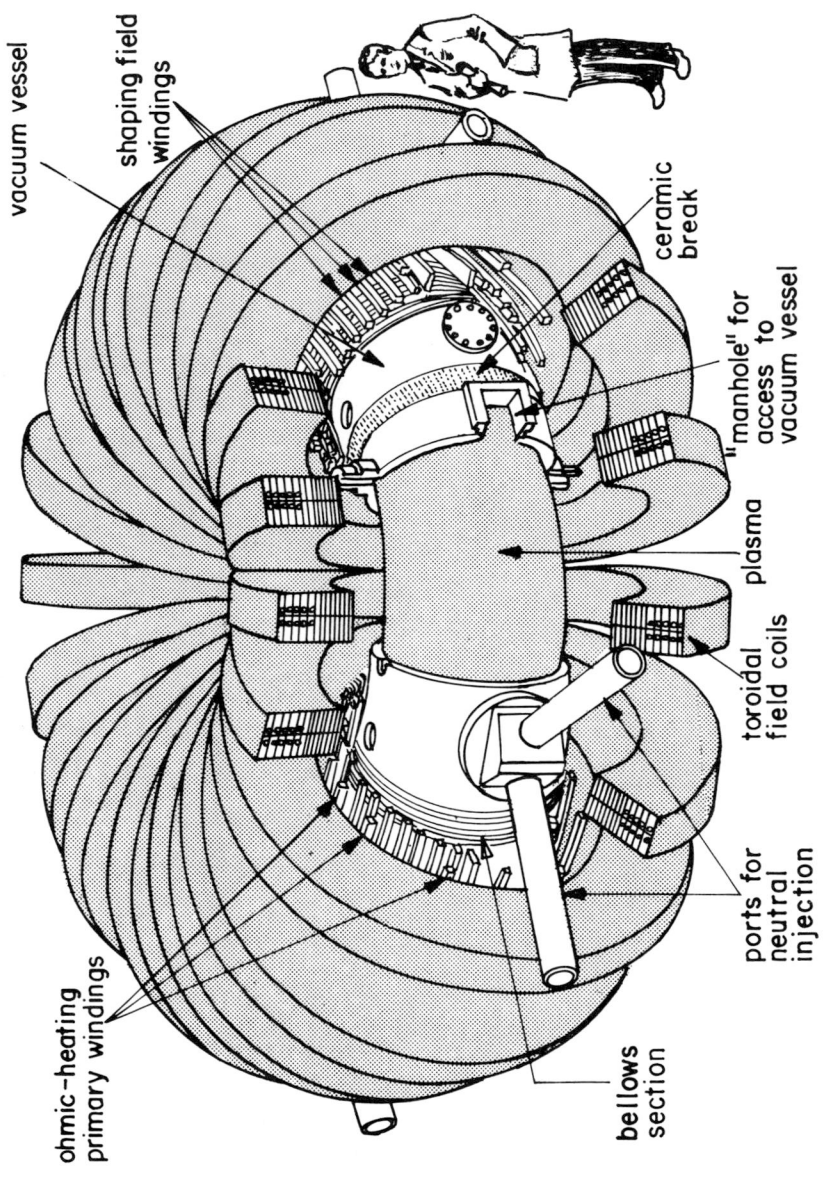

vacuum vessel

shaping field windings

ceramic break

"manhole" for access to vacuum vessel

plasma

toroidal field coils

ports for neutral injection

ohmic-heating primary windings

bellows section

Fig. 23.6-11 Artist's sketch of the large Tokamak at Princeton. Reproduced from Fusion Power by Magnetic Confinement, ERDA 11, Energy Research and Development Administration, Washington, D.C., 1972.

flows in DT, a flux of 10^{13} neutrons/cm^2-sec is expected to strike the wall. To reach the required temperature, the Ohmic heating must be supplemented. Several alternatives are being explored; neutral beam heating, which requires 650 equivalent A at 80 kev, or radio frequency heating, which calls for either 400 Mw at the electron cyclotron frequency, corresponding to 3 cm wave length or, for 200 Mw at the lower hybrid resonance, $\lambda \simeq 30$ cm.

The estimate made for the TFTR are a little less optimistic[24] since the designers in the U.S.A. are seriously concerned about enhanced energy losses due to several types of trapped particle instabilities, and it is believed that, to reach reactor-like conditions, it will be necessary to employ intense neutral beams at 200 kev so that the enhanced nuclear reactions add to the power produced (see Fig. 23.6-12).

(e) Engineering Speculations

It should be clear from this account of the present state of both theory and experiment that any suggestion of a detailed engineering study of a fusion reactor is premature; indeed, in no device yet built have directly detectable amounts of nuclear energy been released; that reactions have occurred can only be inferred from the appearance of neutrons. Furthermore, current experimental devices work at miserably low values of β so that the energy invested in the magnetic field, which is not included in Lawson's energy balance, is used with very poor efficiency. By making more or less plausible extrapolations from present experiments, however, and assuming no surprises are in store, one can construct at least rough models of fusion reactors.[26, 27]

---

[26]E. L. Draper, ed., Technology of Controlled Fusion Experiments and the Engineering Aspects of Fusion Reactors, USAEC, Washington, D.C., 1974.

[27]F. L. Ribe, "Fusion Reactor System," Reviews of Modern Physics 47, 7-41 (1975).

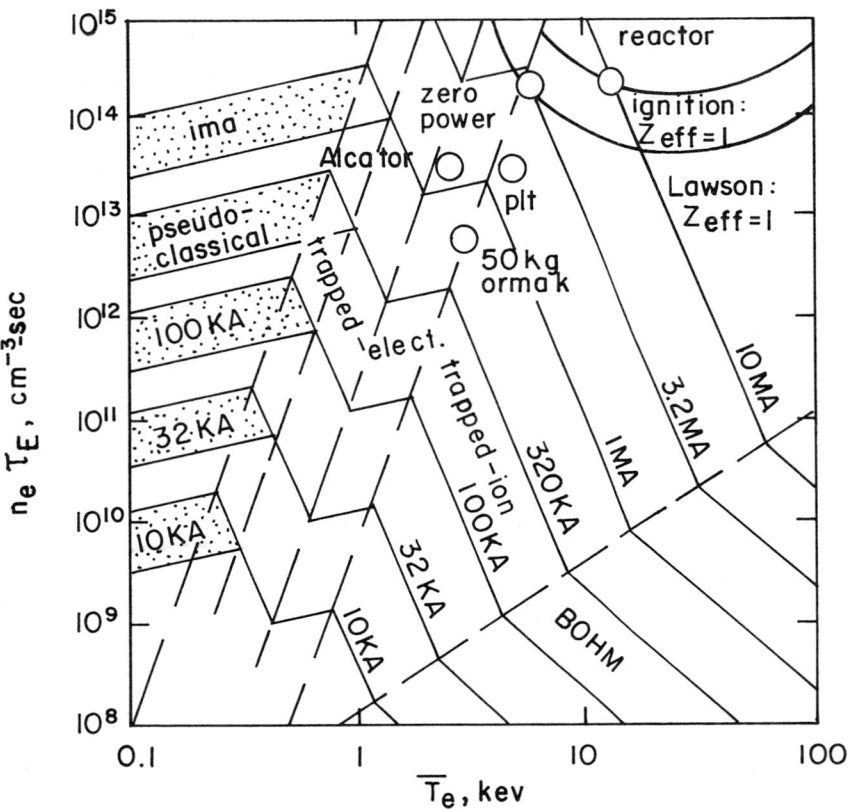

Fig. 23.6-12 Projected and current values of $n\tau$ and T for
              Tokamak devices. Reproduced with modifica-
              tions and permission from Ref. [24].

The initial enthusiasm for thermonuclear fusion was aroused by the vision of a DD reactor, which would use as a fuel the universally plentiful deuterium and produce only energy and helium, hence at one blow solving the fuel and the environmental problems. It is, however, clear by now that the deuterium reactor lies somewhere in the remote future and that present day speculation must be limited to the less attractive DT reactor.

As has been observed, tritium does not occur in nature but must be bred in the reactor by using the $Li^6 + n \rightarrow \alpha + T$ reaction; thus, the reactor must be surrounded by a tritium-breeding lithium-loaded blanket which, in addition, must do some neutron multiplication and, of course, shield the surroundings from radiation. It is the more or less straightforward design of this blanket, not the obscure properties of plasma dynamics, that determines the scale of a fusion reactor. It is fortunate that both Li^7 and beryllium can multiply 14 Mev neutrons by the reactions $Li^7 + n \rightarrow Li^6 + n + n$, $Be^9 + n \rightarrow 2\alpha + 2n$, and that lithium has excellent heat-transfer properties; hence, one blanket design has a first wall consisting of liquid Li pumped parallel to the magnetic field in niobium or stainless steel pipes. Alternatively, a eutectic mixture of lithium and beryllium fluorides, $(LiF)_2 BeF_2$, may be used as a moderator, tritium breeder and coolant. These systems have good high temperature properties and stand up well to neutron fluxes but always require a wall thickness of ~ 1 m to absorb the neutron flux. Moreover, niobium is permeable to hydrogen; hence the tritium produced can be extracted and its concentration kept down to parts per million (see Fig. 23.6-13).

Beyond the breeding blanket, a further layer of insulation is required to protect the main magnetic coils that lie outside. On economic grounds, it appears necessary to make these superconducting so that they must be held at $\sim 5°K$. This requires an extra meter of insulation; thus, the wall surrounding the vacuum vessel must be 2 m in thickness. To give the wall an adequate life

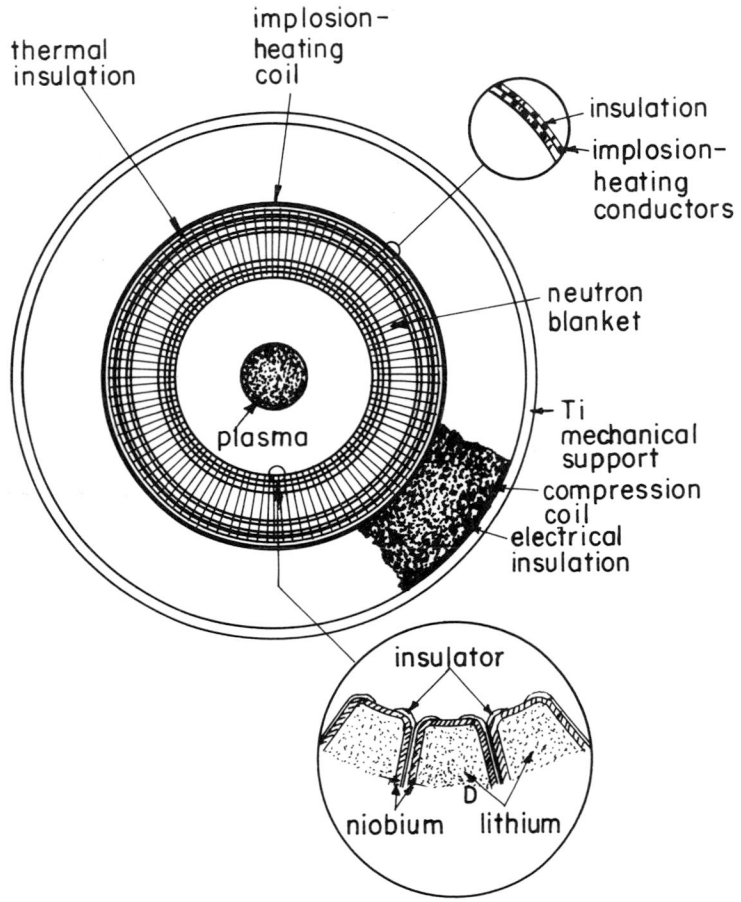

Fig. 23.6-13 A possible wall design for a θ-pinch reactor.
Reproduced with permission from Ref. [27].

time of several years, the power loading is at most a few Mw/m^2 (see Fig. 23.6-14). This requirement defines the minimum size of the system and it appears that fusion reactors become competitive with more conventional fission plants or fossil-fuel plants on the ground of capital cost only for sizes of ~ 1000 Mw/unit. The integration of units of this size into a power network is currently being developed. The economic advantages of large units of any type exerts a continuous pressure toward increases in scale.

The form of the reactor beyond this depends on the particular confinement scheme exploited. A Tokamak can be operated at a steady state but, in that case, must be equipped with a means of eliminating spent fuel and replacing it. The spent fuel can be eliminated by a magnetic diverter, in which field lines near the surface are peeled off and diverted out of the main vacuum vessel into a second heavily pumped dump chamber, so that the outer layer of plasma is lost (see Figs. 23.6-15 to 23.6-17). The surface of the plasma is thus continually removed. Unfortunately, the dump vessel must be as large as the main vacuum system. A second scheme for eliminating spent fuel and protecting the reactor wall is to flush the system with neutral gas. If the plasma density exceeds 10^{15} cm^{-3}, neutral gas is excluded by ionization and again only the outer layers of plasma are swept off. Steady-state systems may be refueled by injecting pellets of solid deuterium or tritium. These must be fired at high speed but even a small pellet could replace the reactor charge.

Low-β systems such as the Tokamak face the problem of thermal stability. If α-particle heating occurs, the plasma should be heated by reaction products so that on ignition its temperature rises; however, since the reaction rate increases so rapidly with temperature, the temperature may rise rapidly to ~ 200 kev and the system must be so designed that the high temperature plasma is also confined. On the other hand, it may be possible to prevent the

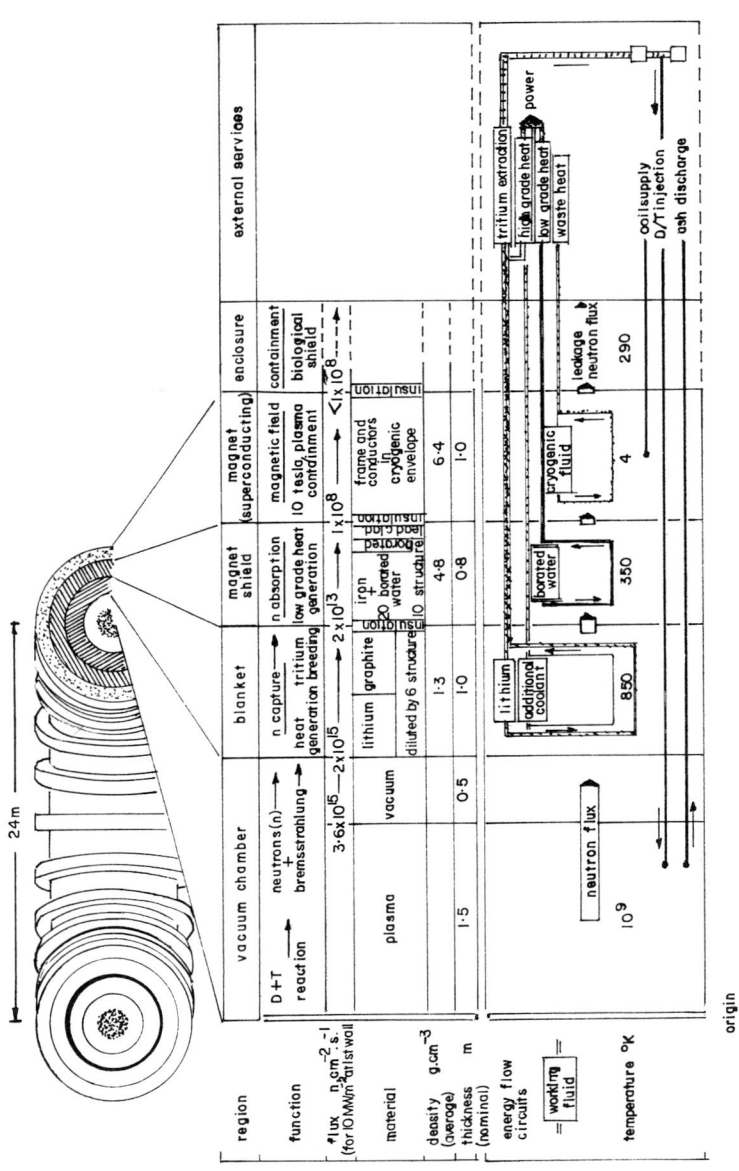

Fig. 23.6-14 Picture-book design for the core of a fusion reactor. Reproduced with permission from the Culham Laboratory Annual Report for 1972, U.K.A.E.A., London, 1973.

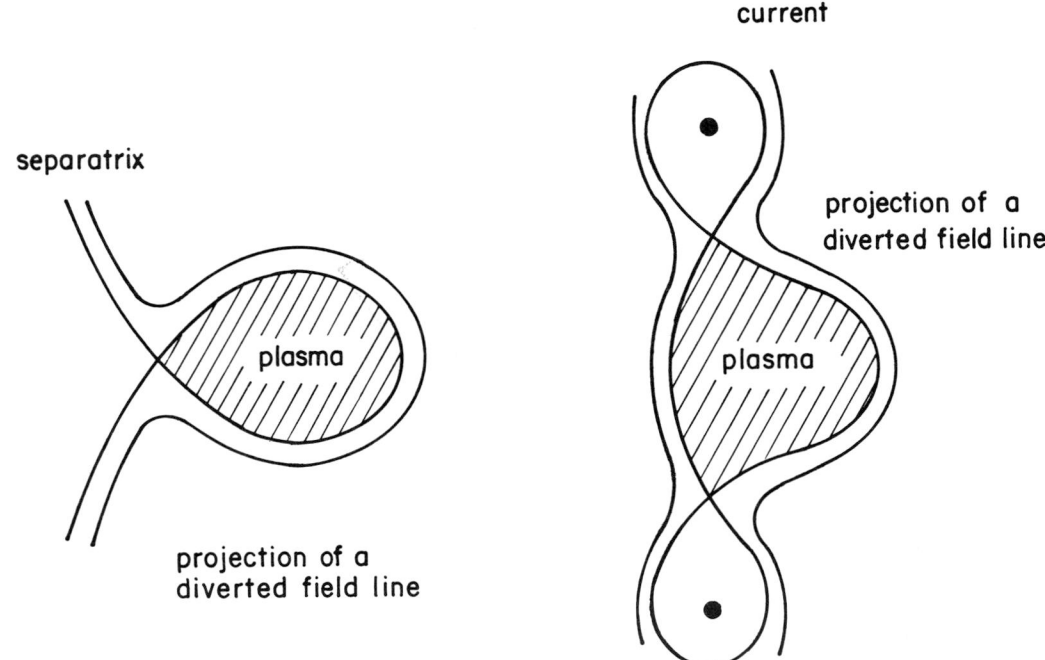

Fig. 23.6-15 Principle of the magnetic divertor. Field lines
              beyond the separatrix are led out of the reacting
              volume into a dump chamber. Reproduced with
              permission from Ref. [27].

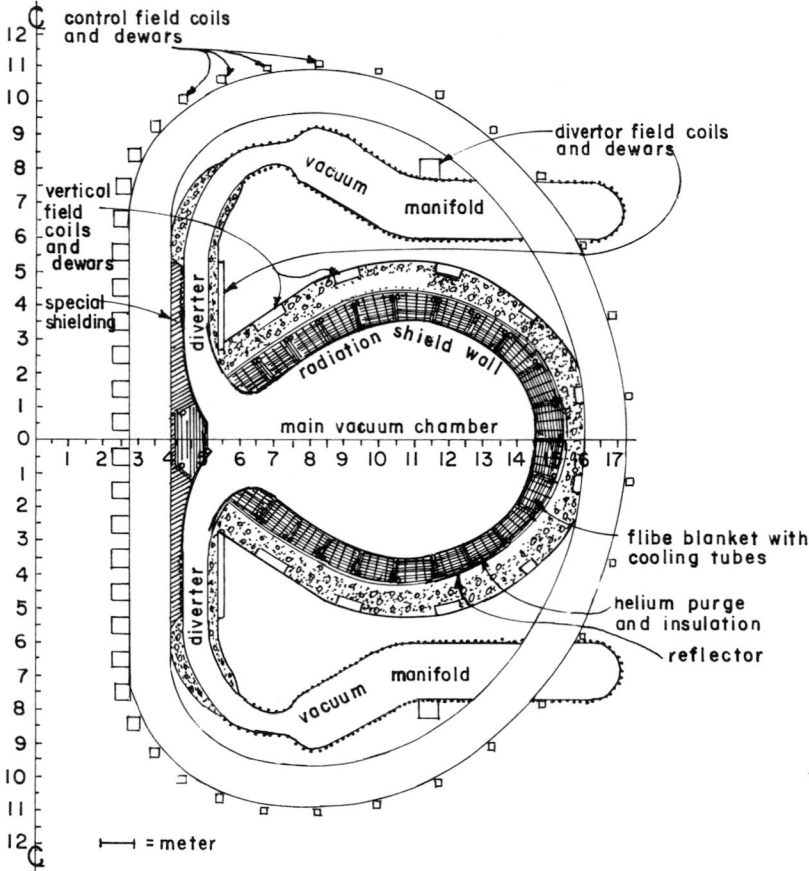

Fig. 23.6-16 Picture-book design for a Tokamak reactor cross section showing one version of a dump chamber. Reproduced from Ref. [27] with the author's permission.

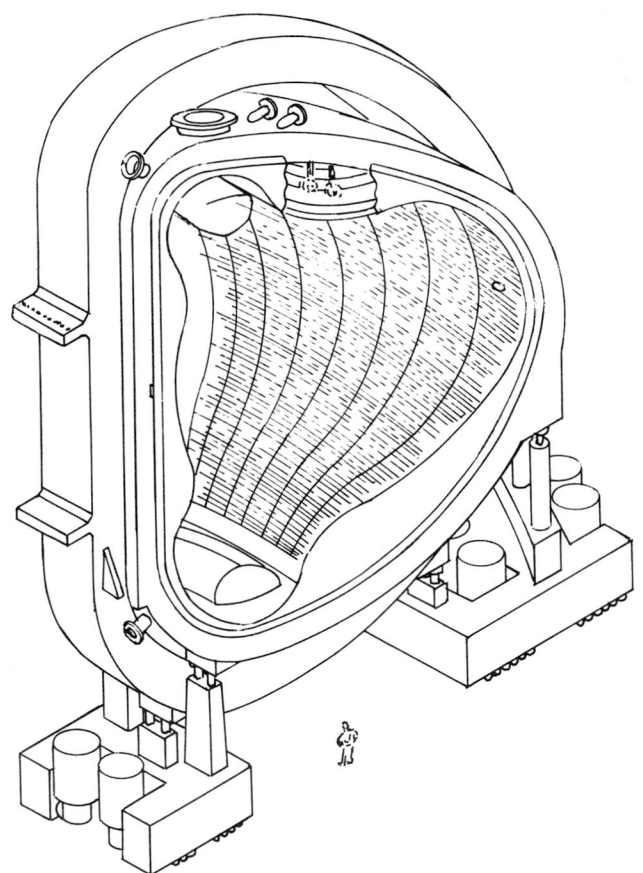

Fig. 23.6-17 A second picture-book design for a Tokamak
              reactor cross section showing the scale. Re-
              produced from Ref. [27] with the author's per-
              mission.

temperature rise by manipulating the plasma, increasing the density, or perhaps the impurity level to enhance radiation losses.

If the high-β, θ-pinch system could be used as a reactor, it would have
a good many advantages. The energy released in the plasma, instead of
causing the temperature to rise, would cause the plasma to expand against the
magnetic field, producing an external e.m.f. so that a reasonable fraction
(about 10%) of the energy produced could be taken off directly as electricity
to avoid the losses associated with a thermal cycle (see Fig. 23.6-18).

The magnetic mirror has special problems. Since the energy gain can
only moderately exceed the energy that must be supplied to the plasma, it is
essential that the energy be circulated in a highly efficient way. Post has suggested a way of doing this (see Fig. 23.6-19). After passing outside the mirror points, the containing field is slowly expanded into a fan. As this happens,
the escaping particles stream parallel to the magnetic field, for the energy
and the magnetic moment, $\frac{1}{2}mv_\perp^2/B$, are both constant; hence, as B decreases,
$v_{\parallel}$ increases. At the same time, the density drops. When the Debye length
exceeds the transverse scale of the system, the ions and electrons may be
separated. The ion beam is now allowed to climb up an electric potential and
a small transverse field forces those ions with small kinetic energy into collector plates so that the energy appears as an electric current. The possibility of making such an ion battery has been demonstrated but, to achieve the
necessary field expansion, the fan must cover a circle of radius > 100 m and
forms the major capital cost of the reactor system (see Fig. 23.6-20).

A possible useful intermediate stage in the development of a fusion reactor may be the fission-fusion hybrid.[28] This scheme, as its name implies,

[28] L. M. Lidsky, "Fission-Fusion Systems: Hybrid, Symbiotic, and Augean,"
Nuclear Fusion 15, 151-173 (1975).

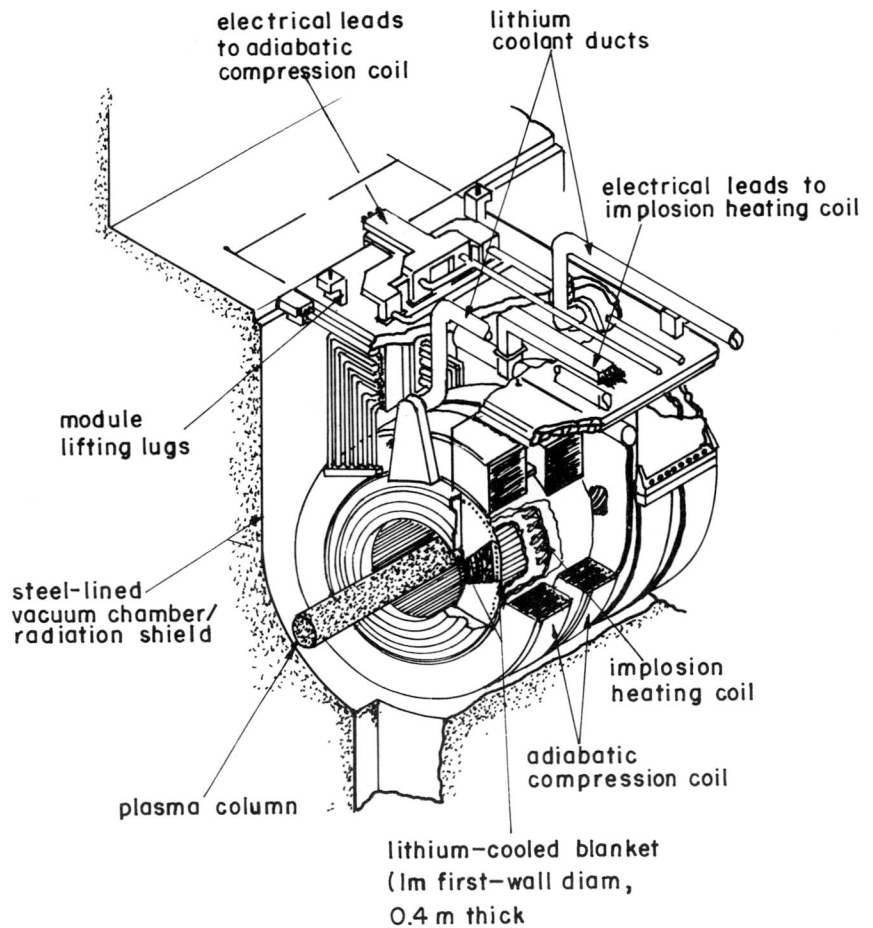

Fig. 23.6-18 Possible cross section of a θ-pinch reactor.
Reproduced with permission from Ref. [27].

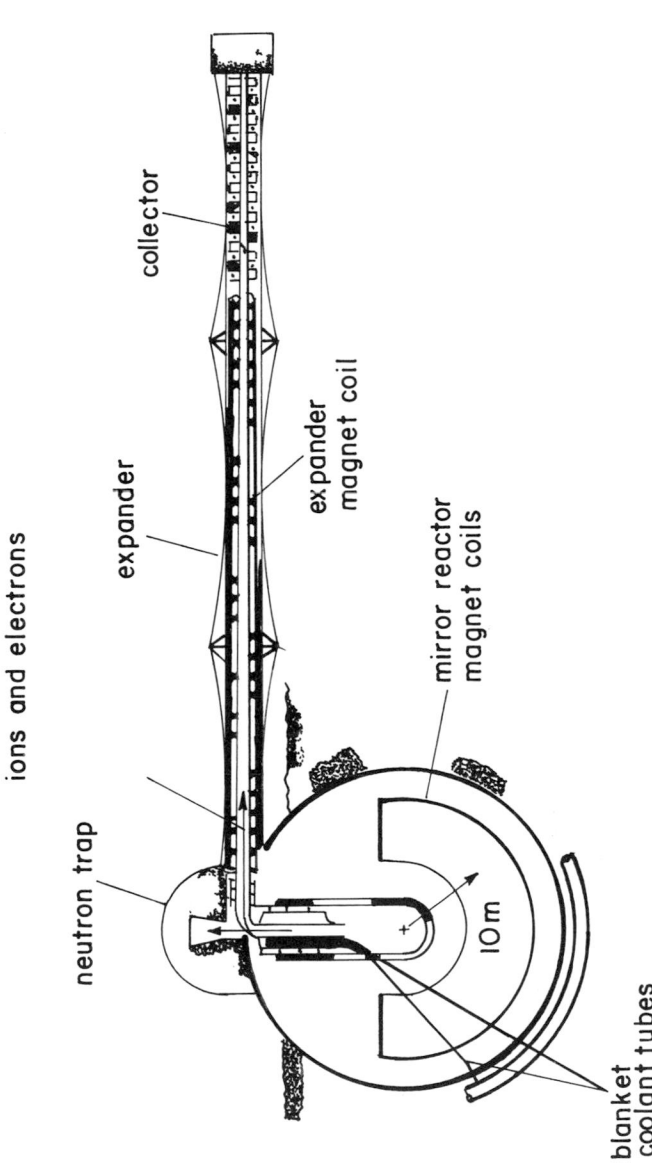

Fig. 23.6-19 Principle of Post's direct conversion system. Charged particles escaping through the loss cone pass into the expander where the field is slowly reduced. Since both the energy $\frac{1}{2}m(v_{\parallel}^2 - v_{\perp}^2)$ and the magnetic moment ($\frac{1}{2}mv_{\perp}^2/B$) are constant, particles end up travelling parallel to B. They then run up an electrostatic potential barrier and are collected with low kinetic energy. Reproduced from Ref. [27] by courtesy of the author.

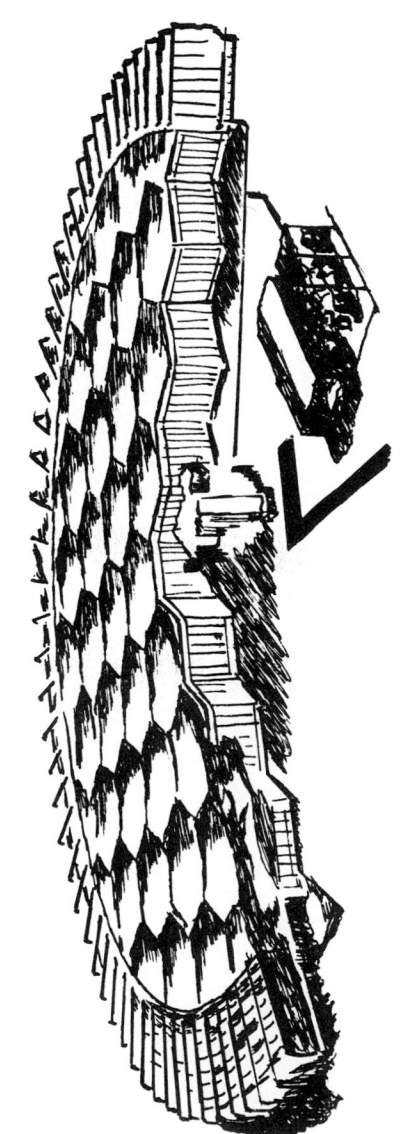

Fig. 23.6-20 Artist's conception of a mirror reactor, showing the reacting core surrounded by the expansion fan of the direct convertor. Reproduced with permission from Ref. [27].

uses a combination of the fission and fusion processes. The thermonuclear reactions are used not so much to produce energy as to produce neutrons. The absorbing blanket, instead of consisting of the light elements, is made of natural uranium or thorium, cooled by liquid lithium. Energy release and neutron multiplication now occur through the fast fission process and, in addition to breeding tritium, the system breeds plutonium as well. Even if a breeder of this kind produced no energy, it would be an important energy source, for the production rate of thermonuclear neutrons per unit of energy produced exceeds the specific rate of production of fission neutrons by a factor of ~ 10 and hence much better breeding rates of plutonium can be expected than in the fast-fission reactor.

In spite of the optimism shown by much of the fusion community, the path ahead is not clear. The fraction of basic plasma theory, the necessary foundation for engineering studies, that is firmly based and experimentally verified is modest. Some unexpected instability may limit the performance of Tokamaks or the expected instabilities may saturate at a higher level than estimated. The low values of β to which Tokamaks are limited may render the entire scheme economically impossible, and uncontrolled synchrotron radiation may display unexpected effects. On the other hand, the "scientific demonstration" lies within an order of magnitude both in T and in $n\tau$ and only a modest extrapolation is required, particularly in view of recent progress, and there seems every hope that it can be realized.

If the "scientific demonstration" reactor, the next generation of fusion experiments, comes up to expectations, the technical possibility of building a fission-fusion breeder will have been demonstrated. Such a system is far from realizing the pollution-free fusion system, but would be a great step toward solving energy problems. On the other hand, although our scientific knowledge and technical ability make a fusion-fission hybrid possible, the question has been raised as to whether or not our standards of morality, political responsibility and social sophistication are adequate for a plutonium-based energy supply.

After a successful scientific demonstration, a modest advance in plasma technology would allow the development of a pure fusion DT system, which represents a much less severe environmental hazard even than a fission reactor, for the tritium charge in the reactor volume amounts to less than a kilogram and the amount in the blanket absorbed on metals need be only a little larger. The blanket design can be selected so that a minimum of neutron-induced reactivity is produced. Beyond these lies the ultimate goal, the DD reactor, whose development will remain as a challenge to the originality, inventiveness, industry and wisdom of the next generation.

23.7 Introduction to Laser-Driven Fusion

If nature does not present unforeseen obstacles, experiments demonstrating the scientific feasibility of laser-driven fusion will be done in the seventies. Major effort will then turn on the larger lasers required for further tests to meet generator requirements and toward the design and economic analyses of reactor configurations needed for power-reactor development.

The classical "breakeven" experiment, which is planned as a critical test of feasibility, is the production of fusion energy in a small pellet of thermonuclear fuel heated and very highly compressed by proper application of energy in a laser pulse, with fusion energy comparable to the incident laser energy. The phenomena which will make this laser-fusion experiment possible are complex and difficult to analyze. Quantitative prediction must be based on calculations with high speed computers which allow simulation of physical processes. The important phenomena which must be included are laser-plasma coupling, electron and ion thermal conduction, electron-ion energy exchange, hydrodynamic motion with shock formation, nuclear reactions, and energy transfer by radiation, α-particles, and neutrons and by the recoil of charged particles from neutron collisions. The calculations are somewhat more difficult for small fusion pellets. Theoretical analyses must also address the problems of laser-plasma coupling instability, the possible production of fast electrons in the laser-deposition region which may subsequently preheat the dense pellet core, the possible hydrodynamic instabilities and the effects of laser-illumination asymmetry which may reduce the symmetry of the implosion and hence reduce the pellet compression. With these complications, the present computations can only be a guide to the necessary experiments. The calculations are of sufficient interest to have stimulated the recent rapid growth of work on laser-driven fusion.

Computations show that the minimum laser energy that is required for the fusion-energy output from the DT pellet to equal the laser-energy output is somewhat less than one kj. The fusion yield rises rapidly with laser energy and is predicted to exceed the laser energy by a factor of 20 at 10 kj of laser energy and by a factor of 60 at 100 kj. Experiments will give lower fusion yields because of departures from the idealized computer studies. The physics may also be inadequate in the codes, giving effects which may degrade or enhance the fusion yield.

A difficulty in the operation of the high-power lasers arises from non-linear effects in the lasing medium, which result from the enormous peak powers in the laser pulse. These effects may amplify laser-intensity fluctuations, modifying the gain or eventually leading to damage in the laser glass. These are not difficulties in principle since they can be solved by reducing the power density in the pulse and compensating by an increase in the laser aperture. Neodymium-doped glass lasers are being constructed in the U.S. to produce the desired laser pulses at the kj level from disc amplifiers of about 10 cm diameter driven by rod amplifiers, with energy extraction of several j/cm^2. Similar developments are also underway in the Soviet Union, France, Germany, and Japan.

Stellar fusion reactions producing energy were correctly analyzed in the thirties. Hans Bethe showed that a sequence of fusion reactions, the "carbon cycle", was the source of energy in the sun. He received the Nobel Prize in 1971 for this work. In the Manhattan project and following the Second World War, attention was given to the release of fusion energy. These studies showed that the fusion reaction of deuterium with tritium is by far the easiest to produce. Reaction of neutrons with lithium is a highly effective source of tritium, the reaction proceeding both by inelastic collisions of the fast neutrons with Li^7 and by capture of slow neutrons in Li^6 with the production of about 1.5 tritons for each initial neutron. Since each DT reaction produces one neutron and an alpha particle, the lithium reactions allow for copious production of tritium from a fusion source. In a mixture of deuterium and tritium, the tritium may be produced by neutron capture in a moderating blanket of lithium surrounding the reaction chamber.

An important approach to controlled fusion power is implementation of a pulsed release of energy on a scale and under conditions suitable for controlled production of power. This approach may be generally characterized as the method of inertial confinement, with the reaction time set by the disassembly of the thermonuclear fuel under the high pressure of the reacting fuel, the expansion being retarded only by the inertia of the fuel.

The method of inertial confinement removes the constraints present in a quasi-stationary system, which arise from the limits on density and externally sustained pressure. Since a thermonuclear fuel, reacting by particle collisions, burns more rapidly at high density with the rate per unit volume increasing as the square of the density, high density fuels are desirable. The penalty of inertial confinement, however, is that the time scale of the process is set by the dynamics of the burning fuel and cannot be controlled by external means. Inertial confinement is, therefore, of interest only if the gain in energy production, which results from the removal of the constraint on density, is not offset by the loss of control of the reaction time.

The characteristics of an energy source suitable for producing the temperature and density required for efficient thermonuclear energy are dependent on the fuel-pellet size and density. A typical pellet radius for achieving useful energy production is approximately 1 mm. A DT sphere of 1 mm at solid density (0.19 g/cm^3) and at 10 kev has a thermal energy of about 1 Mj (megajoule). This energy unit may be compared with 1 kwh (3.6 Mj). If this

energy is to heat the pellet before appreciable expansion has occurred, the
energy must be delivered to the pellet in a time less than the time in which
the central density drops significantly. This "hydrodynamic disassembly"
time is equal to the sound transit time from the pellet surface to the center.
For a DT plasma, the sound velocity at 10 kev is 10^8 cm/sec. Thus, at
10 kev and 1 mm radius, the hydrodynamic disassembly time is 10^{-9} sec.
The power required to supply 1 Mj in 10^{-9} sec is 10^{15} w. For comparison,
the power output of Hoover Dam is about 2×10^9 w and the total installed
electric power in the U.S. is about 10^{12} w. It is, therefore, apparent that a
non-conventional power source of extraordinary characteristics is necessary
to provide the very short pulse of enormous power required to heat an iner-
tially confined pellet. The essential requirement for this pulsed source is
that power be drawn slowly from a conventional source (e. g. , 10 kw for 100
sec) and accumulated, then released in a time 10^{-11} of the accumulation time.
This performance cannot be approached with conventional electronic and
switching circuits, which are limited to times of the order of fractions of
microseconds. A further method of time compression is, therefore, required.
The only method which is now feasible is a pulsed laser, which can extract the
energy stored in a lasing medium in a time as short as 10^{-12} sec. In preparation, the laser pulse can be excited by flash lamps or an electrical discharge in a time commensurate with that used in ordinary electrical circuits.

   The laser also meets a second requirement, which is to produce a
pulse of energy which can be directed and concentrated relatively easily on a
small fusion target.

23.8 Analytic Estimates for Energy Production

    A. Scaling Laws

In this section, we analyze the scaling laws for energy requirements and fusion-energy production using simplified models. This analysis is intended to clarify the dominant phenomena, which are somewhat obscured in computer studies.

The methods used in this section are intended to be only semi-quantitative and we use approximations which are intuitively plausible. More accurate methods may be used if increased precision is desired. Quantitative results cannot be obtained by rather simple methods. Our purpose is pedagogical, to clarify the dominant features of the phenomena.

The thermonuclear fuel which is most readily brought to reaction is DT. It reacts by the process

$$D + T \rightarrow n + \alpha + W, \quad W = 17.6 \text{ Mev.} \tag{23.8-1}$$

About 80% of this energy goes into the neutron and the balance to the 3.6 Mev α-particle. The nuclear cross section has a resonance near zero energy, with the effect of the Coulomb barrier included, and it reaches a peak of about 5 barns at 125 kev. According to Wandel et al,[1] an appreciable reaction rate results at several kev (1 kev = 1.16×10^7 deg K). In contrast to the DT reaction, the DD reaction, producing $T + p$ and $n + He^3$ through two equally

[1] C. F. Wandel, T. Hesselberg and O. Kofoed-Hansen, "A Compilation of Some Rates and Cross Sections of Interest in Controlled Thermonuclear Research," Nucl. Instrum. Methods 4, 249-260 (1959).

probable reactions, has a reaction rate about a factor of 30 lower at several kev and hence is much harder to bring to the condition at which significant fusion energy is produced.

For DT fuel, the dimension, time scale, and laser-energy requirement can be readily estimated for a specified level of multiplication of the laser energy by the fusion process. To obtain the appropriate relationship, we define an energy multiplication factor by

$$M = E_{fusion}/E_{laser}(output)$$

and an efficiency for coupling of laser energy into the thermal energy of the DT by

$$\epsilon_L = E_{plasma}(thermal)/E_{laser}(output). \qquad (23.8-2)$$

The first factor M, which measures overall energy gain in a practical application, depends on the external laser characteristics and on the system for converting laser energy into fusion energy. The second factor ϵ_L is determined by the laser-plasma interaction and by the mechanism of energy transfer into hydrodynamic motion and DT compression. The rate of fusion energy production is

$$(dE/dt)_{fusion} = (4\pi r^3/3)n_D n_T \overline{\sigma v}(\theta_i)W = (4\pi r^3/3)(n^2/4)\overline{\sigma v}(\theta_i)W \qquad (23.8-3)$$

with $n = n_D + n_T$ = total ion density and $\overline{\sigma v}$ the reaction rate averaged over the thermal ion distribution. The initial thermal energy is

$$E_{thermal} = (4\pi r^3/3)(3n/2)[\theta_e(0) + \theta_i(0)], \qquad (23.8\text{-}4)$$

where θ_e and θ_i are, respectively, the electron and ion temperatures times Boltzmann's constant. To proceed, we first need an estimate of the ratio of the electron-to-ion temperature. The laser energy deposition is almost entirely to the electrons, which subsequently heat the ions. In the absence of hydrodynamic and non-linear effects, the energy transfer is by electron-ion collisions and, for Maxwellian distributions, is given by the well-known equation

$$d\theta_i/dt = (\theta_e - \theta_i)/\tau_{th} , \qquad (23.8\text{-}5)$$

with the thermalization time[2]

$$\tau_{th} = \frac{3m_e m_i}{8(2\pi)^{\frac{1}{2}} ne^4 \ell n\Lambda} \left(\frac{\theta_e}{m_e}\right)^{3/2} . \qquad (23.8\text{-}6)$$

For the temperature and density range of interest, we set $\ell n\Lambda = 5$ and m_i equal to the average ion mass, giving

$$n\tau_{th} = 5.0\,\theta_e^{3/2}\,10^{12} \text{ sec cm}^{-3}, \quad \theta_e \text{ in kev}. \qquad (23.8\text{-}7)$$

---

[2] L. Spitzer, Jr., Physics of Fully Ionized Gases, 2nd ed., p. 80, Interscience Publishers, Inc., New York, 1961.

The ions and electrons reach the same temperature if the time available for the reaction is considerably greater than τ_{th}, giving the condition

$$nt \gg 5.0 \times 10^{12} \, \theta_e^{3/2} \, cm^{-3} \, sec = 1.58 \times 10^{14} \, cm^{-3} \, sec, \quad \theta_e = 10 \text{ kev.}$$

$$(23.8\text{-}8)$$

For the moment, we assume that this condition is satisfied and set $\theta_e = \theta_i$ in Eq. (23.8-4). We will at the end of this section determine the conditions for which this assumption is valid.

For the case of weak heating, the temperature does not increase appreciably during the reaction and the reaction constant in Eq. (23.8-3) can be evaluated at the initial temperature. Setting $\theta_e(0) = \theta_i(0) = \theta_0$ in Eqs. (23.8-3) and (23.8-4), we find

$$E_{fusion} = \frac{4}{3} \pi r^3 n^2 \, \overline{\sigma v} (\theta_0) Wt, \quad E_{thermal} = 4\pi r^3 n \theta_0. \qquad (23.8\text{-}9)$$

The condition for energy multiplication is then

$$E_{fusion} = (M/\epsilon_L) E_{thermal}, \qquad (23.8\text{-}10)$$

giving

$$nt = \frac{M}{\epsilon_L} \frac{12\theta_0}{W \overline{\sigma v}(\theta_0)}. \qquad (23.8\text{-}11)$$

Equation (23.8-11) now allows us to estimate the energy requirements. To relate the time to the pellet dimension, we use the time for a rarefaction wave to move from the surface to the center of the pellet,

$$t = r/v_{sound}. \qquad (23.8\text{-}12)$$

The sound velocity is

$$v_{sound} = v_0 \theta^{\frac{1}{2}}, \ \theta \text{ in kev}, \ v_0 = 3.5 \times 10^7 \text{ cm/sec}. \qquad (23.8\text{-}13)$$

The laser energy is then

$$E_L = (1/\epsilon_L) \, E_{thermal} = 4\pi(M^3/\epsilon_L^4 n_0^2)\theta_0 \left[\frac{12 \ \theta_0^{3/2} \ v_0}{W \overline{\sigma v} \ (\theta_0)} \right]^3. \qquad (23.8\text{-}14)$$

If the coupling efficiency ϵ_L is assumed to be independent of temperature, the laser energy given by Eq. (23.8-14) has a broad minimum at 10 kev. At this temperature,

$$nt = 5.7 \, (M/\epsilon_L) \, 10^{13} \text{ cm}^{-3}\text{-sec}, \ E_L = \frac{M^3}{\epsilon_L^4} \frac{n_s^2}{n_0} \ 1.6 \text{ Mj},$$

$$n_s = 4.5 \times 10^{22}/\text{cm}^3 \text{ (solid density)}. \qquad (23.8\text{-}15)$$

The nt product for $M/\epsilon_L \approx 2$ is the often quoted "Lawson condition" for energy

breakeven to be reached.[3]   For $M/\varepsilon_L = 2$, the thermalization condition of
Eq. (23.8-8) is not satisfied and the derivation of Eq. (23.8-11) is, therefore,
not valid.   We turn next to a further discussion of the conditions under which
the assumption of thermalization is correct.

B.   Direct Laser Heating of Uncompressed Spheres

The results of Eq. (23.8-15) show the strong dependence of the laser
energy on the parameters M and ε_L and on the pellet density.   Similar results
have been obtained in Refs. [4] to [11] and were the cause of much discussion

[3] J. D. Lawson, "Some Criteria for a Power Producing Thermonuclear Re-
     actor," Proc. Phys. Soc. London B70, 6-10 (1957).

[4] O. N. Krokhin, "Self-Regulating Regime of Plasma Heating by Laser Radi-
     ation," Z. Angew. Math. u. Phys. 16, 123-124 (1965).

[5] J. M. Dawson, "On the Production of Plasma by Giant Pulse Lasers,"
     Phys. Fluids 7, 981-987 (1964).

[6] N. G. Basov and O. N. Krokhin, "The Conditions of Plasma Heating by the
     Optical Generator Radiation" in Proceedings of the Third International Con-
     gress on Quantum Electronics, pp. 1373-1377, Columbia University Press,
     New York, 1964.

[7] A. Caruso, B. Bertotti and P. Guipponi, "Ionization and Heating of Solid
     Material by Means of a Laser Pulse" Nuovo Cimento B45, 176-189 (1966).

[8] Yu. P. Raizer, "Heating of a Gas by a Powerful Light Pulse," Zh. Eksp.
     Teor. Fiz. 48, 1508-1519 [Sov. Phys. JETP 21, 1009-1017] (1965).

[9] A. Caruso and R. Gratton, "Some Properties of the Plasmas Produced by
     Irradiating Light Solids by Laser Pulses," Plasma Phys. 10, 867-877 (1968).

[10] P. Mulser and S. Witkowski, "Numerical Calculations of the Dynamics of a
     Laser Irradiated Solid Hydrogen Foil," Phys. Lett. A28, 703-704 (1969).

[11] A. F. Haught and D. H. Polk, "Plasmas for Thermonuclear Research Pro-
     duced by Laser Beam Irradiation of Single Solid Particles" in Proceedings
     of the Conference on Plasma Physics and Controlled Nuclear Fusion Re-
     search, Culham 1965, Volume II, pp. 953-968, International Atomic Energy
     Agency, Vienna, 1966.

on realistic values for the parameters. The laser requirement for M = 1 (fusion energy equal to thermal energy) and for ϵ_L = 1 (perfect coupling) and for a pellet at solid density is 1.6 Mj with a pulse length of less than 10^{-9} sec. This is a very formidable technological requirement, the largest present laser in this pulse length range giving about 600 j.[12] The assumption of perfect coupling is, of course, highly optimistic under most conditions since perfect coupling cannot occur unless energy is not reflected at the pellet surface or lost through blow-off in producing hot low-density plasma that expands from the pellet surface. Coupling sufficient to heat the pellet at solid density is, in fact, possible only if the laser energy can easily penetrate into the pellet interior. Penetration can occur if the laser wavelength is sufficiently short so that the critical density of the plasma is slightly above the solid density. In this case, the laser light penetrates directly, heats the solid pellet, and essentially complete energy deposition can result. A laser wavelength of about 1500A is required. For such a laser, the assumptions ϵ_L = 1 and uniform heating of the electrons are reasonable. Lasers in this wavelength region have not operated successfully with a significant energy output. The technological problems, therefore, are even more severe than those at wavelengths of 1 μm or longer; work continues on development of lasers of this type.

Detailed numerical calculations relevant to the model of efficient pellet heating were made by Chu,[13] who assumed an energy source in the electrons of the pellet corresponding to classical laser heating but did not study the actual

---

[12]N. G. Basov, O. N. Krokhin and G. V. Sklizkov,"Heating of Laser Plasmas for Thermonuclear Fusion"in Laser Interaction and Related Phenomena, p. 389, Plenum Press, New York, 1972.

[13]M. S. Chu, "Thermonuclear Reaction Waves at High Densities," Phys. Fluids 15, 413-422 (1972).

problem of energy deposition. He followed correctly the electron-ion energy exchange, the hydrodynamic motion of the pellet, and the fusion reactions. He showed that the estimate based on Eq. (23.8-14) was very optimistic, the correct laser energy requirement for breakeven being increased by a factor of about 600 or to about 1000 Mj. The large increase was due, in part, to numerical factors arising from the actual details of the hydrodynamic expansion but was mainly due to the marked lag in the ion-temperature increase since the time available during expansion is insufficient for thermal equilibration between the electrons and ions. In his calculation, no significant pellet compression could take place because the rapid heating throughout the pellet causes immediate pellet expansion.

C. Summary of Results for Uncompressed Spheres

The uncompressed sphere heated efficiently by a short wavelength laser, which can penetrate the solid density plasma, requires approximately 10^5 Mj of laser input to reach a comparable fusion output. This prohibitive laser-energy requirement is even further increased if the requirement of useful energy production is imposed. The energy multiplication M for practical application is determined by the efficiency of conversion of fusion energy into electrical energy and of electrical energy input to the laser into laser energy. Assuming a thermal-to-electrical conversion of 40% efficiency and 25% efficiency for the laser, the required energy multiplication for overall energy breakeven is equal to 10. Equation (23.8-15) shows that this increases the laser energy requirement by an additional factor of 10^3!

It is therefore clear that any expectation of producing useful fusion energy in the absence of compression is unrealistic. For a compressed sphere, a new problem results since the coupling efficiency ϵ_L to a compressed system becomes much less than unity. Results to be given later show that M/ε_L is, in fact, under the best conditions, equal to approximately 200, requiring that the Lawson condition be exceeded by a large factor. Under these conditions, the thermalization requirement of Eq. (23.8-8) is met. However, the large multiplication of the initial thermal energy by the fusion process invalidates the assumption that the temperature remains constant during the reaction. We therefore turn next to an evaluation of the effects of compression and of fusion-energy deposition on the preceding analysis.

D. Uniform Sphere with Compression and Self-Heating

We now remove the assumption of no heating and return to Eqs. (23.8-3) and (23.8-4), setting $\theta_e = \theta_i = \theta$. The heating equation is then

$$3n(d\theta/dt) = (dE/dt)_{fusion,\ deposited} = (n^2/4)\overline{\sigma v}(\theta)W_{dep}, \qquad (23.8-16)$$

with W_{dep} the fraction of the fusion energy W deposited in the fuel. The time available is determined by the temperature; to allow for the variation of sound velocity as the temperature increases, we change variable from t to r(t) with dr/dt the sound velocity. Equation (23.8-16) may then be written

$$\int_{\theta_0}^{\theta_1} \frac{d\theta v_0 \theta^{\frac{1}{2}}}{\overline{\sigma v}(\theta)W_{dep}} = \frac{1}{12} \int_0^r ndr. \qquad (23.8-17)$$

The ratio of fusion energy produced to initial thermal energy is

$$\frac{E_{fusion}}{E_{thermal}}(0) = \frac{W}{12\theta_0} \int_0^r n\,\overline{\sigma v}\,(\theta)dt$$

which, using Eq. (23.8-16), can be rewritten as

$$\frac{E_{fusion}}{E_{thermal}}(0) = \frac{W}{\theta_0} \int_{\theta_0}^{\theta_1} \frac{d\theta}{W_{dep}(\theta)} . \tag{23.8-18}$$

The analysis may now be completed by an estimate of the energy deposited.

The deposition of energy by neutrons, due to the charged-particle recoils, depends on the ratio of pellet size to neutron range. For 14 Mev neutrons, the neutron range in DT is approximately 50 times larger than the α-particle range at 5 kev electron temperature [see Eq. (23.8-20)]. Consequently, α-particle heating dominates the initial ignition sequence. Neutron heating can become important for large pellets at high density. Simple estimates show that this occurs in the Mj energy-input range for pellets with density of the order of 10^3 g/cm^3. This result may be determined with sufficient accuracy for the present purpose by ignoring possible neutron deposition and accounting approximately for the α-particle energy. The plasma is relatively transparent to the α-particles, particularly at high temperatures. The fraction of the energy deposited is approximately r/R_α for large R_α and unity for $R_\alpha \to 0$. We interpolate roughly between these limits by assuming

$$W_{dep} = W_\alpha \frac{r}{R_\alpha + r} , \quad R_\alpha = \alpha\text{-particle range}, \quad W_\alpha = \alpha\text{-particle energy (3.6 Mev).}$$

$$\tag{23.8-19}$$

The α-particle range is determined by energy loss to electrons and to ions, the loss to electrons being dominant for electron temperatures below approximately 40 kev. In this temperature range,[14]

$$\frac{dE_\alpha}{dx} = - \frac{32\pi^{\frac{1}{2}}}{3} \, ne^4 \ln\Lambda \, \frac{E_\alpha^{\frac{1}{2}}}{\theta_e^{3/2}} \left(\frac{m_e}{m_\alpha}\right)^{\frac{1}{2}},$$

giving[15]

$$R_\alpha = \lambda_0(\theta_e^{3/2}/n), \quad \lambda_0 \cong 2 \times 10^{21} \text{ cm}^{-2}, \quad \theta_e \text{ in kev.} \qquad (23.8\text{-}20)$$

Thus, the integral in Eq. (23.8-18) gives

$$\frac{E_{fusion}}{E_{thermal}(0)} = \frac{W}{\theta_0 W_\alpha} \int_{\theta_0}^{\theta_1} \left(\frac{\lambda_0 \theta^{3/2}}{rn} + 1\right) d\theta. \qquad (23.8\text{-}21)$$

We now make the final approximation of this section and evaluate r and n at their initial values. This approximation neglects the hydrodynamic motion and fuel depletion during the burning of the DT. The depletion effect limits the

[14]D. J. Sigmar and G. Joyce, "Plasma Heating by Energetic Particles," Nucl. Fusion 11, 447-456 (1971).

[15]D. J. Hughes and R. B. Schwartz, Eds., Neutron Cross Sections, Brookhaven National Lab., Report No. 325, p. 58, U.S. Government Printing Office, 1958.

energy multiplication M/ϵ_L to $W/6\theta_0 = 3000$ kev/θ_0. We use Eq. (23.8-2) for $E_{fusion}/E_{thermal}(0)$ and we set $W = 5W_\alpha$. The result is

$$\frac{M}{\epsilon_L} = \frac{2\lambda_0}{n_0 r_0 \theta_0} (\theta_1^{5/2} - \theta_0^{5/2}) + 5 \frac{\theta_1 - \theta_0}{\theta_0}. \tag{23.8-22}$$

A similar approximation in Eq. (23.8-17) yields

$$\frac{v_0}{W_\alpha} (I_1 + \frac{\lambda_0}{n_0 r_0} I_2) = \frac{n_0 r_0}{12} \tag{23.8-23}$$

with I_1 and I_2 defined by

$$I_1 = \int_{\theta_0}^{\theta_1} \frac{\theta^{\frac{1}{2}} d\theta}{\overline{\sigma v}(\theta)}, \quad I_2 = \int_{\theta_0}^{\theta_1} \frac{\theta^2 d\theta}{\overline{\sigma v}(\theta)}. \tag{23.8-24}$$

Equation (23.8-23) expresses $n_0 r_0$ as a function of θ_0 and θ_1 from which Eq. (23.8-22) gives M/ϵ_L. The corresponding laser-energy requirement is

$$E_L = 4\pi \frac{r_0^3}{\epsilon_L} n_0 \theta_0. \tag{23.8-25}$$

From these results, we readily obtain the case of weak heating, as given in Eq. (23.8-15), by going to the limit $\lambda_0 \to \infty$.

The temperature rise for a small temperature increase, for $R_\alpha \gg r$, is

$$\theta_1 - \theta_0 = \frac{12 v_0}{25 W_\alpha \lambda_0} \left(\frac{M}{\epsilon_L}\right)^2 \frac{\theta_0}{\overline{\sigma v}(\theta_0)} \; ; \qquad (23.8\text{-}26)$$

at 10 kev,

$$\theta_1 - \theta_0 = 0.233 \left(\frac{M}{\epsilon_L}\right)^2 \text{ kev.} \qquad (23.8\text{-}27)$$

Thus, appreciable heating does not occur for M/ϵ_L of order of unity. For M/ϵ_L equal to 4, however, the temperature rise is about 4 kev and the reaction rate has about doubled. This result already gives substantial correction to the energy requirement; the correction is clearly very large for large energy multiplication. Under such conditions, evaluation of Eqs. (23.8-22) and (23.8-23) is necessary. The results are given in Fig. 23.8-1 for several values of the initial temperature θ_0. The very large departure of the correct result from the weak heating approximation is apparent. The laser-energy requirement drops by a factor of 400 for the ratio of fusion energy to initial energy M/ϵ_L = 100. This large drop in laser-energy requirement results from the DT temperature reaching 30-50 kev where the DT reaction rate has a maximum, increasing by a factor of about 10 over the rate at 10 kev and a factor of 100 over the rate at 4 kev. The laser, therefore, is required only to provide the relatively small ignition energy of the fuel, which subsequently through self-heating reaches the optimum temperature range for efficient fusion-energy release. The optimum initial temperature changes with energy multiplication; for small values of M/ϵ_L, the optimum is at about 10 kev, as previously pointed out for

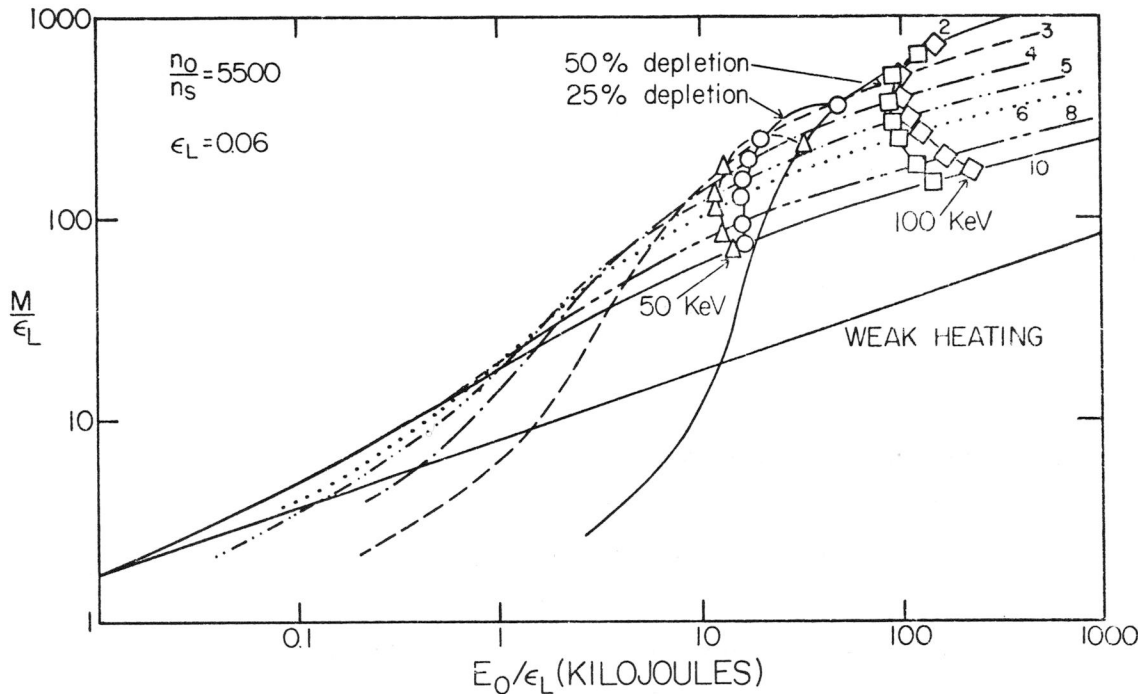

Fig. 23.8-1 Analytic results for laser energy as a function of M/ϵ_L, based
on Eqs. (23.8-25), (23.8-26) and (23.8-27), for a compression
ratio of 5500 and a laser coupling efficiency ϵ_L of 0.06. The
points of 25% and 50% fuel depletion and of 50 and 100 kev final
temperature are indicated. Reproduced from studies at KMSF
performed by K. A. Brueckner et al.

the weak heating case. For larger values of M/ϵ_L, the optimum temperature drops well below 10 kev, reaching 2 kev for M/ϵ_L greater than about 350. The achievable minimum, which is of interest in practice, depends on other loss processes which prevent heating, the most obvious being radiation loss by bremsstrahlung. The total free-free emission rate per unit volume exceeds the α-particle heating rate below about 4 kev, which gives a lower limit for the occurrence of self-heating if the DT is transparent to the radiation. The condition for transparency is that the volume emission not exceed the surface blackbody radiation, or $V\epsilon_{ff} \leqq 4\pi r^2 \sigma T^4$, with ϵ_{ff} the emissivity and σ the Stefan-Boltzmann constant. Using[16]

$$\epsilon_{ff} = 4.9 \times 10^{-24} n^2 \theta^{\frac{1}{2}} (\text{erg/cm}^3\text{-sec}), \ \theta \text{ in kev},$$

$$\sigma T^4 = 1.05 \times 10^{24} \theta^4 (\text{erg/cm}^2\text{-sec}), \qquad\qquad (23.8\text{-}28)$$

we find the transparency condition

$$rn^2 \leqq 6.43 \times 10^{47} \ \theta^{7/2} \ \text{cm}^{-5}. \qquad\qquad (23.8\text{-}29)$$

Combining this with Eq. (23.8-11), using $t = r/v(\theta)$, and evaluating the resulting expression at 4 kev, we obtain for the transparency

$$n/n_s \leqq 7.75 \times 10^4 \ (\epsilon_L/M). \qquad\qquad (23.8\text{-}30)$$

[16] C. W. Allen, Astrophysical Quantities, 2nd ed., p. 100, University of London, The Athlone Press, 1963.

This result shows that the fuel is transparent unless it is very highly compressed under conditions of large energy multiplication, i.e., large M/ϵ_L. However, under the conditions of interest for laser fusion, with high compression and large values of M/ϵ_L, the DT can become opaque to bremsstrahlung and fuel heating can occur for temperatures considerably lower than 4 kev. Thus, the energy requirement given by Eq. (23.8-28) and Fig. 23.8-1 can be used for the lower values given for θ_0. The exact treatment of the radiation loss requires more careful analysis of the radiation problem.

E. Nonuniform Compressed Sphere with Self-Heating and Propagation

In Section 23.8D, the effect of self-heating on fusion energy production was analyzed using a simple model of the processes of energy deposition and hydrodynamic expansion. We next evaluate an extremely important effect, which occurs in a sphere with nonuniform initial conditions. The case of practical interest which arises is a result of shock convergence in a compressed sphere. Convergence produces a sufficiently high central temperature to cause central ignition of the DT fuel, which is surrounded by relatively cold fuel in which the ignition condition is not met. In this case, the energy produced in the ignited and burning central region of the fuel, which is only partially deposited in this fuel region, can heat the surrounding cold fuel sufficiently for laser ignition to occur. This process leads to the formation of a spherically expanding burning wave which can propagate throughout the fuel, causing complete ignition. The energy required to initiate the burning process is then substantially reduced from the requirement for uniform ignition.

We analyze the condition for spherical propagation using a relatively simple model which illustrates the essential features of the process.

   Spherical expansion of the burning region of the fuel results when energy escapes from the burning fuel which is deposited in the adjacent cold layers of fuel. The energy loss can result from hydrodynamic expansion, which follows the rapid pressure build-up in the ignited and burning fuel, from electron thermal conduction or from energy deposition by the reaction products escaping from the fuel. The thermal conduction is subsonic for the temperature ranges of interest in the ignition and propagation phase of the burning front, provided that the fuel density is of the order of magnitude of 10^3 g/cm^3. The latter process usually dominates, the propagation of the burning front being sufficiently supersonic to advance more rapidly into the cold fuel than the hydrodynamic disturbance from the pressure increase of the thermal conduction front. The rate of advance of the burning front then follows from energy conservation, ignoring the hydrodynamic motion. The rate of energy production in a uniformly burning region at density n_0 is

$$\dot{U}_f = (4/3)\pi r^3 \overline{\sigma v} W_\alpha (n_0^2/4) \tag{23.8-31}$$

ignoring the neutron energy which is only weakly deposited in the burning region until the region increases in size and becomes an appreciable fraction of the neutron mean free path. The rate of change of internal energy in the expanding region, assuming negligible energy in the cold fuel and uniform density, is

$$\frac{d}{dt}\left(\frac{4}{3}\pi r^3 n_0 \theta_0\right) = \frac{4\pi}{3} n_0 r^3 \dot{\theta} + 4\pi n_0 r^2 \theta \dot{r}. \tag{23.8-32}$$

Energy conservation therefore gives for the propagation velocity of the burning front

$$\dot{r} = \frac{n_0 \overline{\sigma v} W_\alpha}{12\theta_0} r - \frac{1}{3} r \frac{\dot{\theta}_0}{\theta_0}. \tag{23.8-33}$$

The temperature in the burning fuel increases rapidly from the α-particle heating until the increase in α-particle range with increasing electron temperature causes the α-particles to escape into the surrounding cold fuel. The consequence is that the temperature adjusts itself so that the α-particle range is approximately equal to the radius of the burning region and

$$r \cong R_\alpha = \lambda_0 \theta_0^{3/2}/n_0. \tag{23.8-34}$$

Equation (23.8-34) breaks down if the electron temperature exceeds roughly 40 kev, the contribution of the ions no longer being negligible. The propagation of the burning front is then changed and a more detailed analysis is required. Since

$$\dot{\theta}_0/\theta_0 = 2\dot{r}_0/3r_0, \tag{23.8-35}$$

the ratio of the velocity of the burning front to the sound velocity $v_s = v_0 \theta^{\frac{1}{2}}$ is therefore

$$\frac{\dot{r}}{v_s} \cong \frac{3 \overline{\sigma v} W_\alpha \lambda_0}{44 v_0}. \tag{23.8-36}$$

With the values previously given for v_0 and λ_0, Eq. (23.8-36) gives

$$\dot{r}/v_s = 1.37\ (10^{16}\ \text{sec/cm}^3)\ \overline{\sigma v}. \tag{23.8-37}$$

The burning front therefore advances supersonically ($\dot{r}/v_s \approx 2$) if the reaction rate is about 2×10^{-16} cm^3/sec, which occurs at 15 kev. At a lower temperature, the hydrodynamic expansion dominates and additional energy production in the burning region is required to maintain energy balance. At a higher temperature, the burning front becomes strongly supersonic, $\dot{r}$ reaching a peak of about $10v_s$ at temperatures between 40 and 50 kev. The front also accelerates further as the neutron heating becomes important.

We now use the preceding results as the condition for supersonic propagation of the burning front. To determine the condition for this required temperature of 20 kev or greater to be reached in a central region of radius r_0 and density n_0 initially heated to θ_0, we use the results of Section 23.8C. Equation (23.8-33) gives the condition

$$n_0 r_0 = \frac{6v_0}{W_\alpha} \left[I_1 + \left(I_1^2 + \frac{W_\alpha \lambda_0 I_2}{3v_0} \right)^{\frac{1}{2}} \right]. \tag{23.8-38}$$

Evaluating this at $\theta_0 = 4$ kev and $\theta_1 = 20$ kev, we obtain

$$(n_0/n_s)r_0 = 2.81. \tag{23.8-39}$$

The initial thermal energy to produce central ignition is then

$$E_{th} = 4\pi n_0 r_0^3 \theta_0 = 7.99 \times 10^6 \left(\frac{n_s}{n_0} \right)^2 \text{kj}. \tag{23.8-40}$$

The initial energy in the rest of the fuel, which is ignited by the expanding and burning wave, is determined by the temperature and density. The minimum energy is given by the degeneracy energy of the electrons if the temperature is much less than the degeneracy temperature. The degeneracy energy of fuel with radius R is

$$E_{deg} = (4/3)\pi R^3 n_0 \epsilon_{deg}, \quad \epsilon_{deg} = 2.68 \, (n_0/n_s)^{2/3} \text{ ev/electron,} \qquad (23.8\text{-}41)$$

and

$$E_{deg} = 80.5 \, R^3 \left(\frac{n_0}{n_s}\right)^{5/3} \text{ kj.} \qquad (23.8\text{-}42)$$

After ignition, the fuel burns at a temperature of 40-150 kev, depending on the dimension of the fuel. The fuel depletion is determined in the absence of hydrodynamic motion by

$$\dot{n} = -\overline{\sigma v}\,(n^2/2) \qquad (23.8\text{-}43)$$

giving

$$n/n_0 = [1 + \frac{\overline{\sigma v}}{2} n_0 t]^{-1}. \qquad (23.8\text{-}44)$$

The fusion yield is

$$E_f = \frac{4}{3}\pi R^3 (n_0 - n)\frac{W_{fus}}{2} = \frac{4}{3}\pi R^3 \frac{n_0^2}{4} W_{fus} \frac{\overline{\sigma v}\, t}{1 + n_0 \frac{\overline{\sigma v}}{2} t} \qquad (23.8\text{-}45)$$

For the time t we assume the hydrodynamic disassembly time

$$t = \frac{R}{v_s(\theta_{burn})} \qquad\qquad (23.8\text{-}46)$$

and burning at 50 kev for which $v_s = 3.47 \times 10^8$ cm/sec and $\sigma v \cong 10^{-15}$ cm^3/sec; the result is

$$E_f = \frac{1.75 \times 10^7 \, (n_0/n_s)^2 R^4}{1 + 0.0648 \, R\, n_0/n_s} \text{ kj.} \qquad (23.8\text{-}47)$$

For a fixed compression ratio, Eqs. (23.8-39) and (23.8-42) determine R as a function of the initial internal energy. Equation (23.8-47) then determines the fusion yield. Results are given in Fig. 23.8-2 as a function of initial internal energy for several values of the compression. At low initial energy, the maximum yield ratio occurs for high compression. The optimum drops for large initial energy, for which the ignition energy is small compared with the degeneracy energy, leading to lower compression requirements. For initial internal energies on the 1 to 10 kj, the energy multiplication for compressions of 3×10^3 to 10^4 is several thousand.

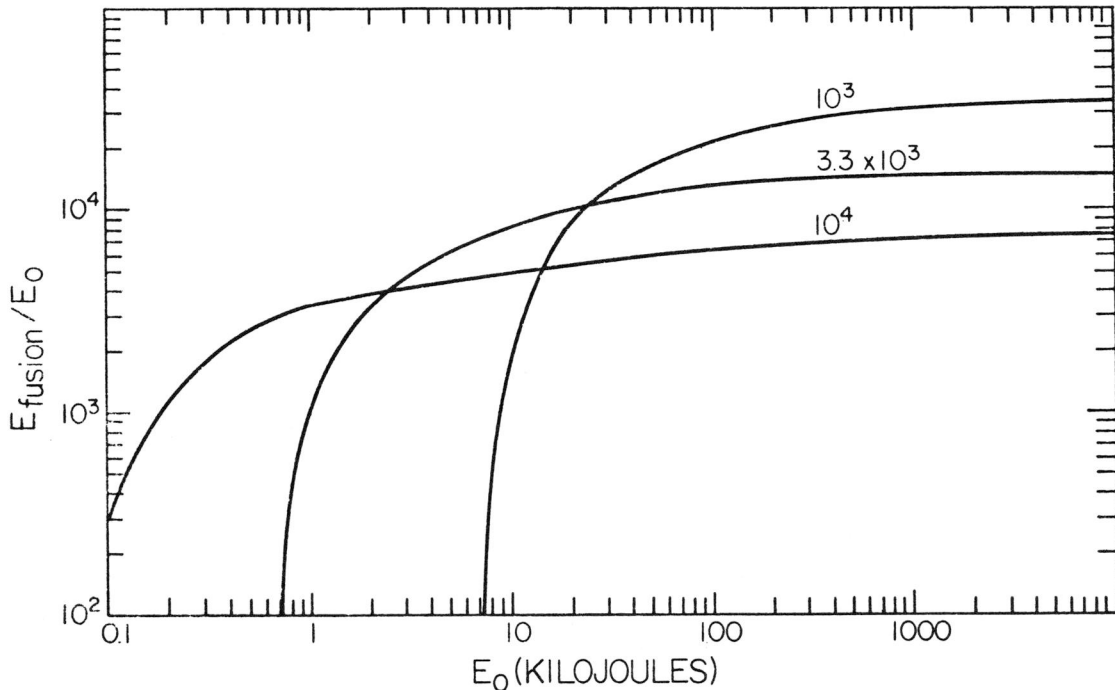

Fig. 23.8-2 Analytic estimate of fusion-energy production in cold DT fuel ig-
            nited by a spherically propagating burning wave. Equation (23.8-
            43) gives the ignition energy, Eq. (23.8-45) the degeneracy energy,
            and Eq. (23.8-47) the fusion energy. The initial energy E_0 is
            the sum of the ignition and degeneracy energy. The laser energy
            required to give this initial energy depends on the efficiency of
            laser-energy deposition and of hydrodynamic transfer to the com-
            pressed fuel, and typically is a factor of ten to twenty larger than
            E_0. The curves are labeled by the compression ratio relative
            to solid DT (0.19 g/cm^3). Reproduced from studies at KMSF per-
            formed by K. A. Brueckner et al.

23.9 <u>Compression by Shocks</u>

A. Introduction

The analysis in Section 23.8 shows that the production of useful amounts of fusion energy can result only if the pellet is highly and efficiently compressed. The compression must also be carried out under conditions which bring the central region of the pellet to ignition temperature and density but which leave the rest of the compressed pellet as cold as possible. The latter condition is essential to minimize the overall energy requirements.

The problem of compression centers around the hydrodynamics of convergent shocks and of providing the pressure at the pellet surface which is required to produce the desired hydrodynamic motion. The pressure variation is determined by the absorbed laser flux and the energy transfer from the deposition region into the ablating pellet surface. These processes thus depend critically on a correct description of the laser deposition process and the energy flow into the dense pellet. Finally, the compression may be limited by departures from spherical symmetry produced by nonuniform laser illumination, intrinsic pellet asymmetries, or by hydrodynamic instability causing amplification of small disturbances in the pellet motion.

B. General Features of the Hydrodynamics

The conditions to be achieved in the compressed pellet are qualitatively clear from the analysis in Section 23.8. The center of the compressed pellet must be brought to the ignition temperature of a few kev while the rest of the pellet is highly compressed at the lowest possible temperature. Very high compression is desirable to maximize the reaction rate after the pellet has

ignited. Too high compression requires excessive work done against the degeneracy pressure of the electrons. Excessive compression is also unnecessary since the pellet yield is eventually limited by fuel depletion.

The driving pressure producing the pellet compression results from
ablation of the dense pellet surface, which is produced by the flow of energy
from the laser-deposition region, the energy being carried by hot electrons.
If the electron distribution is Maxwellian and the electron mean free path short
relative to the depth of the overdense plasma, the electron energy is carried
by classical electron conduction, which can be described by the standard diffusion equation of heat flow. Since the conductivity of the electrons is proportional to $(\theta_e)^{5/2}$, the temperature in the conduction region is a steeply rising
function of distance from the thermal front, with most of the conduction region
being approximately isothermal.

Since the compression is ablation-driven with the velocity of material
removal at the dense pellet surface determined essentially by the local speed
at the thermal front, the time available for the hydrodynamic motion is less
than the sound transit time to the pellet center through the relatively cold
plasma ahead of the conduction front. The motion of the pellet surface must
therefore be supersonic with respect to the cold plasma and shocks necessarily
form. The shock heating, however, must be minimized since the shocks heat
the plasma irreversibly, increasing the final work which is required to reach
the desired compression. The strength of the shocks which form is determined by the pressure at the conduction front, which is fixed by the temperature in the conduction region. The temperature is determined by the laser
flux which is absorbed near the critical density surface.

The problem of minimizing the shock heating while, at the same time,
driving the pellet to high compression is qualitatively clear but the highly
non-linear nature of the process makes analysis difficult. The problem is,

however, ideally suited to computer study, which is the basis of the quantitative results given in Section 23.11. A relatively simple case of a single shock in a spherical geometry will now be considered.

C. Compression and Fusion Yield in a Single Convergent Shock

The first published suggestion of the possibility of obtaining increased fusion yield by producing pellet compression in a convergent shock is due to Daiber, Hertzberg, and Wittliff.[1] They did not analyze the problems of laser-energy deposition, thermal conduction processes and pellet ablation driving the implosion, or energy losses into the ablated plasma. Their results are, however, of considerable interest since they were among the first to suggest using a laser as a hydrodynamic driver of pellet compression rather than as a plasma heater. Their analysis was based on the compression produced by uniform pressure applied to the pellet surface which produces a single shock. This case is of interest since the results for fusion yield and shock energy can be obtained analytically from well-known results of Guderley,[2] provided that the perturbation of the hydrodynamic process by the fusion reaction is ignored. This is approximately valid under the conditions which can be produced by a single shock, if the fusion yield is not appreciably larger than the hydrodynamic energy. The passage of the first converging shock gives a density increase of a factor of four (for a gas with $\gamma = 5/3$) followed by adiabatic compression to a density ratio of about 15. The shock is then reflected at the center and on returning gives a further shock compression to a

---

[1]J. W. Daiber, A. Hertzberg and C. E. Wittliff, "Laser-Generated Implosions," Phys. Fluids 9, 617-619 (1966).

[2]G. Guderley, "Starke kugelige und zylindrische Verdichtungsstösse in der Nähe des Kugelmittelpunktes bzw. der Zylinderachse," Luftfahrtforschung 19, 302-312 (1942).

maximum density ratio of 33. This is the maximum compression that can be achieved in a spherical geometry by passage of a single shock.

The lengthy analysis required to evaluate the shock energy and fusion yield is given by Brueckner and Jorna.[3] The result, which may be expressed in the form of Eq. (23.8-15) with $E_{laser} = E_{shock}/\epsilon_L$, is

$$E_{laser} = 40.9 \left(\frac{n_s}{n_0}\right)^2 \left(\frac{E_{fusion}}{E_{laser}}\right)^3 \frac{1}{\epsilon_L^4} \text{ Mj.} \qquad (23.9-1)$$

In the original paper by Daiber et al,[1] the energy requirement for a fusion energy equal to twice the shock energy and an initial density equal to solid density is given as approximately 2 Mj. Since details of their calculations are not given, we have not been able to determine the source of the large discrepancy between their result and that given by Eq. (23.9-1). They also did not determine the coupling efficiency ϵ_L.

The surprising result of the analysis leading to Eq. (23.9-1) is that the shock compression has not reduced the laser energy requirement from that given by Eq. (23.8-15), assuming that the coupling efficiency is the same. The effect of compression appears to be offset by an inefficient temperature distribution which reduces fusion energy production in most of the fuel. The compressed center, which is strongly heated, has too small a volume and too short a compression time to contribute appreciably to the net fusion yield.

---

[3]K. A. Brueckner and S. Jorna, "Laser Driven Fusion," Rev. Mod. Physics 46, 325-368 (1974).

The effective laser coupling to the shock cannot be readily estimated from the preceding analysis, since the laser absorption, surface-pellet ablation, and the energy lost in the expanding plasma must be separately calculated.

Calculations by Lubin and collaborators[4] give for DD a fusion yield of 3.7×10^9 n for an absorbed laser energy of 927 j. Scaling this result to DT gives a fusion yield of 0.41 j.

Calculations at KMS Fusion for a solid DT sphere with a laser pulse constant in time give $E_{fusion} = E_{laser}$ at a laser energy of about 500 Mj, although the calculation differs from the model described above in that the effect of pellet self-heating is included.

The present analysis and the numerical results given show that pellet compression sufficient to give an interesting fusion yield cannot be produced by a single shock. This result follows because a single shock in spherical geometry cannot give a compression greater than a factor of about 30, which is too small to overcome the other inefficiency factors in the laser coupling and in the hydrodynamic processes.

23.10 Quasi-Adiabatic Compression of Spheres and Shells

A. Compression of Spheres

Compression larger than that which can be reached in a single shock can be produced if the fuel is subject to a rising pressure or, equivalently, to a succession of shocks of increasing strength which are adjusted in time so that the successive shocks do not overtake each other before arriving at the

---

[4]M. J. Lubin, E. B. Goldman and K. Yuan, Laboratory for Laser Energetics, Report No. 5, University of Rochester, Rochester, N.Y., 1971.

center of convergence. In either case, the compression and temperature his-tories after the passage of the first shock follow approximately an adiabat until the shock reaches the center of convergence, where the kinetic energy of motion is converted into internal energy and a reflected shock forms. The final tem-perature necessary to initiate the fusion reaction is determined, for a given final compression, by the first shock strength. It is particularly important to avoid excessive early heating of the DT by the initial shock or by successive shock coalescence, since the final temperature reached by the hydrodynamic compression may be too high. The optimum occurs when the DT, after final compression, reaches the minimum temperature required for ignition. If a proper pressure history is achieved, the achievable compression is limited finally only by the degeneracy pressure of the electrons or possibly by ignition of the fuel before the maximum compression has been reached.

As the initial pressure is applied to a cold sphere, a strong shock is formed which brings the material to a temperature θ_0 moving inward at velo-city v_0. This shock will reach the sphere center at a time t_0. As the sphere compresses, the velocity of pressure waves generated at the radius of applied pressure should increase at the rate required to prevent overtaking of the ini-tial disturbance. This gives the condition limiting the rate of pressure in-crease, viz.

$$v_s(t)(t_0 - t) = r. \tag{23.10-1}$$

If this condition is satisfied, the compression is close to adiabatic and

$$\theta = \theta_0 (r_0/r)^2, \quad p = p_0 (r_0/r)^5, \quad v_s = v_0 (\theta/\theta_0)^{\frac{1}{2}}, \tag{23.10-2}$$

so that Eq. (23.10-1) may be rewritten as

$$r^2 = r_0^2(1 - t/t_0), \quad t_0 = r_0/2v_0. \tag{23.10-3}$$

The compression at a minimum radius r_m is

$$\rho_m/\rho_0 = (r_0/r_m)^3 \tag{23.10-4}$$

at which

$$\theta_m = \theta_0(r_0/r_m)^2. \tag{23.10-5}$$

The maximum temperature θ_m is fixed by the ignition conditions, which de-termine θ_0 for a given compression. The ignition occurs from shock coales-cence within r_m and from heating by the reflected shock. The radius r_m and density ρ_m must also be sufficiently large so that α-particle heating can occur, causing fuel heating and the formation of propagating burning waves from the region of ignition. These processes require numerical analysis for a quantitative determination of θ_m and ρ_m; a reasonable approximation is θ_m = 2 kev, which leads to the 5 kev local maximum within r_m at which ignition occurs.

The absorbed laser flux, allowing for the effects of ablation-energy loss, is

$$\Phi_L(\text{absorbed}) \cong 4\pi r^2 4 v_s p, = 16\pi r_0^2 v_0 p_0 \left(\frac{r_0}{r}\right)^4$$

$$= 8\pi \frac{r_0^3 p_0}{t_0} \frac{1}{(1 - t/t_0)^2} . \tag{23.10-6}$$

The absorbed laser energy to the implosion maximum is

$$E_L(\text{absorbed}) = 8\pi r_0^3 p_0 \frac{1}{(1 - t_m/t_0)} = 8\pi r_0^3 p_0 \left(\frac{\rho_m}{\rho_0}\right)^{2/3} . \tag{23.10-7}$$

Using $p_0 = 2 n_0 \theta_0$ and Eq. (23.10-5) for θ_m,

$$E_L(\text{absorbed}) = 16\pi r_0^3 n_0 \theta_m = 4\frac{4}{3}\pi r_m^3 3\theta_m. \tag{23.10-8}$$

Thus, the absorbed laser energy is four times the energy required to bring the compressed material to θ_m. The excess laser energy is carried off in the internal and kinetic energy of the ablation products.

As an example, we consider a pellet with final compression ratio of 10^4, a final temperature of 2 kev, and a radius of 300 microns for the finally compressed DT. This pellet if burned to 30% of completion gives about a 3 Mj fusion yield. The absorbed laser energy, from Eq. (23.10-7), is

$$E_L(\text{absorbed}) = 43.6 \text{ kj.} \tag{23.10-9}$$

The energy multiplication is about 75, a typical result for an optimal implosion. The absorbed laser flux may be written

$$\Phi_L(\text{absorbed}) = \frac{E_L(\text{absorbed})}{t_0} \frac{1 - t_m/t_0}{(1 - t/t_0)^2} . \qquad (23.10\text{-}10)$$

The initial temperature, from Eq. (23.10-5), is

$$\theta_0 = \theta_m(\rho_0^{2/3}/\rho_m) = 4.32 \text{ ev}, \qquad (23.10\text{-}11)$$

giving

$$v_0 = 1.98 \times 10^6 \text{ cm/sec}, \quad t_0 = 15.2 \times 10^{-9} \text{ sec}. \qquad (23.10\text{-}12)$$

The maximum laser flux at $t = t_m$ is

$$\Phi_L(\text{absorbed}) = \frac{E_L(\text{absorbed})}{t_0} \left(\frac{\rho_m}{\rho_0}\right)^{2/3} = 1.33 \times 10^{15} \text{ w}. \qquad (23.10\text{-}13)$$

The incident laser energy and flux may be much larger than given by Eqs. (23.10-9) and (23.10-13), due to incomplete absorption in the pellet corona. This effect may be estimated from a model for the absorption mechanisms. The peak power density is very high; at the initial radius of 300 μ, the absorbed power density at $t = t_m$ is 1.3×10^{16} w/cm^2 and anomalous absorption and/or anomalous loss processes are probably present. This difficult

problem requires experimental data for a determination of the incident laser
flux, as a function of frequency, to meet the requirements of Eqs. (23.10-9)
and (23.10-13).

B. Other Configurations

The use of a high-density shell filled with initially solid or gaseous
DT changes the laser power requirements markedly. The high-density shell,
which stores kinetic energy during the implosion, produces the final temperature and pressure in the DT by transfer of energy from the shell to the DT.

To allow further reduction in the peak laser flux, a shell of DT within
a dense tamping shell can be used. In this case, for a thin DT shell, the
adiabatic condition within the shell is easily maintained. The result is a very
marked decrease in the peak laser power, which has important practical consequences for laser dosages.

23.11 Confinement in the Implosion of Spherical Shells

A. Introduction

The experimental and theoretical studies of the implosion of spherical
shells provide an excellent test of many of the important features of laser-
driven fusion. The implosion dynamics, symmetry, and stability can be analyzed using lasers in a readily available energy range and standard diagnostic
methods. A number of glass shells of varying diameter and wall thickness
have been imploded at the KMSF neodymium glass-laser facility, giving 1.06 μ
radiation. The experimental data necessary to analyze the implosions have
been obtained from the X-ray spectrum, the velocity distribution of the

expanding plasma, and the overall energetics as determined by monitoring of the laser energy, reflection, and the plasma expansion energy. The structure of the imploding shell has been determined by X-ray pinhole cameras, which record a time-integrated image of the pellet X-ray emission. The details of the measurements will not be given here.

In this section, we shall not discuss the general problems of implosion of shells to thermonuclear burn conditions, of the associated laser requirements, and of pellet-design variants intended to optimize implosion conditions. Our objective is to concentrate on several important features of implosions produced in the (KMSF) laboratory and of the analysis techniques developed for interpreting the results. Comprehensive analyses of more general theoretical aspects of pellet response and thermonuclear burning are given by Brueckner and Jorna[1] and of shell implosions by Clarke et al.[2]

We shall first review the general features of laser-driven implosions and show how the overall implosion dynamics can be inferred from the experimentally determined quantities. The problems of symmetry and shell stability are reviewed in Section 23.11C and some interpretations of the experiments are given in 23.11D. In Sections 23.11E and 23.11F, we correlate the X-ray pinhole observations with a detailed computer simulation of the implosion and determine the consistency of the observations with theory.

B. Simplified Implosion Analysis

We consider a symmetric shell implosion and defer the problem of

---

[1]Reference [3] of Section 23.9.

[2]J. S. Clarke, H. N. Fisher and R. J. Mason, "Laser-Driven Implosion of Spherical DT Targets to Thermonuclear Burn Conditions," Phys. Rev. Lett. 30, 89-92 (1973).

symmetry and stability to Section 23.11C.   The implosion analysis is in-

tended to clarify general features and to establish correlation with experiment.

The results are only semi-quantitative, a more detailed and quantitative ana-

lysis is described in Section 23.11F.

The implosion of a thin shell driven by a constant laser-energy absorp-

tion can be well described by the assumption of a mass shell driven by ablation

pressure.   The ablated material is maintained under approximately isothermal

conditions by rapid electron conduction.   The energy balance may then be

easily determined if the divergence resulting from the spherical geometry of

the flow is ignored; this assumption is not essential but allows a simple solu-

tion for the density and velocity distribution of the plasma.   The solution to

the equations of momentum and continuity, i.e., to

$$\rho \frac{dv}{dt} = -\frac{\partial p}{\partial x}, \quad \frac{\partial \rho}{\partial t} + \frac{\partial}{\partial x}\rho v = 0, \qquad (23.11-1)$$

is easily shown to be

$$\rho = \rho_0 \exp(-x/ct), \quad v = c + x/t, \qquad (23.11-2)$$

where

$$c = [(1 + Z)\theta/m_i]^{\frac{1}{2}} \qquad (23.11-3)$$

is the isothermal sound speed and the ion and electron temperatures have

been assumed to be equal.   We assume that the point $x = 0$ is the ablation point, where the velocity of the ablated material is c. The material is correspond-ingly accelerated from $v = 0$ to $v = c$ at the ablation point, giving a reaction force per unit area

$$F_R = \dot{m}c^2.$$ (23.11-4)

The pressure also acts to accelerate the shell, giving the equation of motion

$$m(t)(dv/dt) = - (p_0 + \dot{m}c^2), \quad p_0 = n_0 \theta(1 + z_i).$$ (23.11-5)

The energy per unit area in the expansion is

$$E = \int_0^\infty \rho\left(\frac{3}{2} \frac{(1 + z)\theta}{m_i} + \frac{1}{2} v^2\right) dx = 4p_0 ct.$$ (23.11-6)

Energy conservation requires that

$$E_L = 4p_0 ct + \frac{1}{2}mv^2,$$ (23.11-7)

with E_L the absorbed laser energy. The mass ablation rate is

$$\dot{m} = \rho_0 c, \quad = p_0/c.$$ (23.11-8)

Using these results, we integrate Eq. (23.11-5) to give

$$v(t) = -2c\ln(m_0/m).$$

(23.11-9)

The distance moved is

$$r(t) = -2ct\left(1 - \frac{m}{m_0 - m}\ln\frac{m_0}{m}\right).$$

(23.11-10)

The ratio of the kinetic energy of the imploding shell to the laser energy is

$$\frac{\frac{1}{2}mv^2}{E_c} = \frac{\left(\ln\frac{m_0}{m}\right)^2}{\left(\ln\frac{m_0}{m}\right)^2 + 2\left(\frac{m_0}{m} - 1\right)},$$

(23.11-11)

where we have used the relation $\dot{m}t = m_0 - m$. The maximum energy transfer occurs at $m_0/m \cong 5$ where $(1/2)mv^2/E_c = 0.25$. For $m > m_0/5$, the velocity of the shell drops; for $m < m_0/5$, the rising shell velocity is compensated by the decrease in mass of the remaining shell.

The energy transfer efficiency of Eq. (23.11-11) results from Eqs. (23.11-5) and (23.11-6) which, in turn, depend on the assumption of isothermal conditions to the ablation point. Detailed numerical calculations of the subsonic conduction zone show, for 1 μ laser radiation, that the drop in temperature between the sonic points and the ablation point modifies the shell acceleration and energy balance appreciably, the net force acting to accelerate the shell being dropped by about a factor of two. We shall now make an ad hoc adjustment of v(t) and r(t) for this factor. The peak energy-transfer efficiency may then be easily shown to be reduced by a factor of about three from the result of Eq. (23.11-11).

As an example, we consider a 100 μ diameter shell with the initial wall thickness of 2 μ. The experimental measurements of the X-ray spectrum show a typical temperature of 1 kev. Figure 23.11-1 gives the implosion of this shell for different levels of absorbed power. The maximum energy in the implosion results from an absorbed power of about 10^{11} w and a pulse length of 325 picosec, giving an absorbed energy of 32.5 j. For these conditions, the kinetic energy of the implosion is 1.85 j, about 80% of the shell mass has been ablated, and the velocity of the imploding shell is 4.4×10^{7} cm/sec.

The results for the example just given are easily generalized to other cases. Generally, one observes that marked implosions are expected to occur with the absorbed laser energy somewhat less than $4m_0c^2$, which for 1 kev temperature is 0.19 j/nanog. The kinetic energy in the implosion is 6 to 8% of the absorbed energy under the optimum conditions and the implosion velocity of the shell is about twice the sound speed in the external isothermal plasma.

The model just given can be placed on a more closely empirical basis by utilizing the experimental measurements of the pellet response. The direct measurement of the distribution of velocity in the plasma expansion determines the sum of the kinetic and internal energy in the plasma at the end of the laser pulse. The ablated mass is also measured. According to Eq. (23.11-6), the sum of the internal and kinetic energy in the plasma is 4pct; thus, the average pressure acting on the imploding pellet is

$$\bar{p} = E_{absorbed}/4ct. \hspace{3cm} (23.11-12)$$

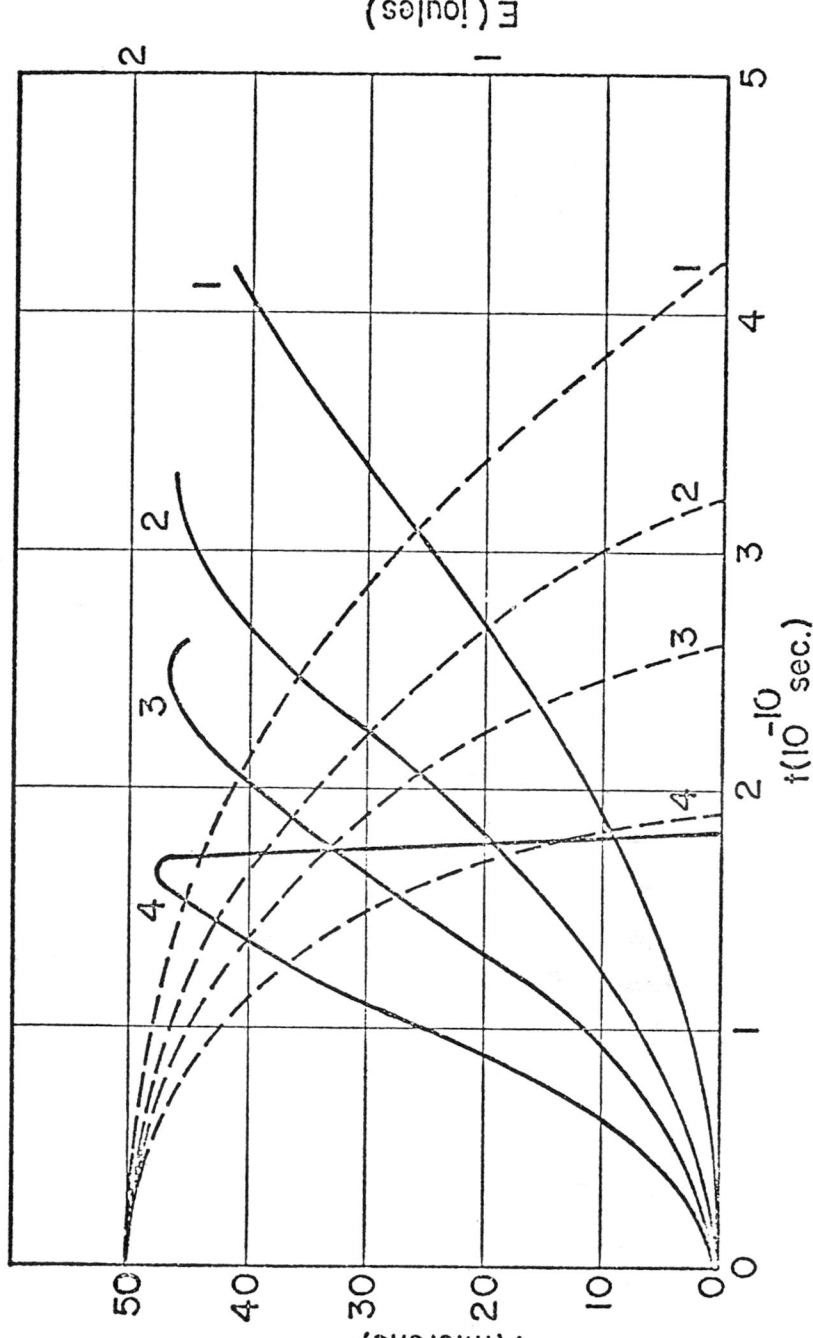

Fig. 23.11-1 Typical implosion pictures for different values of laser flux. The curves marked 1-4 are for total absorbed laser flux of 5×10^{10}, 7.5×10^{10}, 10^{11}, 1.5×10^{11} w, respectively. The dashed curves give the variation of radius with time, up to the collapse time. The solid curves give the kinetic energy for the same cases. Reproduced from studies at KMSF performed by K. A. Brueckner et al.

The model can also be checked by using Eq. (23.11-2) for the velocity distribution and Eq. (23.11-9) for the ablated mass. Assuming that the mass ablation rate is constant, the equation of motion for the shell is fixed by $\bar{p}$ and Eq. (23.11-9). As before, the sound velocity can also be determined by the measurement of the X-ray spectrum, from which the corona temperature is inferred. Finally, the overall description can be confirmed by using the X-ray pinhole determination of the inward motion of the imploded shell.

An example of the analysis described above is given by laser shot No. 1181 on a 70 μ diameter shell with initial wall thickness of 1.2 μ. The laser energy was 75 j and the pulse half-width was 200 picosec. The measured distribution of velocity of expansion of the plasma is given in Fig. 23.11-2. The total kinetic energy of expansion for this case is 9.2 j and the measured ablated mass 32 nanog. The balance of the laser energy was reflected, transmitted by the expanding plasma after the implosion was complete, or emitted in X-rays from the hot, high density glass plasma. Before using these results, we introduce a correction to the energy effectively absorbed in the isothermal plasma, and subtract the energy carried off by the anomalous fast ion flux which does not contribute to the implosion. The origin of this flux is analyzed in a separate study on corona physics; the origin appears to be in strong electric fields produced in the pellet corona by the collisionless motion of fast electrons produced in the laser-deposition region. This is about 30% of the energy; the energy effectively absorbed by the isothermal plasma is therefore about 6.4 j. Equation (23.11-11) then gives for the average pressure 23.6 megabars. Equation (23.11-9) gives an ablated mass of 33 nanog, in good agreement with the measured mass. The initial pellet mass is 42.5 nanog; roughly three-fourths of the pellet is ablated during the implosion. From Eqs. (23.11-3) and (23.11-9), we therefore see that the velocity of the shell

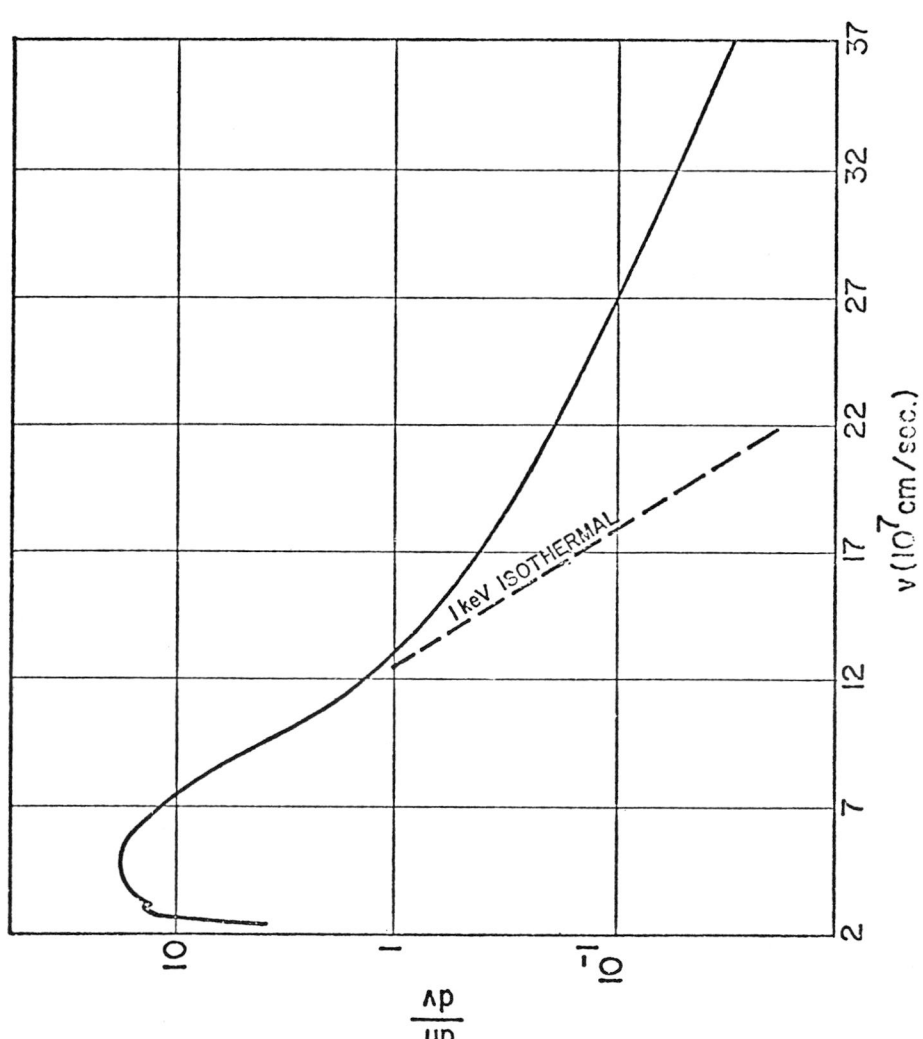

Fig. 23.11-2 Measured plasma velocity spectrum for shot No. 1181. The normalization is arbitrary. Reproduced from studies at KMSF performed by K. A. Brueckner et al.

at the end of the pulse is 2.4×10^7 cm/sec, the kinetic energy in the imploded material is 0.45 j, and the expected wall motion during the pulse is 23 μ. The wall motion can be read from microdensitometer traces of the X-ray pinhole images to a precision of a few μ and is measured to the well defined peak of the X-ray image. The width of the X-ray image is however determined principally by the pinhole aperture and is considerably greater than the precision with which the peak can be determined. The X-ray pinhole photograph shows an apparent inward motion of the shell of about 18 μ, in reasonable agreement with the estimate.

The kinetic energy of 0.45 j in the imploded material is about 5% of the absorbed energy and only 0.6% of the energy in the laser pulse. This very low overall coupling efficiency can be increased by better matching of the laser pulse duration to the implosion time, change in the pellet design to move closer to the maximum transfer efficiency by increased mass ablation, increased absorptivity for a larger pellet with larger scale height in the pellet corona, reduction in laser wavelength to improve absorptivity, reduce the fast ion flux, and improve hydrodynamic transfer efficiency. These problems are under experimental and computational study.

The preceding analysis shows that the experimental measurements of pellet response can be directly correlated with an elementary model of the implosion. Shells can be readily imploded, providing a good semiquantitative test of the basic process of ablation, energy balance, and momentum transfer in the pellet implosion.

Much more interesting and critical questions concerning pellet response arise from considerations of stability and symmetry. We now turn to a study of the theory together with the relevant experimental results.

C. Symmetry and Stability

A pellet can be symmetrically imploded only if (a) the ablation-pro-
duced pressure is symmetric, (b) the pellet is spherical and has uniform wall
thickness, and (c) the implosion is stable against perturbations produced by
the illumination pattern, shell-thickness variations, or other sources of disturbances in pressure, temperature, or density. These three problems are
discussed in the following subsections. Additional sources of asymmetry can
result from phenomena occurring in the pellet corona, such as laser self-
focussing instability, polarization-dependent laser absorption, and magnetic
fields. These are reviewed elsewhere in considerable detail.[1]

The energy flow to the ablation front, producing the ablation pressure,
is determined by the laser deposition as moderated by lateral energy flow due
to electron conduction in the pellet surface. The lateral energy flow is essential since uniform deposition of laser energy is extremely difficult to obtain
because of the structure of the laser beam and/or the optics used to illuminate
the pellet.

The lateral conduction flow produced by flux nonuniformity takes
place over the entire isothermal region of the corona, which we characterize
by a scale height L. The conduction flow under quasi-stationary conditions then
compensates for flux uniformity, giving a temperature difference determined
by

$$L(\Delta K_e \theta / L_{non}^{2}) \approx \Delta \phi, \qquad\qquad (23.11-13)$$

with L_{non} the scale of the nonuniformity in ϕ. The classical conductivity for
a plasma with z = 10 is

$$K_e = 2 \times 10^{27} \, \theta_e^{5/2} \, \frac{erg}{cm \; sec} \; (\theta_e \text{ in kev}). \qquad (23.11\text{-}14)$$

At 1 kev with an absorbed flux of 10^{15} w/cm^2, Eqs. (23.11-13) and (23.11-14) give

$$\frac{\Delta \theta}{\theta} \approx 0.31 \, \frac{L_{non}(\mu)^2}{L(\mu)} \, \frac{\Delta \phi}{\phi} . \qquad (23.11\text{-}15)$$

For a scale height of 50 μ, a flux variation of 50%, Eq. (23.11-15) gives a temperature variation less than 10% only if the scale of nonuniformity is less than 6 μ. This condition may be relaxed if the flux variation is less or if the temperature is higher. At 1.5 kev, for example, with a flux variation of 25%, the temperature variation is less than 10% for a nonuniformity scale of 22 μ.

The expected nonuniformity of temperature and hence of pressure may be determined by a two-dimensional calculation which takes into account the laser flux, refraction in the corona, conduction flow, and the hydrodynamics of the pellet response. The results are, of course, similar to those given above since the general scaling laws of conduction flow are given correctly by Eq. (23.11-13). A typical sequence of temperature and pressure computed with the KMSF 2-D code HYRAD are shown in Fig. 23.11-3 for several times for a 120 μ diameter pellet with 2 μ wall. The parameters of the target shots are given in Table 23.11-1. The pellet illumination for this case involved two f = 0.6 lenses giving high irradiance at the pellet poles ($\theta = 0°$ and $180°$). The calculated energy absorption is shown in Fig. 23.11-4. The numerical calculations show a temperature nonuniformity consistent with Eq. (23.11-15).

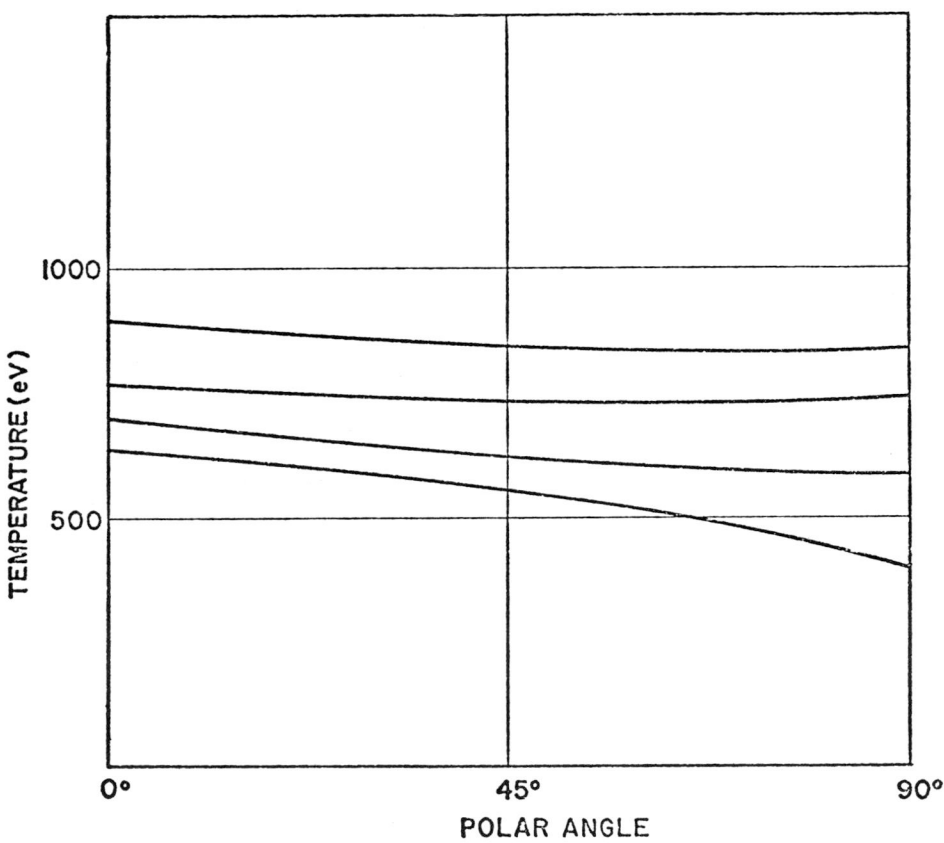

Fig. 23.11-3 Typical computed temperature distributions measured from the
              axis of illumination symmetry (0 degrees) to the pellet equator
              (90°) for the illumination pattern of Fig. 23.11-4.   The succes-
              sive curves are for an increasing laser flux which gives a corona
              temperature increasing with time. The temperature distribution
              becomes more uniform with increasing average temperature.
              Reproduced from studies at KMSF performed by K.A. Brueckner
              et al.

Table 23.11-1 Parameters of the target shots given in Fig. 23.11-3. The
laser pulse half-width is between 400 and 500 picosec. Reproduced from studies at KMSF performed by K.A. Brueckner et al.

Shot No.	Diameter, μ	Wall thickness, μ	Energy on target, j	Final diameter/initial diameter[*]
951	127	3.6	124	0.94
972	160	2.9	242	0.79
1000	125	4.3	209	0.67
1024	102	1.1	195	0.81
1037	102	2.4	164	0.62

[*]The initial diameter is obtained from direct measurement of the glass shell.
The final diameter is measured at the intensity maxima of the densitometer
traces of the X-ray image, with a precision of a few microns.

The actual symmetry of energy flow is readily determined experimentally by the X-ray pinhole photographs. The X-ray emission is a strong function of temperature; thus, measurement of the symmetry of X-ray emission
allows determination of the temperature variation. A number of examples of
X-ray film density as a function of angle are given in Fig. 23.11-5. The corresponding corona temperatures, as inferred from the calibrated film response
and computed X-ray emission, are in the range of 900 to 1000 ev and are approximately proportional to the film density over the indicated range. These
results show that it is possible to obtain temperature uniformity to ± 10%.
We conclude that the symmetry of initial deposition and the conduction flow
are sufficient to give a high degree of symmetry of pressure at the ablating
pellet surface. Further experimental and computational confirmation of this
conclusion are given in Section 23.11D.

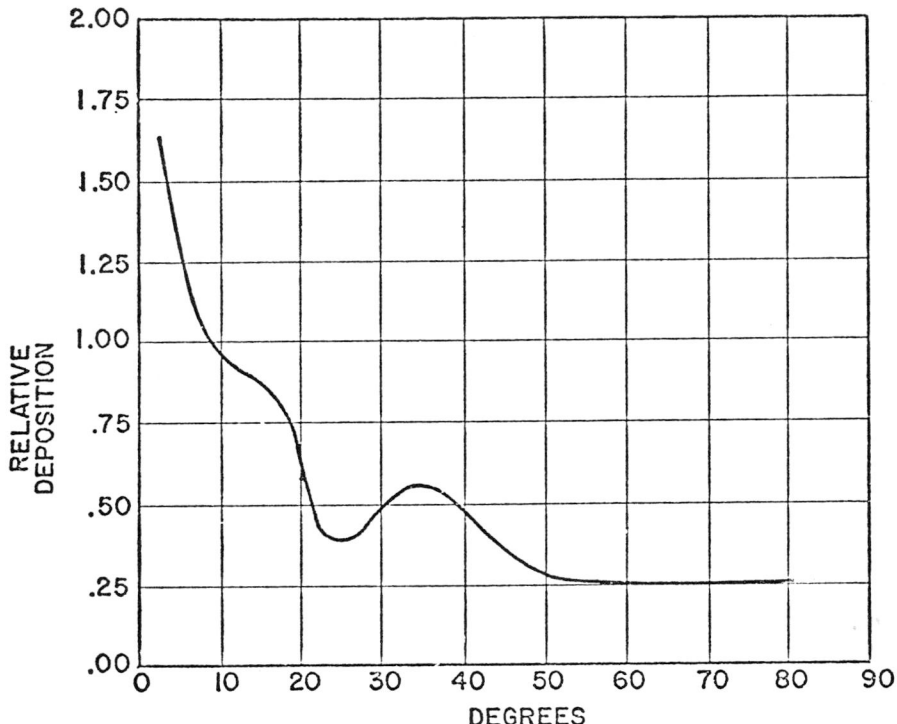

Fig. 23.11-4 Computed variation of energy deposition with angle for a 125 μ
              diameter pellet illuminated by two f = 0.6 lenses. The calcula-
              tion includes the effect of lens aberrations, which lead to the
              variation indicated in the figure. The pellet is located 20 μ from
              the circle of least confusion of the lenses, which has a diameter
              of approximately 70 μ. Reproduced from studies at KMSF per-
              formed by K. A. Brueckner et al.

The glass shells used in these experiments are selected to give excellent external sphericity and uniformity. The configuration of the glass bub-
ble may be represented by two spheres, each very accurately spherical, but
not perfectly concentric. The targets of interest have a wall thickness of the
order of microns or several thousands of atomic layers. The center of the

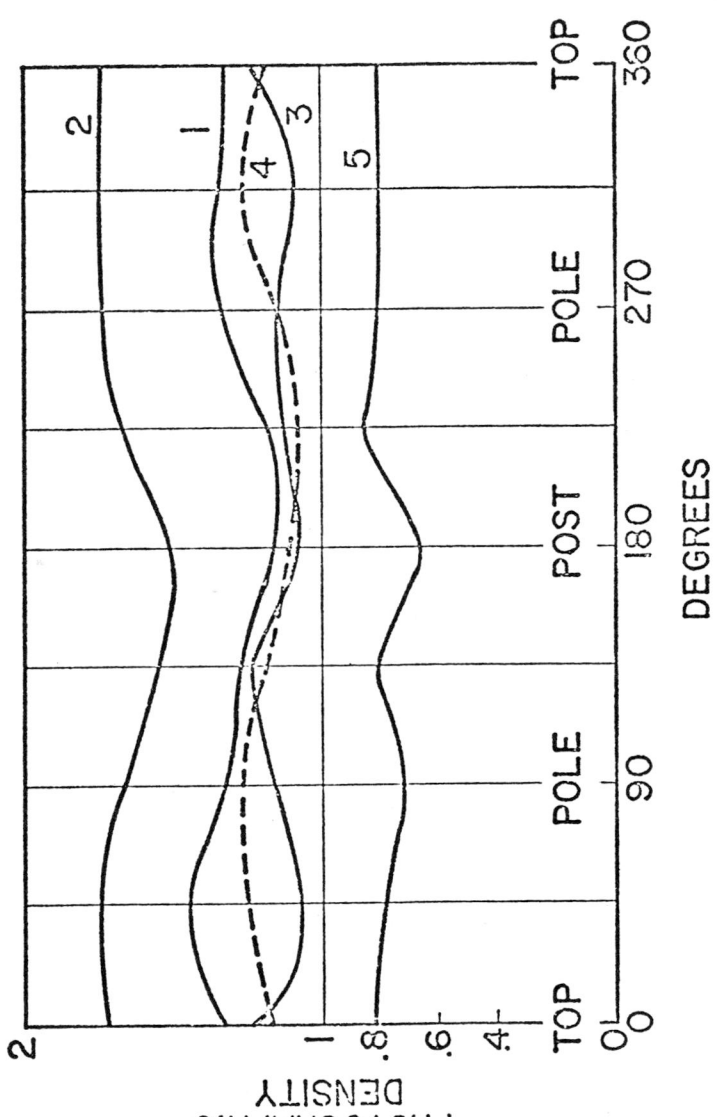

Fig. 23.11-5 Measured film density variation for several target shots. The curves labelled 1-5 are for laser shots 972, 951, 1000, 1024, and 1037. The target size, laser energy, and pulse width are given in Table 23.11-1. Reproduced from studies performed at KMSF performed by K. A. Brueckner et al.

inner spherical surfaces of the shells may therefore be expected to be randomly distributed with respect to the center of the outer surface over a sphere
with radius a large fraction of the average wall thickness.

For two spheres with radii r_1 and r_2 and with centers offset by a distance δ, the shell thickness is given by

$$W(\theta) = r_1 - r_2 + \delta \cos \theta + \theta(\delta^2/r_1). \qquad (23.11-16)$$

The comments above, which are supported by examination of microballoons,
shows that δ is randomly distributed from zero to a maximum which is the
order of $r_1 - r_2$. The probability of δ falling within a fraction f of $r_1 - r_2$ is
of the order of f^3. Thus, a wall uniformity of 10^{-2} has an a priori probability
of 10^{-6}.

In practice, it is possible without great difficulty to select microballoons with wall-thickness variation of less than 10%. The behaviour of these
under laser irradiation is of great interest, since the glass shells simulate
thermonuclear fuels.

The type of shell asymmetry characteristic of the glass shells described
above may be easily seen to have a relatively small effect on implosion symmetry. We consider first a shell with the eccentricity in wall thickness given
in Eq. (23.11-16), accelerated by uniform external pressure. The acceleration of various parts of the shell is inversely proportional to the wall thickness,
so that the various parts of the shell do not all accelerate toward the initial
center of the outer sphere. The variation in displacement, however, is proportional to $\cos \theta$ and hence the accelerating surface remains spherical but
with a shifting center. This configuration persists until convergent motion

increases the wall thickness, lateral flow occurs, or ablation increases the shell nonuniformity.

To analyze quantitatively the implosion symmetry for an eccentric shell, a detailed calculation has been carried out, using the two-dimensional hydrodynamic code HYRAD. The example considered was r_1 = 55.5 μ, r_2 = 58 μ and δ = 0.125 μ, uniformly illuminated with 100 j of 1.06 μ radiation delivered in 0.9 nsec in a linearly rising profile. The results are shown in Fig. 23.11-6, where the inner radius is plotted at various times during the implosion. The implosion maintains fairly good spherical symmetry but implodes to a center displaced from the initial origin, verifying the analysis given above.

The effect of higher order distortions is more serious, causing changes in the spherical convergence. These distortions are, however, expected to occur with much smaller amplitude than the characteristic eccentric offset of the glass shell.

The thin glass shells are susceptible to Taylor instability under the accelerating pressure of the low-density and high-temperature plasma. Convective overturning or the formation of Bénard cells, driven by the thermal gradient, may also occur. The Taylor instability may be inhibited by the viscosity of the plasma and by the ablation flow at the pellet surface. According to the analyses, the convective overturning is less of a problem, being reduced by the viscosity and thermal conductivity of the plasma and by the ablation flow through the unstable zone. We therefore restrict our discussion to the Taylor instability.

The instability growth is limited by convection through the unstable zone due to ablation. The effective growth time is the time to convect through the unstable zone. For the case given in Fig. 23.11-7, the computed mass

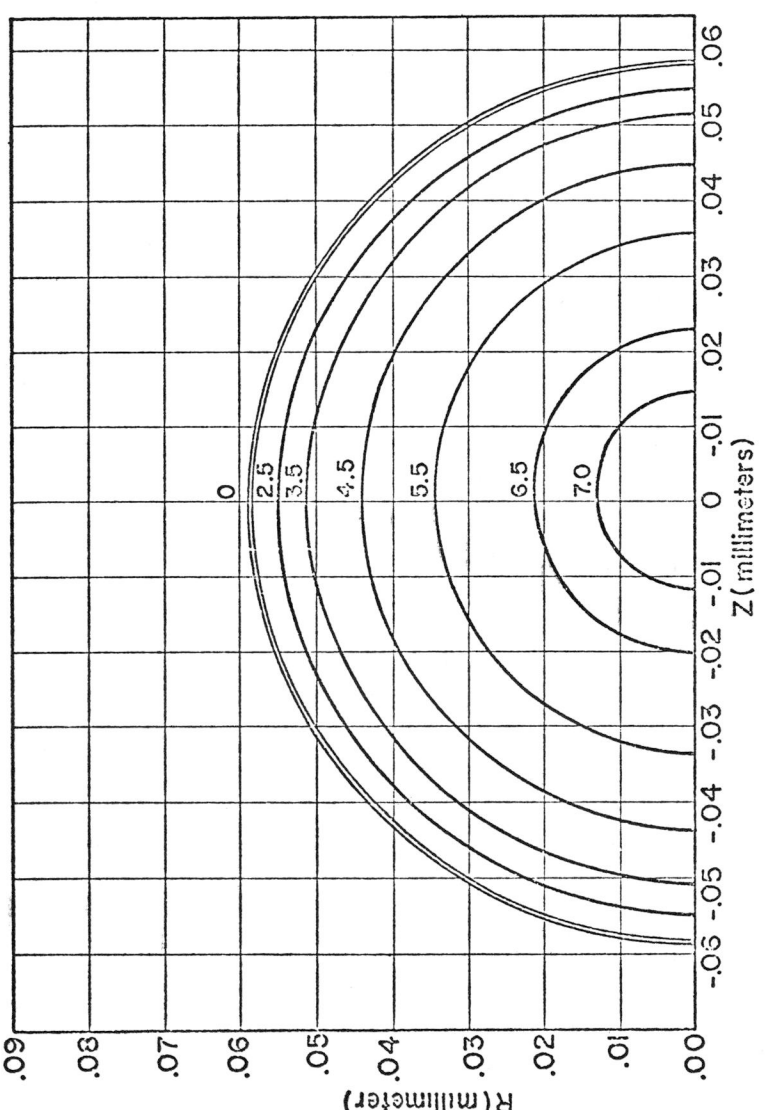

Fig. 23.11-6 Two dimensional calculation of the implosion of an eccentric shell. The position of the inner surface of the shell is shown. The convergence to an off-set center is apparent. The curves are labelled by the time in nsec. Reproduced from studies at KMSF performed by K. A. Brueckner et al.

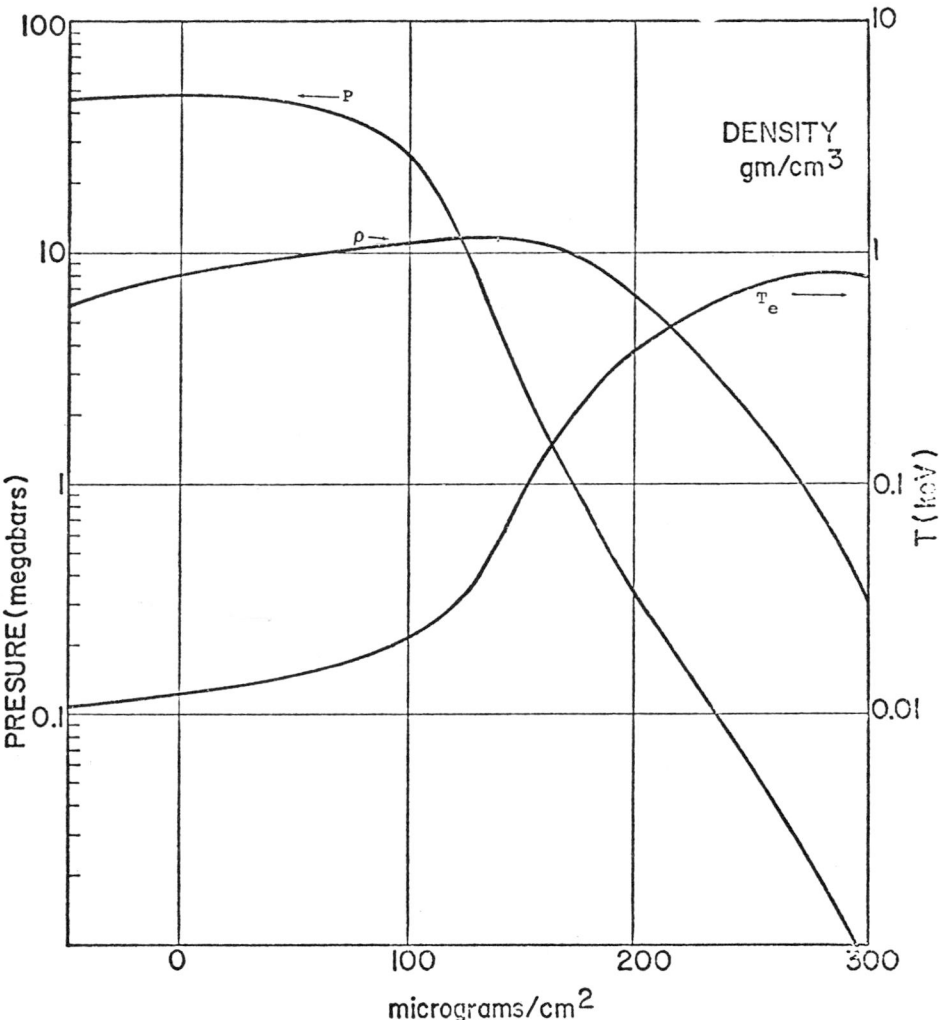

Fig. 23.11-7 Pressure, electron-temperature, and density for a typical
              ablation front.   The shell acceleration is approximately 10^{17}
              cm/sec^2 and the mass ablation rate 2.8×10^5 g/cm^2-sec.
              The zero in the mass scale is an arbitrary reference point in
              the shell. Reproduced from studies at KMSF performed by
              K. A. Brueckner et al.

ablation rate is 2.8×10^5 g/cm^2-sec, corresponding to a convection time through the unstable zone of 50 to 100 picosec. In this time, the instability growth is 1.5 to 3 e-foldings. For this growth to have a small effect on the shell, the initial perturbation in shell thickness, at a transverse scale of the order of μ, should be less than 10^{-2} to 10^{-3} μ. If this perturbation is substantially exceeded, the shell may be expected to develop structure on the scale of μ. The subsequent development of the configuration of the unstable shell can be expected to be very complex, with a disordered zone of the order of several μ in depth developing, centered about the original shell center. This region will be strongly dissipative for the energy flow from the conduction front, which may stabilize extensive non-linear growth.

This problem has been numerically simulated, using the 2-D code HYRAD. The calculation made was of the motion of the first half-wavelength off of the axis of symmetry of an initial ripple in the shell radii with wavelength 4 μ and an amplitude of 10^{-2} of the shell thickness. The mesh is shown in Fig. 23.11-8 at various times during the implosion. The initial disturbance grows, at the rear of the imploding shell, at a rate of about 10^{11} sec^{-1}. It is apparent, however, that communication of this growth to the front surface is markedly delayed. In this problem, the shell has imploded to approximately half of its original radius before noticeable distortion of the front surface occurs.

The possible effect of shell breakup, before the implosion is complete, on the experimental measurements is discussed in the next Section 23.11D.

D. Interpretation of Experiments

Implosions have been observed using the KMSF laser system illuminating the pellet through two large-aperture-lens-elliptical-mirror combinations.

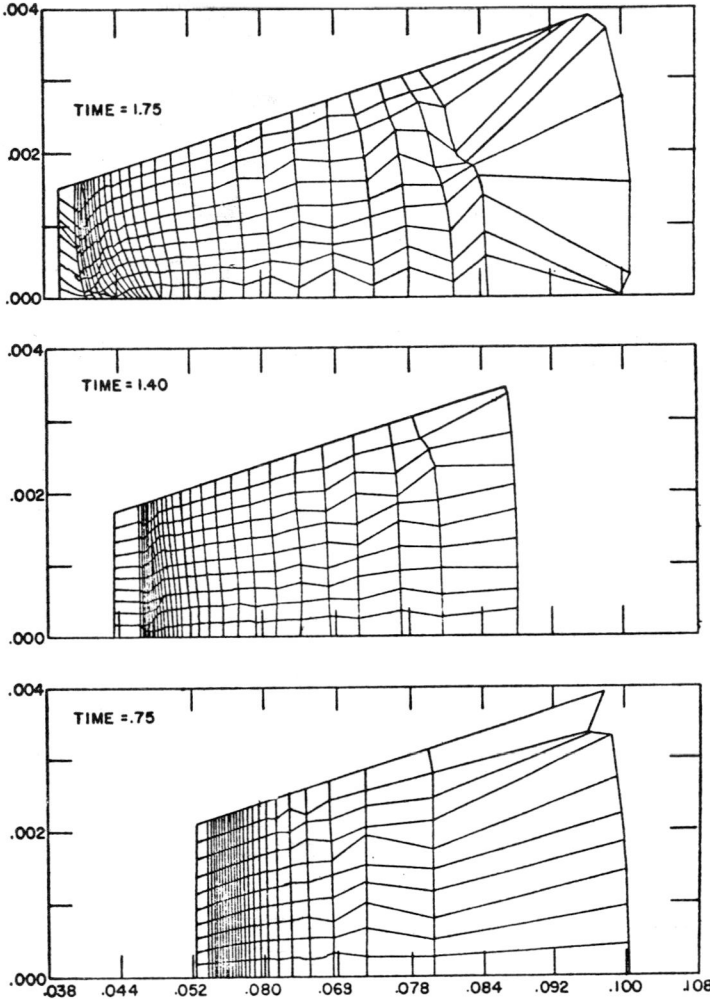

Fig. 23.11-8 Two-dimensional calculation of the implosion of a shell with an
initial ripple. The front surface and a surface near the rear of
the imploding region are shown. The growth of the ripple in the
rear of the imploding shell is shown. The delay in communi-
cating knowledge of the growth to the front surface is apparent.
The curves are labelled by the time in nsec. Reproduced from
studies at KMSF performed by K. A. Brueckner et al.

The illumination pattern used gives a nonuniform intensity pattern with nearly normal incidence at the pellet surface. The energy absorbed by the target is determined by charge-cup measurements of the velocity distribution of the expanding plasma and is confirmed by measurements of the overall energy balance. The implosion characteristics of the pellet are determined with X-ray pinhole cameras, which give time-integrated images of the pellet. The X-ray emission from the pellet is passed through a 0.7 mil Aℓ absorber, which absorbs nearly all the line-radiation from the target. The resulting X-ray spectrum reaching the film is computed to be centered at about 2 kev with a broad spectrum between 1 and 3 kev. The pinhole resolution with a 10 μ pinhole is calculated to be 14 μ, including refraction effects. This resolution has been directly confirmed experimentally.

The symmetry of energy deposition has already been shown in Fig. 23.11-5, which gives examples of the film density measured on different glass shells. The symmetry of X-ray emission, which is achieved in practice, is variable, due to small alignment errors in the laser system, in the illumination optics, or in the pellet positioning. Good symmetry is, however, obtained under the best conditions. From the computer simulation of the process, together with the film calibration, the temperature can be determined. At the ablation front, under good symmetry conditions, the temperature has a variation of less than ± 10%.

The implosion symmetry is readily determined by direct measurement of the X-ray image. The inward motion of the shell is apparent from the inward displacement of the X-ray emission maximum. A quantitative analysis of this correlation is given in Section 23.11F and is based on the computer simulation of the implosion.

The implosion symmetry for several examples is shown in Figs. 23.11-9 to 23.11-12. The observed inward motion of the X-ray image is

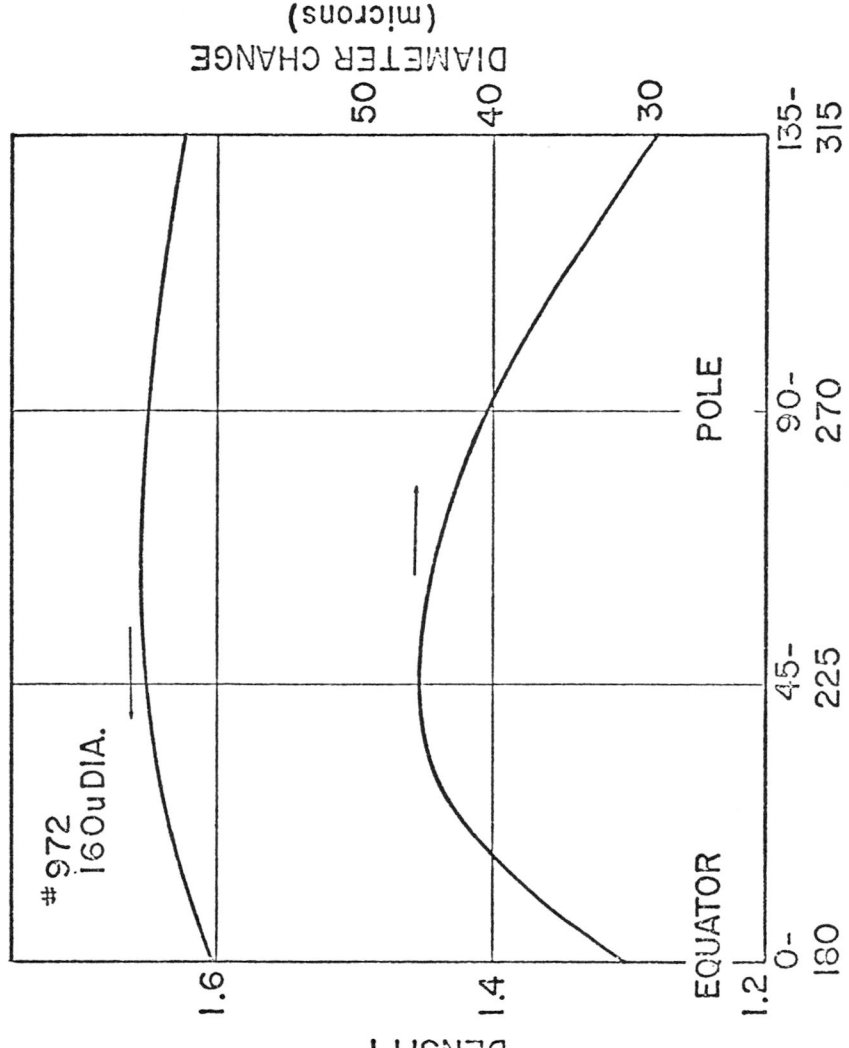

Fig. 23.11-9 Measured peak film densities averaged at θ and $\theta + 180°$, and measured diameter change, for laser shot No. 972 (see Table 23.11-1); 160 µ diameter. Reproduced from studies at KMSF performed by K. A. Brueckner et al.

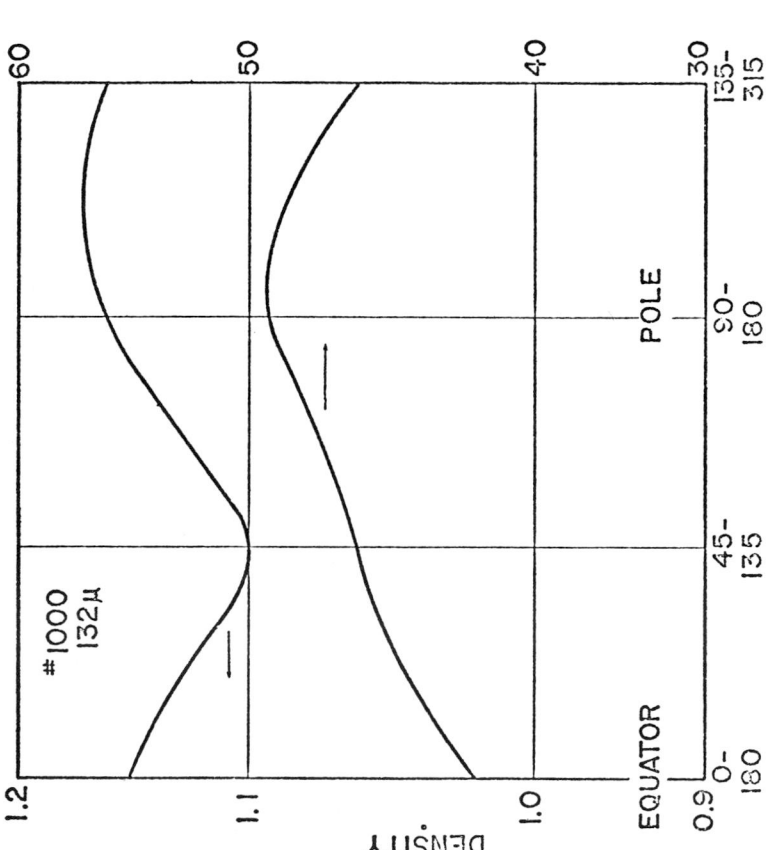

Fig. 23.11-10 Measured peak film densities averaged at θ and θ + 180°, and measured diameter change, for laser shot No. 1000 (see Table 23.11-1); 132 μ diameter. Reproduced from studies at KMSF performed by K. A. Brueckner et al.

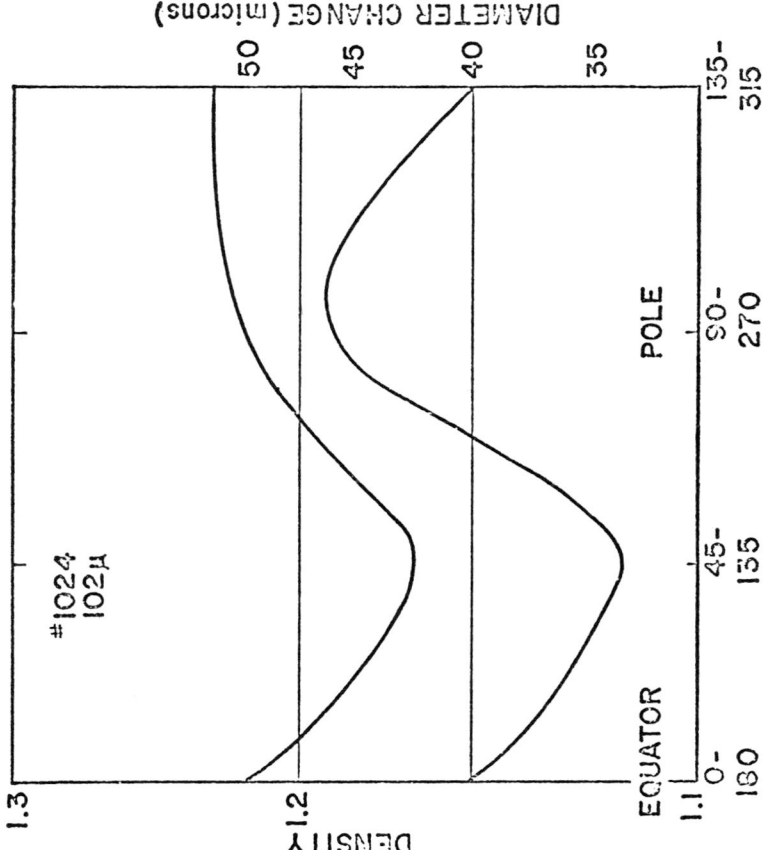

Fig. 23.11-11 Measured peak film densities averaged at θ and θ + 180°, and measured diameter change, for laser shot No. 1024 (see Table 23.11-1); 102 μ diameter. Reproduced from studies at KMSF performed by K. A. Brueckner et al.

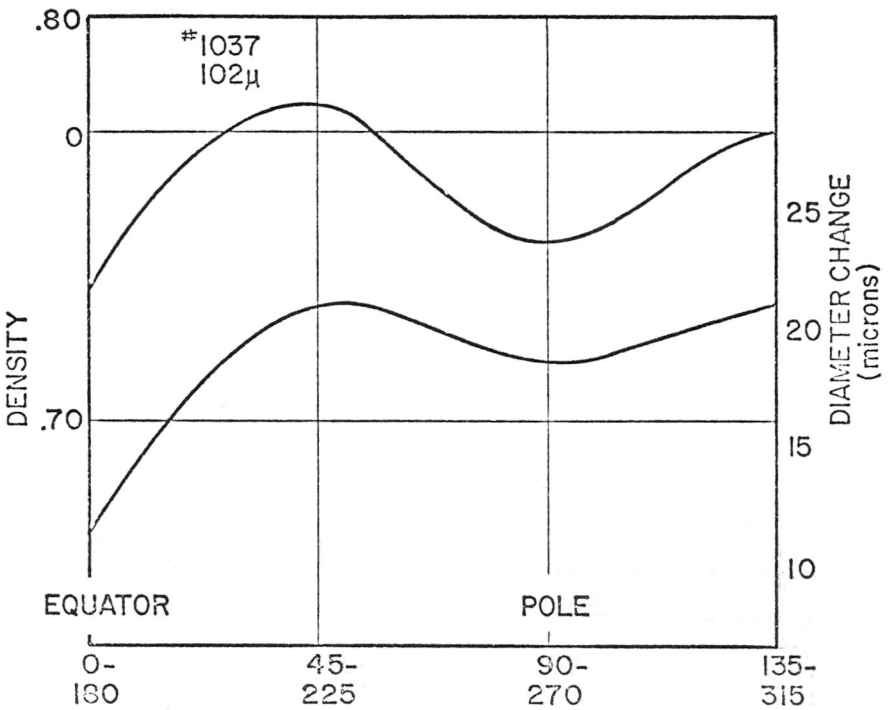

Fig. 23.11-12 Measured peak film densities averaged at θ and θ + 180°, and
measured diameter change, for laser shot No. 1037 (see Table
23.11-1); 102 μ diameter. Reproduced from studies at KMSF
performed by K. A. Brueckner et al.

nonuniform to about ± 10%, showing that the glass shells move inward toward
a region of convergence with a diameter 10 to 20% of the initial diameter. This
conclusion is limited to a scale set by the resolution of the X-ray pinhole camera; instability growth in the glass shell on a much smaller scale would not
be apparent in the film record. However, no appreciable distortion on the
scale of the camera resolution can be detected.

A rough estimate of shell acceleration can be obtained from the laser-pulse length and the shell mass. These confirm the model described in Section 23.11B. To obtain a quantitative check of the implosion process, a more detailed analysis is essential. The computer model used is described and applied in Section 23.11F.

Charge-cup measurements typically show two components of the plasma expansion energy: a slow ion component with velocities less than about 3×10^7 cm/sec, and a well-defined fast ion component extending up to several times 10^8 cm/sec. Measurements of incident and reflected laser energy generally yield absorbed energies which agree within a factor of two with the charge-cup values. As is discussed in Section 23.11F, computer simulation of the X-ray pinhole pictures yields absorbed energies which agree with the slow ion component. It is believed that the fast ions arise from electrostatic acceleration mechanisms in the corona and contribute only weakly to the implosion.

E. X-Ray Pinhole Analysis

In this section, we describe the general features of image formation from a pellet. More quantitative results are given in Section 23.11F. The high energy component of the X-ray emission, selected by the 18μ (0.7 mil) aluminum absorber in the X-ray camera, is not self-absorbed in the target. The aluminum absorption cuts off the lower end of the X-ray spectrum at about 2 kev. The emitted radiation is a combination of bremsstrahlung and recombination radiation and comes principally from a region in the pellet surface in which local thermodynamic equilibration is maintained. The ionization of the glass changes rapidly in the ablation front; this is important in line emission.

For the energetic photons, important in the X-ray picture, the temperature is above a few hundred volts and ionization in the emission region may be assumed to be essentially complete. For the purpose of preliminary analysis, we therefore neglect the z-dependence of emissivity. The bremsstrahlung emission has the following temperature and density variation:

$$h\nu(\text{free-free}) \sim \theta^{-\frac{1}{2}} n_e^2 \exp - E/\theta, \qquad (23.11\text{-}17)$$

with $E = h\nu$ (the photon energy). The recombination radiation has approximately the same dependence on temperature and density; thus, the total emission per unit volume may be assumed to have the dependence of Eq. (23.11-17).

In the ablation region, the temperature rises rapidly away from the ablation front, increasing approximately as $(\text{distance})^{2/7}$. The density drops, first maintaining pressure balance but then decreasing more rapidly as the material accelerates under the pressure gradient in the isothermal region. The emission consequently is peaked in a relatively shallow zone near the ablation front; the detailed structure can be obtained by the computer model described in the next section.

The relatively narrow zone of emission from the shell localizes the X-ray image focused at any instant near the momentary position of the ablation front. The emission is a monotonically increasing function of temperature and hence of laser flux, rising very rapidly as the temperature approaches the characteristic temperature of the X-rays forming the X-ray image; it rises more slowly with further temperature increase.

The photographic density of the X-ray image depends on the velocity of the emitting shell, varying inversely as the implosion velocity. Thus, the X-ray image focused at constant emission rate can be expected to be most intense near the initial radius. The actual distribution of intensity is determined by the emission rate, which is fixed by the energy flow to the ablation front, and by the velocity, which is fixed by the acceleration history. The time-integrated image therefore is a complicated convolution of phenomena which record the evolution of the implosion.

Some features of the integrated X-ray image are apparent. A narrow, well-defined ring near the original pellet radius shows little motion. Inward displacement of the ring shows net inward motion, with the position of the most intense part of the ring determined by the maximum of (emission rate)/shell velocity. The width of the ring is determined by the distance moved by the emitting region during the peak of the laser pulse; thus, a pellet which has been appreciably imploded should show a broadened and displaced emission ring.

The emission from regions near the ablation front is a direct consequence of energy flow from the laser-deposition region. A second emission source can arise from compression and heating at the center of convergence of the implosion. The inner surface of the shell is stopped at the center of convergence by a strong reflected shock which compresses and heats the shell. The temperature reached in the compressed shell is determined by the kinetic energy of the shell, which is momentarily converted to internal energy as the shell is turned around by the reflected shocks. In the final compression, the density in the shell can be increased by a large factor, depending on the effective equation of state of the material. The duration of the emission is determined by the transit time of the reflected shock, which may be estimated from the sound velocity and dimension of the compressed material.

The time-integrated emission from the shocked region can be relatively large, particularly per unit area as recorded by the X-ray pinhole camera.

The density of the compressed material can be of the order of several g/cm^3, which is one to two orders of magnitude greater than in the radiating region of the ablation zone.

At this compression, the dimension of the central region is several μ and the radiating time for a kev is of the order of 10 picosec. The high intensity can, however, easily compensate for the short duration of the emission.

The determining feature of the implosion which controls the central emission maximum or "implosion spike" is the velocity reached in the imploding shell, which determines the stagnation temperature. To reach an average temperature of 1 kev, which fully ionizes glass, requires an average kinetic energy per atom of about 18 kev, corresponding to an implosion velocity of 6×10^7 cm/sec. The central temperature under these conditions is considerably greater due to the convergence of the shock, the work done by the outer layers of the imploding shell providing additional heating and compression of the compressed center.

The preceding remarks are intended only to emphasize the important qualitative features of the X-ray pinhole record of the implosion. In addition to the symmetry and implosion onset which are apparent from the external pellet emission, the appearance of the "implosion spike" gives an excellent indication of the central conditions being achieved in the final stages of convergence of the implosion. We now turn to the quantitative analyses of the implosion history.

F.   Pinhole Analysis by Computer Simulation

The method of analysis is based on a comparison of experimental pinhole traces with those generated by a computer simulation model of the target plasma. An attempt is made to find values of the free parameters in the calculation which yield a pinhole form that matches the experiment. In this way, values can be inferred for those parameters for which the results are sensitive. The most important such parameters are absorbed energy, shell mass, pulse form and the degree to which spherical symmetry was maintained during the implosion.

One of the important parameters in obtaining a match between calculated and experimental pinhole forms is the degree of spherical convergence allowed in the calculation. Two-dimensional computer runs indicate that, when the spherical convergence is only slightly degraded by illumination asymmetry, for example, the effect is equivalent to the addition of a small amount of tangential velocity to the otherwise radially converging material. This effect can be included in a one-dimensional calculation by the addition of a small amount of angular momentum to the material initially. The infinitesimal elements, which compose the spherical shell, are assumed to have their angular momentum vectors oriented uniformly so that the shell always moves spherically but is limited in its minimum distance of approach to the center. In this way, geometrical effects which introduce only a small perturbation on spherical symmetry can be included in the one-dimensional calculations.

We have considered only time-integrated pinhole images with a high degree of circular symmetry. From the qualitative description of Section 23.11E, one expects to observe three different categories of pinhole forms: (a) rings, (b) central spikes, and (c) a combination of ring and spike. Rings generally result from relatively massive shells which do not move very far during the time the laser pulse is on. Shells of this type do collapse but not strongly enough for the collapse phase to be visible in the X-ray picture. Central spikes result from strongly imploded shells, which usually are still receiving energy from the laser pulse at the time of maximum compression. A ring-like form with a central spike is an intermediate case which implodes strongly enough for the collapse phase to be visible at the center.

The breadth of the ring and its position are a measure of how much motion occurred during the high-intensity part of the laser pulse. Since the shell accelerates from rest, it spends a relatively long time at large radius. After it begins to move rapidly, however, it does not contribute appreciably to the time-integrated image until the point of maximum compression. At this point, the kinetic energy of collapse is converted to thermal energy and, if the energy per unit mass is high enough, the central compression will be visible in the X-ray picture. The relative intensities of the ring and central spike depend strongly on the degree of spherical convergence, which is limited by two-dimensional effects.

In Fig. 23.11-13 is shown a sequence of pinhole forms corresponding to different values of absorbed energy for a shell 50 μ in radius and 2μ thick with a 0.75 nsec exponential pulse. As the absorbed energy increases, the ring moves inward, the peak density increases, and the peak-to-center ratio decreases. Very high energy per unit mass generally produces pinhole pictures with a single central spike for an exponential pulse because the peak laser intensity occurs close to the time of maximum compression.

In Fig. 23.11-14 is shown a sequence of pinhole forms illustrating the effect of shell mass for a given laser energy. In these cases, shells 104 μ in radius and of different thickness were heated by 50 j in a 0.75 nsec exponential pulse. These forms differ from those in Fig. 23.11-13 by the presence of both a ring and a central emission spike. For a given energy, a more massive shell produces an emission ring of larger radius and a less prominent central feature.

Figure 23.11-15 shows the effect of pulse length on the pinhole form. The cases illustrated are for 10 j absorbed on a shell 50 μ in radius and 2 μ

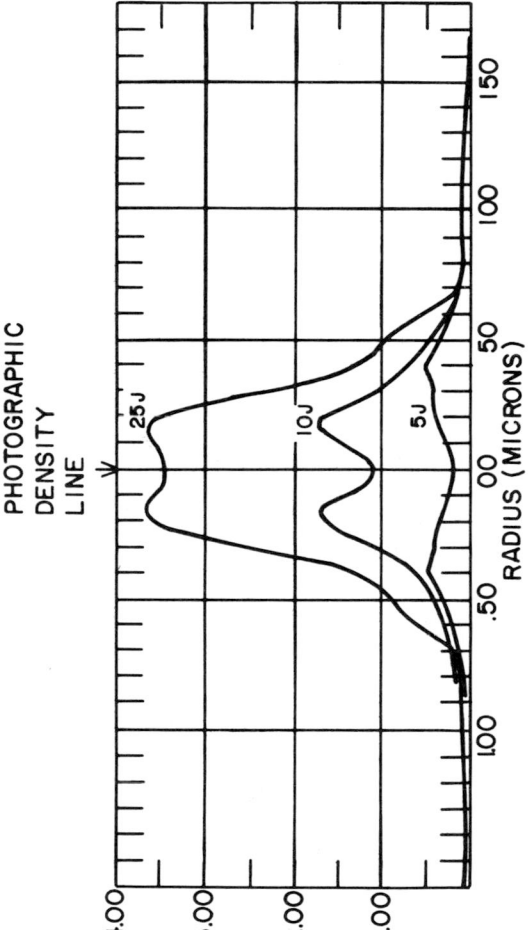

Fig. 23.11-13 Calculated effect on the X-ray pinhole form of absorbed energy. As the energy increases, the ring moves inward, the peak density increases and the peak-to-center ratio decreases. Reproduced from studies at KMSF performed by K. A. Brueckner et al.

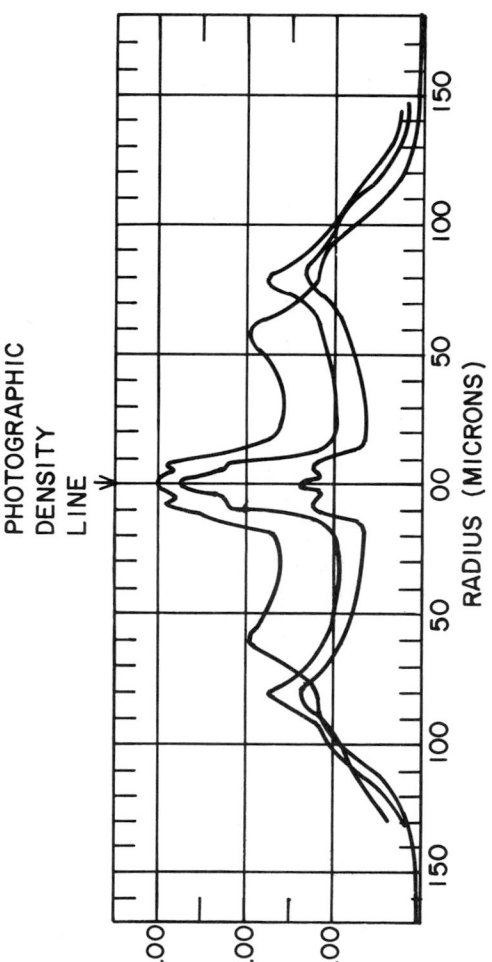

Fig. 23.11-14 Calculated effect on the X-ray pinhole form of shell thickness. The larger the shell mass, the larger the radius of the ring and the less prominent the central emission spike. Reproduced from studies at KMSF performed by K. A. Brueckner et al.

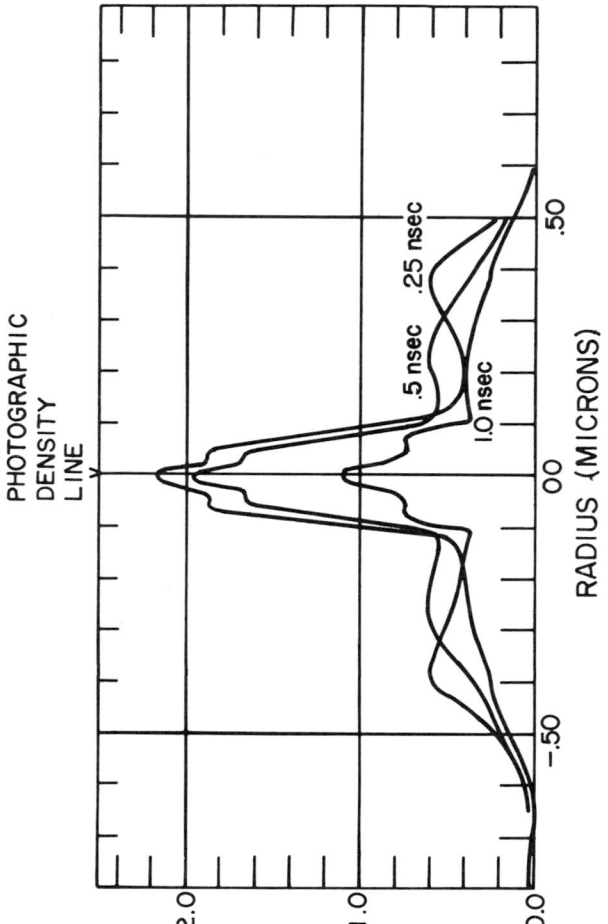

Fig. 23.11-15 Calculated effect on the X-ray pinhole form of pulse length. Short pulses tend to produce rings of large radius with prominent central spikes. As the pulse lengthens, the ring moves inward and blends with the central spike to produce a single central feature. Reproduced from studies at KMSF performed by K. A. Brueckner et al.

thick. In all three cases, the temporal form of the pulse is square. The conspicuous inward motion of the ring as the laser pulse lengthens shows that the emission maximum tends to follow the average position of the shell during the high-power part of the pulse. The collapse time for a shell this size with 10 j absorbed is approximately 0.5 nsec. The 1 nsec case results in a smaller central maximum because only one half of the energy is absorbed by collapse time.

Figure 23.11-16 illustrates the effect of different degrees of central convergence on the height of the central emission feature relative to the ring. In all three cases, 3.3 j were absorbed by a shell 36 μ in radius and 1.3 μ thick. The three different angular momentum values correspond to initial tangential velocities of 2.0 $\times 10^6$ cm/sec, 2.2 $\times 10^6$ cm/sec and 2.5 $\times 10^6$ cm/sec, which are roughly 0.1 of the average radial velocity of 1.7 $\times 10^7$ cm/sec. In these calculations, the inner wall of the shell came in to 4.5 μ, 5.2 μ and 5.7 μ, respectively.

In Fig. 23.11-17, an experimental pinhole form is compared with calculated pinhole forms for three different absorbed energies. In this experiment, the target was a glass shell 42 μ in radius and 1.7 μ thick. The shell was initially supported by a 2 to 3 μ diameter silica fiber. The computer match indicates about 10 j absorbed, which compares with 6 j measured by both blast probes and streak camera. This is a typical case for targets which produce a well-collapsed ring.

The computer calculation, which matched the experimental form, indicates a maximum temperature and density of 150 ev and 44 g/cm^3 and a collapse time of 0.8 nsec. There were, however, no independent measurements of these quantities.

In Fig. 23.11-18 is shown the computer match to a pinhole form typical of the class of single central spikes. The target was a glass shell 30 μ in

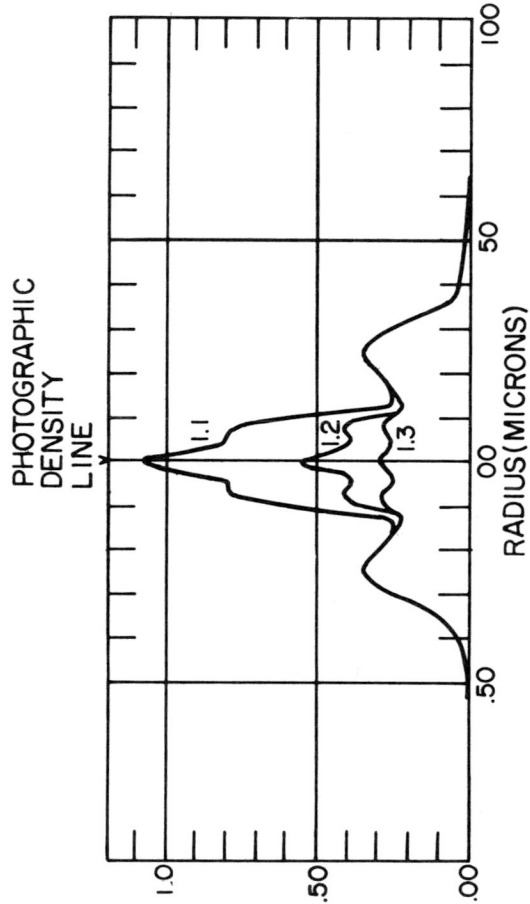

Fig. 23.11-16 Calculated effect on the X-ray pinhole form of different degrees of spherical convergence. The relative height of the central spike is very sensitive to the presence of small amounts of tangential velocity in the material. The indicated parameters L_1, L_2 and L_3 correspond to initial tangential velocities of 2.0×10^6, 2.2×10^6 and 2.5×10^6 cm/sec. Reproduced from studies at KMSF performed by K. A. Brueckner et al.

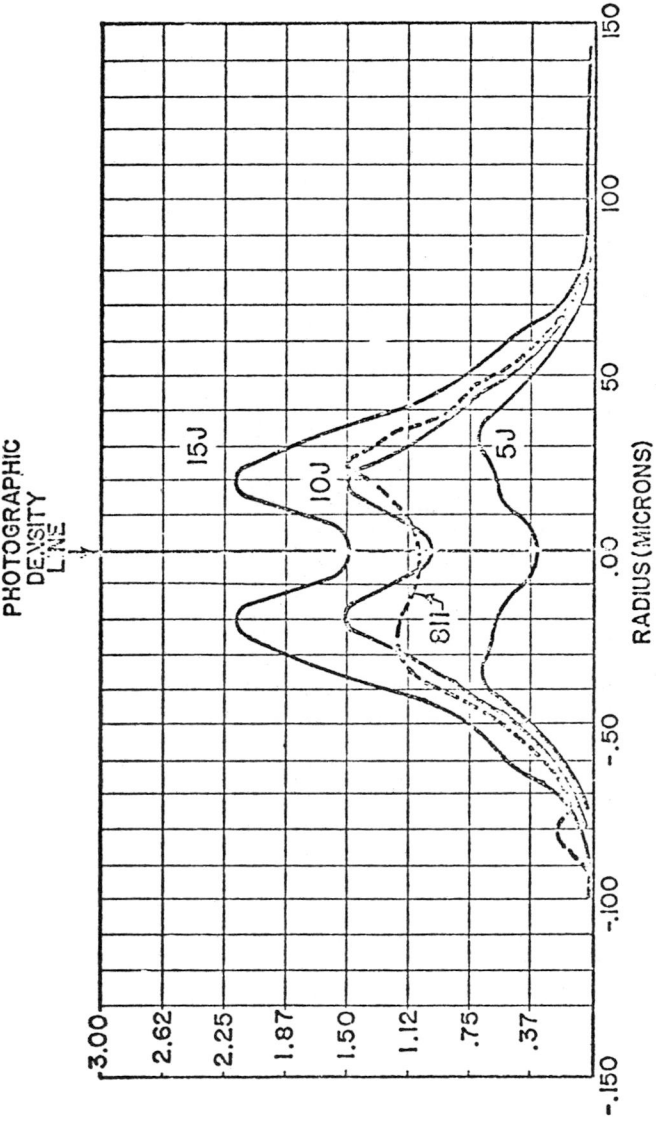

Fig. 23.11-17   Comparison of microdensitometer trace for the X-ray pinhole photograph on shot No. 811 with forms calculated by computer simulation.   The 10 j match compares favorably with a measured energy of 6 j by both blast probes and streak camera.   Reproduced from studies at KMSF performed by K. A. Brueckner et al.

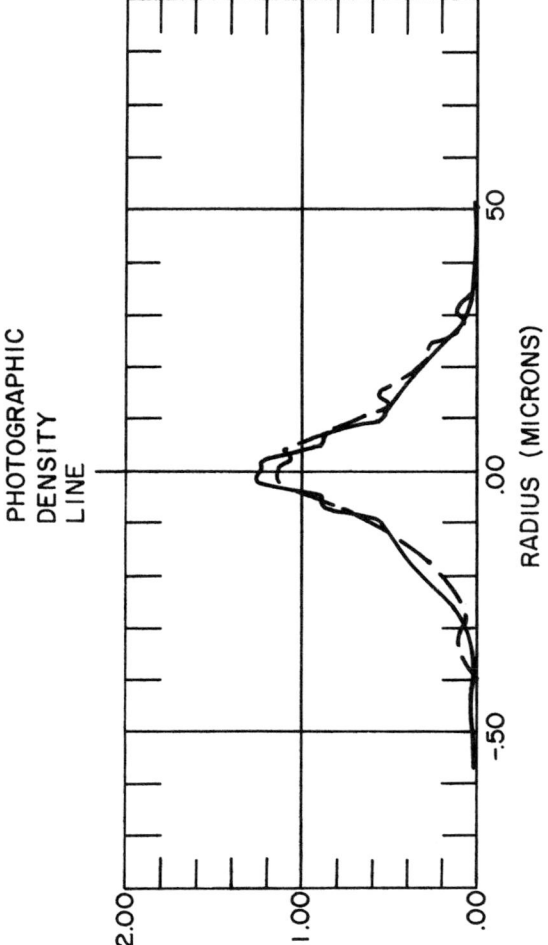

Fig. 23.11-18 Computer match to pinhole form on shot No. 1036. This case, which is typical of single central spikes, was matched with 6 j absorbed. The experimental values of absorbed energy were 7 j in slow ions and 5 j in fast ions.
Reproduced from studies at KMSF performed by K. A. Brueckner et al.

radius and 2.5 μ thick. The computer match was obtained for 6 j absorbed, which compares with 7 j measured in the slow ion component by the charge cups. This shot had an additional 5 j in fast ions, which apparently contribute only weakly to compression. In this case, the computer match indicates a collapse time of 0.4 nsec with a maximum temperature and density of 160 ev and 55 g/cm^3.

In both of the above cases, a computer match to the experimental pinhole form was obtained, without decreasing the spherical convergence, by adding angular momentum to the shell. The sensitivity of the computer match to the degree of symmetry of the implosion was tested by variation of the initial angular momentum of the shell, which roughly simulates the effects of nonuniform convergence. The best match was obtained with negligible angular momentum, indicating nearly symmetric final convergence. The computer simulation shows marked sensitivity to slight non-convergence since the central temperature and density, while only slightly affected, have a very pronounced effect on the X-ray emission. An indication of loss of symmetry was, however, obtained in the pinhole form given in Fig. 23.11-19. In this case, an experimental pinhole form is compared with the computational sequence shown in Fig. 23.11-16. The position of the outer ring was obtained with 3.3 j absorbed. The experimental values for absorbed energy were 4 j in slow ions and 2 j in fast ions.

In order to match the central spike, a considerable amount of nonconvergence had to be assumed. The best fit to the experimental pinhole form was obtained for the calculation which yielded a minimum radius of 5.2 μ. This value, which is 14% of the initial radius, is consistent with observed asymmetries of 10% to 20%.

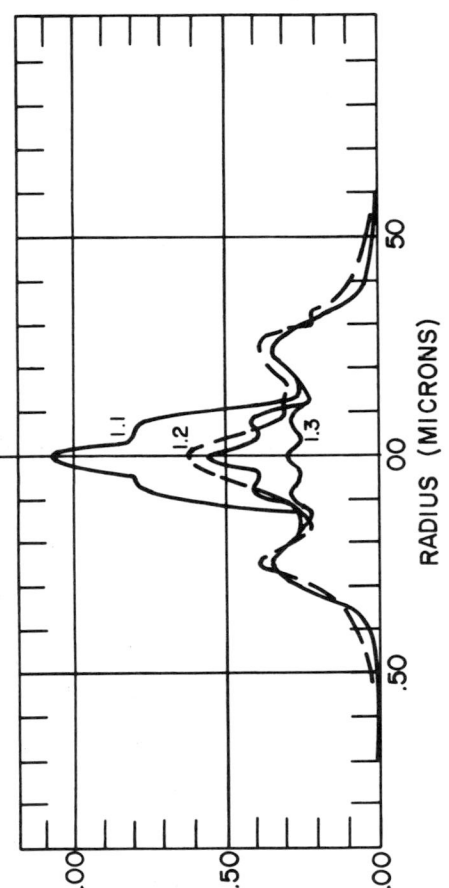

Fig. 23.11-19 Comparison of experimental pinhole form for shot No. 1189 with calculated forms for different degrees of spherical convergence. This case is typical of a well defined ring with a central spike. The indicated parameters L_1, L_2 and L_3 correspond to initial tangential velocities of 2.0×10^6, 2.2×10^6 and 2.5×10^6 cm/sec. Reproduced from studies at KMSF performed by K. A. Brueckner et al.

G.   Conclusions

The dynamics of shell implosions described by relatively elementary analyses of ablation and momentum balance give a semi-quantitative model, which
is in good agreement with more complete computer simulation. The problems
of symmetry and energy flow in the pellet surface are also readily analyzed and
the order-of-magnitude conditions for good implosion symmetry determined.
Computer simulation again confirms the nature of the phenomena.

The X-ray pinhole formation is shown to give a highly useful signature
of the pellet implosion history, which is readily correlated with the computer
predictions of the expected images. This process of "pattern recognition"
allows quantitative inferences of transient pellet conditions, although the results so obtained are, of course, subject to uncertainties which may arise from
the accuracy of the computer model.

The implosion experiments, in correlation with the analytic estimates
and the results of computer simulation, show that successful implosion with
symmetry not degraded by two-dimensional effects, can be obtained within the
resolution of the measurement, in good quantitative agreement with the predictions. Although Rayleigh-Taylor instability is predicted to be marginally
present, depending on the initial shell perturbation, no growth is apparent in
the recorded pinhole images. Some degradation from perfect convergence can,
however, be detected for some implosions, and is evident from a reduction in
the strength of the "implosion spike" predicted for perfect implosion symmetry.
This effect may be due to the occasional loss of illumination symmetry due to
laser alignment or pellet-positioning errors or to occasional substantial eccentricity in the glass shell.

A more sensitive test of final symmetry of convergence and of possible
late instability showing mixing in the pellet center is given by the details of the
neutron production.

23.12 Semi-Empirical Estimates of Neutron Production in Shell Implosions

A. Introduction

The implosion of glass shells filled with high pressure DT gas has given a first crucial experimental test[1] of several aspects of laser-driven fusion. We shall now present a simple model consistent with experiment, which accounts for the observed neutron production and which incorporates the phenomena known to be important in the process. Much more complete numerical calculations are available; these, however, tend to obscure the phenomena and are not readily obtainable for the wide range of parameters of possible interest.

The basic models for laser absorption and the hydrodynamic motion of the shell are give in Sections 23.12B and 23.12C. The effects of ion-electron equilibration and of conduction loss on the reaction rate and the maximum DT temperature are estimated in Sections 23.12D and 23.12E. Neutron production is determined in Section 23.12E. In Section 23.12F, the results of the earlier sections are combined to obtain estimates for neutron yield as a function of laser flux and pellet parameters.

B. Laser Coupling and Shell Hydrodynamics

The absorption of energy in the plasma formed from a glass shell, for a laser flux density less than roughly 5×10^{15} w/cm^2, may be estimated

---

[1]K. A. Brueckner, "Symmetry and Stability in Laser-Driven Fusion," invited talk, American Physical Society Meeting, Anaheim, California, February 1, 1975.

from classical absorption.[2] The free-free absorption coefficient at 1μ is

$$K_{ff} = \frac{K_0 (n/n_c)^2}{\theta^{3/2} (1 - n/n_c)^{1/2}}, \quad K_0 = 24Z \, cm^{-1}, \tag{23.12-1}$$

with θ in kev. The absorption coefficient in a pellet with critical radius r_c and an exponential density distribution

$$n = n_c \exp - (r - r_c)/L \tag{23.12-2}$$

is

$$A = 1 - \exp - 2 \int_{r_c}^{\infty} K_{ff} dr = 1 - \exp - \frac{8}{3} \frac{K_0 L}{\theta^{3/2}}. \tag{23.12-3}$$

The scale height L is typically of the order of the pellet radius; consistent with experiment and with numerical estimates, we assume $L = r/2$, with r_0 = initial pellet radius.

To obtain the temperature and hence the absorption coefficient, we use

---

[2]K. A. Brueckner, "Interpretation of the Interaction of Lasers with Spherical Targets," Part 2 of a series of four lectures on "Laser-Induced Nuclear Fusion" given at the summer school on The Physics of Quantum Electronics, Santa Fe, New Mexico, June 23-July 4, 1975. To be published by Addison-Wesley Publishing Company, Inc., and edited by Stephen Jacobs and Marlan O. Scully.

the result for an isothermal, one-dimensional plasma expansion:[3]

$$E(\text{expansion}) = 4p_0 c, \quad c = \text{isothermal sound velocity} = v_0 \theta^{\frac{1}{2}},$$

$$v_0 = 3 \times 10^7 \text{ cm/sec } (\theta \text{ in kev}). \tag{23.12-4}$$

In Eq. (23.12-4), p_0 is the pressure at the sonic point, which is close to the ablation front, with the plasma assumed to be isothermal outside the ablation front. The pressure, for equal electron and ion temperatures, is $p_0 = n_i \times (1 + z)\theta$. The ionization state and the density at the sonic point can be accurately determined only from more detailed numerical calculation. For temperatures of the order of a kev, the plasma is fully ionized with $z \cong 10$. The density n_i is expected to be of the order of $10^{22}/\text{cm}^3$.

The energy balance is

$$\varphi \left(1 - \exp{-\frac{4}{3} \frac{K_0 r_0}{\theta^{3/2}}}\right) = 4 n_i (1 + z) v_0 \theta^{3/2}. \tag{23.12-5}$$

This equation is most conveniently used to determine $\varphi_L r_0$ as a function of $D/\theta^{3/2}$, and hence θ as a function of φ_L and r_0. The pressure is then determined.

The equation of motion for implosions of the shell may be approximated, consistent with implosion calculations for a shell, by

[3]K. A. Brueckner, P. M. Campbell and R. A. Grandey, "Laser-Driven Implosion of Spherical Shells," Nuclear Fusion 15, 471-486 (1975).

$$m(dr/dt) = - p_0, \tag{23.12-6}$$

with

$$m = m_0 - \dot{m}t \tag{23.12-7}$$

and the mass ablation rate

$$\dot{m} = p_0/c. \tag{23.12-8}$$

The motion of the shell is then given by

$$v = - c \ln(m_0/m), \tag{23.12-9}$$

$$r_0 - r = (c^2/p_0)\left(m_0 - m + m \ln \frac{m_0}{m}\right). \tag{23.12-10}$$

The shell mass per unit area remaining after the ablation in the implosion is determined by Eq. (23.12-10) with r = 0. The implosion time is

$$t = (m_0 - m)/m = (m_0 - m)c/p_0. \tag{23.12-11}$$

The efficiency of energy transfer to the implosion is

$$\frac{1}{2} \frac{mv^2}{E_L \text{(absorbed)}} = \frac{1}{8} \frac{m}{m_0 - m} \ln^2 \frac{m_0}{m} .$$ (23.12-12)

C. Gas Compression

The ablated glass shell transfers energy to the DT gas, compressing and heating the gas to the maximum conditions for thermonuclear reactions. The final gas conditions may be affected by incomplete energy transfer from the shell to the gas and by conduction loss from the gas to the shell.

The kinetic energy of the relatively dense shell cannot be completely transferred to the gas, for the moment ignoring thermal conduction, since the shell retains some internal kinetic and thermal energy as the shell is stopped by the back pressure of the gas. The transfer efficiency may be estimated by a consideration of the conditions as the shell reaches the minimum radius. As the shell is turned around, the gas and shell momentarily reach pressure balance so that the internal energy in the shell is (we assume an ideal gas equation of state)

$$\Delta E_s = (3/2) p_m 4\pi r_m^2 \Delta r_m,$$ (23.12-13)

with r_m the minimum inner shell radius and Δr the shell thickness. From mass conservation,

$$4\pi r_m^2 \Delta r_m \rho_s = 4\pi r_0^2 \rho_{s,0}) M_s / M_0), \quad M_s = \text{shell mass at } r = r_m,$$

$$M_0 = \text{initial shell mass.}$$ (23.12-14)

The pressure in the gas, with conduction-cooling ignored, is

$$(4/3)\pi r_m^3 (3/2)p_m = f_H E_S,$$ (23.12-15)

with f_H the hydrodynamic transfer efficiency and E_S the shell kinetic energy. We assume that the internal energy and kinetic energy of relative motion with the shell are equal. Combining these equations, we find

$$f_H = 1 \Big/ \left(1 + 6 \frac{\Delta r_0}{r_0} \frac{m}{m_0} \frac{\rho_{S,0}}{\rho_{g,0}} \frac{\rho_g}{\rho_S}\right) = 1 \Big/ \left(1 + 2 \frac{M_s}{M_g} \frac{\rho_g}{\rho_S}\right).$$ (23.12-16)

The gas temperature, neglecting conduction, is found from Eqs. (23.12-15) and (23.12-16),

$$\theta_m = m_a v_S^2 / 6 \left(\frac{M_g}{M} + \frac{2\rho_g}{\rho_S}\right).$$ (23.12-17)

Thus, if the shell mass is much larger than the gas mass and the final shell density is much larger than the gas density, the gas temperature θ_m is much higher than the stagnation temperature $\theta = m_a v_S^2 / 6$ corresponding to the implosion velocity. Equation (23.12-17) shows that the gas temperature θ_m rises monotonically with decreasing gas mass. The optimum conditions for neutron production are, however, also affected by conduction loss and by the gas volume and density, as will be discussed in the next sections.

The gas density is more difficult to estimate, since the details of the shock heating of the gas depend on the acceleration of the shell. If, however, the laser flux is constant or slowly increasing with time, the gas is initially heated by the passage of an incoming and reflected shock, with the average compression ratio in the range of 20 to 30.[3] The initial outgoing shock is again reflected at the incoming shell surface and the gas is further heated. After the first strong shock traversal, the subsequent compression is approximately adiabatic unless the shell is strongly accelerated by a rapid rise in the applied pressure. In the absence of such a pressure increase, the final compression is

$$\rho_m \cong \rho_1 (\theta_m / \theta_1)^{3/2}, \quad \rho_1 = (20 \text{ to } 30)\rho_0. \qquad (23.12\text{-}18)$$

The gas is heated from θ_1 to θ_m by the transfer of energy from the shell to the gas, as discussed in the preceding paragraphs. The temperature θ_1 is approximately 1/4 of the stagnation temperature[3] of the gas, corresponding to the shell velocity at the time of return of the first reflected shock, giving

$$\theta_1 \cong m_a v_1^2 / 24, \qquad (23.12\text{-}19)$$

while θ_m is given by Eq. (23.12-16). Thus, neglecting the difference between v_S^2 and v_1^2, we find

$$\frac{\rho_m}{\rho_0} \cong 200 \left(\frac{M_g}{M_S} + \frac{2\rho_g}{\rho_S} \right)^{3/2} . \qquad (23.12\text{-}20)$$

For a typical case with $M_g + (M_S/10)$ and $\rho_g = \rho_S/5$, $\rho_M/\rho_0 = 570$, which is consistent with complete implosion calculations. Equation (23.12-20) shows that the compression ratio drops rapidly with increasing gas mass, the final damping action of the shell becoming less effective as the gas mass increases.

Equation (23.12-20), based only on the qualitative derivation of this section, can be considered only a rough estimate, although the scaling with M_g/M_S is correct. The numerical coefficient should be adjusted by complete implosion calculations. The maximum compression can also, in principle, be markedly increased by the use of an optimized shell-acceleration history.

D. Conduction Loss

The conduction cooling is particularly important for small shells and is a strong limiter on the achievable gas temperature. The thermonuclear reaction rate is a very strong function of temperature in the one-to-two kev temperature range so that a relatively small drop in temperature can have a pronounced effect on neutron yield.

We use an approximate method to estimate the gas temperature as a function of time. The gas is heated by compression in the implosion and cooled by conduction. The combined effect is given by

$$3\frac{d\theta}{dt} = -\frac{1}{r^2}\frac{\partial}{\partial r}r^2 k\frac{\partial\theta}{\partial r} + \frac{2\theta}{p}\frac{d\rho}{dt} . \qquad (23.12-21)$$

The thermal conductivity for DT is

$$k = k_0\theta^{5/2}/n, \quad k_0 = 5\times10^{27} \text{ cm}^2/\text{sec} \; (\theta \text{ in kev}), \qquad (23.12-22)$$

where n = ion number density.

We approximate these terms by

$$\frac{1}{r^2} \frac{\partial}{\partial r} r^2 \frac{k_0 \theta^{5/2}}{n} \frac{\partial \theta}{\partial r} \cong \frac{-k_0 \theta_0^{3/2}}{R^2 n} \ ,$$

$$\frac{2\theta}{\rho} \frac{d\rho}{dt} = - \frac{6\theta_0}{R} \frac{dR}{dt} \ , \tag{23.12-23}$$

with θ_0 the central temperature and R the inner radius of the glass shell. Equation (23.12-9) may then be rewritten as

$$\frac{6}{5} \frac{d}{dt} \left(R^2 \theta_0 \right)^{-5/2} = \frac{-k_0}{nR^7} = \frac{-k_0}{n_0 r_0^3} \frac{1}{R^4} \ . \tag{23.12-24}$$

Equation (23.12-12) may be integrated to give

$$\theta_0 = \frac{\theta_m r_m^2}{R^2 \left(1 + \frac{5}{6} \frac{k_0 \theta_m^{5/2} r_m}{n_0 r_0^3} \int_{-\infty}^{t} dt' \left[\frac{r_m}{R(t')} \right]^4 \right)^{2/5}} \ , \tag{23.12-25}$$

with θ_m the temperature at $R = r_m$ in the absence of conduction. To evaluate Eq. (23.12-25), we need an estimate of R as a function of time, which we obtain from the equation of motion of the shell near the implosion maximum, viz.

$$(M_g + M_s)(d^2 R/dt^2) = -4\pi r^2 p,$$

(23.12-26)

giving

$$R \cong r_m [1 + (t - t_m)^2 / \tau_m^2],$$

(23.12-27)

with the characteristic time τ_m of the implosion maximum given by

$$\frac{1}{\tau_m^2} = \frac{1}{2} \frac{4\pi r_m}{M_s + M_g} p_{0,m} = 3 \frac{M_g}{M_s + M_g} \frac{\theta_{0,m}}{m_a r_m^2} .$$

(23.12-28)

The temperature $\theta_{0,m}$ is the central temperature maximum, to be determined by the following equations.

Equation (23.12-25) may be rewritten

$$\theta_0 = \frac{\theta_m}{\left[1 + \left(\dfrac{t - t_m}{\tau_m}\right)^2\right]^2} \frac{1}{\left[1 + \lambda_m f\left(\dfrac{t - t_m}{\tau_m}\right)\right]^{2/5}} ,$$

$$\lambda_m = \frac{5}{6} \frac{\theta_m^{5/2} k_0 r_m \tau_m}{n_0 r_0^3} , \quad f(x) = \int_{-\infty}^{x} \frac{d^4}{(1 + y^2)^4} .$$

(23.12-29)

The temperature is sharply peaked near $t = t_m$; at this time, the temperature has been depressed by conduction by the factor

$$\theta_{0,m} = \theta_m \Big/ \left(1 + \frac{5\pi}{32}\lambda_m\right)^{2/5} .$$ (23.12-30)

To determine the neutron production, we need the spatial distribution of temperature in the compressed gas, which we obtain from the constant-flux solution of the conduction equation,

$$\theta(r) = \theta_0[1 - (r/R)]^{2/7},$$ (23.12-31)

which we assume to hold as a function of time as R varies near the implosion maximum at $R = r_m$.

E. Neutron Production

The neutron yield is given by

$$N_n = 4\pi \int_{-\infty}^{\infty} dt \int_0^R r^2 dr \frac{n^2}{4} \overline{\sigma v}(\theta) = \frac{V_0 n_0 n_m}{4} \int_{-\infty}^{\infty} \frac{dt}{R^3} r_m^3 \int_0^R \frac{3r^2 dr}{R^3} \overline{\sigma v}(\theta).$$

(23.12-32)

This radial integral may be evaluated by using Eq. (23.12-31) for the radial variation of temperature and by approximating the temperature variation of the reaction rate for θ near $\theta_{0,m}$ by

$$\overline{\sigma v}\,(\theta) = \overline{\sigma v}\,(\theta_0)(\theta/\theta_0)^k, \tag{23.12-33}$$

with

$$k = \theta_{0,m}\left(\frac{d}{d\theta}\,\ell n \overline{\sigma v}\,(\theta)\right)_{\theta = \theta_{0,m}}. \tag{23.12-34}$$

The radial integral then gives

$$\int_0^R \frac{3r^2}{R^3}\,dr\,\overline{\sigma v}\,(\theta_0)(1 - r/R)^{2k/7} = \frac{6\overline{\sigma v}\,(\theta_0)}{\left(\frac{2k}{7} + 1\right)\left(\frac{2k}{7} + 2\right)\left(\frac{2k}{7} + 3\right)}. \tag{23.12-35}$$

The factor $6\!\left/\!\left(\frac{2k}{7} + 1\right)\left(\frac{2k}{7} + 2\right)\left(\frac{2k}{7} + 3\right)\right.$ gives the reduction of the effective re-action volume from $4\pi R^3/3$, due to the temperature gradient in the compressed gas. At 1 kev, $k \cong 6$ and this factor is 0.13.

The remaining integral over time in Eq. (23.12-32) may similarly be evaluated to give

$$\int_{-\infty}^{\infty} dt\left(\frac{r_m}{R}\right)^3\,\overline{\sigma v}\,(\theta_0)$$

$$= \overline{\sigma v}\,(\theta_{0,m})\tau_m \int_{-\infty}^{\infty} \frac{dx}{\left(1 + x^2\right)^{3+2k}}\left(\frac{1 + \lambda_m f(0)}{1 + \lambda_m f(x)}\right)^{2k/5}. \tag{23.12-36}$$

The integral over x gives the correction to the effective interaction time τ_m due to the rapid temperature change and cooling near the implosion maximum. The integrand is very strongly peaked near x = 0; we therefore neglect the relatively slow variation of f(x) which we evaluate at x = 0. The result for k = 6 is

$$\int_{-\infty}^{\infty} \frac{dx}{\left(1 + x^2\right)^{3+2k}} = \frac{\sqrt{\pi}}{2} \frac{\Gamma(2k + 5/2)}{\Gamma(2k + 3)} \cong 0.24. \tag{23.12-37}$$

Collecting these results, we find

$$N_n = 0.031\, V_0 (n_0 n_m/4)\, \overline{\sigma v}\, (\theta_{0,\,m})\, \tau_m. \tag{23.12-38}$$

F. Applications

For convenience in application, we first summarize the required equations. As a function of external plasma temperature, the laser flux per unit area is given by [Eq. (23.12-5)]

$$\varphi_L = \frac{4n_i(1 + z)\theta}{1 - \exp - 160r_0/\theta^{3/2}}. \tag{23.12-39}$$

The absorbed laser flux is [Eq. (23.12-5)]

$$\varphi_{L(abs)} = \varphi_L \left(1 - \exp - 160r_0/\theta^{3/2}\right). \tag{23.12-40}$$

The ablated mass fraction $x = m/m_0$ is [Eq. (23.12-10)]

$$1 - x + x \ln x = \frac{\varphi_{L(abs)} r_0}{4 m_0 c^3} = n_i r_0 m_i / m_0 .$$

(23.12-41)

The implosion time is [Eqs. (23.12-11) and (23.12-10)]

$$t_{imp} = r_0 / c (1 - x + x \ln x) .$$

(23.12-42)

The energy transferred to the gas is [Eqs. (23.12-12) and (23.12-16)]

$$E_{gas} = \frac{M_0 c^2}{2} \frac{x \ln^2 (1/x)}{\left(1 + \frac{x M_0}{M_g} \frac{\rho_g}{\rho_\tau} \right)} , \quad M_0 = 4 \pi r_0^2 m_0 ,$$

(23.12-43)

giving the maximum gas temperature in the absence of conduction cooling as

$$\theta_m = E_{gas} m_a / 3 M_{gas} .$$

(23.12-44)

The maximum temperature corrected for cooling loss is [Eq. (23.12-30)]

$$\theta_{0,m} = \theta_m \bigg/ \left(1 + \frac{5\pi}{32} \lambda_m \right)^{2/5} , \quad \lambda_m = \frac{5}{6} \frac{k_0 r_m \tau_m \theta_m^{5/2}}{n_0 r_0^3} ,$$

(23.12-45)

with the characteristic time [Eq. (23.12-28)] and compression [Eq. (23.12-20)]

$$\tau_m = \left(\frac{1}{3}\frac{M + M_g}{M_g}\frac{m_a}{\theta_{0,m}}\right)^{1/2} r_m, \quad \left(\frac{r_0}{r_m}\right)^3 = Q \bigg/ \left(\frac{M_g}{M_T} + \frac{2\rho_g}{\rho_T}\right)^{3/2}, \qquad (23.12\text{-}46)$$

where $Q = 200$.

The neutron production is

$$N_n = \frac{0.031}{4} n_0 n_m V_0 \overline{\sigma v}(\theta_{0,m}) \tau_m. \qquad (23.12\text{-}47)$$

The variables in these equations, which determine the neutron yield, are the laser flux φ_L, the pellet radius r_0, the initial shell mass per unit area m_0, and the initial gas density n_0. The parameters of the model, which cannot be accurately determined without numerical study of the implosion, are the external electron density $n_i(1 + z)$ at the isothermal conduction front, the ratio of gas to shell density ρ_g/ρ_S at the implosion maximum, and the numerical factor of 200 in Eq. (23.12-46) giving the maximum compression. We will choose reasonable values for these parameters, consistent with experiment and numerical calculations, and use the above equations to determine the scaling laws. We use the values

$$n_e = n_i(1 + z) = 2.5 \times 10^{22}/\text{cm}^3, \quad \rho_T/\rho_g = 5, \qquad (23.12\text{-}48)$$

and the numerical value of $Q = 200$ and $Q = 800$ in Eq. (23.12-46). Some typical results are given in Table 23.12-1 for shells of radius $30\,\mu$ and $100\,\mu$. For

Table 23.12-1 Implosion examples. Reproduced from calculations done by
K. A. Brueckner.

Variable	Units	Results for various examples				
shell radius	μ	30	30	30	100	100
shell thickness	μ	1.35	1.35	1.35	4.50	4.50
DT pressure	atm	30	60	60	90	90
plasma temperature	keV	1.70	2.00	2.00	1.40	1.40
laser power	Tw	0.50	0.79	0.79	1.30	1.30
laser flux/unit area	10^{15} w/cm^2	4.41	7.02	7.02	1.036	1.036
absorptivity	%	19.6	15.6	15.6	62	62
implosion time	picosec	142	131	131	521	521
final/initial shell mass	dimensionless	0.10	0.10	0.10	.10	.10
laser energy	j	70.8	103.9	103.9	678	678
energy transferred to gas	j	0.45	0.71	0.71	20.8	20.8
maximum gas temperature, neglecting conduction	kev	5.76	4.54	4.54	2.38	2.38
maximum gas temperature at r = 0, including conduction	kev	1.28	1.73	2.53	2.08	2.25
compression parameter Q [Eq. (23.12-46)]	dimensionless	200	200	800	200	800
compression ratio	dimensionless	434	283	1133	203	813
maximum gas density	g/cm^3	2.60	3.39	13.60	3.65	14.6
time of implosion maximum	picosec	25.4	19.2	10.0	56.9	34.5
neutron yield	neutrons	5.24 $\times 10^5$	5.53 $\times 10^6$	7.23 $\times 10^7$	2.48 $\times 10^9$	8.72 $\times 10^9$

a 30 μ shell with initial shell thickness of 1.35 μ with 60 atm DT pressure and a laser power of 0.79 Tw, the plasma absorptivity is 15.6%, the DT central temperature maximum is 1.73 kev for a compression of 283 and a final DT density of 3.39 g/cm^3, giving a neutron yield of 5.53×10^6. For a higher compression of 1,133 giving a final DT density of 13.6 g/cm^3, the central temperature maximum is increased to 2.53 kev and the neutron yield to 7.23×10^7.

For a 100 μ radius shell with thickness of 4.5 μ filled with 90 atm of DT, a laser power of 1.30 Tw and a compression ratio of 813 give a maximum central temperature of 2.25 kev and a neutron yield of 8.72×10^{9}.

Other examples are easily computed; sets of tabulations are available.

23.13 Lasers for Fusion

The general characteristics of the laser follow from the peak power. The characteristics of neodymium-doped glass are summarized in Table 23.13-1. A safe operating level for a neodymium glass laser is between 5 and 10 Gw/cm^{2} for a beam with good spatial uniformity. We will return later to the problem of meeting adequate uniformity requirements. We assume optimistically 10 Gw/cm^{2} as a safe operating level. The laser configuration to provide a given peak power depends on the choice of rods (circular or rectangular) which are edge-pumped by flashlights, or of disks (at Brewster's angle) which are face-pumped. If rods are used, the absorption of pump energy in the glass limits the diameter of a circular rod or the thickness of a rectangular rod to 5 to 8 cm, depending on the doping level of the glass. One of the largest rod amplifiers in present use is an 80 mm, 2% doped CD-2 rod at KMS Fusion. The large 9-beam laser at the Lebedev Institute uses 9 output rods with 45 mm diameter. Many lasers built by CGE have 64 mm diameter output rods. A rectangular rod amplifier with 40×240 mm cross section has recently been tested at the Lebedev Institute in Moscow. We assume a diameter of 60 mm as a reasonable limit for rod amplifier.

The useful aperture of a single amplifier can be increased if the glass is arranged in disks at Brewster's angle to minimize reflection losses and allow face-pumping (see Fig. 23.12-1). In this case, the aperture can be increased

Table 23.13-1 Neodymium glass characteristics. Reproduced from studies
at KMSF performed by K. A. Brueckner et al.

safe power level (limited by surface damage or self-focusing damage)	5 to 10 Gw/cm^2
gain for Owens-Illinois ED-2 glass; the gain is slightly affected by beam spectral structure and pulse length	0.16 per cm per j/cm^3 of stored energy
energy storage as determined by pumping intensity and spontaneous emission amplification	0.3 to 0.6 j/cm^3
maximum glass dimension to limit parasitic amplification	50 to 60 cm
maximum glass thickness to give uniform pumping, assuming two-sided illumination normal to the surface	40 to 60 mm

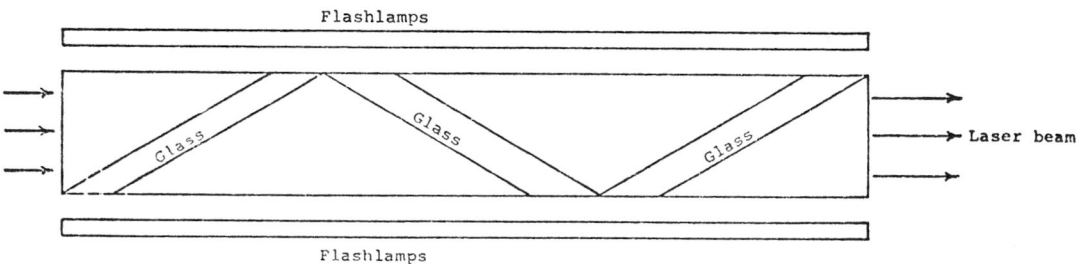

Fig. 23.13-1 Typical disk amplifier; 10 cm clear aperture, 2.5 cm slab
thickness; effective aperture = 122 cm^2; optical slab thickness
= 3 cm; energy storage = 440 j (0.4 j/cm^3); low signal gain
= 1.78. Reproduced from studies at KMSF performed by
K. A. Brueckner et al.

until amplification of spontaneous emission along the disk becomes too large. Present experimentally tested designs show that an effective area in the glass disk of about 10^5 cm^2 is possible corresponding to a length along the elliptical disk of about 60 cm and an open circular aperture of 30 cm diameter.

An optimized amplifier arrangement uses edge-pumped rods up to a diameter of 60 mm and then disk amplifiers up to a diameter of 300 mm. If additional energy is required, the system is paralleled.

The overall configuration of a laser depends on the gain per unit length and the final energy requirement. The gain of strongly pumped neodymium glass depends on the dimension of the glass and the details of the flash-lamp geometry. A gain of 0.064 to 0.080 per cm in ED-2 glass, corresponding to an energy storage of 0.4 to 0.5 j/cm^3, is reasonable. The overall gain requirement is determined by the oscillator input, typically of the order of 10^{-3} j, and the level of saturation reached in the amplifiers. Typical laser configurations to give a kj pulse with peak powers of 2×10^{12} w and 2.5×10^{13} w are shown in Tables 23.13-2 and 23.13-3. The amplifiers required are well within the present state-of-the-art. The largest disk amplifier presently operating is at KMS Fusion with 100 mm aperture, a low signal gain of about 20, and a gain of about 8 at 1000 j output. A disk amplifier with 300 mm aperture is being designed at the Lawrence Livermore Laboratory.

The analysis just given indicates only approximately the characteristics of a glass laser designed for feasibility experiments. A laser must typically be substantially over-designed to provide for additional losses in mirrors, lenses, electro-optic shutters, etc. These effects are, however, readily compensated by an additional stage of amplification or increased gain per amplifier stage and hence are not the major sources of practical difficulty in achieving the desired laser performance.

Table 23.13-2 Laser amplifier to give 10^3 j with peak power of 2×10^{12} w.[*]

Amplifier type	Energy output, j	Diameter, mm	Path length, cm	Low signal gain
rod	0.18	4	50	24.4
rod	3.36	16	50	24.4
rod	16.7	33	30	6.8
rod	76.3	64	30	6.8
disk	324	87	30	6.8
disk	1000	128	30	6.8

Table 23.13-3 Laser amplifier to give 10^3 j with peak power of 2.5×10^{13} w.[*]

Amplifier type	Energy output, j	Diameter, mm	Path length, cm	Low signal gain
rod	0.043	4	30	6.8
rod	0.297	12	30	6.8
rod	2.03	30	30	6.8
disk	13.9	79	30	6.8
disk	99	156	18	3.16
disk	307	260	18	3.16
disk(3)	3×333	260	18	3.16

---

[*]Reproduced from studies at KMSF performed by K. A. Brueckner et al.

As mentioned earlier, the laser pulse to be fully useful must be accurately controlled in its time variation. This cannot be achieved with electro-optic shutters of the Kerr cell or Pockel cell type. These give a pulse form which at low power is roughly Gaussian in form with a full width at half maximum of 1 to 1.5×10^{-9} sec. This pulse is distorted by saturation effects in the amplifiers; as a result, the amplifier output pulse has a considerably more rapid rise than the oscillator output. Other methods of pulse forming are, therefore, necessary to deliver the required pulse form at the high power of the laser, allowing for possible distortion through the laser. A relatively simple design now operating at KMSF uses a "pulse-stacker" and starts with a pulse of about 30 picosec duration from a mode-locked YAG oscillator, which is divided by beam splitters, selectively delayed and attenuated, and reformed to give a composite pulse of the desired form. The characteristics of the pulse-stacker are shown schematically in Fig. 23.13-2. A short pulse of 20 picosec is produced by a mode-locked oscillator. The pulse is divided into five pulses by a series of partially reflecting mirrors. Each pulse is given a predetermined time delay and attenuation and the pulses are brought back to a single optical path by a second succession of partially reflecting mirrors. The "stacked" pulses consequently can be given any desired composite pulse form. The number of pulses used in practice is determined by the overall pulse length desired, by the pulse length from the oscillator, and by the acceptable intensity ripple in the stacked pulse sequence.

In operation, the pulse stacker is adjusted to compensate for the measured nonlinearity of the amplifier train, to correct for the pulse distortion which arises from saturation, possible index variation with energy depletion of the lasing medium, and relaxation effects in the lasing action.

The next requirement is the control of the laser-intensity distribution through the laser amplifiers so that good spatial uniformity is maintained.

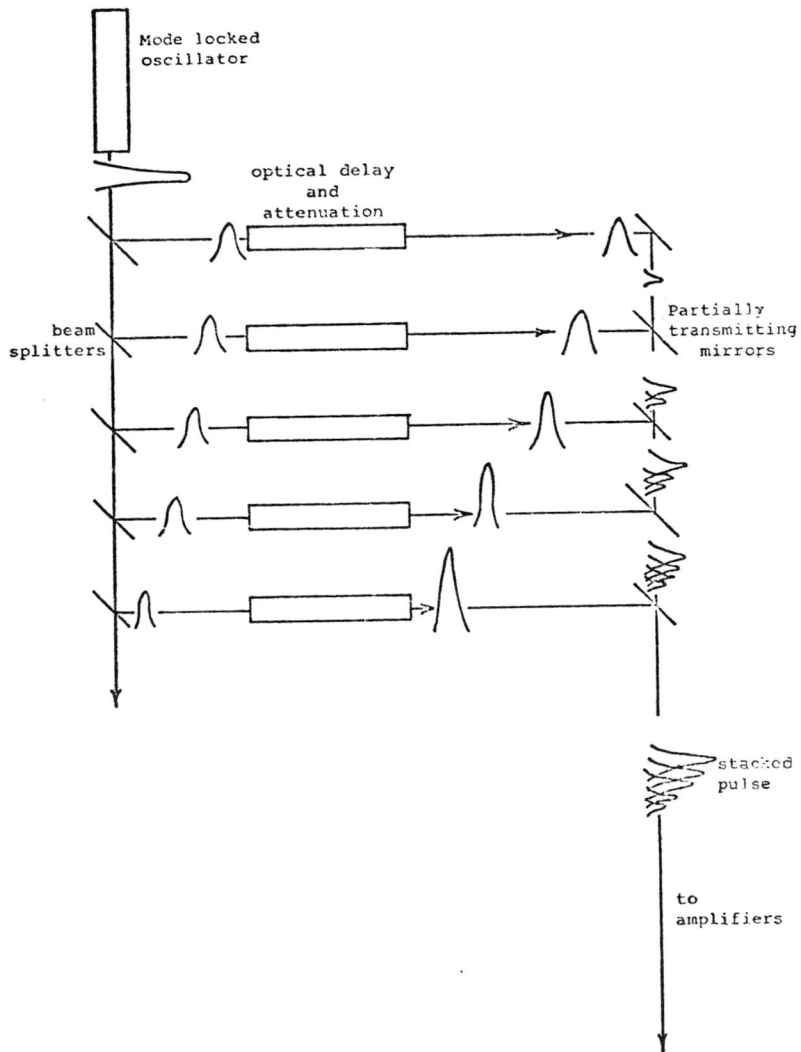

Fig. 23.13-2 Schematic diagram of the pulse-stacker. Reproduced from
studies at KMSF performed by K. A. Brueckner et al.

This is very important since the maximum power limit is set by the onset of catastrophic self-focusing in the glass, which is initiated by the effect of intensity nonuniformity on the index of refraction of the glass. The index of refraction increases with power in the glass according to

$$n = n_0 + n_2 E^2, \quad n_2 = 1.3 \times 10^{-13} \text{ (E in esu).}$$

A simple analysis shows that a beam nonuniformity ΔP with scale b gives rise to a large field increase which in turn leads to damage, with a self-focusing length f of

$$f = (b/4)\sqrt{n_0/n_2 E^2} \cong 400 \, b/[\Delta P(Gw/cm^2)]^{\frac{1}{2}}.$$

This equation shows that self-focusing can be avoided in an amplifier with 50 cm length, with field fluctuations of 5 Gw/cm^2, only if the scale of the nonuniformity is greater than 2.8 mm. If the fluctuations are greater or the amplifier length greater, the scale of the nonuniformity must be correspondingly increased. The importance of this problem is well-known in work with high powered lasers, damage being easily produced if a laser is operated with poor control of the intensity patterns.

In practice, with high quality laser glass, the intensity variations in the laser beam which initiate self-focusing are results of Fresnel rings produced by apertures in the amplifier train. Other sources of interference, which can cause severe damage, are improperly aligned prisms and internal reflection from rod surfaces. These problems can all be avoided if the laser is designed to have an aperture considerably larger than the useful energy-producing aperture. If this is done, the laser beam is apertured only at very low intensity and

the resulting Fresnel rings are very weak. This requirement is most important in the smaller laser rods, which fortunately are also the lower cost elements of the amplifier train. The output-disk amplifiers can be more fully
filled, the diffraction patterns forming late in the system in the widely spaced
disks giving relatively little self-focusing problems. The problems of Fresnel
ring formation can also be alleviated by the use of "soft apertures" rather than
typical apertures with sharply defined radii. Soft and improved apertures are
being tested. Tests at KMSF show that 5 Gw/cm^2 is a safe operating range.
It is expected that 10 Gw/cm^2 will be reached with better intensity control by
improved aperture design.

The next requirement is the control of the laser to deliver spatially uniform energy to the pellet. This requires good laser-beam uniformity, together
with an optical system near the diffraction limit to focus the energy onto the
target. The laser should deliver energy to the pellet with about 100 μ radius
with a flux variation due to laser intensity variation of less than 20%. With an
f/1 focal-ratio illuminating lens, this requires a beam divergence of less than
about 10^{-4} radian and corresponds, at one μ, to a diffraction-limited beam
at 1 cm aperture and to ten times diffraction limit at 10 cm aperture. These
requirements are met fairly easily with uniformly pumped small diameter laser rods with a laser beam only partially filling the laser rods and hence only
weakly apertured. In practice, the beam divergence can be increased by the
nonuniform pumping and thermal distortion in larger diameter rods, and by
edge effects from the apertures limiting the beam diameter. Closely associated with the beam divergence in nonuniformly pumped and thermally distorted rods is strong birefringence which depolarizes the laser beam. Since
polarizing elements are usually present in the laser (as for example, in the
disk amplifiers set at Brewster's angle), the depolarization leads to substantial energy rejection.

The problems of thermal distortion of the laser rods can be alleviated by increasing the laser-cycle time, by using small diameter rods, by reducing the doping level of the rods to improve uniformity of the pumping, by filtering the flash-lamp light to reject energy ineffective in pumping the rods, and by maintaining a stable firing and cooling cycle. This problem, however, remains a major source of difficulty with a glass laser system.

The laser must also be protected against damage due to amplification of laser energy reflected from the target or from other reflecting elements inadvertently placed in the laser beam. The successive laser stages must be isolated to prevent excessive amplification of spontaneously emitted light, which depletes the stored energy in the glass. The target must also be protected from energy amplified through the laser before the main laser pulse, which can be the result of spontaneous emission or of oscillator energy leaking through the pulse-forming network before the main pulse. The desired isolation can be provided by both passive and active shutters. A Faraday rotator with polarizing plates before and after the rotator allows nearly loss-less forward passage of the laser beam and many tens of db reverse attenuation. Pockel and Kerr cells can be opened by short electrical pulses, allowing fairly good transmission in either direction while open. A Pockel cell can work effectively down to times of about 1.5 nsec and a Kerr cell to 0.5 to 1 nsec. Low-level isolation can also be provided by saturable dye cells which have low transmission at low power and transmit with relatively low loss at the high powers characteristic of the full laser pulse. These techniques can adequately protect the laser system.

Particular care must be taken to protect the target from very low power coming from amplified spontaneous emission ahead of the main pulse. A few millijoules of energy arriving many microseconds ahead of the main pulse can destroy the target. In addition to fast shutters open for only several nsec around the main pulse, further protection can be provided with saturable dye cells. If necessary, final protection can be provided with very thin foils placed in the beam; these are vaporized and made transparent by the leading edge of the main laser pulse.

With a laser designed to provide a suitable pulse, the final problem is the symmetric delivery of energy on the target. This can be done with multiple beams formed with beam splitters and mirrors combined with one or more output stages. Large diameter precision optics can also deliver the energy with adequate symmetry. Fortunately, the analysis of the physics of implosion shows that highly precise uniformity of illumination is not essential, a considerable variation being acceptable due to smoothing effects in energy flow in the pellet surface [see Fig. (23.13-3)].

In summary, a kj laser for fusion experiments is within the present state-of-the-art, although careful attention must be given to critical elements of the system. These are: (a) careful uniformity control to reduce self-focusing damage, (b) pulse shaping to give the required steeply rising output-pulse form, (c) control of pumping levels and thermal distortion to avoid beam degradation and rod birefringence, (d) laser isolation to prevent damage from reflected energy, (e) target protection against energy arriving before the main laser pulse, and (f) symmetric target illumination.

In addition, the requirement of laser reliability should be emphasized. A kj laser is a complex instrument and careful design with good engineering attention to detail is essential.

Other lasers are now being developed, for experiments in the 1 to 10 kj range, for studies of laser-driven fusion feasibility. Of particular interest is the N_2-CO_2 laser being built at Los Alamos to give 1 kj output at $10.6\,\mu$. The N_2-CO_2 laser is pumped by a strong electric field applied to the lasing medium

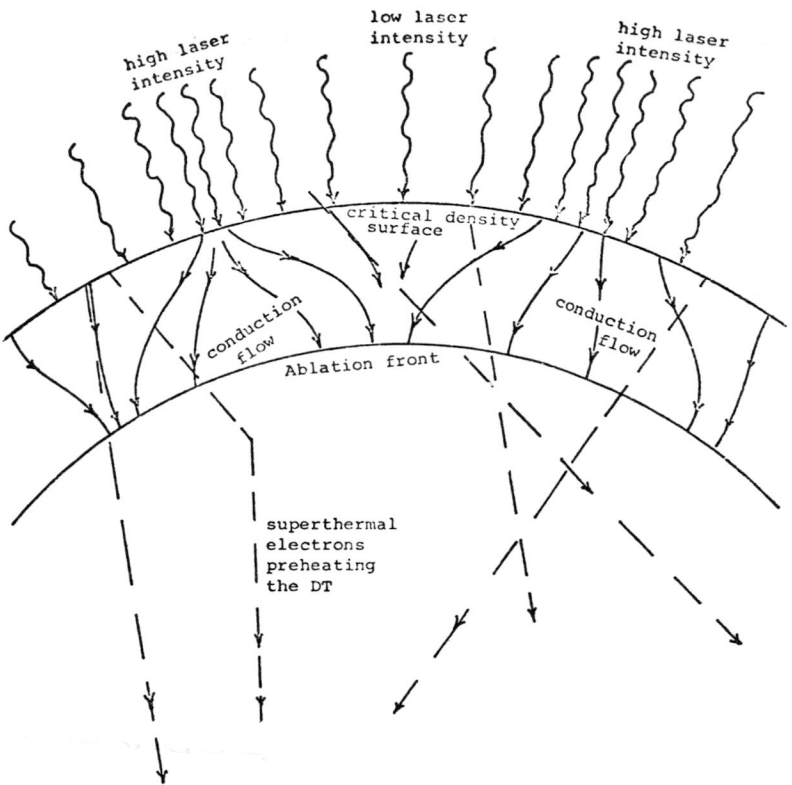

Fig. 23.13-3 Schematic diagram showing smoothing effects in energy flow at a pellet surface. Reproduced from studies at KMSF performed by K. A. Brueckner et al.

in which ionization is maintained by an electron beam. This laser offers the possibility of relatively high (3-5%) efficiency of conversion of electrical to laser energy, compared with 0.1 to 0.3% for a glass laser. The long wavelength of this laser presents considerable theoretical problems in the laser-fusion application, which can be resolved only by experiment. The slow relaxation process of energy transfer in the lasing medium may also make pulse-form control difficult.

Another interesting possibility is a lasing medium of a gaseous alkylio-
dide, in which flashlight pumping produces excited iodine atoms in the $5\,^2P_{\frac{1}{2}}$
state.   This laser operates at 1.3 μ and is expected to have 0.5 to 2% efficiency,
depending on the flash lamp characteristics.   A kj laser of this type with an
nsec pulse is being built at the Max Planck Institute near Munich.

CHAPTER 24

ENVIRONMENTAL ASPECTS OF NUCLEAR POWER APPLICATIONS

24.1   Introduction

Energy use is hazardous. It must be discussed in the context of direct and social costs paid for its use. Our exposition is deliberately repetitive with progressively more detailed introduction of technological facts.

A.   Fatalities Associated With Occupational and Environmental Hazards

The U.S. population is exposed to significant occupational and environmental hazards. The most damaging of these has been associated with the use of motor vehicles. A 1974 categorization[1] of occupational and environmental hazards is reproduced in Table 24.1-1. These data serve to emphasize the fact that historically the occurrence of nuclear-reactor accidents has not been a hazard to the U.S. population, because of relatively limited reactor deployment

---

[1] "An Assessment of Accident Risks in U.S. Commercial Nuclear Power Plants," a study directed by N. C. Rasmussen, U.S. Atomic Energy Commission, Report WASH-1400, Washington, D.C., draft report, August 1974; final report, November 1975. Available from NTIS, Springfield, Va. 22161.

Table 24.1-1.   Estimated numbers and individual probabilities of occurrence
of fatal accidents by various causes in the U. S. during 1969.
Reproduced from Ref. [1].

Accidents involving	Number of accidents during 1969	Approximate individual risk of suffering an acute fatality during the year[a]
motor vehicles	55,791	3×10^{-4}
falls	17,827	9×10^{-5}
fires and hot substances	7,451	4×10^{-5}
drowning	6,181	3×10^{-5}
poison	4,516	2×10^{-5}
firearms	2,309	1×10^{-5}
machinery (1968)	2,054	1×10^{-5}
water transport	1,743	9×10^{-6}
air travel	1,778	9×10^{-6}
falling objects	1,271	6×10^{-6}
electrocution	1,148	6×10^{-6}
railways	884	4×10^{-6}
lightning	160	8×10^{-7}
tornadoes	91[b]	4×10^{-7}
hurricanes	93[b]	4×10^{-7}
all others	8,695	4×10^{-5}
all accidents	111,992	6×10^{-4}
nuclear accidents for 100 reactors	0	2×10^{-10}

(a) Based on the total U.S. population.
(b) Average for the period 1953 to 1971.

and because of relatively safe operation. Our task in the following discussion is to estimate likely future dangers from sources of this type. Many of our factual inputs will be taken from a recently completed study performed under the direction of N. C. Rasmussen.[1]

A compilation of occupational and environmental hazards has been given by Jordan[2] and is reproduced in Table 24.1-2. Reference to Table 24.1-2 shows that people have willingly accepted risks with an average death rate of 0.95 to 2.4 per 10^6 hours of exposure in driving private cars or flying in airplanes. Smaller segments of the population have engaged in motorcycling with a death rate of 6.6 per 10^6 hours and in rock climbing with a death rate of 40 per 10^6 hours. An alternative view of voluntary and involuntary risks is provided by the composite curves shown in Fig. 24.1-1, where the death risks per hour for 10^9 people are shown as a function of average annual benefit (or cost) expressed in $/p-y for voluntary (e.g., skiing, mountain climbing) and involuntary (e.g., environmental pollution, mass transportation) risks. We note that, at a given level of dollar benefits (or costs), voluntary activities are customarily accepted with death risks that are larger by factors of 10^3 to 10^4 than involuntary activities.

The frequency of occurrence of man-caused calamities decreases rapidly with increasing number of fatalities, as is illustrated by the data plotted in Fig. 24.1-2. As causes of deaths, nuclear power plants are seen to follow a curve about four orders of magnitude below that for persons on the ground killed by air crashes. They are, in fact, far safer than any of the devices included in Fig. 24.1-2.

The use of fossil fuels as an environmental hazard has not been included

---

[2]W. H. Jordan, "Nuclear Energy: Benefits Versus Risks," Physics Today 23, 32-38 (1970).

Table 24.1-2 Estimated death rate per million hours of exposure; from W. H.
Jordan (Ref. [2]), PHYSICS TODAY, American Institute of Physics.

Type of risk	Death rate per 10^6 hours of exposure
driving a private car	0.95
transportation in railroads or buses	0.08
air line travel	2.4
driving a motorcycle	6.6
disease contracted in old age	1.0
rock climbing	40.0
incremental exposure* to 1 mrem/y	10^{-5}
incremental exposure* to 10^{-2} mrem/y	10^{-7}

*The radiation exposure units rem and mrem are defined in
Section 24.2A.

in the compilation of Table 24.1-1. It has been estimated[3] that about one
fatal respiratory infection per year occurs for every 10^2 persons-parts-per-
million (p-ppm) of continuous SO_2 exposure. Thus, for an SO_2 concentration
of 10^{-4} g/m^3 $\simeq$ 0.1 ppm (which is often observed in large metropolitan areas),
we might expect among a population of 10×10^6 p the following number of annual
deaths: $10 \times 10^6 \times 0.1/10^2 = 10^4$ deaths per year. Relative hazard assessments for coal and uranium use differ widely[4] and it is not possible to compare their deleterious health effects with confidence.

[3]R. Wilson, "Tax the Integrated Pollution Exposure," Science 178, 182-183
(1972).

[4]See, for example, the discussion and referenced articles by E. A. Martell,
A. Elkeles and J. S. Wiley in American Scientist 63, 618-620 (1975).

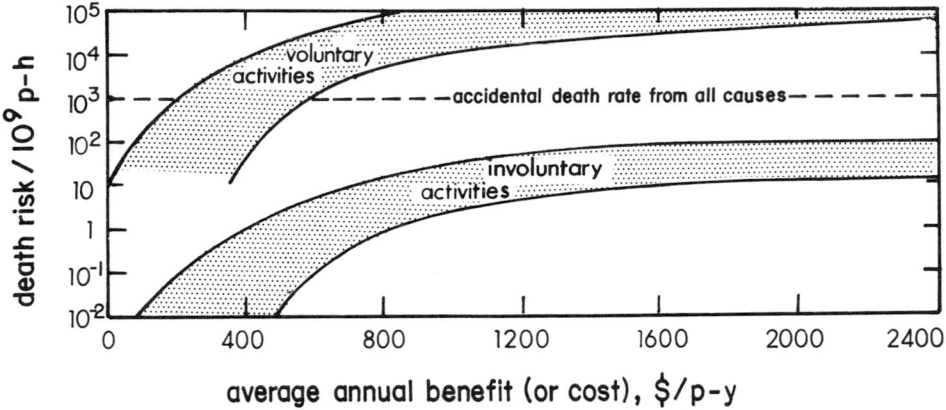

Fig. 24.1-1 The death risk per 10^9 people per hour as a function of average
annual benefit (or cost) in dollars per person per year for voluntary and involuntary activities. The accidental death rate from
all causes is about 1 death per million people per hour. Reproduced from C. L. Comar, "Technology Assessment with Special
Reference to Energy," Cornell Energy Project Paper No. 70-1,
Cornell University, Ithaca, New York, October 1970.

Mining operations are hazardous. Examination of the data compiled in
Table 24.1-3 shows that coal mining produced about ten times as many disabil-
ities as uranium mining and milling, per 10^6 Mw_eh of energy recovered, while
the number of injuries per 10^6 man-hours of work was roughly comparable
for these two types of occupations. These estimates reflect the relatively high
level of sophistication and safety achieved in uranium mining and milling. We
recall that typical energy densities for uranium ores (5.7 bbl_e of petroleum
per metric ton of ore, see p. 17 in Volume I) and for coal (4.8 bbl_e per mt,
see Table 1.2-2) are not greatly different. Furthermore, since the number of
injuries per 10^6 man-hours of work is similar, we must conclude that the
principal reason for greater safety in the extraction and milling of uranium is
associated with greater speed of processing of these ores through utilization

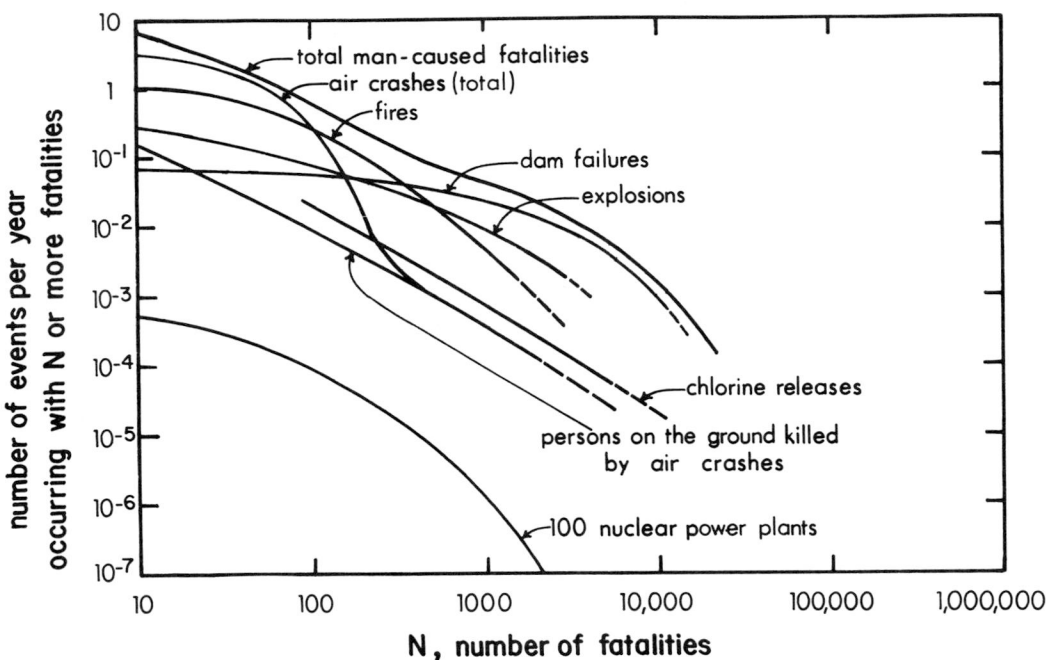

Fig. 24.1-2 Frequency of man-caused events with fatalities greater than N. Automobile accidents cause about 50,000 deaths per year; reproduced from Ref. [1].

of improved technological procedures. The hazards associated with coal mining have been significantly decreased since passage of the Coal Mine Safety Act of 1969. Reference to Table 24.1-3 also shows that oil recovery and refining were somewhat less hazardous than uranium mining and milling. The potential health hazard of Th-230 (with a radioactive half-life of 80,000 years) and hence the importance of careful handling and deposition of uranium tailings have recently been emphasized. Comey,[5] using estimates by R. O. Pohl of

_____

[5]D. D. Comey, "The Legacy of Uranium Tailings," Bulletin of the Atomic Scientists 31, 43-45, September 1975. See also, ibid. 32, No. 2, 61-64.

Table 24. 1-3 Accidents associated with extraction processes in the U.S.
during the period 1965 to 1969. Reproduced from L. B. Lave
and L. C. Freeburg, "Health Effects of Electricity Generation
from Coal, Oil and Nuclear Fuel," Nuclear Safety 14, 409-428
(1973).

Process	Accidents per year		Injuries per 10^6 man-h	Disability per 10^6 man-h	Disability per $10^6 Mw_e h$, 1969
	fatal	nonfatal			
coal mining	246	10,251	43.5	8441	1545
uranium mining	8	272	39.8	8702	
uranium milling	1/3	59	17.0	1091	157
oil drilling and production	1104*		10.2	1176	
oil refining	1060*		5.5	793	135

*Includes both fatal and nonfatal accidents.

(lung-cancer) deaths, has calculated 440 to 982 deaths (depending on the world
population) from untreated tailings per gigawatt-year of electrical energy
produced.

24.2 Radiation Protection Standards; Radiation Exposures; Radiation Emitted from Nuclear Reactors Under Normal Operating Conditions

A. Measures of Radiation Exposure

Radiation dosages and effects are measured in various units, depending
on the context. The basic source unit is the curie (1 millicurie = 10^{-3} curie,
1 microcurie = 10^{-6} curie) which corresponds to 3.7×10^{10} nuclear transfor-
mations per second. The rad is a measure of absorbed radiation dose and

equals the radiation exposure that corresponds to an energy deposition of
0.01 joule per kg (10^{-5} joule per g) of material.

The biological effect produced by the absorption of radiant energy depends not only on the energy deposited but also on the quality factor, which is defined to measure the biological changes produced by the ionizing radiation [e.g., alpha (α) particles, beta (β) particles, neutrons (n), and gamma (γ) -rays] that are responsible for energy deposition. The rem (from Roentgen-equivalent-man) is the unit dose equivalent and is defined as the product of the absorbed dose measured in rad and the quality factor.

The roentgen refers specifically to γ-radiation (X-rays) and is defined as the radiant energy which produces in 1 cm^3 of dry air, under standard conditions of temperature and pressure, total ionization in the form of positive ions, negative ions and electrons with a combined charge of 1 electrostatic unit.

The mass of radioactive material that produces a curie is a function of the half life of the isotope. Thus, U^{238} has a very long half life and several tons of material are required to produce one curie; a curie of Th^{232} would be produced in about one half of a cubic meter of material, while a curie of the short-lived Po^{212} would contain only about 3.7×10^{10} atoms. The relation between the curie and the rad is not readily quantified because it depends on both the detailed nature of the source radiation and the atomic and molecular structure of the absorbing material in which energy deposition occurs. In practice, it is often useful to use the unit 1 millirem = 1 mrem = 10^{-3} rem. The quality factor for γ-rays and β-particles is unity (1); the quality factor for α-particles is 10. Thus, the biological damage produced by absorption of a given amount of energy in the form of α-particles is ten times greater than the corresponding damage produced from equivalent energy absorption by β-particles or γ-rays. The exposure doses measured in rem are additive for people. A person exposed to 2 rad of α-particles, 7 rad of γ-rays, and 2 rad of β-particles suffers a total exposure of $2 \times 10 + 7 \times 1 + 2 \times 1 = 29$ rem.

B.   Biological Effects of Radiation Exposure on Humans

There are extensive empirical data on the effects suffered by humans as the result of radiation exposure. The most dramatic source of data for fatal, acute and chronic exposures comes from experience following the atomic bomb blasts at Hiroshima and Nagasaki near the end of World War II. Other data are derived from measures of occurrence of lung cancer among uranium miners, bone cancer among painters of luminous (radium-containing) watch dials, chronic tissue damage following therapeutic exposure to heavy X-ray doses of selected patients (e.g., for spondylitis), animal experiments, etc. Limited information is available also on genetic damage and mutations.

Scientists at UNSCEAR (United Nations Committee on the Effects of Atomic Radiation) have collected and disseminated information on radiation damage. Exposure standards have been determined by ICRP (International Commission on Radiological Protection). The following are examples of radiation-damage effects: leukemia is sometimes induced by exposure to 100 rad or more, cataracts often develop after long-term exposure to 500 rad of β-particles or γ-rays or to 200 rad of mixed γ-rays and neutrons, hereditary damage requires chronic exposures to more than 100 rad. Generally, the observed biological effects are highly variable for different populations in different locations.

C.   Normal Background Radiation and Dose Limits

Natural radiation-exposure levels are highly variable but normally fall between 100 and 200 mrem/y. Exposure levels increase with elevation; thus, people living in Denver may be subjected to 50 mrem/y more than residents of the Midwestern plains. Threshold levels below which exposure produces no measurable damage are not known to occur with certainty but are believed by many to exist (see Fig. 24.2-1). If hazard estimates are made by treating radiation damage as a linear function of exposure, excessively large damage estimates result. [*]

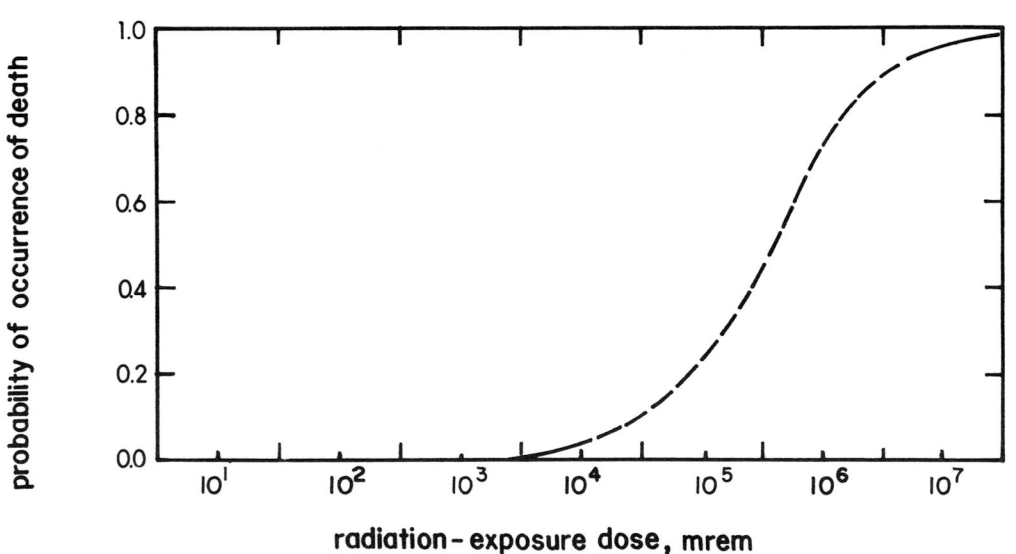

Fig. 24.2-1 Schematic diagram showing a logistics curve for the probability of occurrence of death within 1 year as a function of radiation-exposure dose in mrem/y. If this type of representation is correct, then there is an effective threshold level (between 100 and 200 mrem/y) below which the probability of fatal exposure in a large population becomes negligibly small.

---

[*] For a readable summary of quantitative information relating to the biological effects of radiation exposure, see B. L. Cohen, Nuclear Science and Society, pp. 45-73, Anchor Books, Garden City, N. Y., 1974.

Maximum allowed dose limits on the peripheries of nuclear reactors used to be fixed at 170 mrem/y but have recently been set "as low as practicable", generally below 5mrem/y. Gofman and Tamplin[1] have argued that supplementary radiation exposures at a level of 170 mrem/y would be responsible for 20 additional leukemia victims per rem exposure per 10^6 people. This estimate actually applies for an increase in radiation exposure from 50 to 51 rem and is far too large at the much lower dosage levels to which people are normally exposed (compare Fig. 24.2-1). If we use this unreasonably large estimate, we may determine the additional number of leukemia victims in a population of $200 \times 10^6 p$, which is produced by incremental exposure to 170 mrem/y, as follows: supplementary number of leukemia victims per year = 170×10^{-3} (rem/y) $\times$ ($200 \times 10^6 p$) $\times$ (20 leukemia victims/rem-$10^6 p$) = 680 additional leukemia victims per year. Cancers of all types are sometimes assumed to be produced by radiation exposure about five times more frequently than leukemia. Hence, the indicated procedure leads to the conclusion that the total additional number of cancer cases of all types per year is 3,400/y if $200 \times 10^6 p$ in the U. S. were exposed to an incremental radiation dose of 170 mrem/y. It is appropriate to reemphasize the fact that the preceding numbers are unreasonably high and have been derived by using unreasonable input data.

Assuming that the linearity between radiation dose and victims continues to hold at the much lower exposure levels allowed by current regulations, namely, 5 mrem/y, the estimated numbers of victims of leukemia and of cancers of all types in a population of $200 \times 10^6 p$ are reduced to 20 and 100 per year, respectively. These lower estimates are also unreasonably high,

---

[1] J. W. Gofman and A. R. Tamplin, Poisoned Power: the Case against Nuclear Plants, Rodall Press, Emmaus, Pennsylvania, 1971.

especially since only incidences of thyroid cancers and leukemia appear to scale linearly with radiation doses.

Ball[2] made (in 1970) a much more optimistic evaluation of radiation risks and arrived at estimates of 0.02 deaths per year associated with a general population exposure of 0.01 mrem/y, which he considered to be the proper level for 17 reactors in the U.S., each yielding as much as 170 mrem/y; the corresponding leukemia risk was stated to be 0.002 cases/y. Ball's estimate of the decrease in life span for humans was 2 hours at an incremental dose of 5 mrem/y or 24 seconds at 0.01 mrem/y.

All of the preceding numerical data are so doubtful that it is not reasonable to use these numbers as a basis for a discussion of what the allowed radiation dose should be. We note that an incremental change of 5 mrem/y is certainly much less than the normal increase of ~50 mrem/y associated with a move from a location such as Albany, New York, to Laramie, Wyoming, or the increase of ~20 mrem/y associated with relocation from a single-dwelling wood-frame residence to a condominium in a concrete structure.

In view of the greatly different sensitivities of human organs or tissues to radiation exposure, the ICRP has recommended a maximum total exposure of 5 rem/y, as well as the following differential exposure limits: 0.5 rem/y for gonads and red marrow; 1.5 rem/y for single organs; 3.0 rem/y for skin, bone and thyroid (1.5 rem/y for thyroid exposure in children up to 16 years of age); 7.5 rem/y for hands, forearms, feet, ankles. These basic radiation standards are seen to be far higher than those produced in connection with current or anticipated nuclear-reactor developments.

---

[2]R. H. Ball, "Nuclear Power and the Environment," pp. 1-20, December 1970.

Genetic effects are generally assumed to be proportional to the gonad-exposure dosages.

D. Natural and Man-made Sources of Radioactivity

The average sea-level dose rate is 28 mrad/y at middle latitudes, where most of the world population resides; it decreases toward the equator by about 10% and doubles with elevation for every 1.5 km above sea level up to about 5 km.

Cosmic radiation incident on the atmosphere produces multiple nuclear reactions that are accompanied by the production of radioactive materials, which are collectively responsible for a dose of 28.7 mrad/y. In decreasing order of abundance, ^{14}C (1.3 to 1.6 $\times 10^{-2}$ curie/m^3), ^{7}Be, ^{3}H (tritium), and ^{32}P are radionuclides present in air that are produced by reactions associated with the incidence of cosmic rays. There are long-lived radionuclides present on earth which were formed during its creation and contribute 50 mrad/y (e.g., uranium and thorium rocks which produce ^{226}Ra during decay, ^{40}K, ^{87}Rb, ^{50}Va, ^{115}In, etc.).

Internal radiation within the human body is provided by food ingredients. Thus, the isotope of potassium with an atomic weight of 40 (^{40}K) contributes about 20 mrad/y; it is a normal component of the human body and is taken in with food and eliminated in such a manner that a steady level is maintained. The radionuclides ^{14}C and ^{87}Rb are also ingested with food and contribute about 1 mrad/y. In addition, ^{222}Ra and its daughter product ^{210}Po are taken in from air during breathing and are responsible for an average exposure level of 0.6 mrad/y. Evidently, such activities as dancing, attending a crowded theatre, or sleeping with one's spouse involve far greater radiation hazards than are currently produced by nuclear reactors.

A 1971 estimate of man-made radiation exposures is reproduced in Table 24.2-1. The heavy effect provided by diagnostic X-rays is especially noteworthy and suggests that improvements in this category should receive special emphasis. An average aircraft crew exposure level of 670 mrem/y has been estimated and may suggest the desirability of added shielding, rapid crew rotation, and short life span for this type of occupation.

E.   Estimated Radiation Exposure Following a Major Nuclear Accident

We shall consider the likelihood of occurrence of a major nuclear accident in the following Section 24.3. Here we note the following principal conclusions:[3] heavy release of radioactive isotopes from a 1,000 Mw$_e$ nuclear reactor leading to 2.5×10^6 man-rem of total body dose and 30×10^6 man-rem of thyroid dose, under favorable meteorological conditions in a metropolitan area with a well-run emergency evacuation system, would probably not be responsible for more than 500 deaths. On the other hand, more credible accidents involving nuclear reactors should not produce fatalities.

F.   Effluent Releases from Nuclear Reactors[4]

The U.S. Atomic Energy Commission has officially adopted the following regulation concerning radioactive effluents: these must be "as low as practicable." Hull[4] has reviewed a number of studies which indicate that the air-dilution volumes required to achieve allowed concentration levels for air

---

[3]Nuclear Power and the Environment, p. 16, International Atomic Energy Agency, Vienna, 1973.

[4]A. P. Hull, "Comparing Effluent Releases from Nuclear and Fossil-Fueled Power Plants," Nuclear News 17, 51-55 (1974).

Table 24.2-1 A 1971 estimate of annual per capita radiation doses from man-
            made sources in the U.S.; reproduced from a release by the
            U.S. Environmental Protection Agency.

Man-made radiation source	Average dose in mrem/y
medical activities	
diagnostic radiography	103
therapeutic radiography (e.g., treatment of malignant and non-malignant diseases)	6
radiopharmaceuticals	2
occupational sources[*]	0.8
environmental sources	
global fallout (mostly from ^{90}Sr with a half-life of 28.5y)	4
worldwide ^{3}H and ^{85}Kr	0.05
AEC installations	0.01
nuclear power reactors	0.002
nuclear fuel reprocessing	0.0008
miscellaneous sources[**]	2.6
total from man-made sources	118
estimated total from all sources	130

[*]About 0.4% of the U.S. population was classified as radiation workers in 1969.

[**]Color television sets, luminous watch dials, radiation gauges, etc., air-craft transport (which adds about 1 mrem/y on the average and 670 mrem/y for the crew).

pollutants other than radioactive compounds, <u>as well as for radionuclides</u>, are larger for conventional fossil-fuel plants than for nuclear reactors. Starr et al[5] have noted that the allowed concentration levels for non-radioactive pollutants are much closer to levels at which health hazards occur than are the allowed levels for radionuclides.

A nearly ten-fold decrease in the release rates of radionuclides from the BWR (boiling water reactor) was achieved between 1967 and 1972. Projected release rates, based on past experience, are shown in Table 24.2-2 for 1,000 Mw$_e$ nuclear power plants operating at an 80% load factor. We note that both the PWR (pressurized water reactor) and HTGR (high temperature gas-cooled reactor) have substantially lower emission rates than the BWR. Furthermore, a theoretical projection made by workers at the AEC and based on estimated fuel-leakage rates and transfer coefficients actually yielded[4] larger values than the experimentally-based projections shown in Table 24.2-2.

Estimated effluents that will occur for a plant reprocessing fuel for a 1,000 Mw$_e$ reactor are summarized in Table 24.2-3. Comparison of the data summarized in Tables 24.2-2 and 24.2-3 shows that the radioactive outputs associated with fuel-reprocessing are somewhat greater than those produced during normal plant operation.

The dose rates as functions of distance from a BWR and a PWR are shown in Fig. 24.2-2 for reactors normalized to yield 500 mrem/y at a distance of 500 m (0.31 mile) from the reactors. Under normal operating conditions, the exposure level to radioactivity is seen to decrease rapidly with distance.

---

[5]C. Starr, M. A. Greenfield, and D. F. Hausknecht, "A Comparison of Public Health Risks: Nuclear vs. Oil-Fired Power Plants," Nuclear News 15, 37-45 (1972).

Table 24.2-2 Projected radioactive effluent releases from a 1,000 Mw_e reactor with an 80% load factor; reproduced from Ref. [4].

Airborne radioactivity in curies/year in			Radioactivity of liquid effluents in curies/year for	
reactor type	gaseous effluents	halogens and particles	fission and corrosion products	tritium
BWR	1.66×10^4	5.31	49.6	1.04×10^2
PWR	9.65×10^3	0.17	30.2	5.75×10^3
HTGR	2.76×10^3	<0.02	0.27	8.35×10^2

Table 24.2-3 Projected release rates of radioactive effluents from a plant reprocessing fuel from a 1,000 Mw_e plant with a total output of 8.72×10^5 Mw_ed; reproduced from Ref. [4].

Airborne radioactivity in curies/year in the		Radioactivity of liquid effluents in curies/year for	
gaseous effluents	halogens and particles	fission and corrosion products	tritium
2.68×10^5 as ^{85}Kr	0.40 (<0.07 as ^{131}I)	53.5 (8.3 as ^{90}Sr)	4.83×10^3

G. Some Views Concerning Exposure Hazards to Radiation from Nuclear Reactors

It has been suggested that a total supplementary exposure over 30 years of 10 rad represents a genetically "safe" limit[6] in the sense that this exposure level would increase the natural mutation rate by not more than 30%. At

[6]I. Asimor and T. Dobzliansky, "Genetic Effects of Radiation," U.S. AEC, Understanding the Atom Series, Oak Ridge, Tennessee, 1966.

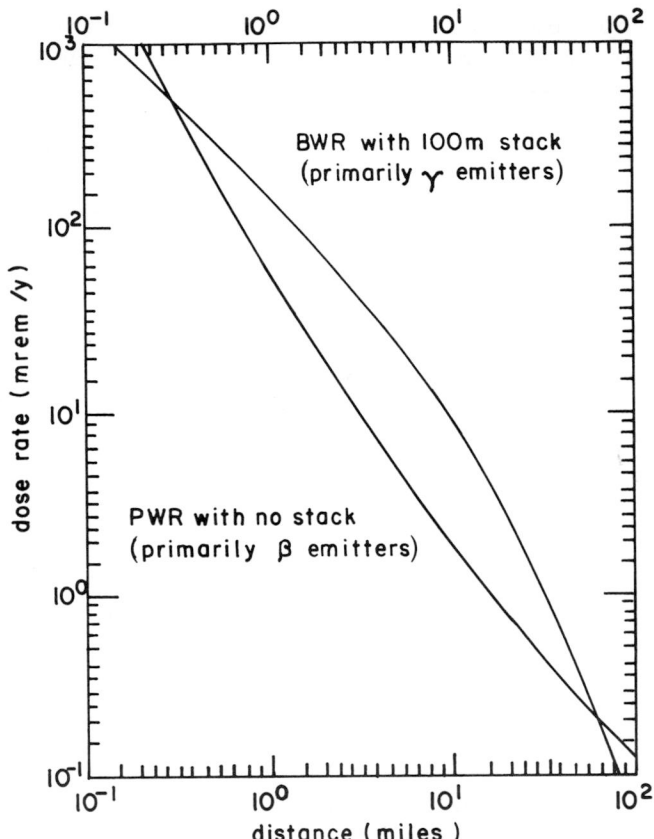

Fig. 24.2-2 Dependence of dose on distance for LWRs yielding 500 mrem/y
            at 500 m (0.31 mile); reproduced from T. Rogers and C. C.
            Gamersfelder, Conference on Environmental Effects of Nuclear
            Power Stations, pp. 127-145, IAEA, Vienna, Austria, 1971.

an average quality factor of 3, this cumulative dose is $\sim 10^3$ mrem/y and much
higher than the exposure with currently operative reactor controls.

Pauling[7] has estimated that atmospheric explosion of a single large

---

[7]L. Pauling, <u>No More War</u>, Dodd and Mead Co., New York, 1958.

nuclear bomb of the type still being tested during 1975 by the French, Indians and Chinese could be responsible for a cumulative total of 10,000 leukemia and bone-cancer deaths and perhaps could lead to as many as 90,000 other fatalities. Assuming a latency period of 10 years and a life expectancy of 70 years, Pauling[8] has concluded that a supplementary exposure of 170 mrem/y would increase the incidence of all cancers by about 480×10^{-6}. He then computes the number of supplementary cancers in the U.S. with a population of 200×10^6 as 10^5 per year for a supplementary dose rate of 170 mrem/y.

Workers at the Atomic Bomb Casualty Commission (ABCC), which has been charged with collection and evaluation of data in Hiroshima and Nagasaki, have given an exposure level of about 10^6 man-rads per year as responsible for about one case of leukemia per year. Thus, the ABCC analysis suggests that supplementary exposure of 200×10^6 people to 170 mrem/y would lead to about 30 cases of leukemia per year, which is seen to be very much less than Pauling's value of 10^5 cases of cancers of all types per year for the same supplementary exposure level.

Computations by Gofman[9] have been discussed in Section 24.2C and are believed by many[10-13] to yield excessively large estimates for cancer incidence.

[8] L. Pauling, "Genetic and Somatic Effects of High-Energy Radiation," Bulletin of Atomic Scientists 26, No. 7, 3-5 (1970).

[9] J. Gofman, as quoted by R. Wilson and W. J. Jones, in Energy, Ecology and the Environment, p. 240, Academic Press, Inc., New York, N.Y., 1974.

[10] K. Z. Morgan and J. E. Turner, Principles of Radiation Protection, John Wiley and Sons, New York, N.Y., 1967.

[11] ICRP Publication No. 9, Pergamon Press, Ltd., Oxford, 1966.

[12] L. A. Sagan, "Human Costs of Nuclear Power," Science 171, 487-493 (1972).

[13] Various authors quoted by Wilson and Jones in Ref. [9]

H.   Our Views on Radiation Hazards of Safely Operating Reactors

At presently operative control levels for nuclear reactors, a safely
operating reactor appears to constitute a lesser hazard than any other currently available and economically competitive energy source.   The operative constraint "as low as practicable" for radiation output should be interpreted to
mean that all reasonable effort will be exerted to reduce supplementary radiation levels from nuclear reactors to still smaller values provided other economic and social costs do not become prohibitively high.

We must not forget that all energy use implies environmental degradation and health dangers.   It is therefore improper to single out a particular
energy source and discuss its hazards out of context.   Within context, we must
learn to think in terms of total hazard and benefit per unit of energy produced.

24.3 Accidents Involving Nuclear Fission Reactors

The probable catastrophic failure rate of nuclear reactors cannot be
estimated with any reasonable degree of certainty.   We are here dealing with
an event for which fortunately no meaningful statistical background exists.   In
the absence of this information, it has become customary to engage in the following three separate types of activity:

i.   We attempt to estimate the number of reactor-years before a major
failure will occur.   Estimates are contained in such statements as "the time to
failure is more than 100 reactor-years because no catastrophic failure has as
yet occurred," "somewhere between 10^4 and 10^5 reactor-years will correspond
to the expected lifetimes for untested and carefully tested pipes of the type used
in reactor-coolant systems," etc.   We note that if the average expected period
before a catastrophic failure occurs were 10^5 reactor-years, and 10^3 reactors

each producing $10^3 Mw_e$ (or about one half of our projected year 2000 electrical generating capacity) were in operation, then the average failure rate over a very long period of time (i.e., a period of time much longer than 10^2 years) would be expected to be 1 reactor every 100 years. The preceding statement does not rule out the possibility that a catastrophic failure will actually occur tomorrow or next year or not before 374 years have elapsed, etc.

ii. We construct failure scenarios of the worst possible sequence of events that we are able to invent for a given reactor. This type of analysis requires deep and intimate understanding of all types of reactor designs. It is doubtful even then that an actual failure will follow one of our scenarios in detail. Accidents are by definition unpredictable. However, construction of conceivable failure scenarios is a useful activity because it provides us with better understanding of weaknesses in design details where improvements or changes should be made.

iii. On the assumption that a credible catastrophic failure does occur, we investigate required countermeasures and procedures to minimize loss of life and property. Explosion of a nuclear reactor in a manner analogous to that which occurs in a nuclear fission bomb is not a credible catastrophic failure because it violates well understood physical laws. A loss of coolant accident (i.e., a LOCA accident), followed by failure of the emergency core coolant system (ECCS) and by a rapid temperature rise in the reactor core, melting of fuel rods and other solid reactor components, radioactive contamination of the surrounding atmosphere, and radioactive contamination of populated areas under the worst possible meteorological conditions, may be a credible accident for some types of reactors operating under conceivable conditions.

Before proceeding with a discussion of events leading to catastrophic

failures, we must reemphasize that presently available data indicate that
energy generation per unit of output by nuclear reactors has exacted in the
past lower environmental, human-health, and associated costs than any other
process of energy generation used by man. We shall find that this enviable
record is likely to continue in the future in the absence of deliberate sabotage,
diversion of materials and other planned human efforts to utilize nuclear
power for nefarious purposes. The price we must expect to pay for nuclear
energy is an economically competitive charge for capital investment, operation and distribution, as well as eternal vigilance to prevent deliberate misuse
or misappropriation. In this sense, the commitment to guard all components
of the nuclear power industries is analogous to the commitment made by the
Dutch to maintain their dikes "forever" to prevent flooding of reclaimed land,
or the commitment made by the builders of dams not to use their reservoirs
to inundate the lower-lying countryside, or the decision made by civilized man
to continue food production by efficient methods as a mandatory alternative to
mass starvation, etc.

Analysis of the consequences of a nuclear accident must properly begin
with a quantitative description of nuclear-reactor characteristics. We have
reproduced in Table 24.3-1 an estimate from Ref. [1] of the performance
characteristics and contents of important fission-reactor types.

Reference to Table 24.3-1 shows that the steady-state charges and discharges of the 1,000 Mw_e (nominal) fission and breeder reactors contain many
hundreds of kilograms of U-233, U-235 or fissile Pu. At refueling intervals
which are slightly longer than 1 year, the steady-state discharges have

---

[1] "High-Level Radioactive Waste Management Alternatives," Frank K. Pittman,
    ed., U.S. Atomic Energy Commission, Report WASH 1297, May 1974.
    Available from NTIS, U.S. Dept. of Commerce, Springfield, Va. 22151.

Table 24.3-1 Nuclear-reactor-plant characteristics; reproduced from Ref. [1].

Plant characteristic	LWR-U Total power or mass	LWR-U, Pu[a] Contributions from U	LWR-U, Pu[a] Contributions from Pu	LWR-U, Pu[a] Total from U and Pu	HTGR[b] Charge of Th and U	HTGR[b] Fresh Makeup of U-235	HTGR[b] Recycled Makeup of U-235	HTGR[b] Total power or mass	General Electric, follow-on LMFBR[f] Contributions from core	General Electric, follow-on LMFBR[f] axial blanket	General Electric, follow-on LMFBR[f] radial blanket	General Electric, follow-on LMFBR[f] Total power or mass
electric power, Mw$_e$ (net)	1000	676	324	1000	-	-	-	1160	-	-	-	1011
thermal power, Mw	3077	2081	996	3077	-	-	-	3000	2081	195	141	2417
average specific power, Mw/mt[b]	37.5	37.5	37.5	37.5	-	-	-	80.65	155.6	13.0	8.5	53.76
average burnup, Mw$_e$d/mt	32,873	32,873	32,873	32,873	-	-	-	94,264	104,542	8725	9051	41,792
refueling interval, days[c]	365.25	365.25	365.25	365.25	365.25	365.25	365.25	365.25	385	385	385	385
steady-state charge, kg												
Th	-	-	-	-	8434	-	-	8434	-	-	-	-
U-233	-	-	-	-	217	-	-	217	-	-	-	-
U-235	875.2	592.0	59.8	651.8	29.6	373.0	30.4	433	-	20	14	34
total U	27,350	18,500	8409	26,909	357.9	403.0	104.5	865.4	5038	6884	4798	16,720
fissile Pu[d]	-	-	270.3	270.3	-	-	-	-	786	-	-	786
total Pu[e]	-	-	441.0	441.0	-	-	-	-	1093	-	-	1093
U + Pu + Th	27,350	18,500	8850	27,350	8792	-	-	9299.4	6131	6884	4798	17,813
steady-state discharge, kg												
Th	-	-	-	-	7819	-	-	7819	-	-	-	-
U-213	-	-	-	-	219.3	-	-	219.3	-	-	-	-
U-235	243.4	164.6	26.4	191.0	30.7	30.8	2.6	64.1	-	14	10	24
total U	26,137	17,679	8190	25,869	366.1	105.3	70.0	541.4	4439	6580	4583	15,602
fissile Pu[d]	180.1	121.8	151.2	273.1	-	-	-	2.1	714	234	163	1111
total Pu[e]	254.9	172.4	273.1	445.5	-	-	-	10.0	1051	245	171	1467
U + Pu + Th	26,572	17,851	8463	26,315	8185	105.3	70.0	8370	5490	6825	4754	17,069

(a) PWR with self-sustaining Pu recycle; (b) based on full power and fuel charged;
(c) at 80% load factor; (d) Pu-239 + Pu-241; (e) Pu-238 + Pu-239 + Pu-240 + Pu-241 + Pu-242.
(f) General Electric plant typical of advanced units (later than 1990).

similarly heavy loads of fissile materials. Thus, an analysis of the safety of nuclear reactors must include proper consideration of the implications of storage and discharges of materials of this type.

A. Accidents Involving Water-Cooled Nuclear Reactors

As we have discussed at length in Chapter 21, the reactor fuel for a 1,000 Mw_e, water-cooled nuclear reactor consists of about 30 to 100 t of uranium. This nuclear fuel is contained in cylindrical rods which are about 3 to 4 m long and about 1 cm in diameter. An assembly of 50 to 200 rods constitutes a fuel bundle. The reactor core contains hundreds of fuel bundles. The fuel bundles are surrounded by water which serves as coolant, moderates the fission reaction and, in the case of the BWR, is also used as working fluid for electricity generation in a secondary power cycle of conventional design.

Under normal operating conditions, the radiation hazard associated with nuclear reactors has been shown to be relatively small. An accident at a nuclear power station is highly unlikely to occur but, if it does occur, will lead to significant augmentation of the levels of radioactivity. Many of the fission products are radioactive and undergo subsequent nuclear reactions. Some of the disintegration products are stable, others are not and remain radioactive on time scales varying from seconds to thousands of years. The radioactive fission products which accumulate in the fuel rods are gaseous, liquid or solid under existing conditions of temperature and pressure. Important components are isotopes of iodine, krypton, xenon, cesium, and strontium.

Dangerous releases of radioactive materials can only occur if the fuel in the core is dispersed. The stored radioactive waste products cannot produce short-term, catastrophic radiation levels. The sequence of unlikely events leading to reactor-core melting (at about 5,000° F) has been examined in the Rasmussen report.[2] Because the designs of nuclear reactors (compare Chapter 21) include such safety features as the ECCS (emergency core cooling system) and the reactor containment vessel or building, a series of sequential failures must occur before radioactive releases to the environment will take place. A LOCA (loss of coolant accident) followed by failure of the ECCS is an important preliminary to hazardous reactor failures.

The analysis in Ref. [2] was carried out by using event trees and fault trees. The event tree is constructed by postulating an initial failure within the system and noting the sequence of subsequent events which will lead to core melting and radioactive releases. A fault tree is then defined in order to assess the likelihood of failures along the event tree paths that are responsible for the accident. The quantitative study requires engineering data on failures of components (e.g., valves, pumps, pipes), operator errors, maintenance deficiencies, and systems failures. In the absence of data, plausible scenarios are constructed. Assessments of the environmental impact after failure require detailed studies of meteorological conditions and changes, evacuation procedures, population distributions, etc. The final output[2] may then be presented in terms of probability estimates for various categories of accidents.

Following a LOCA and failure of the ECCS, heat released in radioactive decays will heat the reactor core to the melting point within 30 to 80 seconds. After melting, the containment vessel may fail promptly due to pressure or after some hours due to melting. If the failure is prompt, the plant facilities designed to contain radioactive releases will not function and all radioactive gases and perhaps as much as one half of the core will be released. If melt-through of the containment structure occurs, most of the solid radioactive elements will be trapped in the soil.

---

[2]Reference [1] of Section 24.1.

Estimates from Ref. [2] of the consequences of a most likely accident, which is expected to occur on the average once for every 20×10^3 plant-years of operation, are summarized in Table 24.3-2 and show that this average nuclear-reactor accident does not constitute a major disaster of the type that is associated, for example, with the crash of a commercial aircraft.

The probabilities of failures and their long-term consequences are evaluated in Table 24.3-3 for two classes of failures, namely, an average failure with probability of occurring once in 20,000 plant-years of operation and a failure causing more than 10^2 fatalities with probability of occurring once in 8×10^5 plant-years of operation. Accidents with more than 10 fatalities are predicted to occur once in 350,000 plant-years; those with more than 4,500 fatalities have an a priori probability of occurring once during 10^9 plant-years of operation.

Estimates concerning property damage are summarized in Table 24.3-4 and should be contrasted with the following calamities: There are on the average in the U.S. three fires per year causing property damage in excess of $\$10 \times 10^6$; once every seven years, a fire will cause 100 or more fatalities; four hurricanes during the last ten years have caused property damage between \$0.5 and $\$5 \times 10^9$ and one hurricane in five years causes 100 or more fatalities.

The indicated damage levels in Tables 24.3-2 to 24.3-4 have been determined for the present generation of nuclear reactors. Reactors constructed in the future are likely to be safer. Policy decisions concerning nuclear reactor safety cannot be made without comparing data of the type assembled here with comparable information for non-nuclear energy sources.

Table 24.3-2 Consequences of a most likely reactor failure leading to melting of the reactor core; reproduced from Ref. [2].

a. The numbers of fatalities, injuries, latent (delayed) fatalities, and genetic defects are all expected to be less than 1.

b. The number of thyroid nodules caused by radiation is about 4.

c. Property damage <u>outside the reactor</u> will be less than $\$10^6$.

Table 24.3-3 Latent health effects expected during a 20-year period following a reactor failure that produced 100 fatalities; reproduced from Ref. [2].

Effect	Number of plant years of operation before the indicated accident will occur on the average		Normal incidence rate for people living near the site
	20,000 (average accident)	10^6	
latent cancers	< 1	170	17,000
thyroid illness	< 1	1,400	8,000
genetic effect	< 1	25	8,000

B. Aspects of Nuclear-Safety Analysis

We shall now take a more detailed look at the technical background that leads to the conclusions stated in Section 24.3A.

Safety analysis deals with the prediction of consequences of deviations from desired performance. The principal engineering functions that must be examined are (i) control and (ii) cooling and containment. Two general

Table 24.3-4 Average number of plant-years of operation before occurrence
of reactor failures associated with various levels of property
damage; compiled from data in Ref. [2].

Average number of plant-years of operation before occurrence of an accident with the indicated level of property damage	Level of property damage (not including damage done to the reactor)
2.0×10^4	$\geq \$10^6$
5.5×10^4	$\geq \$10^7$
1×10^6	$\$2$ to $\$3 \times 10^9$
1×10^9	$\$4$ to $\$6 \times 10^9$

approaches are employed for safety assessments. These are (a) forming a
consensus on the maximum credible accident or design basis accident (DBA)
for a particular reactor and providing assurance that its consequences will not
endanger the public; (b) calculating and tracing the probabilities of the sequences of faults and events by assessing the probability of failure of any component or function and then determining the probabilities of resulting hazards
to the public. The second approach has been widely used in Europe and has
been applied also in the U.S. in a study of the safety of LWRs.[2]

    i. Control

An assessment of nuclear safety may begin with an analysis of the kinetic behavior of nuclear fission reaction and of the stability and reliability of
their control systems. We have shown in Appendix 21A that the time dependence of the number density and flux of neutrons in a reactor is directly related to neutron sources (fission), absorption and diffusion. In Section 21A.1,
we defined a multiplication constant k for the chain reaction in a nuclear reactor

as representing the ratios of the populations of succeeding generations of neutrons. A generation time may be thought of as the average time elapsed from
birth in fission to death by absorption in a fissile nuclide; this time includes
the time during which a neutron migrates randomly in the reactor and the time
for emission of delayed neutrons. If k equals unity, the population and flux of
neutrons (including the delayed contributions) are constant and the reactor is
said to be critical. This condition of criticality may be maintained by controlling neutron absorption through the positions of control rods. If $k > 1$, the
neutron population and flux grow at a fractional rate proportional to the excess
of k ($\equiv$ k - 1) divided by the effective generation time which, for $(k - 1) \simeq 0$,
is the average time for appearance of the longest delayed neutrons (about 56
sec for $^{235}_{92}$U). As the excess k increases, fewer of the delayed neutrons are
required and the effective generation time will become quite short. If delayed
neutrons are not required for multiplication, the condition is referred to as
"prompt criticality." This condition is strictly avoided, except in some research reactors. In thermal neutron reactors, the slowing and migration of
neutrons may lead to a prompt generation time of 0.1 to 1.0×10^{-3} sec. In
fast neutron reactors, it may be as short as 1×10^{-6} sec. Nuclear weapons
require the shortest times and largest excess k-values achievable.

The neutronic time behavior of a reactor is described by a set of linear
differential equations. The rates of production of six emitters of delayed neutrons are related to flux. Their rates of decay and fractional contribution are
known. Thus, the coupling between these equations and that for the neutron
flux is linear. The specified or designed time dependence of the control-system sensors and drives is a required input. Finally, the effects of temperature, density and dimensional changes associated with the production and flow
of heat may be incorporated. Either the time or frequency dependence of the
entire system may be determined by the use of modern control theory; here

included are all of the parameters required for safe start-up, shutdown, operation and response to demands for power from the system.

The coupling of changes of power and temperature with other parameters is described most simply through coefficients of reactivity, which are defined as the fractional change in k produced by unit change of an independent variable (e.g., power or temperature). Safe reactor design requires that these coefficients be negative to assure that a positive fluctuation in power is self-limiting. We note a number or special control constraints. A power coefficient of reactivity refers to a process which operates in such a manner that, as power changes, all temperatures in the system become dependent variables and change to their steady-state values. A fuel or prompt coefficient measures the effect of an adiabatic change in fuel temperature without heat flow, as in a rapid change of power. A moderator coefficient defines the overall change determined by a change in the moderator. In LWRs, water temperatures may change independently of the fuel temperatures. Isothermal coefficients refer to reactivity changes when the temperature is uniform throughout, for example, in a reactor operating at low power or in a critical assembly at zero power when heat is supplied from the outside. Each of these coefficients may be calculated from basic data, confirmed by measurements on reactors and used to predict reactor dynamics under a real condition.

In order to depend on negative feedback, the engineer must make certain that there is no way in which the control system or any change in the reactor core can "introduce reactivity," i.e., increase fission or decrease absorption or leakage in excess of the total negative change in the reactivity from the feedback loops. While the moderator coefficient is almost always negative in the reactors we have discussed, the moderator heats slowly compared to the fuel. Some reactors are given large negative temperature coefficients by mixing fuel and moderator so that they change temperature together.

We catalogue ways in which the reactivity k is controlled: (a) control rods cannot be accidentally withdrawn or ejected; (b) fuel cannot, in any probable way, become more dense, e.g., by movement of fuel elements toward the center of the reactor; (c) absorbing coolants, such as sodium, cannot be lost or pushed from the core, e.g., by boiling; (d) moderation cannot suddenly increase by flow of cold moderator into a reactor or by fall or flow of a scattering material into an under-moderated or a fast reactor. During the conceptual and final design phases, careful and detailed inspection is made of all conceivable ways in which the neutron-multiplication constant may change.

The control of reactors requires systems whereby the operator and control-system devices may change reactivity as needed. At start-up, the operator brings the reactor from subcritical to critical condition and then to the operating power level by withdrawing the control rods. This operation is perhaps the most hazardous of all and is highly proscribed by both administrative procedures and programmed electro-mechanical devices and systems. Devices to measure neutron, γ-ray and heat flows, as well as other sensors, are located in the core. A neutron source provides a detectable flux before criticality is reached. This control is necessary in order to avoid large fractional statistical fluctuations in neutron population and to provide instrument readings that may be readily monitored. Mechanical arrangements prevent the operator from withdrawing the control rod too rapidly. Instruments indicate both the level and rate of change of flux, power, etc. Limiting devices automatically and promptly insert reserve control or safety rods if either the level or the rate of change exceed prescribed levels. Since operation of the first chain reaction in December 1942, the operation of a safety device for quickly inserting rods and causing shutdown of a reactor has been called a "scram". Redundant paths and devices reduce the probability of control-system failure. As mentioned earlier, if all other means fail, gravity forces safety rods into the core. Alternative independent backup procedures for introducing reactor poisons are provided, for example, by injecting a soluble boron compound into the water in the core of an LWR and by releasing boron-containing spheres from a hopper into the core of an HTGR.

A number of limiting signals scram a reactor during operation. Among these are low-level seismic signals, changes beyond carefully selected limits and at chosen locations of flow, pressure, temperature, neutron flux, γ-ray intensity, and other properties of the system. Signals indicating malfunction anywhere in the power-conversion system can also shut the reactor down. Some reactor systems are "load following," that is to say, changes in demand for electrical power signal the reactor-control system to adjust control-rod positions accordingly. The LWR is inherently somewhat load following through moderator-temperature feedback.

The theory and practice of control is a highly developed branch of engineering. Applications of control systems provide both the methods of mathematical analysis and the electro-mechanical equipment essential for the design, construction and operation of stable devices containing sensors, differentiators, integrators, discriminators, logic gates, decision systems, and computer-programmed electro-mechanical drives to move and position control rods. In particular, reactor control is a highly specialized branch of nuclear engineering.[3] Control engineers have the responsibility of seeing to it that there is no built-in procedure which will lead to a reactor power level exceeding the capability of the cooling and containment systems. In addition, operator errors are eliminated.

---

[3] M. A. Schultz, Control of Nuclear Reactors and Power Plants, McGraw-Hill Book Co., Inc., New York, 1961.

ii.   Cooling, Containment and Modes of Failure

After a large reactor is shut down, the containment barriers will fail if temperatures rise excessively because insufficient heat is transferred from the reactor.   The consequences of this type of failure, including each of the following topics, have been introduced in Section 24.3: (a) analysis of human attitudes toward risk; (b) analysis of the nature of nuclear power-plant accidents with reference to the levels and dispersal of radioactivity, loss-of-coolant accidents, reactor transients, and accidents involving stored and spent fuel; (c) development of methods for assessing risks by tracing accidents through event and fault trees and associating magnitudes of releases and consequences with the probabilities of occurrence of branches of the trees; (d) definitions of release categories and their associated probabilities of occurrence; (e) development of perspectives concerning the risks of nuclear accidents and other societal risks.

The fission-product inventory, which constitutes most of the radioactivity, is within Zircaloy-clad fuel elements.   This cladding is the first of three successive barriers to the release of radioactivity.   Of the fission products listed in Table 24.3-5, the rare gases, iodine and, to a lesser extent, tellurium, cesium, rubidium, and strontium may escape from the high melting uranium dioxide.   The steel pressure vessel surrounds the core and confines radioactivity in the cooling water and other core components, the piping, and the pressurizing and heat-exchange system.   This pressure vessel is inspected and monitored because it represents the second important barrier for confining radioactivity.   The final barrier is provided by the containment building or shell that surrounds the reactor and serves to confine or absorb gases, liquids and radioactivity; released steam may be condensed here while heat produced by decaying radioactive nuclides is transferred to an external sink.   In

Table 24. 3-5 Approximate yields of chemical families of fission products for the fission of ^{235}U. The nuclides listed live long enough to grow and persist at significant concentrations in fuels. They are listed in order of increasing retention in uranium oxides and other oxygen-containing solids. The tendency for release of the most radiologically significant nuclides is indicated in Tables 24. 3-7 and 24. 4-1.

Chemical family	Fractional yield of persistent or stable element	Radiologically significant nuclides		
		mass numbers	half life	yield
inert gases				
Kr	0.039	85	10.3y	0.003
Xe	0.35	133	5.27d·	0.0066
halides				
Br	0.005	83	2.3h	0.005
I	0.01	129	$1.7 \times 10^7 y$	0.01
		131	8. d	0.03
		133	21h	0.066
		135	6.7h	0.061
chalcogenides				
Te	--	127	105d	0.0002
		129	33.5d	0.0034
		131	30h	0.0044
		132	77h	0.044
		133, 134	≤63min.	0.117
alkali metals and alkaline earths				
Rb	0.038	87	$6 \times 10^{10} y$	0.025
Cs	0.19	135	$2.6 \times 10^6 y$	0.064
		137	30y	0.061
Sr	0.14	89	51d	0.048
		90	27.7y	0.059
Ba	0.057	140	12.8d	0.064
transition metals				
Zr, Mo, Tc, Ru	0.65	Tc 99	$2.12 \times 10^5 y$	0.05
		Ru 103	39.7d	0.03
		105	4.5h	0.009
		106	1.02y	0.004
rare earths and other trivalent metals Yt, La, Ce, Nd, Sm, Pm, Eu	0.56	La 140	42.2h	0.064
		141	3.8h	0.060

the event of partial or complete failure of the three barriers, each escaping chemical species must be followed during dispersal in the atmosphere, hydrosphere, lithosphere, and biosphere.

Figure 24.3-1 shows the barriers that limit the consequences of the most serious kinds of malfunction of a PWR system. We refer to these as engineered safety features (ESFs). The ESFs include the following components: (a) a fail-safe control system that trips the safety-rod insertion and scrams the reactor; (b) the emergency-core cooling system (ECCS) that will promptly inject water from accumulators and storage tanks into the core and flood the containment structure adequately to submerge the fuel elements (see Section 21.13E); (c) the steam-pressure suppression system described in Section 21.13E and shown in Fig. 21.13-8; (d) the post-accident radioactivity removal (PARR) system, which is designed to circulate the atmosphere within the containment vessel through filters and aqueous sprays for the removal of iodine, particulate matter and other vectors of radioactivity; (e) the post-accident heat removal (PAHR) system designed to circulate water in the containment vessel through a heat exchanger capable of removing heat produced by decaying fission products; (f) the containment building integrity (CI) that, except for very small and known leakage, retains radioactivity released from the core until it is absorbed by PARR, the water or the walls; pressure suppression is essential to CI; (g) emergency sources of electrical power that can be supplied promptly in case plant power or power from the outside fail or are cut off; usually, a rapidly starting Diesel motor-generator set supplies emergency requirements while batteries are used for instrument power.

An event-fault tree analysis may be outlined by selecting initiating events and tracing the consequences.[2] Examples of initiating events are the following: (a) production of excessive neutronic reactivity without proper

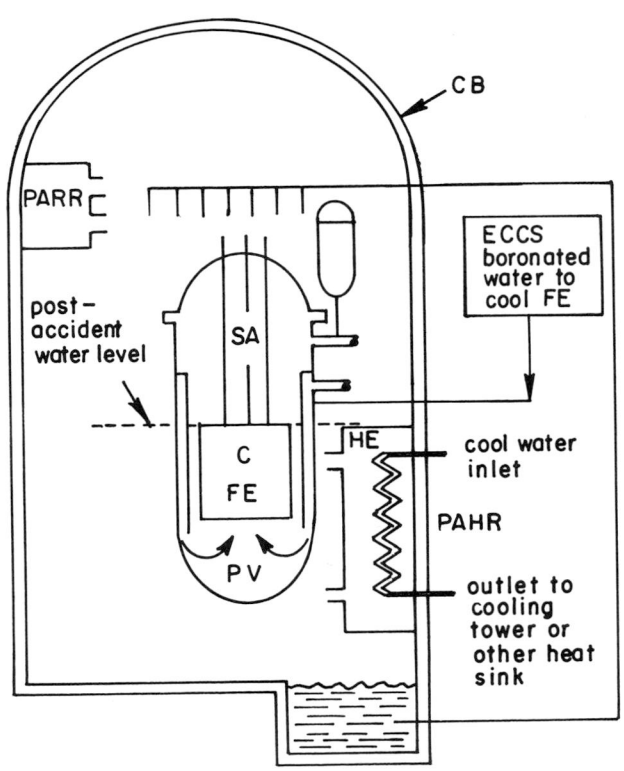

Fig. 24.3-1 The essential engineered safety features of a PWR.

Legend C: reactor core;
         FE: clad fuel elements;
         PV: pressure vessel;
         SA: safety rod neutron absorbers;
         ECC: emergency core cooling; injects water if the cooling loop
               fails;
         PAHR-HE: post accident heat removal-heat exchanger, pressure
                    suppression;
         PARR: post accident radioactivity removal; sprays and filters
               remove radioactivity from the atmosphere inside the
               containment building;
         CB: containment building; has very low leakage to the outside,
         withstands ~ 3 atm.

response of the control and safety systems; (b) failures in the cooling system, e.g., pressure-vessel failure, pump failure, or a break in the pipes going to and from the steam generators; of these events, the last is the most important. Six events have been identified in Ref. [2] as leading to a loss of coolant. Excessive neutron production is highly unlikely to occur. Pressure-vessel failure entered into a decision in England not to install an LWR. Inspection and surveillance procedures in the U.S. make pressure-vessel failure much less likely than a break in the pipes. Occurrence of pump failure has been made unlikely by redundancies in pumps, power and paths for removing heat.

An event or fault tree tends to define sequences of complete success or total failure, whereas partial failures are perhaps also possible. Event- and fault-tree analyses are made more realistic by the assignment of probabilities and also yield estimates for the worst outcomes. A complete break of a pipe in the cooling loop, which allows water and steam to escape through the full cross section of the pipe, may be taken as the initiating event (A) in Fig. 24.3-2. Detectable leaks will normally occur prior to the propagation of a large crack, but no allowance is made for the possibility of anticipating the break and shutting the reactor down. Four subsequent events will occur in our simplified illustration: electric power is or is not available (B), the ECCS does or does not flood the core in time (C), fission products are or are not removed from the atmosphere in the containment (D), the containment building does or does not confine the accident (E). The probabilities to be estimated are assigned to the branches of the event tree in Fig. 24.3-2. A complete analysis would include many additional events and would require the use of computer programs to enumerate the resulting alternatives.

The determination of probabilities is a difficult part of the analysis. The event tree provides a systematic way of selecting the necessary and sufficient sequences for calculation of the probabilities for various releases. After

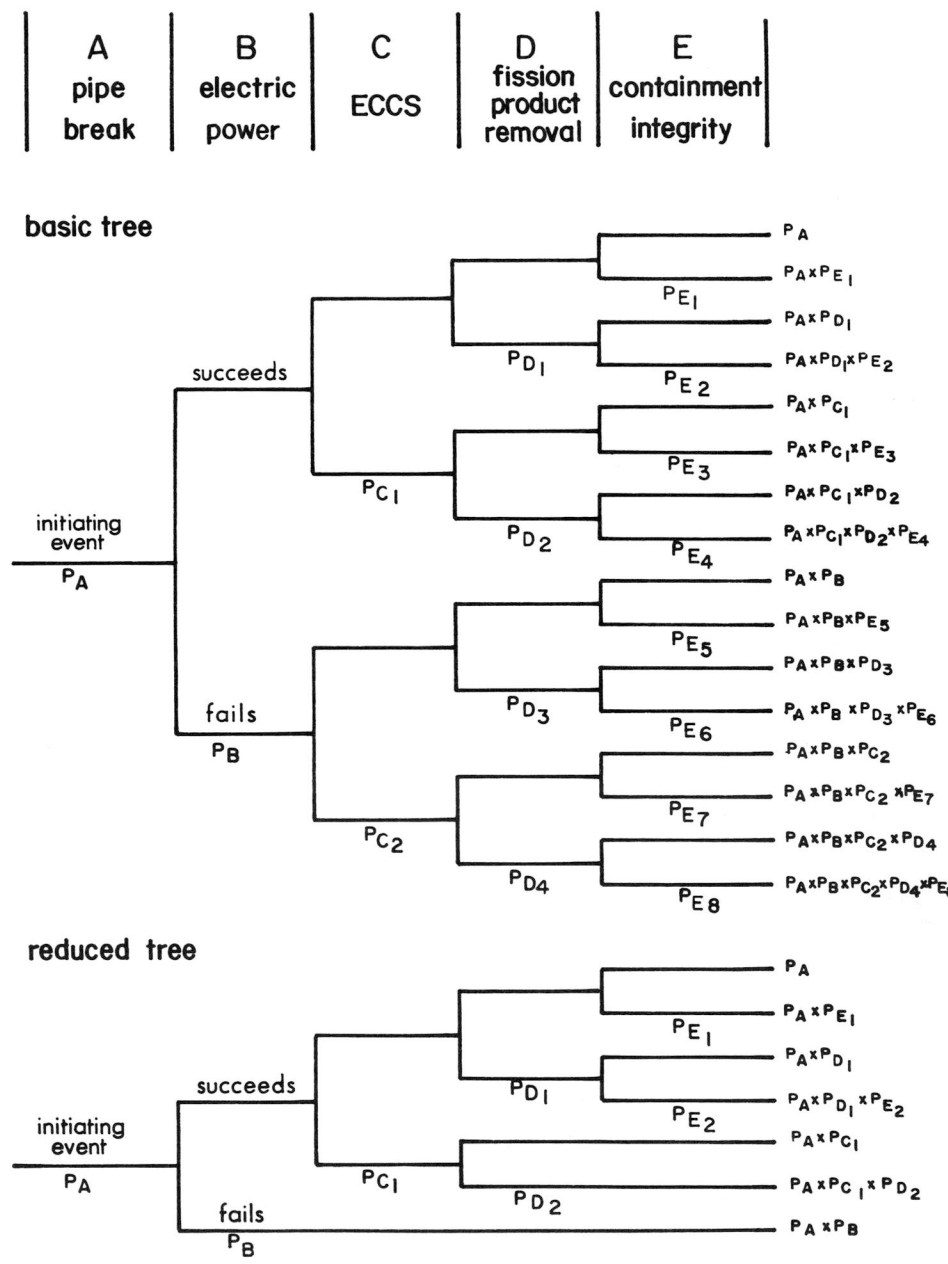

Fig. 24.3-2 Simplified event trees for a large LOCA; reproduced from Ref. [2].

selection of initiating events and definition of event trees, failure modes must be determined for each of the components and processes. This part of the analysis is divided into two tasks in Ref. [2], viz. identification of all the ways or modes in which the containment might fail and an examination of common-mode failures.

In the determination of subsequent failure probabilities, all combinations and sequences of failures leading to a given failure must be included. Both "and" and "or" types of branches must be studied. The "and" sequence refers to faults leading to complete failure while the "or" sequences refer to faults any one of which will cause the specified failure ultimately. As an illustration, [2] we show in Fig. 24.3-3 three stages in a fault tree leading to the loss of electrical power, which will vitiate the ESF. The "and" symbol demands multiplication while the "or" symbol signifies addition of probabilities in the computations.

The probabilities of occurrence of a given fault are taken from industrial experience or from a direct analysis of the equipment or function. Common modes for failure may be examined and the interdependence of probabilities may then be included in the calculations. Common features shared among failure modes may include an outside event (e. g. , an earthquake), a defective manufacturing process or defective materials for classes of components, a common principle of operation, or a human operator.

Many probability estimates may be known only within a factor of ten. In this case, the computer may be instructed to repeat each calculation many times after selecting from random numbers within the range of certainty for each fault. The final result has a probability distribution of known width. Such a procedure is called a Monte Carlo calculation.

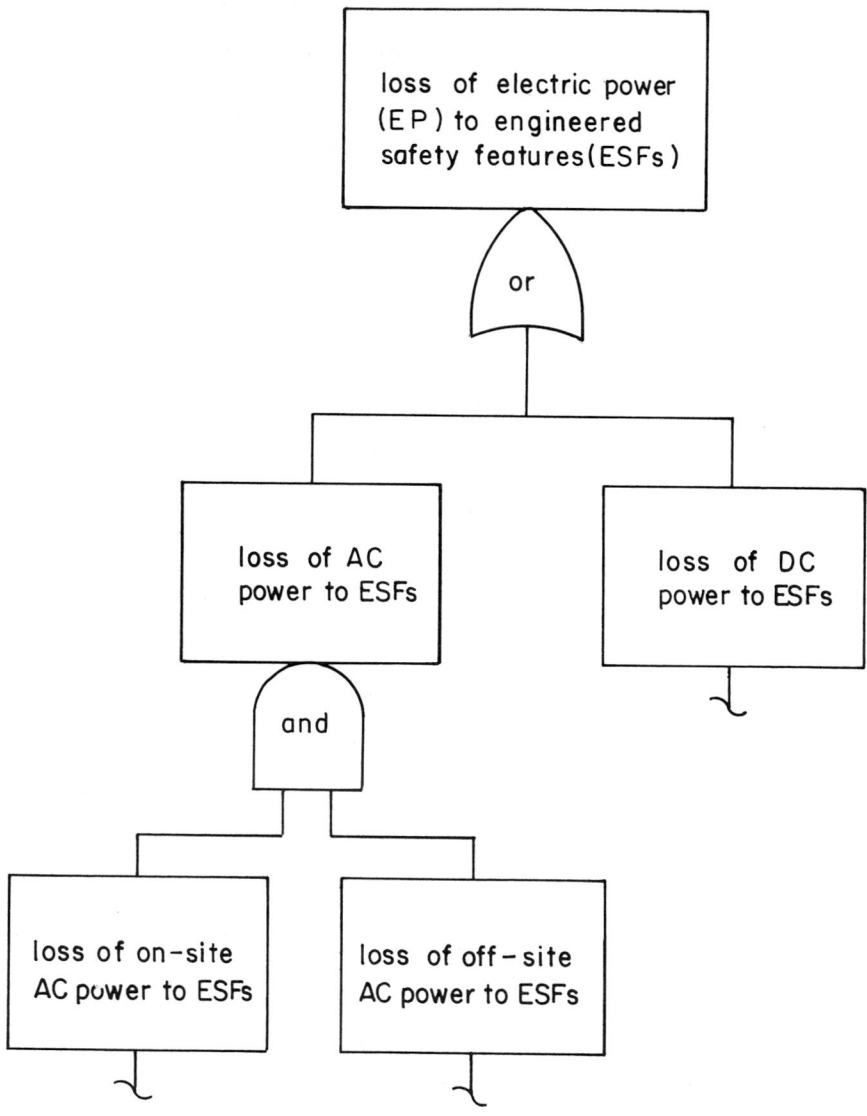

Fig. 24.3-3 Illustration of fault tree development; reproduced from Ref. [2].

iii. Radioactivity Release

In order to estimate the magnitudes of the released radioactivity, use is made of experimental work on the escape of 45 biologically significant radioactive nuclides from the fuel. Detailed experimental data on the behavior of the entire reactor core and its nuclides following a LOCA and activation of the ECCS are not available. However, extensive knowledge has been accumulated from smaller scale research. Unfortunately, available evidence is sometimes selected to support partisan positions. Programs to obtain required full-scale engineering data have been unreasonably delayed.

Following a complete break in a coolant pipe three feet in diameter, the water in the core flashes to steam very promptly with 80% conversion in 10 seconds. Removal of the moderator makes the reactor subcritical and stops the chain reaction. At the same time, the required heat transfer from the fission products is large, as may be seen from Fig. 24.3-4 and from Table 24.3-6, which show the energy and radioactivity of the fission products as functions of time and stage in the fuel cycle, respectively. The ECCS is required to inject water containing a neutron absorber within about 20 seconds in order to prevent the temperature of the zircaloy cladding from rising above 1200° C. It is known that zircaloy reacts rapidly with water above about 1270° C, thus adding heat while removing this containment barrier and permitting the very hot UO_2 to be redistributed by turbulent flows in the vaporizing, water-steam mixture within the vessel. Furthermore, the chemical reactions liberate hydrogen that could, if mixed with air, react explosively. During flashing and injection of coolant, the fuel rods are subject to hydrodynamic and thermal forces that are not easily predicted. The prediction that adequate heat transfer will be reestablished between the emergency cooling system and the hot rods is uncertain because surface boiling may occur (see Appendix 21A).

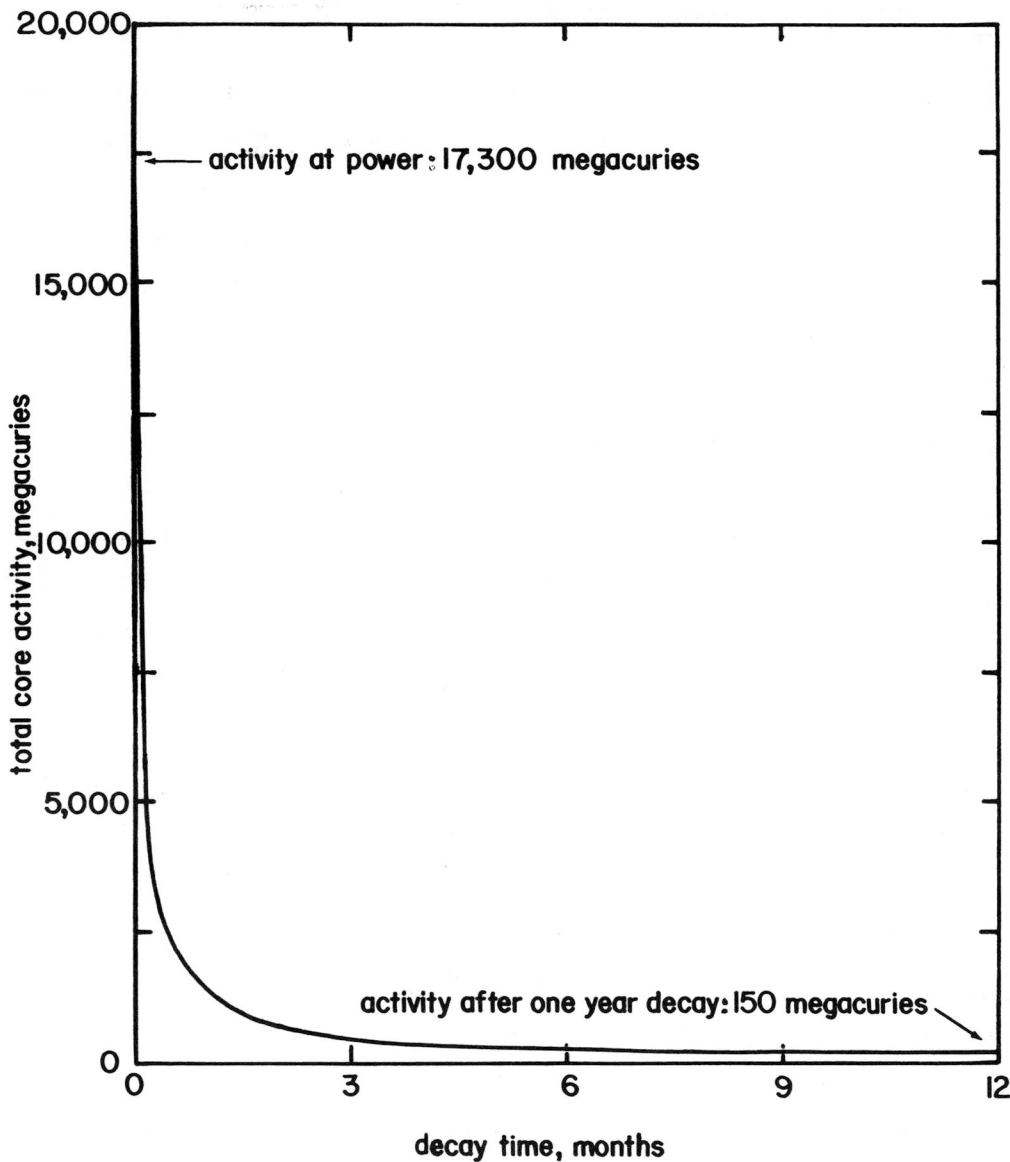

Fig. 24.3-4 The radioactivity in a 1,000 Mw$_e$ LWR core as a function of time
after shutdown. Fission products, actinide elements and radionuclides formed by neutron activation are included. This figure
has been reproduced from Ref. [2].

Table 24.3-6 Typical radioactivity inventory for a 1,000 Mw$_e$ nuclear power reactor; reproduced from Ref. [2].

Location	Total inventory (curies)			Fraction of core inventory		
	fuel	gap	total	fuel	gap	total
core(a)	8.0×10^9	1.4×10^8	8.1×10^9	9.8×10^{-1}	1.8×10^{-2}	1
maximum in the spent fuel storage pool(b)	1.3×10^9	1.3×10^7	1.3×10^9	1.6×10^{-1}	1.6×10^{-3}	1.6×10^{-1}
average in the spent fuel storage pool(c)	3.6×10^8	3.8×10^6	3.6×10^8	4.5×10^{-2}	4.8×10^{-4}	4.5×10^{-2}
shipping cask(d)	2.2×10^7	3.1×10^5	2.2×10^7	2.7×10^{-3}	3.8×10^{-5}	2.7×10^{-3}
refueling(e)	2.2×10^7	2×10^5	2.2×10^7	2.7×10^{-3}	2.5×10^{-5}	2.7×10^{-3}
waste-gas storage tank			9.3×10^4	-	-	1.2×10^{-5}
liquid-waste storage tank			9.5×10^1	-	-	1.2×10^{-8}

(a) Core inventory based on the activity 1.2 hours after shutdown; (b) inventory of 2/3 core loading; 1/3 core with three-day decay and 1/3 with 150 days of decay; (c) inventory of 1/2 core loading; 1/6 core with 150-day decay and 1/6 core with 60-day decay and 1/6 core with 60-day decay; (d) inventory based on 7 PWR or 17 BWR fuel assemblies with 150-day decay; (e) inventory for one fuel assembly with three-day decay.

However, experimental studies have now established the conditions for adequate convective heat transfer, the heat transfer relations and the extent of reaction between zirconium and water for different times of injection. Timing is crucial if the integrity and location of the fuel elements are to be preserved, the more so the higher the reactor power.

To carry the analysis beyond the failure of the ECCS, we must use a calculated probability that emergency cooling will not be adequate and determine the consequences of the reaction and melting of the cladding, thus permitting UO_2 at very high temperatures to fall into water at the bottom of the reactor vessel where cooling is inadequate to prevent rapid steam formation. This rearrangement of the fuel may generate steam pressures that are sufficient to rupture the containment building, either directly or by flying fragments. In this manner, the escape of radioactivity may be estimated from molten UO_2 (m.p. $\sim 2865°C$) moved about by expanding steam.

Early estimates of the dispersal of a reactor core by disruptive forces were made by workers at the Brookhaven National Laboratory.[4] The calculated serious consequences to life and property ($\$10^9$ to $\$10^{11}$) discussed in the 1957 report have biased past public debate. A major difference between the recent study[2] and the earlier study[4] is the conclusion that, while breeching of the containment is more likely to occur than had been estimated earlier, the release of radioactivity is reduced and occurs more slowly. The new, experimentally verified facts are the following: dispersal of the fuel as particles has extremely low probability; melt-through of the containment has higher probability but requires several hours during which many of the fission products

---

[4]Staff of the Brookhaven National Laboratory, "Theoretical Possibilities and Consequences of Major Accidents in Large Nuclear Power Plants," AEC WASH 740 (1957); available from the NTIS, Springfield, Va. 22151.

not bound in the UO_2 (approximately 2%) will be absorbed in the PARR, water, etc. Most of the fission products are not released, either to the air or to ground water.

Review of approximately 1,000 accident sequences releasing radioactivity from the system has resulted in the grouping of these sequences into a relatively small number of categories with decreasing severity and increasing probability of occurrence for PWR and BWR, as will now be described.

iv. PWR Release Categories Assuming Failure of the ECCS

(a) A steam explosion may be caused by a large amount of very hot UO_2 falling into a pool of water. If sufficiently dispersed, the UO_2 will transfer heat rapidly, thus causing an explosion large enough to rupture the pressure vessel and possibly also the containment. About one half of the UO_2 might be ejected and dispersed by the steam into the atmosphere. The PARR and other fission-product removal mechanisms will not have time to function. When in contact with air, UO_2 oxidizes to U_3O_8, thus exposing additional surface and releasing more radioactivity.

(b) Following melting of the core, the mechanisms for removing radioactivity fail to operate. Hydrogen from the zirconium-water and steel-water reactions burns and the containment fails because of excessive pressure. This event is slightly more probable than category (a), but the sequence occurs more slowly.

(c) Events similar to those in categories (a) and (b) occur but the PARR removes radioactivity while the containment remains intact.

(d) The core melts, the PARR fails, the containment fails to isolate the system while extensive deposition and retention of radioactivity occur within the system.

(e) The core melts and finally penetrates containment. The PARR functions properly. The upper portion of the containment fails.

(f) The core melts and penetrates the containment; the PARR fails. The upper portion of the containment remains intact.

(g) This sequence proceeds as in (f), except the PARR functions.

(h) The core does not melt but the cladding is destroyed and only the radioactivity outside of the UO_2 is released to the containment, which does not isolate the system.

(i) This sequence proceeds as in (h) but the containment isolates the system and only slow leakage occurs.

v.   BWR Release Categories

Because the engineered safety features are somewhat different for the BWR (see Section 21.14D), the paths for release of radioactivity are different. Six release categories are listed in Ref. [2]. While the water pool in the vapor-suppression system has the capability of absorbing radioactivity, it is unlikely to be effective if the core melts through the bottom of the primary vessel. Overpressure may rupture the low-pressure containment shell. The most effective remaining removal mechanism is absorption on all of the surfaces within the confinement building, from which matter may escape only through filters and a stack. At the same time, the reliability of the ECCS is greater than for the PWR. The following six failure categories have been identified.

(a) The core melts and is dispersed in a pool of water, thus causing a steam explosion that ruptures all containment and further distributes about one half of the core into the atmosphere, with consequences similar to those in the PWR category (a).

(b) The core melts and causes a steam explosion but is not ejected from the containment. Less fuel is oxidized by the air and less activity is released. Little deposition of radioactivity takes place within the building.

(c) The core melts and breeches the containment but air does not reach the fuel. Radioactivity not in the fuel has time to deposit on the inside surfaces of the containment structures.

(d) The core melts but the containment functions until melting occurs through the bottom into the soil. The walls, water and soil absorb much of the radioactivity.

(e) The core melts but the containment remains integral. Only the radioactivity not absorbed in the building and filters escapes.

(f) The core does not melt. Radioactivity not in the UO_2 is released to the containment; only that not absorbed by surfaces and filters escapes.

vi.   Summary of Failure Probabilities and Radioactivity Releases

The probabilities and properties[2] of various failure categories for the PWR and BWR are summarized in Table 24.3-7.

vii.   The Transport of Radioactivity

The existence of radioactivity presents us with both the opportunity and the need to learn how different chemical species are introduced, transported, and removed from the atmosphere, hydrosphere, biosphere, and lithosphere. Radioactive nuclides may be detected with a sensitivity and specificity at levels far below those of background cosmic radiation or naturally occurring radioisotopes such as $_{19}^{40}K$. The production of large quantities of radiation has emphasized the need for measurements of this type. As a result, more has been learned about the paths taken by radionuclides than about any other potential hazard. This work was begun with the operation of plutonium-producing

Table 24.3-7 Summary of accidents involving the core; reproduced from Ref. [2].

Release category	Probability, y^{-1}	Time of release, h	Duration of release, h	Warning time for evacuation, h	Elevation of release, m	Fraction of core inventory released							
						Xe-Kr	Org-I	I-Br	Cs-Rb	Te	Ba-Sr	Ru [a]	La [b]
PWR a	7×10^{-7}	1.5	0.5	1.5	25	0.8	6×10^{-3}	0.6	0.4	0.4	0.05	0.4	3×10^{-3}
PWR b	5×10^{-6}	2.5	0.5	1.5	0	0.9	7×10^{-3}	0.7	0.5	0.3	0.06	0.02	4×10^{-3}
PWR c	5×10^{-6}	2.0	1.0	1.5	0	0.8	6×10^{-3}	0.2	0.2	0.3	0.02	0.02	3×10^{-3}
PWR d	5×10^{-7}	2.5	3.0	1.5	0	0.5	2×10^{-3}	0.09	0.04	0.03	5×10^{-3}	3×10^{-3}	4×10^{-4}
PWR e	1×10^{-6}	2.5	4.0	1.5	0	0.2	2×10^{-3}	0.03	9×10^{-3}	5×10^{-3}	1×10^{-3}	6×10^{-4}	7×10^{-5}
PWR f	1×10^{-5}	12.0	10.0	1.5	0	5×10^{-3}	2×10^{-3}	8×10^{-4}	7×10^{-4}	1×10^{-3}	9×10^{-5}	7×10^{-5}	1×10^{-5}
PWR g	6×10^{-5}	10.0	10.0	1.5	0	2×10^{-3}	2×10^{-5}	2×10^{-5}	1×10^{-5}	2×10^{-5}	1×10^{-6}	1×10^{-6}	2×10^{-7}
PWR h	4×10^{-5}	0.5	0.5	N/A	0	2×10^{-3}	5×10^{-6}	1×10^{-4}	5×10^{-4}	1×10^{-6}	1×10^{-8}	0	0
PWR i	4×10^{-4}	0.5	0.5	N/A	0	3×10^{-6}	7×10^{-9}	1×10^{-7}	6×10^{-7}	1×10^{-9}	1×10^{-11}	0	0
BWR a	9×10^{-7}	3.0	2.0	2.5	25	1.0	7×10^{-3}	0.50	0.40	0.70	0.05	0.5	5×10^{-3}
BWR b	2×10^{-6}	3.0	0.5	2.5	0	1.0	7×10^{-3}	0.60	0.30	0.10	0.04	0.07	2×10^{-3}
BWR c	1×10^{-5}	28.0	5.0	2.5	0	1.0	7×10^{-3}	0.08	0.05	0.20	0.03	0.06	3×10^{-3}
BWR d	3×10^{-5}	9.0	0.5	2.5	0	1.0	7×10^{-3}	0.10	0.07	0.07	9×10^{-3}	6×10^{-3}	9×10^{-4}
BWR e	1×10^{-5}	5.0	2.0	2.5	0	0.6	3×10^{-3}	0.05	0.02	0.05	2×10^{-3}	3×10^{-3}	6×10^{-4}
BWR f	1×10^{-4}	30.0	5.0	N/A	0	4×10^{-4}	3×10^{-8}	6×10^{-12}	4×10^{-11}	8×10^{-14}	8×10^{-16}	0	0

(a) Includes Mo, Rh, Tc; (b) includes Nd, Y, Ce, Pr, Pm, Np, Pu, Zr.

reactors in the Columbia River Basin in 1944 and was followed by studies of fission products in the air, ground water, soil, the river, fish, and other biota. As the result, we now understand the paths, rates of transport, and points of concentration. No measurable hazard can be ascertained from either the intentional or the accidental releases (e.g., leaks from storage tanks) that have occurred during the years from 1944 through 1974 in the U.S. On the basis of information obtained from low-level accidents in England and Canada, the studies mentioned above, and general knowledge of meteorology, we conclude that good predictions can now be made of the consequences of severe releases of radioactive materials.

viii. Consequences of Accidental Releases

Computer modeling is an indispensible tool for predicting damage by released radioactivity to life and property. Extensive information on the dispersal of radioactivity as a function of height and rate of release for different chemical species, particulate matter, wind velocity and direction, distribution of population, diffusive and convective transport in the moving atmosphere, adsorption and desorption from surfaces, modes of biological uptake, and damage or health effects to humans and other living creatures, may be stored in data banks and programmed to yield the required predictions. Also included in the analyses are the effects of natural phenomena such as earthquakes, floods and windstorms, as well as appropriate preventive measures such as evacuation of populated areas. The results of extensive calculations are shown in Table 24.3-8.

ix. Criticism of Safety Analysis

Since publication of a draft of Ref. [2], three kinds of comments have appeared in the public media: skepticism has been expressed and the report

Table 24.3-8 Immediate consequences of reactor accidents with various probabilities for a single reactor; reproduced from Ref. [2].

Chance per year	Consequences					
	acute fatalities	acute illness	total property damage, $\$10^6$	total evacuation area, mi^2	decontaminated evacuation area, mi^2	
one in 20,000[a]	<1.0	<1.0	<0.1	<0.1	<0.1	
one in 1,000,000[b]	70	170	2750	175	6	
one in 10,000,000[c]	450	900	4000	320	25	
one in 100,000,000[d]	1200	2500	5000	390	30	
one in 1,000,000,000[e]	2300	5600	6200	400	31	

(a) This is the predicted chance of core melt.

(b) About two core melts out of 100 would produce the consequences in this row.

(c) About two core melts out of 1000 would produce the consequences in this row.

(d) About two core melts out of 10,000 would produce the consequences in this row.

(e) About two core melts out of 100,000 would produce the consequences in this row.

has been labeled as biased; higher release probabilities with reduced impact (relative to the 1957 study) have been noted with surprise; EPA staff members have labeled some event probabilities as too low by a factor of 10 while optimistic rates of evacuation were used. In an APS study,[5] integration over population distributions was modified and 16,000 delayed cancers were estimated to occur after an accident with a probability of $10^{-6}/y$.

C. The Safety of Fast Breeder Reactors

Analyses of the ways in which control, cooling or containment of FBRs might fail have been underway since 1945 and have been substantiated by experience with at least 10 steady-state reactor experiments or demonstrations and 2 pulsed experimental reactors. Design basis accidents (DBA) have been agreed upon, but no probabilistic analyses comparable to the Rasmussen study of commercial nuclear plants have been completed.

As was mentioned in Chapter 22 and Appendix 21A, stable control of the FBR (in spite of quicker time constants for neutron multiplication) is assured by the negative prompt or fuel (Doppler) coefficient of (neutron) reactivity and by increased neutron leakage as the density of the core decreases. Temperature coefficients have been measured and calculated and are well understood for use in the engineering analysis and design of system dynamics and control.

Loss of coolant from LMFBR vessels may be given a vanishingly small probability because the pressure inside the reactor vessel is no more than a few atm, redundancy of sodium containment may be provided, the integrity of the reactor vessel may be ascertained periodically and no physical mechanism

---

[5] "Report to The American Physical Society by the Study Group on Light-Water Reactor Safety," Supplement to Rev. of Mod. Phys. 47 (1975).

for rapid failure exists. Analyses show that propagation of an incipient void (because of boiling or formation of bubbles by entrained gas) throughout the core will probably not occur. In the event that the circulating pumps fail, natural thermal convection of sodium will remove heat generated after shutdown. Loss of helium from a GCFR may occur with non-negligible probability because of the relatively high system pressure ($\sim$70 atm). However, the time for loss of coolant cannot be reduced below 30 sec, which is ample for fail-safe shutdown of the reactor, as well as for activation of the auxiliary cooling systems if this is necessary. As long as one helium circulator is operating, post-shutdown heat will be removed even after loss of some pressure. The pressure cannot fall below 2 atm because the reactor vessel is surrounded by a pressurized, gas-tight containment building. Redundancies of circulators and the associated motor drives may be made whatever are required for safety.

Thus, it appears likely that for both LMFBR and GCFR core melting will be avoided. In spite of this judgement, the selected DBA that must be contained without release of radioactivity and undue hazard to the public remains melting and supercritical reassembly of the core. Less than 0.01 of the reactor core might constitute a critical mass if it were accidentally made fully dense and approximately spherical. Careful analysis of the energy released in the event of occurrence of such "criticality" shows that the energy release in a "core disruptive accident" is no more than that from 65 to 70 kg of TNT dissipated among 1×10^4 kg of inert matter and will therefore be contained within the reactor vessels. [6]

More recently, completed experimental work and design studies have shown that both reassembly of a critical mass and melt-through of the con-

---

[6]H. A. Bethe, "Testimony on the California Nuclear Initiative," Energy, Volume 1 (in press, 1976).

tainment are either contrary to physical principles or preventable by the use of dependable cooling systems.

In spite of the foregoing assurances, the hazards of an accidental re-lease from an FBR are sufficiently large to cause at least as great controversy as in the case of the LWR, largely because of the higher energy and fissile mass density of the core and shorter time constants for changes in heat pro-duction.

24.4 Fuel Reprocessing

Fuel elements removed from a reactor are stored for about 3 months at the reactor site and then shipped for reprocessing to prepare fissile and fertile materials for reuse. High- and low-level wastes are solidified and shipped to repositories. Reprocessing costs are probably minimal for a single plant serving 40 to 50 reactors. With large throughput of fluid ma-terials, the potential for radioactive release is large.

At the reprocessing plant, reactor-fuel elements of the type described in Chapters 21 and 22 are opened mechanically and the UO_2 contents dissolved in nitric acid. The released fission-product gases (Kr and Xe) must be trans-ferred to adsorption trains and tanks. Heretofore, dilution with air to toler-able levels of radiation was followed by disposal through tall stacks. The ra-dionuclide tritium may escape with aqueous effluents to the biosphere during reprocessing. Fortunately, zircaloy cladding in the LWR fuel elements re-tains most of the tritium as zirconium hydride. The nuclides requiring special attention are listed in Table 24.4-1.

Table 24.4-1 The listed fission products require special attention in environmental and waste management.

Nuclide	Yield, %	half life	Particle and its energy after decay, Mev
^{85}Kr	0.3	10.76 y	β 0.67 γ 0.514
^{90}Sr	5.8	27.7 y	β 0.546
^{131}I	2.9	8.05 d	β 0.606, 0.25, 0.81 γ 0.364, 0.080, 0.723
^{129}I	1	1.6×10^7 y	β 0.15 γ 0.040
^{137}Cs	5.9	30 y	β 0.66
^{3}T	*	12.26 y	β 0.0186

*Tritium is produced with low yield from rare ternary fission events.
In addition, it may be formed from $(n\gamma)^2 D \rightarrow {}^3T$, $(np) {}^3He \rightarrow {}^3T + H$,
$(n\alpha) {}^6Li \rightarrow {}^3T + {}^4He$, and $(n2\alpha) {}^{10}B \rightarrow {}^3T + {}^4_2He$. The amount formed
must be determined for specific situations.

The halides have significant vapor pressures at the dissolution stage
in the presence of the oxidizing acids that are used. However, retention and
concentration of iodine and bromine may be readily accomplished.

All of the other fission products, uranium, plutonium, and a variety
of isotopes of neptunium, curium and other actinide elements, formed by neutron capture and transmutation, remain in the aqueous solutions that are processed through several stages. Isolation and decontamination of the uranium
and plutonium are accomplished by using valence states which form compounds

that may be transferred selectively in solvent-extraction columns to an immiscible organic liquid phase; by readjusting the valence states, transfer back to an aqueous phase may be accomplished. This selectivity leaves less than 10^{-7} of the fission products with the uranium and plutonium while 0.1 to 0.5% of the latter, as well as all other actinides, remain with the fission-product stream that must be handled at the waste-disposal stage.

Other procedures have been developed for taking the reactor fuel through appropriate reaction steps in shielded facilities adjoining the reactor. All of the heavy elements and minor amounts of the fission products are returned to the reactor. The wastes from these processes may contain very little uranium, plutonium or other actinides. Unfortunately, these procedures are excessively costly for oxide fuels. In the Molten Salt Breeder concept, which is still under development, the fission products are continuously removed from the liquid fuel.

Fragmented tubing, referred to as cladding hulls, may be compacted, packaged, and stored as low-level waste. Other low-level wastes result from processing side streams, decontamination of equipment, and some items of discarded equipment.

Reprocessing plants are now located in South Carolina, Western New York, and Illinois. None will operate during 1975. Four or five plants may be needed after 1985.

A. Transportation of High-Level Radioactive Materials

Highly radioactive fuel elements must be transported from reactors to reprocessing plants or to temporary storage sites. The high-level wastes must be transported from the reprocessing plants to retrievable or permanent storage sites.

The amount of fuel shipped from a standard 1,000 Mw_e LWR to a processing plant may be estimated from the typical exposure required of the fuel (to minimize power costs), which was given in Chapter 21 as 3×10^4 Mw_td/mt. The thermal power of such a reactor is approximately 3,000 Mw_t. Thus, 3,000 Mw_t/30,000 Mw_td/mt = 0.1 mt/d or 36.5 mt/y of fuel must be shipped, corresponding to about one-third of a core. A single fuel reprocessing plant serving 50 reactors receives approximately 1,630 mt of fuel per year. During the year 2000, about 6,500 mt might be shipped to 4 plants if the high AEC reactor projections prove to be correct.

For the FBR, an exposure of at least 100,000 Mw_td/mt is desirable. With heat-power efficiencies of 40%, the weight shipped per 1,000 Mw_e reactor core corresponds to only 9 to 10 mt/y. Inclusion of the blanket material raises the mass shipped to about 18 mt. Ultimate construction of nearly 4,000 reactors of this type has been projected after the year 2020; the corresponding shipments of high-level radioactive materials amount to 36,000 mt/y. Co-location and separate location of the FBR and processing plants are under study.

Steel tanks are used as shipping casks for LWR fuel elements. These are equipped with external fins for air cooling and internally located, lead-shielded, water-cooled, sealable steel tubes. Typical casks weigh about 50 t and are approximately 5.5 m long and 1.5 m in diameter. The casks may be transported either by truck or by railroad. They are built to withstand a drop from 30 ft and a 30 minute fire at 800°C, followed by total immersion in water for 8 h. With these capabilities, potentially hazardous accident rates are comparable for either mode of transportation and have been estimated from experience[1] to be less than 10^{-10} per car-mile.

---

[1] W. A. Brobst, "Transportation Accidents: How Probable? Nuclear News 16, 48-54, July 1973.

HTGR fuel blocks may be transported in large containers made of steel and aluminum. The aluminum serves to conduct heat to the outside fins.

After elapse of the 90-day cooling period before shipping, a typical cask containing 1.5 mt of LWR fuel will dissipate approximately 130 kw_t, which is equivalent to 2.1×10^7 ci. The power from decaying fission products decreases slowly with time. For the fission of ^{235}U, the power follows the empirical relation $P(t) = P(0) \times 0.066t^{-0.2}$, where t = elapsed time ($\geq$ 10 sec) if P(0) is the power at t = 0.

Casks weighing 114 mt and containing about 0.7 mt of fuel or blanket elements, while dissipating up to 13 kw_t, have been proposed for the shipment of FBR elements. Railroads will probably be preferred because of the need for rigorous monitoring and protection during shipping.

Solidified high level wastes may be transported in air-cooled steel-and-concrete containers that are less destructible than fuel-element casks. Failures releasing radioactivity are less likely to occur and would be less severe than for unprocessed fuel. Borosilicate glass is preferred; it would contain 50% by weight of the fission products. By the year 2000, 10^3 Gw_e of nuclear-power capacity might require shipment of approximately 300 mt/y to storage sites. Shielded canisters will probably be used. The internal space for the glass may have a diameter and length of 30 and 300 cm, respectively. Heat dissipation will be about 5 kw_t. Shipments by rail will probably be preferred.

24.5 Nuclear Waste Disposal

An indication of the rate of decay of waste toxicity is given in Fig. 24.5-1, where the relative waste toxicity (after 99.5% removal of uranium and plutonium) from the fuel element of a light-water reactor is shown as a function of time. We note that the rapid decrease in toxicity to a level of about 10^{-3} of

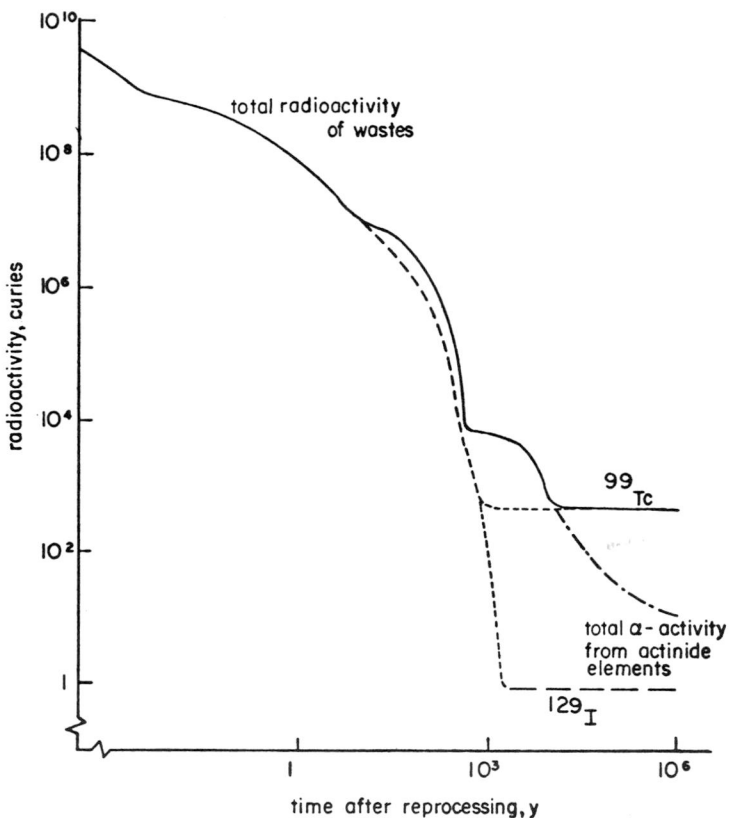

Fig. 24.5-1 The approximate radioactivity of high level wastes, from which
99.5% of the Pu has been removed, is shown as a function of time
after reprocessing of the fuel taken each year from a 10^3 Mw$_e$
LWR. The average residence time in the reactor is about 3 y.
The solid line shows the total radioactivity. Emissions from fission products dominate for ~100 y. After ~600 y, the toxicity of
α-emissions from actinides exceeds that of the fission products.
The dashed line (— —) shows that β-decays of ^{137}Cs and ^{90}Sr dominate to ~800 y, after which low-energy βs from ^{99}Tc (t½ = 2.12
× 10^5 y) are the primary radioactivity. The shortest dashed line
(- - -) represents the final decay of rare earth elements. The
longest dashed line (— — —) corresponds to ^{129}I (t½ = 1.7 × 10^7 y).
The long-and-short-dash line (—·—) represents the αs for long-
lived actinides and their daughters. Adapted from T. H.
Pigford and K. P. Ang, "The Plutonium Cycles," Health Physics
29, 451-468 (1975); by permission of the Health Physics Society.

the initial level requires nearly 1,000 years. This relatively rapid initial decrease in toxicity level results from the decay of the fission products Sr and
Cs which provide most of the radioactivity for 30 to 800 years. Thereafter,
the total toxicity level tends to be dominated by members of the longer-lived
actinide series of elements. After about 10^5 years, the total activity builds
up again because of the formation of radioactive daughter products. Nevertheless, after decay of the fission products, the radioactivity of the wastes without
actinides is less than that of an equivalent amount of ore. Chemical separation
of the wastes into a long-lived actinide fraction, an intermediate-lived cesium
and strontium fraction, and a shorter-lived residue would simplify their use,
management and storage.

The radiological hazard or toxicity of effluents from the fuel cycle may
be represented by an index which is equal to the logarithm to the base 10 of the
dilution with water that is required to yield a maximum permissible concentration[1] for discharge to the environment. On this basis, the initial value of
the index is 16, dropping to 12 after 10^3 y. Also, a relative toxicity[2] is
defined for comparison of the dilution factor with that measured for a standard
uranium ore. On this basis, the wastes have a relative toxicity near 900 after
10^4 y which rises to 8500 in 10^6 y, before finally diminishing monotonically.
Without the actinides, the relative toxicity would fall to 70 in about 900 y. Because of the possibility of concentrating specific elements in the biosphere and

---

[1]National Council on Radiation Protection and Measurements, Basic Radiation Protection Criteria, NCRP Report No. 39, 7910 Woodmont Avenue,
Washington, D.C., 20014 (1974).

[2]A. S. Kubo and D. J. Rose, "Disposal of Nuclear Wastes," Science 182,
1205-1211 (1973).

in organs in man, [3] objections have been raised to using a single index as a measurement for the specified maximum permissible concentrations of some elements.

For 10^6 kw$_e$h generated, we produce 2×10^{-3} to 4×10^{-3} m^3 of high-level liquid wastes holding several hundred to several thousand curies per gallon. Furthermore, 2×10^{-4} m^3 of solidified waste is generated per 10^6 kw$_e$h from enriched nuclear fuel. For a year 2000 nuclear-reactor capacity of 1×10^6 Mw$_e$ operating at a 75% load factor, the industry will generate $\sim 6.6 \times 10^{12}$ kw$_e$h/y and produce 1.3×10^4 to 26×10^4 m^3 of high-level liquid and 2.6×10^3 m^3 of high-level solid wastes per year. The largest of these volumes can be accommodated in a room 200 ft long, 200 ft wide, and less than 20 ft high. A distinguishing feature of the problem of nuclear-waste disposal is the small volume of highly radioactive material produced.

An authoritative 1974 estimate of the projected accumulation of high-level solid wastes to the end of the twentieth century is summarized in Table 24.5-1. Recent projections suggest that the given estimates may be too large.

A. Waste Management

A schematic diagram[4] indicating options in the management of

---

[3]E. A. Martell, "Actinides in the Environment and their Uptake by Man" Report NCARTN/STR-110, National Center for Atmospheric Research, Boulder, Colorado (May 1975); E. A. Martell, "Tobacco Radioactivity and Cancer in Smokers," American Scientist 63, 404-412 (1975).

[4]High-Level Radioactive Waste Management Alternatives, Frank K. Pittman ed., U.S. Atomic Energy Commission Report WASH 1297, May 1974. This report is available from the NTIS, Springfield, Va. 22151.

Table 24.5-1 Projected accumulation of solidified, high-level waste to the year 2000; reproduced from Ref. [1].

Year	Volume[a] of waste, m^3	Actinide[b] mass, mt	Radio-[b,c,d] activity, mci	Thermal[b,c,d] power, Mw_t	Toxicity indices[b,c,d,e] for inhalation	Toxicity indices[b,c,d,e] for ingestion
1974	5	---	200	1	18.08	13.34
1975	30	3	700	3	18.78	14.04
1976	90	9	2,300	10	19.30	14.58
1977	190	18	4,600	20	19.62	14.91
1978	300	29	7,000	30	19.85	15.15
1979	410	40	8,300	40	19.97	15.26
1980	550	50	10,200	50	20.26	15.38
1981	720	70	12,900	60	20.54	15.50
1982	930	100	16,100	80	20.79	15.59
1983	1,170	120	19,600	100	20.99	15.69
1984	1,420	150	22,600	120	21.15	15.76
1985	1,720	190	26,300	140	21.30	15.85
1986	2,060	220	30,200	160	21.40	15.91
1987	2,440	260	34,200	180	21.46	15.98
1988	2,870	310	39,000	200	21.52	16.04
1989	3,350	360	44,300	230	21.57	16.11
1990	3,900	410	50,300	250	21.59	16.18
1991	4,510	470	56,700	280	21.62	16.26
1992	5,190	530	64,400	310	21.63	16.30
1993	5,930	600	72,300	350	21.66	16.36
1994	6,750	680	81,000	380	21.68	16.43
1995	7,650	760	90,500	420	21.71	16.48
1996	8,630	850	101,100	470	21.74	16.53
1997	9,700	950	112,500	510	21.77	16.58
1998	10,820	1,050	123,400	560	21.80	16.61
1999	12,040	1,160	136,200	610	21.83	16.66
2000	13,340[f]	1,270	149,000	660	21.86	16.70
Time elapsed after the year 2000, y						
10^2			5,700	20	21.29	15.55[g]
10^3			30	< 1	20.19	12.74[g]
10^4			10	< 1	19.70	12.14[g]
10^5			4	< 1	18.79	12.38[g]
10^6			1	< 1	18.60	11.86[g]

(a) The volume is based on 0.057, 0.170, and 0.085 m^3 of solidified waste per mt of heavy metal for LWR, HTGR, and LMFBR fuels, respectively.

(b) It is assumed that 0.5% of the product (U and Pu in the LWR and LMFBR and Th and U in the HTGR) is lost; all other actinides remain in the waste.

(c) The waste is initially generated 150, 365, and 90 days after the spent fuel has been discharged from LWR, HTGR, and LMFBR units, respectively.

(d) All tritium and noble gas fission products and 99.9% of iodine and bromine fission products are excluded.

(e) Base 10 logarithms refer to the quantity in m^3 of air for the inhalation index, or of water for the ingestion hazard index, required to dilute radioactive material to limits stipulated elsewhere.[1]

(f) The volume of waste generated through the year 2000 results from the reprocessing of almost 2×10^5 mt of irradiated fuel, about 80% of which is associated with LWR plants.

(g) Beyond the year 2000, fission products (primarily strontium) and transplutonium elements (primarily americium) are the controlling potential hazards in drinking water up to about 350 and 2×10^4 years, respectively. Radioactivity from plutonium losses during reprocessing then becomes controlling until about 10^6 years. Finally, the radioactivity remaining as the result of uranium losses during reprocessing becomes the predominant contribution to the ingestion toxicity index.

high-level radioactive wastes is shown in Fig. 24.5-2. Immediate disposal of slightly enriched reactor fuels from an LWR is not economically advantageous. For this reason, storage until a later date is necessary. Solidification of high-level waste involves an established procedure and is required within 5 years of waste production. Partition or separation of actinide elements from the wastes is not now practiced but may be implemented. Many different feasibility studies and cost estimates have been made of other options shown in Fig. 24.5-2. Most appear to be economically acceptable. As has already been emphasized, interim storage of insoluble solid wastes does not involve large volumes of material. Wastes must now be moved to a federal repository within 10 years. Transmutation of partitioned actinides appears to be a technically and economically reasonable procedure. The storage of fission products for 600 to 800 years is also entirely feasible.

B.   Characteristics of High-Level Liquid Wastes

The characteristics of high-level liquid wastes are summarized in Table 24.5-2.

C.   Characteristics of High-Level Solid Wastes

High-level solid wastes are described in Table 24.5-3.

The purposes of the processes indicated in Table 24.5-3 are removal of water and production of stable solids, which remain sealed inside a corrosion-resistant metallic can. The simplest process is to seal the dried and calcined solid in a pot with a welded closure. Spray drying and fluidized bed calcining increase the throughput of the processes. Phosphate glasses may be made by melting the dried mixture with phosphoric acid. The least soluble

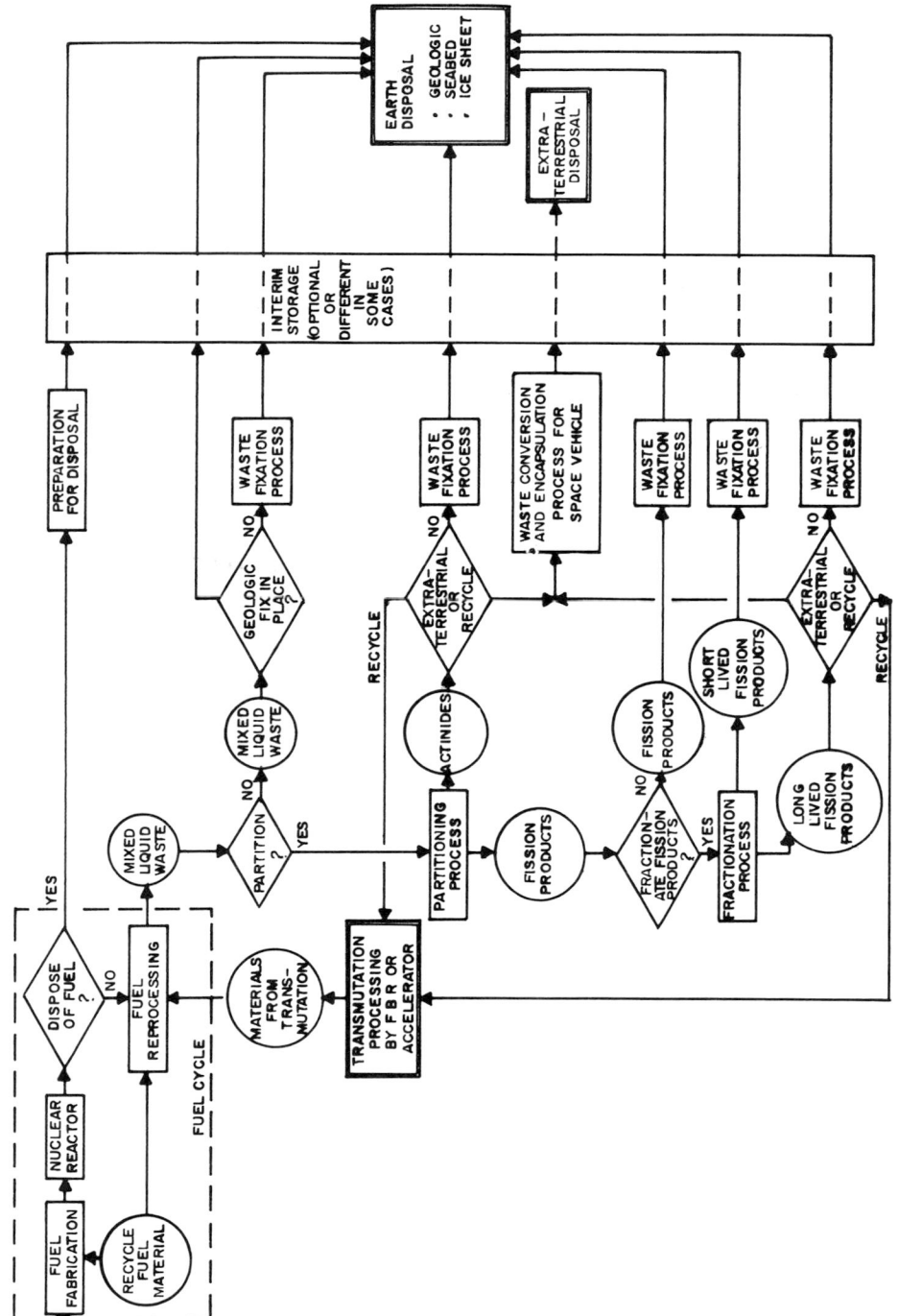

Fig. 24.5-2 Schematic diagram showing options in high-level waste management; reproduced from Ref. [1].

Table 24.5-2 Typical materials in high-level liquid waste; reproduced from Ref. [1].

	Material[b]	Grams/mt from[a]		
		LWR[c]	HTGR[d]	LMFBR[e]
chemicals that are reprocessed	hydrogen	400	3,800	1,300
	iron	1,100	1,500	26,200
	nickel	100	400	3,300
	chromium	200	300	6,900
	silicon	--	200	--
	lithium	--	200	--
	boron	--	1,000	--
	molybdenum	--	40	--
	aluminum	--	6,400	--
	copper	--	40	--
	borate	--	--	98,000
	nitrate	65,800	435,000	244,000
	phosphate	900	--	--
	sulfate	--	1,100	--
	fluoride	--	1,900	--
	sub-total	63,500	452,000	380,000
fuel-product losses[f,g]	uranium	4,800	250	4,300
	thorium	--	4,200	--
	plutonium	40	1,000	500
	sub-total	4,840	5,450	4,800
transuranic elements[g]	neptunium	480	1,400	260
	americum	140	30	1,250
	curium	40	10	50
	sub-total	660	1,440	1,560
other actinides[g]		< 0.001	20	< 0.001
total fission products[h]		28,800	79,400	33,000
	total	103,000	538,000	419,000

(a)   The water content is not shown.

(b)   Most constituents are present in soluble, ionic form.

(c)   U-235 enriched PWR, using 378 liters of aqueous waste per mt, 33,000 $Mw_e d$/mt. The integrated reactor power is expressed in $Mw_e d$ per unit of fuel in mt.

(d)   Combined waste from separate reprocessing of "fresh" fuel and fertile particles, using 3,785 liters of aqueous waste per mt, 84,200 Mwd/mt exposure.

(e)   Mixed core and blanket, with boron as soluble poison. 10% of cladding dissolved, 1,249 liters per mt, 37,100 Mwd/mt average exposure.

(f)   0.5% product loss to waste.

(g)   Losses are given at the time of reprocessing.

(h)   Volatile fission products (tritium, noble gases, iodine and bromine) are excluded.

Table 24.5-3 Characteristics of solidified high-level waste; reproduced from Ref. [1].

form	Pot calcine	Spray phosphate, ceramic	Phosphate glass	Borosilicate glass(a)	Fluidized bed, calcine
form	scale	monolithic	monolithic	monolithic	granular
description	calcine: cake to friable	ceramic: hard to tough	glass: hard to brittle	glass: hard to brittle	calcine, mean particle diameter 100 to 500 μ
bulk density, g/cm³	1.2 to 1.4	2.7 to 3.3	2.7 to 3.0	3.0 to 3.5	1.0 to 1.7
weight % of fission-product oxides, maximum	90	30	25	50	50
thermal conductivity, $w/(m^2)$-$(°C/m)$	0.3 to 0.4	1.0 to 1.4	0.8 to 1.2	1.0 to 1.4	0.2 to 0.4
leachability in cold water, g/cm²-d	1 to 10^{-1}	10^{-3} to 10^{-5}	10^{-4} to 10^{-6}(b)	10^{-5} to 10^{-7}	1.0 to 10^{-1}

(a) Produced by either spray or fluidized bed calcining, followed by melting or by in-canister vitrification processing.

(b) Devitrified phosphate glass exhibits increased leachability (leach rates = 10^{-2} to 10^{-3} g/cm-d).

glasses are made by adding borosilicate compounds (the basis of Pyrex glasses) before melting.

D.   Long-Term Disposal Options

The options shown in Fig. 24.5-2 include combinations of processing methods and of types of repositories.   Present practices yield wastes that require isolation for many thousands of years.   High-level wastes may be partitioned at modest cost into two fractions.   One of these decays with a half life of about 30 years and requires isolation for 20 half lives or 600 years.   The other fraction contains actinides and iodine and requires permanent isolation or disposal.   It is probably possible, at reasonable cost, to partition the wastes further according to nuclear and chemical behavior, thus reducing the amounts of material and increasing the options for uses and special disposal methods.

The engineering feasibility of supervised retrievable storage on the surface of the earth has been determined.   The technical feasibility of storage in salt beds has been established.   Either retrieval after a limited period of time or indefinite salt-bed storage may be accomplished if care is taken not to expose regional ground water.   The possibility of using deep, stable granitic rock formations has been considered; implementation must be deferred until after verification of the stability of the geological strata involved.

Many of the actinides fission with the addition of 1 to 3 neutrons, giving back nearly as many neutrons as were used and thus may be recycled to reactors at reasonable cost.   Disposal of selected fission products in specially designed fission or fusion reactors or particle accelerators may not prove to be too costly.

Disposal in space of the longest-lived nuclides will probably not increase the cost of electric power by more than a few percent.   Adequate reliability and vessel integrity on reentry in case of failure are judged to be attainable but cost estimates are not yet reliable.

Disposal in deep ocean trenches could be permanent and very safe because there is no transport from the sediment in these trenches to the rest of the ocean.   Furthermore, since the underlying tectonic plate is subducted at the rate of 1 to 2 cm/y, solid material in the trench is entrained under the overlying plate.   The difficulty of knowing the fate of the disposed material interferes with acceptance of this option.

E.   Plutonium Toxicity

We have seen that plutonium production constitutes an integral component of fission-reactor technology.   The extreme toxicity of this material has been stressed in the public debate of nuclear-reactor safety. [5]   A great deal remains to be learned about the dangers of exposure to minute concentrations of plutonium and of man-made radioactive substances in general.   It is interesting to note that a recent study has shown that localization of α-emitting particles in the lungs does not appear to raise the level of hazard above that for an equivalent concentration of uniformly distributed material. [6]

---

[5] "The Plutonium Economy: A Statement of Concern, " National Council of Churches of Christ in the U.S.A., 475 Riverside Drive, New York, N.Y. 10027, September 1975.

[6] Alpha-Emitting Particles in Lungs, National Council on Radiation Protection and Measurement, 7910 Woodmont Ave., Washington, D.C. 20014, Report No. 46, August 1975.

F.   Some Quantitative Aspects of Waste Disposal

A highly informative analysis of the nuclear waste-disposal problem has been published recently by B. L. Cohen.[7] Cohen has shown that a reasonable upper limit to the deaths occurring as the result of normal leakage of radioactive wastes buried randomly in the U.S. at a depth of 600 m is likely to be as much as 100 times smaller over a very long time period than the number of deaths normally attributable to the naturally occurring uranium that is burned up in nuclear reactors.

In arriving at his conclusions, Cohen introduces a number of assumptions for electrical energy generation of 4×10^8 kw$_e$-y (about double the total 1970 U.S. electrical energy consumption):

i.   The production of radioactive waste materials was calculated by using the Oak Ridge National Laboratory computer code ORIGEN.

ii.   Cancer-causing dosages were then obtained by using ICRP Publication-2 and the BEIR report. The number of deaths scales linearly with the number of cancer-producing doses and is independent of the number of people ingesting radioactive waste materials.

iii.   With random burial (to obtain upper limits for the hazard in the absence of a direct path such as an artesian well) at depths of 600 m in the U.S., radioactive wastes will not reach the surface in less than about 9×10^3 y because water flows in aquifers at very low speeds (~ 0.03 m/d) and the average distance to the surface is very long (~ 100 km). Furthermore, the radioactive

---

[7]B. L. Cohen, "Environmental Hazards in Radioactive Waste Disposal,"
   Reviews of Modern Physics (in press, 1976); for an abstract of this work,
   see Physics Today 29, 9-14 (1976).

materials are held up by ion exchange with rock and soil material and may not reach the surface for millions of years.

iv.   Irradiation of people after ingestion of radioactive waste is assumed to occur according to the known exposure to radium from uranium ores, which corresponds to a probability of about 4×10^{-13}.

v.   The total number of deaths attributable to the wastes produced during generation of 4×10^{8} kw_e-y is the integral of lives lost over the time elapsing after burial and is found to be less than 0.4 during the first 10^{6} y, less than 1.2 during the first 10×10^{6} y, and less than 4 during 100×10^{6} y.

vi.   The exposure to uranium and its daughter products is reduced for each year of nuclear-reactor operation in an LWR. During the first 10×10^{6} y, it is estimated that 120 lives will be saved by uranium clean-up, as compared with an upper limit of 1.2 lost by ingestion of radioactive waste material. Thus, Cohen's study[4] suggests that the use of nuclear reactors may actually decrease the number of cancer deaths associated with exposure to radioactivity.

24.6 <u>Estimates of Environmental and Safety Aspects of Fusion Power</u>

Environmental and safety aspects of fusion power involve thermal pollution (Section 24.6A); material and land requirements (Section 24.6B); chemical pollutants (Section 24.6C); radiological impact including activity inventories, radioactive effluents, and waste disposal (Section 24.6D); nuclear afterheat (Section 24.6E); operational safety (Section 24.6F); and fuel diversion (Section 24.6G). Although fusion power is a relatively clean and practically inexhaustible solution to our energy-supply problems, significant quantities of radioactive materials must be handled and controlled during the operation of fusion reactors. Waste-disposal schemes similar to those required for fission products will be needed for fusion reactors employing niobium as the structural material. Development of vanadium as a primary structural material offers considerable advantages with respect to management and disposal of long-lived radioactive wastes produced by fusion reactors. A preliminary comparison of the radiological implications of fusion and fission power indicates that reduced environmental and safety problems will be associated with the development of fusion power.

Analyses of environmental and safety aspects of fusion power are, of course, intrinsically related to the specified engineering design of the reactor. Since fusion-reactor designs are not yet defined in detail, environmental and safety analyses are preliminary assessments of the implications of fusion power. We shall focus on deuterium-tritium fusion reactors employing electromagnetic-confinement designs because (a) they are presently the most advanced in terms of engineering design, (b) they are likely to be the earliest application of fusion power, and (c) they are potentially the most environmentally damaging. Environmental and safety analyses of laser-fusion reactors and advanced fuel-cycle (e.g., deuterium-deuterium, deuterium-helium 3) reactors are not discussed.

A. Thermal Pollution

The thermal pollution problem is common to all power plants that operate on thermal cycles. In preliminary design studies,[1,2] thermal-to-electrical energy conversion efficiencies were estimated to be 50 to 56% for Tokamak reactors. At these efficiencies, the thermal-energy dissipation from fusion reactors should be approximately 30% less than that from current fossil-fuel or nuclear-fission power plants. However, design options with higher thermal efficiencies may be available for all types of power plants by the time fusion reactors are fully developed.

B. Material and Land Requirements

Fusion reactors use fuel and structural materials. Since tritium occurs naturally in negligible quantities and must be bred from lithium, the basic fuels for a deuterium-tritium fusion reactor are deuterium and lithium. An estimate[3] of the natural supply of tritium in the world is about 10 kg, which is

[1] A. P. Fraas, "Conceptual Design of the Blanket and Shield Region and Related Systems for a Full-Scale Toroidal Fusion Reactor," USAEC Report ORNL-TM-3096, Oak Ridge National Laboratory, Oak Ridge, Tennessee, 1973.

[2] P. J. Persiani, W. C. Lipinski, and A. J. Hatch, "Some Comments on the Power-Balance Parameters Q and ε as Measures of Performance for Fusion Power Reactors," USAEC Report ANL-7932, Argonne National Laboratory, Argonne, Illinois, 1972.

[3] H. Postma, "Engineering and Environmental Aspects of Fusion Power Reactors," Nuclear News 14, 57-62, April 1971.

only sufficient for the production of approximately 2.5×10^4 Mwd$_e$ by a deuterium-tritium fusion power plant.

Deuterium may be obtained from the ocean where it occurs as one part in every 6,500 parts of hydrogen. Thus, 1.029×10^{25} deuterium atoms with a total mass of 34.4 g are contained in each cubic meter of seawater. The deuterium requirements should be relatively easily met with negligible impact on the processed seawater.

The availability of lithium is the limiting feature in the deuterium-tritium fuel cycle. Lithium occurs in seawater with a mass ratio of about 1 part in 10^7, while its abundance in the crust of the earth is 20 to 32 ppm. However, only 7.4% of natural lithium is lithium-6. The remainder is lithium-7, which is less desirable but may also be used in a fusion reactor. The consumption of natural lithium in fusion reactors, optimized for the production of energy per gram of lithium, has been estimated[4] as 1 g/Mwd$_t$. Minable lithium deposits in the U.S. amount to 5 to 10×10^6 t of Li$_2$O, which correspond to a potential energy production of 2 to 4×10^{12} Mwd$_t$ (≈ 160 to 320 Q or roughly 1,000 to 2,000 times the estimated total U.S. energy consumption in the year 2000). Thus, the environmental degradation for lithium procurement should also prove to be minor.

The procurement of structural materials for fusion reactors will be more demanding than fuel procurement. The high operating temperatures, the alloys and compounds necessary for magnetic confinement, as well as the shielding materials required for high-energy neutron containment, indicate large material demands. Furthermore, fusion reactors with energy-generation

---

[4]W. Häfele and C. Starr, "A Perspective on Fusion and Fission Breeders," Journal of the British Nuclear Energy Society 13, 131-139 (1974).

densities of 1 to 10 Mw_t/m^3 of plasma will make relatively larger material demands for the same power output than fission reactors with energy densities of 100 Mw_t/m^3 of core.[5]

The maximum material requirements by element and the corresponding U.S. and world reserves are summarized in Table 24.6-1. Reference to the data of Table 24.6-1 indicates that an installed fusion-power capacity of 10^6 Mw_e may require more beryllium, chromium, manganese, nickel, niobium, tin, and vanadium than the domestic reserves contain. Table 24.6-2 shows the ore requirements for supplying the structural metals necessary for an installed fusion-power capacity of 10^6 Mw_e. If, on the average, 0.3 m^3 of land per ton of ore is disturbed by mining, the total ore requirements of 5×10^9 t may be obtained from a mining volume of 1.5×10^9 m^3 or, equivalently, 150 km^2 of land mined to a depth of 10 m. This mining area is less than 4% of the average mining area required to supply strip-mined coal for an equivalent electrical-generating capacity for one year.

Land requirements for power-generating facilities consist of plant-site and transmission-line facilities. A summary of plant-site land requirements for 1,000-Mw_e stations utilizing various fuels is shown in Table 24.6-3. Because of lowered environmental hazards for fusion reactors, these plants may be located relatively closer to the load centers.[5] In 1970, the average transmission-line land requirements in the U.S. were 27 mi^2 per 1,000 Mw_e.[6] If fusion plants may be located closer to population centers, the associated transmission-line right-of-way requirements may be reduced as much as 50%.

---

[5]G. L. Kulcinski, "Fusion Power - An Assessment of Its Potential Impact in the USA," Energy Policy 2, 104-125 (1974).

[6]Council on Environmental Quality, "Energy and the Environment: Electric Power," U.S. Government Printing Office, Washington, D.C., 1973.

Table 24.6-1 Estimated maximum material requirements for fusion reactors operating with magnetically-confined deuterium-tritium reactions; based on the data of Ref. [5].

Element	Requirement, mt/Mw_e	U.S. reserves, 10^6 mt	World reserves, 10^6 mt
aluminum	0.6	13	3,000
beryllium	0.12	0.018	0.38
boron	0.8	33	66
chromium	2	~0	370
copper	2	74	310
fluorine	1	9	62
graphite	2	10	large
helium	0.3	1.2	1.2
iron	10	8,500	180,000
lead	11	39	85
lithium	0.95	6	180
manganese	0.2	0	590
molybdenum	0.2	2.5	6.5
nickel	1.5	~0.14	24
niobium	0.8	0.005	7.8
potassium	0.02	42	30,000
tin	0.2	0.009	6
titanium	0.8	23	134
vanadium[*]	0.5	0.1	26
zirconium	0.002	0.06	25

[*]Vanadium is required if it is used as a structural substitute for niobium. In this case, the niobium requirement must be correspondingly reduced.

Table 24.6-2 Metal and ore requirements for a fusion-power capacity of 10^6 Mw_e; based on the data of Ref. [5].

Element	Metal requirements, 10^6 mt	Metal yield from ore, %	Ore requirements, 10^6 mt
aluminum	0.6	10	6
beryllium	0.12	2	6
chromium	2	5	40
copper	2	0.9	220
iron	10	45	22
lead	11	1.5	733
molybdenum	0.2	2	10
nickel	1.5	1	150
niobium	0.8	2	40
tin	0.2	10	2
vanadium*	0.5	5	10
subtotal			1,239
total**			4,956

*Vanadium is required if it is used as a structural substitute for niobium.

**A multiplication factor of four to account for overburden, tunnels, and pillars is included.

C. Chemical Pollutants

Fusion power plants will not emit chemical pollutants such as SO_2, NO_x, CO, CO_2, particulates, hydrocarbons, and aldehydes. Kulcinski[5] has estimated that if fusion power plants are installed entirely in place of fossil-fuel power plants after the year 2000, the 3.6×10^6-Mw_e generating capacity

Table 24.6-3 Estimated land requirements for power-plant sites; based on the data of Ref. [6].

Type of power plant	Land requirement, $mi^2/10^3 Mw_e$
coal*	1.5
oil	0.4
gas	0.25
fission**	0.49
fusion†	0.49

*Two hundred and twenty-five acres are required for the disposal of solid waste generated by air-pollution equipment while 10 acres are needed for the cooling towers.

**Fourteen acres are included for the cooling towers.

†The land requirement for a fusion plant has been conservatively assumed to be equal to that of a fission plant.

---

required in the U.S. by the year 2020 will be supplied by 60% fission, 29% fusion, 9% fossil fuel, and 2% hydroelectric power plants. For this scenario, Kulcinski[5] estimates that the total pollutants emitted by electrical-generation facilities in the U.S. by 2020 will be below 1973 levels.

D.   Radiological Impact

Three major concerns in assessing the radiological impact of a fusion reactor are[7] (a) the biological-hazard potential in the event of an accident, (b) the radioactivity of the effluents under normal operating conditions, and (c) the management and disposal of long-lived radioactive wastes. We shall now

[7]D. Steiner, "The Radiological Impact of Fusion," New Scientist 52, 168-171, December 16, 1971.

consider briefly each of these topics.

i.   Activity Inventories

The principal radioactive materials in a fusion reactor will be the tritium fuel and the activated blanket structure.   Activation of the structure of a deuterium-tritium reactor results from the intense neutron flux of 14.1-Mev neutrons. The neutron flux carries about 80% of the energy that is available from the fusion of deuterium and tritium.   For a 1,000-Mw$_t$ reactor, the neutron flux is about 8×10^{14} n/cm^2-sec.[2]   In a fusion reactor, there will be roughly 5 times more neutrons produced per unit of time, each with 7 times greater energy than in a fission reactor of equivalent thermal power.[4]   Preliminary fusion-reactor designs utilize niobium as the structural material with vanadium as a preferred alternate for minimizing radioactive-waste disposal.   Although vanadium structures are advantageous from the viewpoint of radioactive-waste disposal and nuclear afterheat (see Section 24.6E), vanadium alloys have strength- and oxygen-corrosion limitations when used as structural materials in a high-temperature ($\geqslant 900°$K) environment.[8]   For this reason, the neutron-activation chains of niobium and vanadium are of particular interest (see Figs. 24.6-1 and 24.6-2).

The tritium inventory will include contributions from the blanket tritium-recovery system, the plasma tritium-recovery system, the tritium-storage

[8]D. J. Dudziak and R. A. Krakowski, "A Comparative Analysis of D-T Reactor Radioactivity and Afterheat," Proceedings of the First Topical Meeting on the Technology of Controlled Nuclear Fusion, San Diego, California, 1974; published by the U.S. Government Printing Office, Washington, D.C., 1974. See, also, D. J. Dudziak and R. A. Krakowski, "Radioactivity Induced in a Theta-Pinch Fusion Reactor," Nuclear Technology 25, 32-55 (1975).

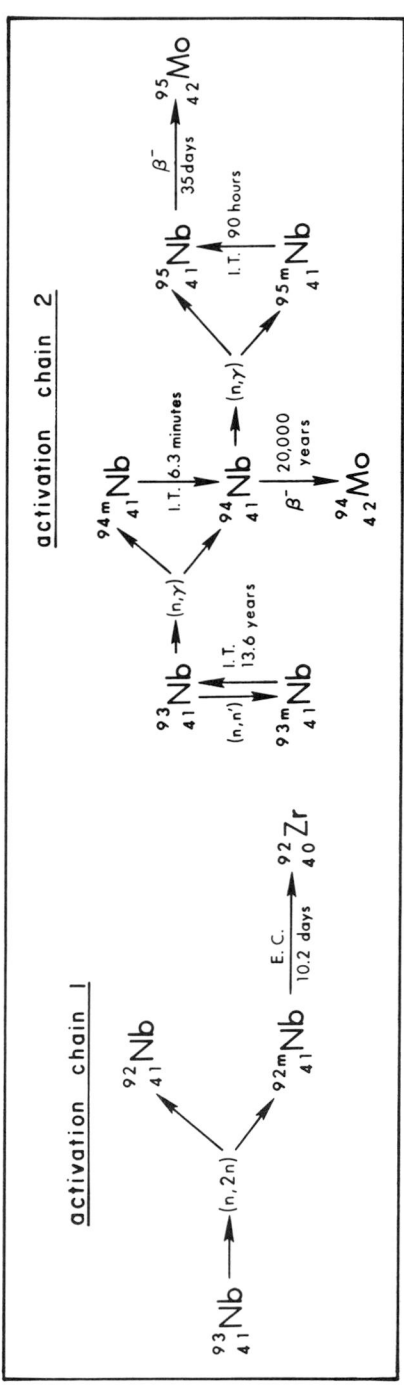

Fig. 24.6-1 Niobium activation chains. Decay by electron capture is indicated by E.C. while decay by isomeric transition is identified by I.T. The symbol m identifies metastable atomic states; based on the data of Ref. [9].

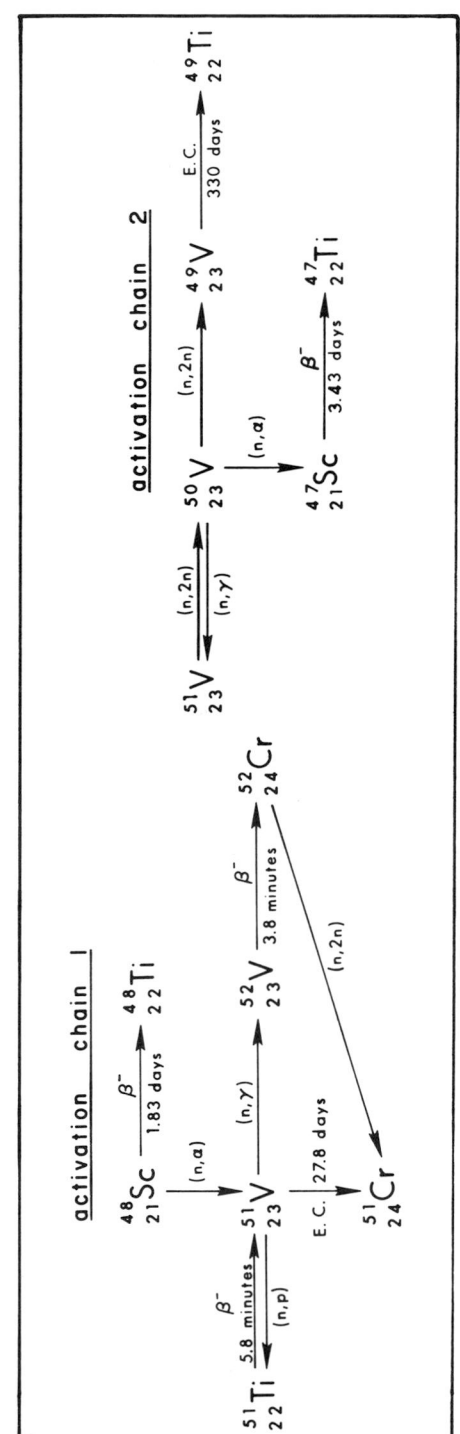

Fig. 24.6-2 Vanadium activation chains. Decay by electron capture is indicated by E.C.; based on the data of Ref. [9].

and processing systems, and the fuel-delivery system.[9] Estimates for the total tritium inventory of a fusion-reactor site vary from 1.2 to 6 g/Mw$_t$.[4,7,9] The specific activity of tritium is 9.6×10^3 Ci/g[2] and the half-life is 12.3 years. The maximum allowable concentrations of waterborne and airborne tritium are 3×10^{-3} µCi/cm^3 and 2×10^{-7} µCi/cm^3, respectively.[10] Tritium is normally considered to be a low-hazard isotope because of its low beta energy ($E_{average}$ = 5.7 kev) and its relatively short half residence time in man of 8 to 10 days.[10] Anspaugh et al[11] have calculated that men living in an environment containing a tritium concentration of 10^3 pCi/m^3 in the air would acquire an annual dose of 3.7 mrem/y. The data of Anspaugh et al[11] indicate that approximately 40% of the total dose to man from tritium is acquired by inhalation and skin absorption while the remaining 60% comes from ingestion.

The radioactive inventory of a fusion reactor as a function of time after shutdown is given in Table 24.6-4. A comparison of the total induced activities of Tokamak and theta-pinch fusion reactors with fission reactor plants as functions of time after shutdown is shown in Fig. 24.6-3. The biological-hazard potential of radioactive releases is defined as the activity divided by the maximum permissible airborne concentration (MPC). Biological-hazard potentials

---

[9] D. Steiner and A. P. Fraas, "Preliminary Observations on the Radiological Implications of Fusion Power," Nuclear Safety 13, 353-362 (1972).

[10] J. E. Draley and S. Greenberg, "Some Features of the Environmental Impact of a Fusion-Reactor Power Plant," paper presented at the Symposium on the Technology of Controlled Thermonuclear Fusion Experiments and the Engineering Aspects of Fusion Reactors, Austin, Texas, 1972.

[11] L. R. Anspaugh, J. J. Koranda, W. L. Robinson, and J. R. Martin, "Dose to Man Via Food-Chain Transfer Resulting from Exposure to Tritiated Water Vapor," USAEC Report UCRL-73195, Lawrence Livermore Laboratory, Livermore, California, 1971.

Table 24.6-4 Radioactive inventory of a fusion reactor after shutdown; reproduced with modifications from Ref. [4].

Element or isotope	Activity, Ci/w_t				
	time after shutdown				
	10^3 seconds (~20 minutes)	10^7 seconds (~4 months)	10^8 seconds (~3 years)	10^{10} seconds (~300 years)	10^{11} seconds (~3,000 years)
tritium					
^{3}H	2×10^{-2}	2×10^{-2}	1.7×10^{-2}	-	-
niobium*					
^{94m}Nb	0.17	-	-	-	-
^{95}Nb	1.1	0.11	-	-	-
^{95m}Nb	1.1	-	-	-	-
^{92m}Nb	0.3	-	-	-	-
^{93m}Nb	0.4	0.4	0.33	-	-
^{94}Nb	7×10^{-4}	7×10^{-4}	7×10^{-4}	7×10^{-4}	$\approx 7 \times 10^{-4}$
niobium total	3.07	0.51	0.33	7×10^{-4}	$\approx 7 \times 10^{-4}$

*The niobium activities refer to an operational lifetime of 20 years. The values for an operational lifetime of 10 years are lower by insignificant factors, except for ^{94}Nb which would contribute only about one-half of the indicated activity. The symbol m identifies metastable atomic states.

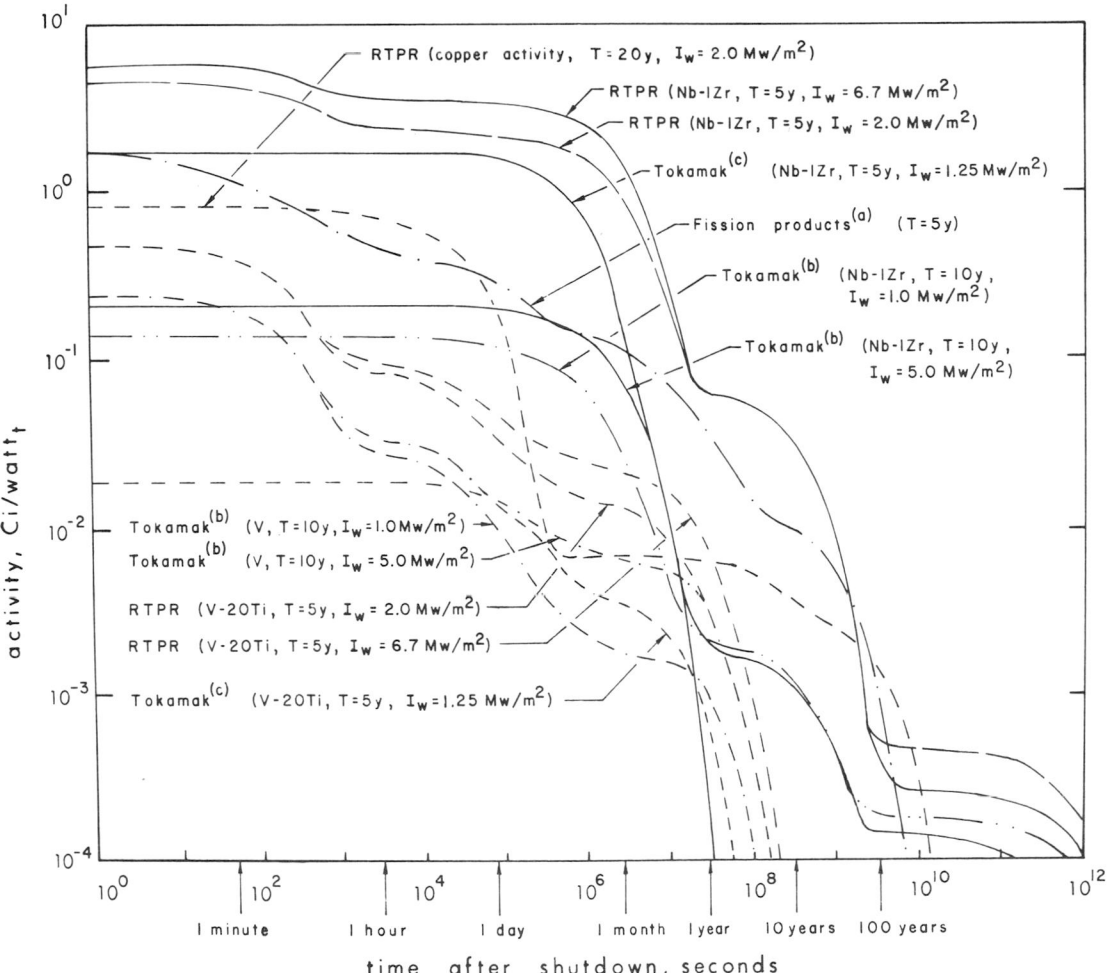

Fig. 24.6-3 Total induced activities of Tokamak and theta-pinch reactors as functions of time after shutdown; based on the data of Ref. [8]. Abbreviations used: T for operating time, I_w for 14-Mev neutron wall loading, RTPR for reference theta-pinch reactor. Additional sources: (a) W. B. Cottrell and A. W. Savolainen, "U. S. Reactor Containment Technology," USAEC Report ORNL-NSIC-5, Oak Ridge National Laboratory, Oak Ridge, Tennessee, 1965; (b) D. Steiner, "The Nuclear Performance of Vanadium as a Structural Material in Fusion Reactor Blankets," USAEC Report ORNL-TM-4353, Oak Ridge National Laboratory, Oak Ridge,

Tennessee, 1973; (c) W. F. Vogelsang, G. L. Kulcinski,
R. G. Lott, and T. Y. Sung, "Transmutation Effects in CTR
Blankets," Transactions of the American Nuclear Society 17,
138-139 (1973) and B. Badger et al, "UWMAK-I, A Wisconsin Toroidal Fusion Reactor Design," USAEC Report
UWFDM-68, University of Wisconsin, Madison, Wisconsin,
1973.

of fusion (1,000-Mw$_t$ Tokamak) and advanced fission reactors are given in Table
24.6-5.   A more detailed analysis, including four alternative structural materials, is given in Table 24.6-6.   The relative biological-hazard potentials of
induced activities of Tokamak and theta-pinch fusion reactors as functions of
time after shutdown are compared with the activities of fission reactors in
Fig. 24.6-4.   An interesting comparison may be drawn between fusion and fission reactors by reference to Tables 24.6-5 and 24.6-6 and Fig. 24.6-4.   Fusion-reactor systems have biological-hazard potential values of one to four
orders of magnitude lower than fission reactors.   Furthermore, the most important radioactive isotopes produced by fission reactors are volatile elements,
while the majority of the radioactive isotopes produced by fusion reactors are
nonvolatile metallic elements.   We refer to the literature for more complete
discussions of this topic.[4, 12, 13]

ii.   Radioactive Effluents

Tritium represents the major potential source of radioactive effluents

[12] F. von Hippell, "Discussion: A Perspective on Fusion and Fission Breeders by W. Häfele and C. Starr," Journal of the British Nuclear Energy Society 14, 119-120 (1975).

[13] J. P. Holdren, "Discussion: A Perspective on Fusion and Fission Breeders by W. Häfele and C. Starr," Journal of the British Nuclear Energy Society 14, 120-122 (1975).

Table 24.6-5 Principal radioactive inventories of fusion and advanced fission reactors; based on the data of Ref. [9].

Inventory	Activity, Ci/kw_t	Maximum permissible airborne concentration, $\mu\ Ci/cm^3$	Biological-hazard potential, km^3 of air/kw_t
fusion reactors			
tritium			
^3H	60*	2×10^{-7}	0.30
niobium structure			
^{95}Nb	155	3×10^{-9}	52
total	714		240
vanadium structure			
^{48}Sc	4.2	5×10^{-9}	0.84
total	55.1		0.86**
fission reactors			
^{131}I, inhalation[‡]	31.6	1×10^{-10}	330
^{131}I, ingestion[†]	31.6	1.4×10^{-13}	230,000
^{239}Pu[††]	0.06	6×10^{-14}	1,000
all plutonium isotopes	18.6		8,300

*The specific activity of tritium is approximately 10^4 Ci/g.

**Impurities within the vanadium may increase this value by a factor of two.

‡See, for example, U.S. Atomic Energy Commission, "USAEC Rules and Regulations," Title 10, Atomic Energy Supplement No. 37, December 10, 1969.

†Exposure through the food-chain (primarily milk) is assumed; see, for example, T. J. Burnett, "Derivation of the 'Factor of 700' for ^{131}I," Health Physics 18, 73-75 (1970); see, also, the discussion given by Häfele and Starr in Ref. [4].

††The plutonium inventory refers to a liquid-metal fast-breeder reactor.

Table 24.6-6 Principal radioactive inventories of fusion and advanced fission reactors; reproduced from Ref. [5].

System or material	Isotope	Half-life	Activity, Ci/kw_t	Maximum permissible concentration, $\mu Ci/cm^3$	Biological-hazard potential, km^3 of air/kw_t
fusion reactors					
	^3H	12.3y	60	2×10^{-7}	0.30
	^{49}V	331d	0.67	1×10^{-10}	6.7
structural materials* 316 stainless steel					
	^{55}Fe	2.94y	140	3×10^{-8}	4.6
	^{58}Co	72d	29	2×10^{-9}	14.5
	^{57}Ni	36h	1.1	1×10^{-10}	11
	^{54}Mn	310d	24	1×10^{-9}	24
	^{60}Co	5.25y	4.7	3×10^{-10}	15.6
total					~77
Nb-1Zr (niobium alloy containing 1% zirconium)	^{92m}Nb**	10.1d	152	1×10^{-10}	1,520
	^{95m}Nb	3.75d	50	1×10^{-10}	500
	^{95}Nb	35d	43	3×10^{-9}	14
	^{89}Sr	51d	38	3×10^{-10}	126
total					~2,200
V-20Ti (vanadium alloy containing 20% titanium)	^{48}Sc	1.81d	12.1	5×10^{-9}	2.5
	^{45}Ca	165d	2.6	1×10^{-9}	2.6
	^{46}Sc	84d	1.87	8×10^{-10}	2.3
total					7.4
Al	^{24}Na	1.5h	630	4×10^{-8}	15.8
	^{26}Al	7.5×10^5y	0.004	1×10^{-10}	0.04
total					~15.8
fission reactors					
	^{131}I	8.04d	31.6	1×10^{-10}	330
	^{239}Pu	24,100y	0.06	6×10^{-14}	1,000
	all Pu isotopes	-	18.2	-	8,300
	^{90}Sr	25y	0.64	3×10^{-11}	21
	^{137}Cs	33y	0.94	5×10^{-10}	2
	all other fission products	>1d	-	-	18,000

*Refers to isotopes produced in the first wall and blanket region of a fusion reactor except in the case of stainless steel, which refers to the first wall only. A 10-year operating life is assumed.

**The symbol m identifies metastable atomic states.

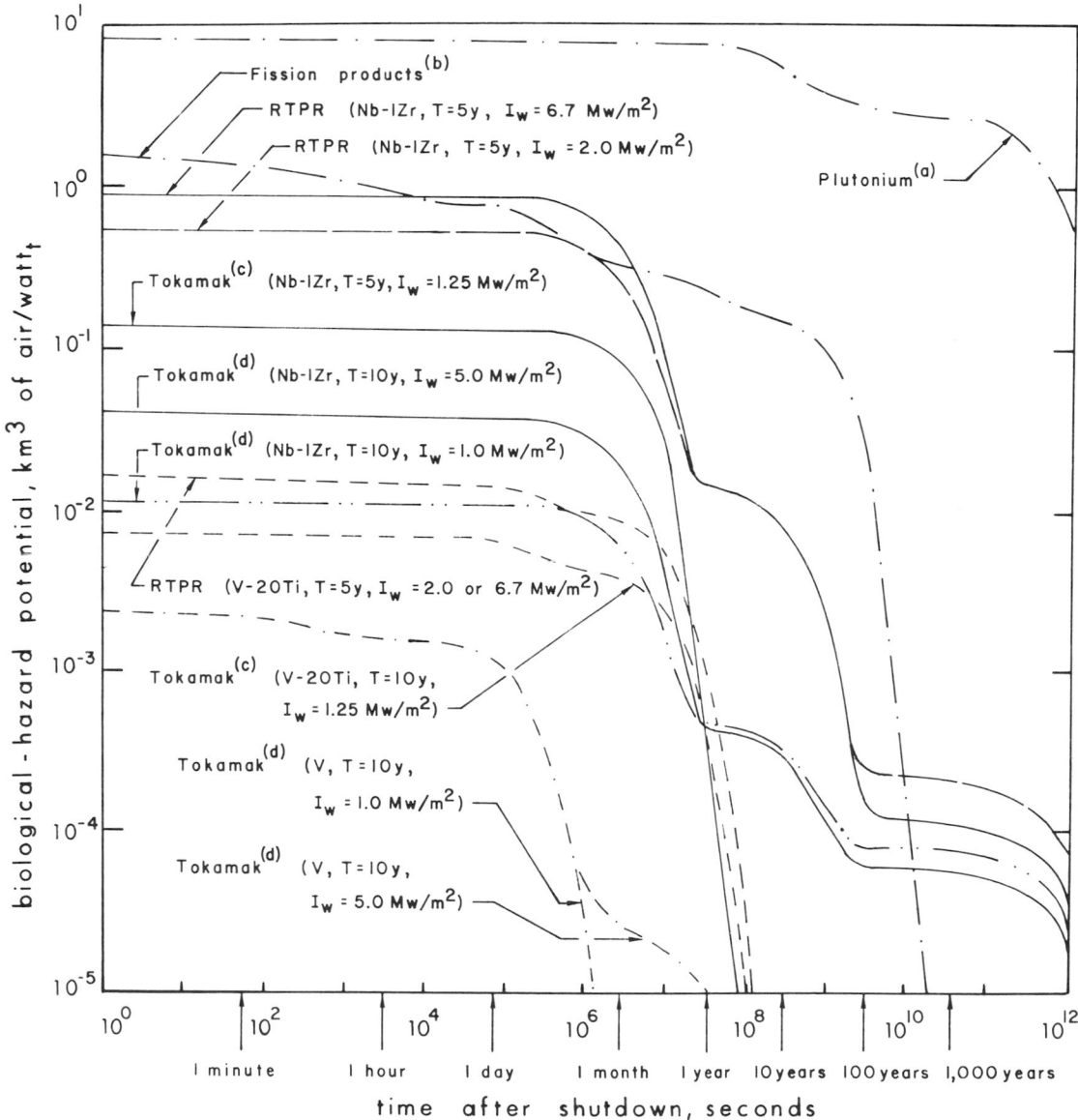

Fig. 24.6-4 Relative biological-hazard potentials of induced activities of Tokamak and theta-pinch fusion reactors and of fission reactors as
functions of time after shutdown; based on the data of Ref. [8].
Abbreviations used: T for operating time, I_W for 14-Mev neutron

wall loading, RTPR for reference theta-pinch reactor. Additional sources: (a) "Final Report, Project Definition Phase," 4th Round Demonstration Plant Program, Westinghouse Electric Corporation, Pittsburgh, Pennsylvania, 1970; (b) W. B. Cottrell and A. W. Savolainen, "U.S. Reactor Containment Technology," USAEC Report ORNL-NSIC-5, Oak Ridge National Laboratory, Oak Ridge, Tennessee, 1965; (c) W. F. Vogelsang, G. L. Kulcinski, R. G. Lott, and T. Y. Sung, "Transmutation Effects in CTR Blankets," Transactions of the American Nuclear Society 17, 138-139 (1973) and B. Badger et al, "UWMAK-I, A Wisconsin Toroidal Fusion Reactor Design," USAEC Report UWFDM-68, University of Wisconsin, Madison, Wisconsin, 1973; (d) D. Steiner, "The Nuclear Performance of Vanadium as a Structural Material in Fusion Reactor Blankets," USAEC Report ORNL-TM-4353, Oak Ridge National Laboratory, Oak Ridge, Tennessee, 1973.

from a fusion reactor because tritium is not only volatile but also tends to diffuse through high-temperature metal walls. The tritium fuel will be bred from lithium with breeding ratios from 1.1 to 2.[14] Draley and Greenberg[10] have made an extensive study of tritium release during the normal operation of a Tokamak fusion reactor using potassium to extract thermal energy from the lithium blanket. The potassium system includes potassium turbines, heat exchangers, and cold traps. A detailed analysis[10] of tritium losses from a 1,000-Mw$_t$ fusion reactor operating with a thermal-to-electrical energy conversion efficiency of 47% indicates that 2.6 Ci/d of tritium will be discharged. Approximately 2 Ci/d of tritium will be released to the air while the remaining 0.6 Ci/d will be discharged in the cooling water. These radiation releases are based on a total on-site tritium inventory of 2 kg and may be compared to releases of 11 Ci/d for current pressurized-water reactors and 0.13 Ci/d for

[14]R. F. Post and F. L. Ribe, "Fusion Reactors as Future Energy Sources," Science 186, 397-407 (1974).

boiling-water reactors, both augmented by approximately 34 Ci/d at fission-fuel reprocessing plants.[10] The radioactive effluents from fusion plants will originate only from the power-plant site since no external fuel cycle will be required as in the case of fission reactors.

The release of 2.6 Ci/d of tritium leads to a calculated[10] maximum individual off-site dose of about 0.17 mrem/y. This dose may be compared with the natural background dose of about 130 mrem/y and the standard for light-water fission reactors of 5 mrem/y. Tritium release may be reduced by a factor of 1,000 or more by second-generation deuterium-deuterium fusion reactors.[10]

Steiner and Fraas[9] have given less optimistic estimates of tritium leakage rates. The tritium recovery, storage, and delivery systems are estimated to contain 5.6 kg of the total 6-kg tritium inventory and are expected to be responsible for only negligible tritium releases. The remaining tritium inventory of 0.4 kg may leak from the lithium and potassium systems into the air and from the potassium system into the steam system. Steiner and Fraas[9] estimate that the leakage rate from the lithium and potassium systems to the atmosphere may be held to 0.0001% per day, which would result in a tritium leakage rate of 4 Ci/d. An additional tritium leakage rate of 3 Ci/d from the steam system has been estimated,[9] which would be discharged from the plant in the condenser cooling water.

iii. Radioactive-Waste Disposal

The primary source of radioactive waste from fusion reactors will be the activated structural material of the blanket since structural materials will have a reduced lifetime because of radiation damage. The magnitude of the waste-disposal problem is directly related to the choice of structural materials. A comparison of long-lived (i.e., ≥ 10 year half-life) activities produced in niobium and vanadium blanket structures of a 1,000-Mw$_t$ Tokamak fusion reactor with the major long-lived activities associated with spent fuel of advanced fission reactors is shown in Table 24.6-7. The data of Table 24.6-7 indicate that the biological-hazard potential associated with the accumulated radioactive waste from niobium is about 1.5% of that associated with spent fuel from an advanced fission reactor. The magnitude of the niobium related waste is, however, not insignificant. On the other hand, the use of vanadium as the blanket structural material reduces the biological-hazard potential of radioactive wastes to a negligible value.

A summary of the radioactive wastes from fusion and fission reactors after 100 years of decay is shown in Table 24.6-8. Comparison of the data in Tables 24.6-6 and 24.6-8 indicates that 100 years after the generation of spent fission fuel the biological-hazard potential has decreased by one order of magnitude. In the same time interval, the biological-hazard potentials of fusion wastes have declined by about two to six orders of magnitude.

E.   Nuclear Afterheat

Closely related to the hazard potential of radioactive inventories is the energy-release rate from the decay of radioactive material (i.e., the afterheat) which may lead to subsequent vaporization and dispersal. The activated blanket structure is the only significant source of afterheat in a fusion reactor.[15] Afterheat considerations would be important in the event of a loss-of-coolant accident.[14] Figure 24.6-5 shows the afterheat as a fraction of the

_____

[15] D. Steiner, "The Neutron-Induced Activity and Decay Power of the Niobium Structure of a D-T Fusion Reactor Blanket," USAEC Report ORNL-TM-3094, Oak Ridge National Laboratory, Oak Ridge, Tennessee, 1970.

Table 24.6-7 Long-lived activities in the blanket structure of fusion reactors and in the spent fuel of advanced fission reactors; based on the data of Ref. [9].

System or structural material and isotope	Mean life, years	Activity generation rate, Ci/kwy_t	Accumulated activity after 1,000 years* of operation, Ci/kw_t	Maximum permissible concentration in water,** $\mu Ci/cm^3$	Biological-hazard potential after 1,000 years of operation, km^3 of water/kw_t
niobium structure					
^{93m}Nb	19.6	8.8	173	4×10^{-4}	4×10^{-4}
^{94}Nb	2.9×10^4	2.9×10^{-3}	2.9	3×10^{-6} †	1×10^{-3}
vanadium structure††					1.4 to 14×10^{-7}
fission: spent fuel					
^{90}Sr	40.4	0.67	27	3×10^{-7}	9×10^{-2}
^{137}Cs	43.4	0.93	40	2×10^{-5}	2×10^{-3}
^{99}Tc	3×10^5	1.2×10^{-4}	0.12	2×10^{-4}	6×10^{-7}
^{238}Pu	128	9×10^{-4}	0.12	5×10^{-5}	2.4×10^{-5}

* The accumulated-hazard potential will approach a steady-state value in 1,000 years.

** The maximum permissible concentration in water is appropriate for underground waste disposal. The values stated are for continuous exposure, as specified by the U.S. Atomic Energy Commission, "USAEC Rules and Regulations," Title 10, Atomic Energy Supplement No. 37, December 10, 1969.

† This value is recommended for a radionuclide which is not listed with a decay mode other than alpha emission or spontaneous fission and with a half-life greater than 2 hours; see the U.S. Atomic Energy Commission, "USAEC Rules and Regulations," Title 10, Atomic Energy Supplement No. 37, December 10, 1969.

†† A niobium impurity present at an atomic concentration of 100 to 1,000 ppm is assumed.

Table 24.6-8 Radioactive wastes from fusion and fission reactors after 100 years of decay; reproduced from Ref. [5].

System or structural material	Isotope	Half-life years,	Activity, Ci/kw_t	Maximum possible concentration, $\mu\ Ci/cm^3$	Biological-hazard potential, km^3 of air/kw_t
fusion reactor					
316 stainless steel	^{60}Co	5.25	8.7×10^{-6}	3×10^{-10}	3×10^{-5}
	^{63}Ni	85	1.3×10^{-4}	2×10^{-9}	6.6×10^{-5}
Nb-1Zr	^{94}Nb	5×10^4	8×10^{-3}	1×10^{-10}	8×10^{-1}
	^{93}Zr	5×10^6	6×10^{-3}	4×10^{-9}	1.5×10^{-3}
V-20Ti	-	-	-	-	negligible
Al	^{26}Al	7.5×10^4	4×10^{-3}	1×10^{-10}	4×10^{-2}
fission reactor	^{239}Pu	2.4×10^4	6×10^{-2}	6×10^{-14}	1×10^3
	^{90}Sr	25	4×10^{-2}	3×10^{-11}	1.3
	^{137}Cs	33	0.115	5×10^{-10}	0.23

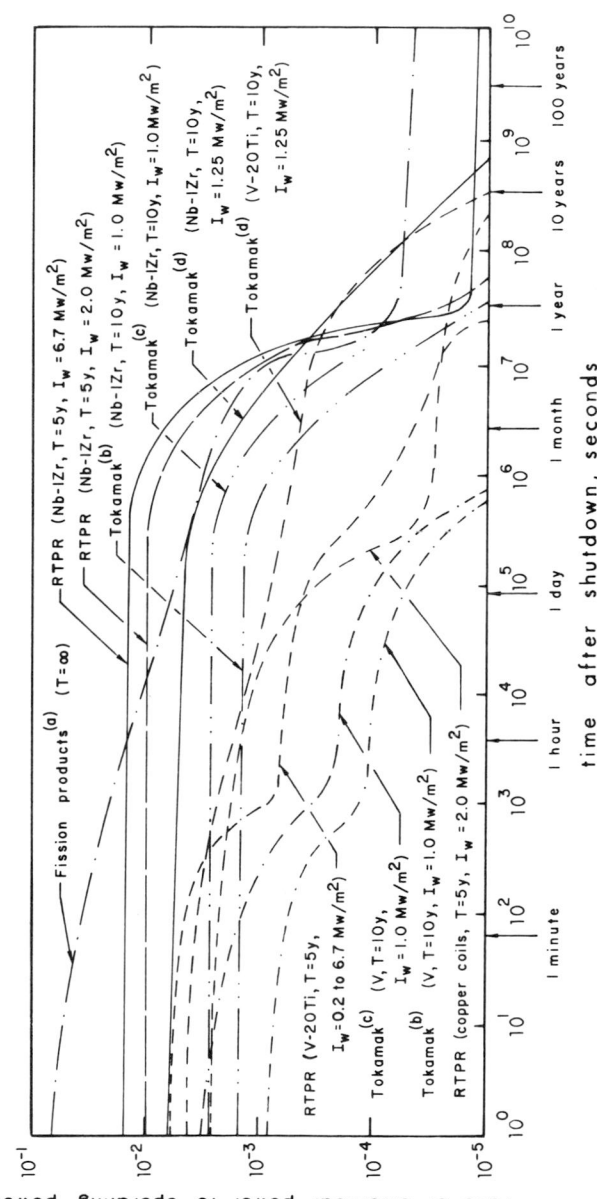

Fig. 24.6-5   Relative afterheat powers of Tokamak and theta-pinch fusion reactors and of fission reactors as functions of time after shutdown; based on the data of Ref. [8]. Abbreviations used: T for operating time, I_w for 14-Mev neutron wall loading, RTPR for reference theta-pinch reactor. Additional sources: (a) K. Shure and D. J. Dudziak, "Calculating Energy Released by Fission Products," Transactions of the American Nuclear Society 4, 30 (1961); (b) D. Steiner, "The Nuclear Performance of Vanadium as a Structural Material in Fusion Reactor Blankets," USAEC Report ORNL-TM-4353, Oak Ridge National Laboratory, Oak Ridge, Tennessee, 1973; (c) Ref. [9]; (d) W. F. Vogelsang, G. L. Kulcinski, R. G. Lott, and T. Y. Sung, "Transmutation Effects in CTR Blankets," Transactions of the American Nuclear Society 17, 138-139 (1973) and B. Badger et al, "UWMAK-I, A Wisconsin Toroidal Fusion Reactor Design," USAEC Report UWFDM-68, University of Wisconsin, Madison, Wisconsin, 1973.

reactor thermal power as a function of time after shutdown for fusion and fission reactors. Reference to Fig. 24.6-5 shows that, for short times, the
afterheat values for fusion reactors with niobium structures and fission reactors differ by about an order of magnitude or more. The corresponding values
for reactors with vanadium structures are smaller by one or two orders of
magnitude. Detailed analyses by Conn, Sung, and Abdou[16] of five potential
fusion reactor designs have shown that afterheat values vary from 0.5 to 5%
of the normal thermal operating power, depending on design and choice of the
structural material. However, the afterheat power densities of the five designs
have a maximum of only about $0.2 \, w/cm^3$ at shutdown, which is considerably
lower than the corresponding power densities of light-water or breeder fission
reactors.[16]

The afterheat power densities associated with the niobium structure of
a 1,000-Mw_t Tokamak reactor and of the fuels of pressurized-water and liquid-
metal fast-breeder reactors are compared in Table 24.6-9. Since the specific
heats of niobium and fission fuels are comparable, the data of Table 24.6-9
imply that the rise rate of the niobium temperature will be substantially less
than the rise rate of the fission-fuel temperature following a loss-of-coolant
accident.[9] For example, the adiabatic temperature rise at shutdown, based
on average afterheat power densities, would be ~0.03° F/sec for vanadium,
~0.1° F/sec for niobium, ~10° F/sec for pressurized-water reactor fuel, and
~50° F/sec for liquid-metal fast-breeder reactor fuel.[9] The data of Fig.
24.6-5 and Table 24.6-9 suggest that the problem of afterheat removal will be

---

[16]R. W. Conn, T. Y. Sung, and M. A. Abdou, "Comparative Study of Radioactivity and Afterheat in Several Fusion Reactor Blanket Designs," Nuclear
Technology 26, 391-399 (1975).

Table 24.6-9 Afterheat power densities of fusion and fission reactors; reproduced from Ref. [9].

Power density	Reactor type		
	fusion[*]	pressurized-water	liquid-metal fast-breeder
average afterheat power density at shutdown, w/cm^3	0.15	19[**]	105[**]
ratio of maximum to average power density	~2	~2	~2

[*]A niobium structure is assumed.

[**]The power densities given here are about one order of magnitude higher than those given by Conn, Sung, and Abdou in Ref. [16].

less significant for niobium fusion reactors and negligible for vanadium fusion reactors as compared to fission reactors.

F. Operational Safety

Serious accidents during the operation of fusion reactors will not result in large-scale uncontrolled nuclear reactions because of the limited amount of fuel present in the reactor (e.g., ~2 g for a 1,000-Mw$_t$ reactor). Furthermore, a fusion reactor will have little radioactive fuel or fuel-related material available for release in the event of a serious accident. Draley and Greenberg[10] have outlined the effects of several possible accidents during the operation of a fusion reactor.

Two accidents may occur in the operation of an electromagnetic-containment system. If the magnetic field increased accidentally, the plasma

would be compressed and the reaction rate would increase, which might severely damage the reactor containment vessel and increase the blanket temperature.[10] For example, adiabatic addition of one-half of the energy contained
in 2 g of fuel has been estimated[10] to increase the total blanket temperature
by 100°C. If a loss of refrigerant caused the superconducting magnets to go
normal, a large quantity of stored electrical energy ($\sim 3.3 \times 10^8$ joules) might
be released suddenly.[10] However, proper design may limit the rate of
energy release to an acceptable value.[10]

Increases in the fuel-injection rate would lead to increased reaction
rates and may trigger plasma instabilities which would serve to alleviate any
serious effects of excess energy production.[10] Any changes in fuel composition would lead to reduced power-production rates.

The rupture of a potassium-water heat exchanger would create a significant hazard. Massive injection of water or steam into molten or gaseous
potassium would result in rapid reactions producing hydrogen and solid cor-
rosive materials. Serious hydrogen explosions and/or fires might cause a
loss of system security and the subsequent release of the tritium contained in
the blanket structure.

A loss-of-coolant accident is probably not a serious accident for fusion
reactors since about 12 minutes of full-power operation of a 1,000-Mw_t reactor would only burn about 1 g of fuel.[10] The afterheat at shutdown would
be[9] only 0.25% and 0.07% of the thermal operating power for niobium and
vanadium Tokamak reactors with 14-Mev neutron wall loadings of 1 Mw/m^2,
respectively (compare Fig. 24.6-5).

The maximum credible accident is the sudden exposure to air of all
the lithium, potassium, and tritium in the operating reactor. Liquid-metal
fires are assumed to occur and burn for approximately three days. The

tritium release would be 760 g or 7.3×10^6 Ci for a $1,000$-Mw$_t$ reactor.[10] For unfavorable meteorological conditions, a dose of 15 rem would be received at a distance of 600 meters.[10]

G. Fuel Diversion

Since the fuel cycle of a fusion power plant will not extend beyond the boundaries of the reactor site, the availability of fusion fuel for diversion is relatively less than for fission fuels. Furthermore, the diversion of fusion fuels for weapon manufacture will also require diversion of fission fuel to trigger the device unless a fission-free thermonuclear weapon is developed in the future. Häfele and Starr[4] suggest that fission-free thermonuclear weapons may be a reality by the time fusion reactors are widespread; thus, the fusion-fuel cycle may require safeguards similar to those for fission-fuel cycles. However, von Hippell,[12] Holdren,[13] and Post and Ribe[14] conclude that the diversion of fusion fuels should not pose a significant safety problem.

H. Epilogue

Fusion reactors may operate with slightly higher thermal-to-electrical conversion efficiencies than future generation fossil-fuel or nuclear-fission power plants. Structural material requirements for fusion reactors will be larger than for either fission or fossil-fuel plants. Fuel and land requirements for fusion plants will be considerably less demanding than for future fossil-fuel (i.e., coal) power plants. Chemical pollutants will not be emitted from fusion power plants; thus, replacement of fossil-fuel plants by fusion plants will reduce chemical emissions from stationary power-generation facilities.

In terms of biological-hazard potentials, fusion reactors appear to offer important advantages as compared to fission reactors. Furthermore, the most dangerous radioactive isotopes produced by fission reactors are volatile elements, while the majority of the radioactive isotopes produced by fusion reactors are nonvolatile. Maintenance of acceptable tritium release rates during the normal operation of a fusion reactor will require sophisticated engineering designs comparable to those of fission reactors. Disposal of long-lived radioisotopes produced in fusion reactors employing niobium as the structural material should not be as difficult as for fission reactors. However, the magnitude of long-lived waste produced in the niobium-blanket structure is not insignificant and disposal schemes similar to those employed for fission products may be required. The use of vanadium as the structural material offers considerable advantages, one of which may be ease in recycling of the blanket structure. Afterheat values for fusion reactors with niobium-blanket structures and for fission reactors differ by less than one order of magnitude for short times after shutdown. Again, vanadium structures offer important advantages in afterheat production and dissipation.

Safe operation of fusion reactors will require safety of engineering and human responsibility comparable to that for fission reactors. However, fusion reactors appear to have a lower accidental damage potential than fission reactors because of the small amount of fuel present in a fusion reactor at any given time and because of the tendency of containment failures to extinguish the fusion reaction relatively more rapidly. Diversion of fusion fuels should not pose the significant safety problem that is associated with fission reactors.

Finally, a preliminary assessment of environmental and safety aspects of fusion power indicates that the major problems will be largely determined by sophistication of the final engineering designs rather than by the laws of physics, which severely limit the design flexibility of fission reactors.

24.7 Licensing of Nuclear Reactors

The practice of nuclear engineering properly includes procedural and legal matters, as well as the conception, design, calculation and prediction of performance and cost. Engineers prescribe the basic safety features and prepare the documentation that demonstrates their efficacy. Using this documentation, they then participate in governmental procedures that are required to obtain licenses for construction and operation of nuclear reactors, as well as for reprocessing, enrichment, mining, refining, and use of other facilities involved in the fuel cycle (see Fig. 21.18-1).

Between 1954 and the end of 1974, the U.S. Atomic Energy Commission regulated and financed most of the developments, operations and facilities connected with nuclear energy. The Atomic Energy Act of 1954 prescribed the manner of regulation. The Nuclear Regulatory Commission (NRC) was established during 1975 and has assumed all regulatory functions. The nuclear energy research and development function has been made one of five kinds of energy development charged by Congress to the Energy Research and Development Agency (ERDA). This separation of federal function, authority and budget is undoubtedly desirable.

The NRC now incorporates a division of the old AEC called the Division of Licensing and Regulation (DOL). A sketch of procedures used for regulation will indicate the continuing form if not the precise terminology of future nuclear regulatory functions. In addition to DOL, an independent Advisory Committee on Reactor Safeguards (ACRS) advises the commission on policy, performance and novel aspects.

Federal regulatory practice is defined in detail by the Code of Federal Regulations (CFR 10), which belongs to a set of public documents in which each agency has a chapter. There is a standard practice for formulating, publishing, public review, and enforcement of all items in CFR. In addition, guides are formulated, publicly reviewed and issued to instruct all participants on the what, how, when, who, and why of meeting the requirements associated with licensing and regulation. Two guides apply directly to safety and environmental matters.

Documents, standards, procedures and organizations are changed and updated, as is deemed advisable. The NRC is evolving faster implementation procedures. Safety and public access are being stressed. These last two objectives are, of course, in opposition.

Some flavor of the standards and requirements for licensing may be had from general statements in 10CFR50, a part of which is devoted to the general terms of the standards. The standards require reasonable assurance that the facility (for which a license is sought) will comply with all 10 of the CFR regulations; that the applicant is technically and financially qualified to operate the proposed facility; that issuance of a license will not be inimical to the common defense and security or the health and safety of the public. "The description of the process should be sufficiently detailed to permit evaluation of the radioactive hazards involved."

Two very detailed safety analyses and two environmental analyses are required prior to licensing. The owner of the facility proposes a Preliminary Safety Analysis Report (PSAR) with the help of the (reactor) supplier, an architect-engineer and consultants. The PSAR is followed by a final safety analysis report (FSAR). The PSAR costs \$1 to \$3 $\times 10^6$. They are reviewed by DOL and by a board appointed to conduct a public hearing. Requests for licenses have been carried to courts of appeal.

In addition, an Environmental Impact Statement (EIS) and an Environmental Impact Report (EIR) are prepared by the owner and the NRC, respectively. Several national and local government agencies review these documents.

Court review and decision occurs on environmental matters only if some party
instigates a legal action.

24.8 Nuclear Safeguards: Diversion of Nuclear Materials, Sabotage and Subversion

Successful contravention of safeguarding measures depends on the technical sophistication, competence and determination of inimical parties. Willrich and Taylor have published analyses[1] of possible ways for preventing diversion and use of fissile materials in the making of bombs. Analyses have also been made of the potentials of sabotage to cause reactor accidents and to effect scattering of toxic materials (particularly Pu and Sr) from reprocessing plants and storage sites. Several recently published works of fiction are based on dramatic acts involving nuclear fission reactors. How safe is any highly organized activity in the face of increasing terrorist activities? A general conclusion is the following: power plants and reprocessing facilities could be shut down and the public badly frightened by sabotage but, unless nuclear bombs are produced or extensive paralysis of a major electrical network is accomplished, the terrorist activity will not be especially damaging. We shall now inspect some technological aspects of the complex human problem of nuclear safeguards.

A. Amounts of Fissile Material Required for Bombs

Taylor and his colleague[1] conclude that, given the opportunity to divert ^{235}U, ^{239}Pu or ^{233}U from the fission-fuel cycle, reasonably competent

---

[1]Mason Willrich and Theodore B. Taylor, Nuclear Theft: Risks and Safeguards, Ballinger Publishing Co., Cambridge, Mass. 1974.

terrorists could make bombs. The approximate amounts that must be diverted in order to make one unsophisticated device will vary with the isotopic composition and chemical form of the material available and to be used and are indicated by data listed in Table 24.8-1.

B.   Points of Diversion

A number of conspicuous points of diversion exist.

i.   The feed to the fuel-fabrication plant may be sensitive to sabotage. Slightly enriched (2 to 4%) uranium, as fed from an isotope-enrichment plant to the LWR fuel cycle, cannot be made into a bomb. Highly enriched ^{235}U for the HTGR is usable for bombs and this material must therefore be regulated. Enrichment and fuel preparation are likely to occur at separate locations. Plutonium, recycled and fed to LWR or FBR fuel-element fabrication, must be regulated. The recycling and fuel-element fabrication could be accomplished within a single enclosed facility.

ii.   Shipments of fuel elements to and from reactors may serve as entries for subversion. As was noted above, fresh LWR fuel would be useful only to terrorists with access to enrichment plants. The HTGR fuel elements may be processed to ^{235}U but the required bulk would be large, as may be seen from Table 24.8-2. The FBR elements and irradiated LWR or FBR elements contain Pu that may be separated by relatively simple chemical processes which could be performed by small, organized, clandestine groups. Here, it is surmised that the radioactivity of irradiated fuel might not deter determined criminals from performing the chemical separations. Therefore, the shipping of fuel must be thoroughly regulated by national and international agencies. The difficulties involved in highjacking and processing sufficient

Table 24.8-1 Approximate critical masses* of materials occurring in nuclear
fission-reactor fuel cycles; private communication from
T. B. Taylor.

Element	Percentage: fissile isotope	Principal, non-fissile isotope(s)	Chemical state	Critical mass, kg
U	93:235	238	metal	17
			oxide or carbide	~30
	95:233	236	metal	5.8
			oxide or carbide	~10
Pu	70:239, 241, 238	240, 242	metal	6.7
			oxide or carbide	~11
	50:239, 241, 238	240, 242	metal	9.6
			oxide or carbide	~16

*As is discussed in Appendix 21A, a critical mass is the amount of fissile
material that will just sustain the fission when surrounded by an efficient
neutron reflector.

material may be judged from the data in Table 24.8-2, where the approximate
mass of fuel that would have to be diverted for the production of a single weapon
is listed (column b). During shipment from reactors, the likelihood of diversion increases in the order FBR, LWR, HTGR. Safety considerations have been

Table 24.8-2 Critical masses, reactor-fuel elements per critical mass, and numbers of critical masses in nuclear reactors; private communication from T. B. Taylor.

	(a)	(b), kg	(c)
BWR, Pu recycle	~0.5	500	~50
PWR, Pu recycle	~1	500	~50
HTGR, ^{233}U recycle	0.067	2,100	~80 ^{233}U ~12 ^{235}U
LMFBR	2.3	100	~330

(a) Approximate number of chemically separable critical masses in a single reactor-fuel element. (b) The total weight of reactor-fuel elements per critical mass. (c) The number of critical masses flowing in the reactor-fuel cycle per 1,000 $Mw_e y$.

used to argue for locating FBRs in centers which house all of the other facilities required in the complete fuel cycle.

iii. The reprocessing plants may be vulnerable. Conceivably, insiders could systematically divert small quantities over a period of time and thus accumulate a critical mass of material without detection. An effective countermeasure involves accurate assay and accounting of mass flows.

iv. The weapons stockpile is clearly a potential location for subversion. The amounts of fissile material stored and handled by military forces around the world are larger than those flowing in the nuclear fission-power fuel cycles.

24.9 Nuclear Energy and Trade Deficits

In a recent article,[1] R. Philip Hammond has emphasized the economic hazard involved in failure to develop nuclear power. He states that "the real risk of nuclear power is in not having enough of it" and emphasizes that the costs of adequate safeguards against nuclear sabotage and subversion are about 1% of the costs of nuclear power ($\simeq 0.13$ mill/kwh$_e$). The costs of an adequate nuclear waste-disposal program are also estimated to be 1% of nuclear-power costs.

Hammond[1] emphasizes the price of not having nuclear power by quoting calculations from a 1975 article in The Economist of London. Cost estimates are expressed in terms of a new currency, the COPEC-year (C-y) which equals the total surplus cash accumulated by members of the Organization of Petroleum Exporting Countries (OPEC) at the 1975 export rate. One C-y $\simeq \$60 \times 10^9$; similarly, one COPEC-day = 1 C-d $\simeq \$164 \times 10^6$ and one COPEC-hour = 1 C-h $\simeq \$6.8 \times 10^6$. The rate at which the assets of the world could be bought if all oil income were used exclusively for this purpose is illustrated in the summary reproduced in Table 24.9-1.

Fortunately, the needs of the people living in the OPEC have thus far precluded the exclusive use of foreign currency acquisitions for the purposes assumed in Table 24.9-1. However, the compilation should serve to reemphasize the thesis that cost-benefit factors must be carefully weighed in making decisions on energy-resource developments and applications.

---

[1]R. Philip Hammond, "The Real Risk of Nuclear Power," Supplement to Energy (Volume 1, in press, 1976).

Table 24.9-1 A 1975 estimate of the economic power of OPEC, expressed as
time required to purchase the world's major assets; abstracted
from Ref. [1].

Asset	Time required for purchase by OPEC if all oil income were applied for the purchase of assets
all companies listed on the world's major stock exchanges	15.6 y
the entire personal wealth of Britains	15.5 y
an annuity of $232 per week for all Arab males	12.8 y
all listed stocks on the New York Stock Exchange	9.2 y
all of Britain's industrial assets	6.0 y
all U.S. foreign investments	1.8 y
IBM Corporation	143 d
Exxon Corporation	79 d
Bank America Corporation	10 d

CHAPTER 25

NUCLEAR STRATEGIES

The term "strategies" referring to nuclear-reactor development has been traditionally associated with efficient fuel utilization in conventional fission reactors. Programming of reactor schedules is essential if we are to optimize the use of scarce, highest grade, uranium and thorium (see Section 2.15 in Volume I for resource estimates). This classical topic will be addressed in Section 25.2, together with a general formulation for nuclear-reactor growth rates.

More recently, questions have been raised about the efficacy of nuclear-reactor construction of any kind. Insofar as conventional fission reactors are concerned, the issue of "net energy return" has surfaced almost as prominently as questions connected with safety and environmental impact, which have been discussed in Chapter 24. Construction schedules for fission reactors designed to produce desired net outputs of electrical energy will be defined in Section 25.1, where it will be shown that the estimated net energy ratios for converting primary fossil-fuel energy to electricity by using nuclear reactors are such that the use of nuclear reactors represents a highly efficient methodology for fossil-fuel utilization. Reactor construction schedules are easily defined for any desired electrical-energy output.

25. 1 Programming of Nuclear-Reactor Development to Produce Desired
       Electricity-Generating Capacity[*]

It is the purpose of the following discussion to determine schedules for
nuclear-reactor development in view of desired electricity-generating capacity.
We shall note, in particular, that net energy ratios for nuclear-energy production from high-grade ore are sufficiently favorable to allow scheduled developments of reactor scenarios that imply large savings in fossil-fuel resources for
desired electricity-generating capacity.

A. Introduction

The development of any new resource requires an energy investment of
existing resources. Depending on the new energy technology involved, the annual and total energy returns may be negative or positive. A continuous time-
dependent energy-investment schedule is shown schematically in Fig. 25.1-1
for a successful new resource development. We have defined the following three
times in this resource development schedule: the time Δt corresponding to the
minimum completion time for the new resource, the time t_b after which there
is a net annual return of energy, and the time t_r after which the time-integrated
energy investment first becomes positive.

Figure 25.1-1 represents an idealized energy-development schedule. In
practice, both the annual rates of energy investment and the annual rates of energy return may be step-functions of the time because new energy-resource developments are initiated and completed in incremental units. A schematic diagram depicting this more realistic situation is shown in Fig. 25.1-2. The times

---

[*]This Section has been reproduced from a paper by S. S. Penner, Energy,
   Volume 1 (in press, 1976).

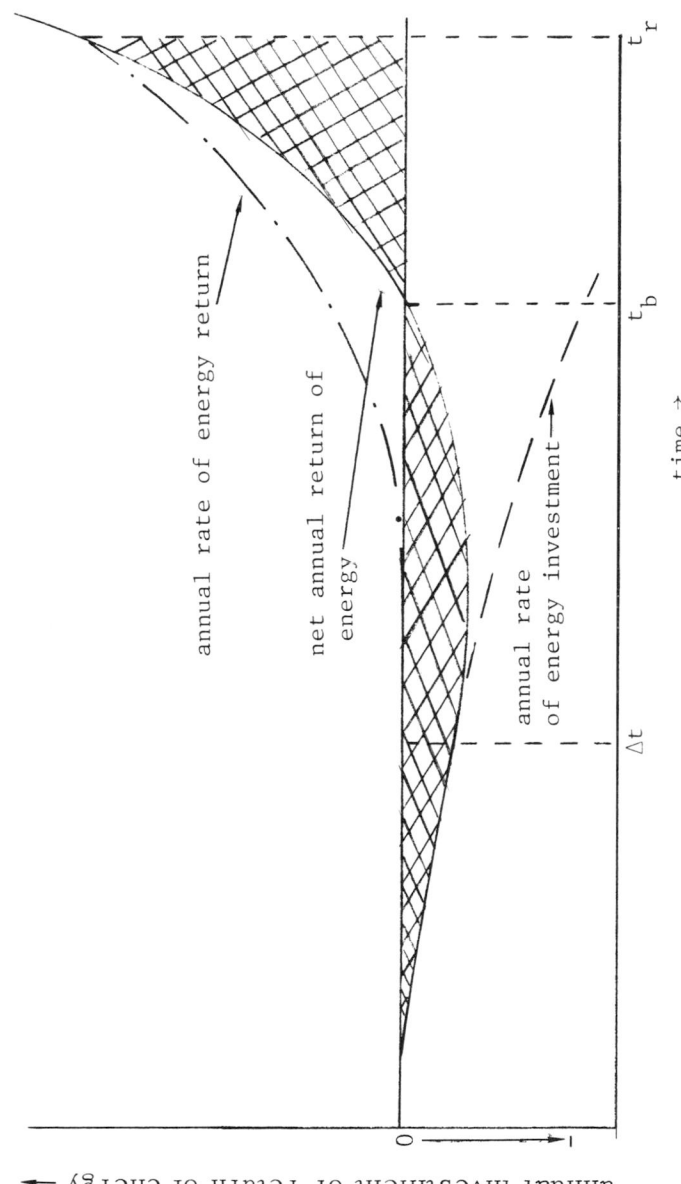

Fig. 25.1-1 Schematic diagram showing continuous annual rates of energy investment, re-turn, and net energy available as functions of time for a successful new resource development. The minimum construction time for the new resource is Δt, the break-even time t_b corresponds to the time when the net annual return of energy reaches zero, and the time t_r is the time required to achieve a time-integrated energy return that equals zero. The entire enterprise becomes a positive con-tribution to energy availability after t_r years when the total area of the hashed region just equals zero.

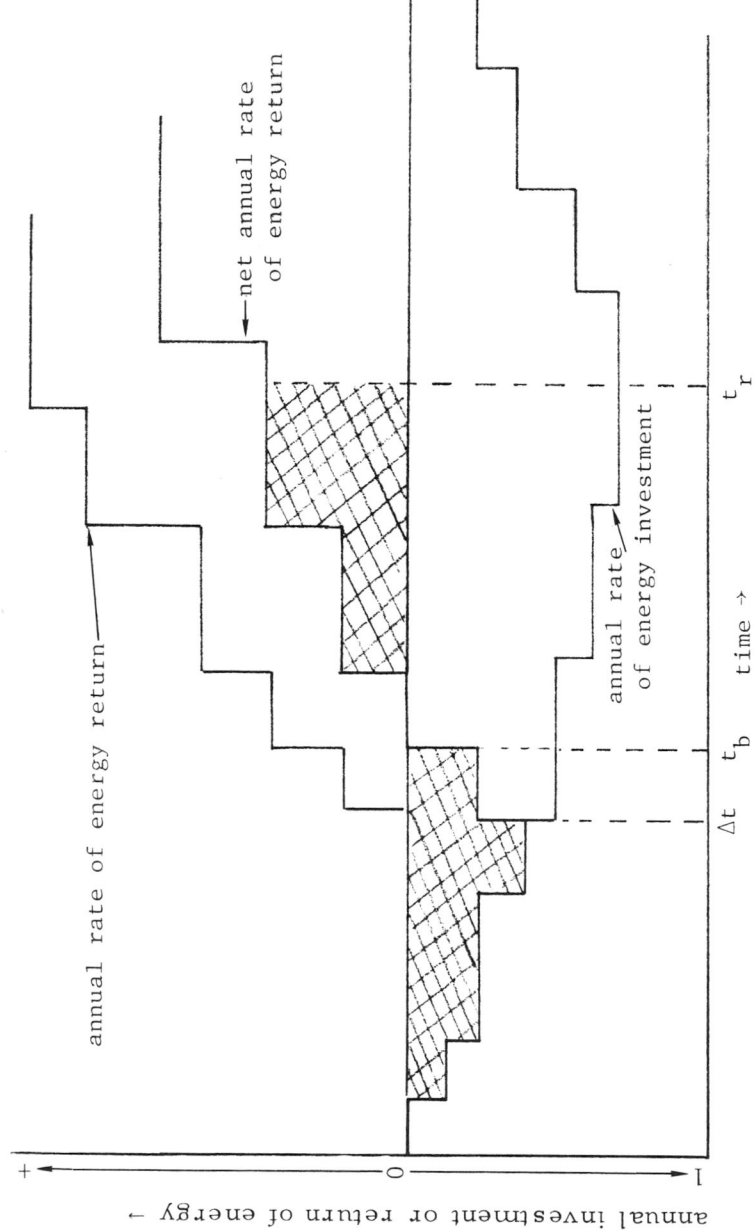

Fig. 25.1-2   Schematic diagram showing incremental annual rates of energy investment, return, and net energy available as functions of time for a successful new resource development. The minimum construction time for the new resource is Δt, the break-even time t_b corresponds to the time when the net annual return of energy reaches zero, and the time t_r is the time required to achieve a time-integrated energy return that equals zero. The entire enterprise becomes a positive contribution to energy availability after t_r years when the total area of the hashed region just equals zero.

Δt, t_b, and t_r are defined as before. The ratio of the height of the first positive increment of the annual return of energy to that for the first negative annual investment of energy depends critically on the net energy ratio, R, for the
technology under development and on the life of the new energy-producing plant
(see Section 25.1C for details). For a sufficiently large value of R, t_r will generally occur within a year or two of t_b.

B. Simplified Schemes for Determining Reactor-Development
   Scenarios; Evaluations of t_b and t_r

We make no distinction between the times in starting a resource development within a given year, i.e., we assume in effect that all energy-resource
development is initiated at the beginning of the year and completed at the end
of the year. Let N(i) be the number of reactors started during the ith year
after the current year; the current year is identified by the label 0 and the number of reactors started during the current year is N(0).

The parameter E_C is defined to be the total energy cost for the construction of a new reactor which requires Δt years for completion. Assuming that
uniform energy expenditures are made during capital construction and capital
investment, the average yearly energy cost is $E_C/\Delta t$ for each new reactor.
The total net energy return from the reactor (with proper allowance for all operating charges) is designated as E_n during the plant life of t_ℓ years. Again
assuming a uniform rate of energy return, the yearly energy return for each
new reactor after completion becomes E_n/t_ℓ. The net energy ratio is defined as

$$R = E_n/E_C \qquad (25.1-1)$$

for the construction of a new reactor. As already noted, E_C is invested uniformly over Δt years while E_n is returned at a uniform rate during t_ℓ years

after Δt years have elapsed from initiation of construction.

   i. Net Annual Rate of Return of Energy

   With the specified definitions, the total energy investment per year in construction after t years is seen to be

$$(E_C/\Delta t)[N(t) + N(t - 1) + N(t - 2) + \ldots + N(t - \Delta t + 1)],$$

while the energy returned per year is given by the expression

$$(E_n/t_\ell)[N(0) + N(1) + \ldots + N(t - \Delta t)] ,$$

where all time indeces must be positive for physically meaningful definitions of terms.

   In view of the preceding relations, the net energy that becomes available per year as the result of reactor development is

$$\epsilon_a(t) = (E_n/t_\ell) \sum_{t'=0}^{t-\Delta t} N(t') - (E_C/\Delta t) \sum_{t'=0}^{\Delta t-1} N(t - t') \qquad (25.1\text{-}2)$$

or, in terms of the net energy ratio R,

$$\epsilon_a(t)/(E_C/\Delta t) = R(\Delta t/t_\ell) \sum_{t'=0}^{t-\Delta t} N(t') - \sum_{t'=0}^{\Delta t-1} N(t - t') . \qquad (25.1\text{-}3)$$

As already noted, t', t and Δt only take on integral values in the present simplified treatment.

   It is apparent from Eq. (25.1-3) that the energy break-even year, t_b, corresponds to

$$R = (t_\ell/\Delta t)\left[\sum_{t'=0}^{\Delta t-1} N(t_b - t')/\sum_{t'=0}^{t_b-\Delta t} N(t')\right].$$ (25. 1-4)

The shortest possible break-even time occurs after $t_b = \Delta t$ years when the initial reactor development has been completed and useful energy generation first takes place; for $t_b = \Delta t$, Eq. (25.1-4) leads to the result

$$N(0) = (t_\ell/R\Delta t)\sum_{t'=0}^{\Delta t-1} N(\Delta t - t')$$ (25. 1-5)

$$= (t_\ell/R\Delta t)[N(\Delta t) + N(\Delta t - 1) + \ldots N(1)].$$

For an accelerated nuclear-reactor development using high-grade fuel, $\Delta t = 5y$, $t_\ell = 25y$, $R = 30$; hence

$$N(0) = (1/6)[N(5) + N(4) + N(3) + N(2) + N(1)],$$

i.e., a positive energy return becomes first possible after five years provided the number of reactors started during the initial year equals one sixth of the total number of reactors started during the following five years. It is apparent from Eq. (25.1-5) that smaller values of R dictate heavier initial program scheduling. For example, for R = 10,

$$N(0) = (1/2)\sum_{t'=1}^{5} N(t')$$

if the annual energy investment is to be returned after $t_b = \Delta t = 5$ years following the year of start-up of construction.

In practice, we shall generally be concerned with the evaluation of ϵ_a for values of t appreciably larger than Δt but normally shorter than t_ℓ ($\sim$25y). It is convenient to use Eqs. (25.1-2) or (25.1-3) for a parametric representation of ϵ_a. We shall now examine several representative reactor-development schedules.

a.   Uniform Scheduling of Reactor Development

For uniform scheduling of reactor development, all of the $N(t') \equiv N_0$ are equal.   Hence Eq. (25.1-3) becomes

$$\epsilon_a(t)/(E_C/\Delta t) = N_0[R(\Delta t/t_\ell)(1 + t - \Delta t) - \Delta t] \qquad\qquad (25.1-6)$$

and $\epsilon_a(t)/(E_C/\Delta t)$ will be positive for

$$R(\Delta t/t_\ell)(1 + t - \Delta t) > \Delta t \; ;$$

thus, for $R = 30$, $\Delta t = 5y$, $t_\ell = 25y$, a positive energy return will first be realized during the fifth year after the present year; for $R = 5$, $\Delta t = 5y$, $t_\ell = 25y$, t will be longer than 9y before a positive energy return occurs.

b.   Sequential Doubling of Reactor Development

We assume that N_0 reactors are developed during the current ($t' = 0$) year, $2N_0$ during the next ($t' = 1$) year, 2^2N_0 during the $t' = 2$ year, etc. Equation (25.1-3) becomes

$$\epsilon_a(t)/(E_C/\Delta t) = [R(\Delta t/t_\ell)(1 + 2 + 2^2 + \ldots 2^{t-\Delta t})$$

$$- (2^t + 2^{t-1} + \ldots 2^{t-\Delta t+1})] N_0. \qquad (25.1-7)$$

The dimensionless parameter $\epsilon_a(t)/(E_C/\Delta t)N_0$ will now assume positive values only for large ratios of $R\Delta t/t_\ell$. In particular, for $R = 5$, $\Delta t = 5y$, $t_\ell = 25y$, a positive energy return will never occur. The minimum value of $R(\Delta t/t_\ell)$ for which a positive energy return becomes possible for $\Delta t = 5y$ and $t = 10y$ is easily found to be $R(\Delta t/t_\ell) = 31.49$ from Eq. (25.1-7), i.e., $R = 157.46$ for $t_\ell = 25y$ and $\Delta t = 5y$; this value of R is unrealistically large. Even for $t = 20y$, $t_\ell = 25y$, R must be at least 155 before a positive energy return occurs. We conclude that continuous doubling of the rate of reactor development is not a realistically defensible posture for a program that is designed to increase the total U.S. energy availability.

    c.   Sequential Doubling of Reactor Development for t_1 Years, Followed by Termination of Construction

For the specified model, Eq. (25.1-3) becomes

$$\epsilon_a(t)/(E_C/\Delta t) = N_0[R(\Delta t/t_\ell)(1 + 2 + 2^2 + \ldots + 2^{t-\Delta t})$$

$$- (2^{t_1} + 2^{t_1-1} + \ldots + 2^{t_1-\Delta t+1})] \qquad (25.1-8)$$

for $t_1 \le t \le t_1 + \Delta t$; for $t_1 + \Delta t < t < t_\ell + \Delta t$, reactor development is complete and maximum energy output occurs. As an example, we choose a 10-year reactor-development schedule, i.e., $t_1 = 10y$. Setting again $\Delta t = 5y$, $t_\ell = 25y$, $R = 30$, $R(\Delta t/t_\ell) = 6$, we may determine the dimensionless ratio $\epsilon_a(t)/(E_C/\Delta t)N_0$

as a function of t from Eq. (25.1-8). Some representative numerical results are listed in the following table.

Table 25.1-1   Numerical values of $\epsilon_a(t)/(E_C/\Delta t)N_0$ as a function of t for $R = 30$, $t_1 = 10y$, $\Delta t = 5y$, $t_\ell = 25y$.

t, y	$\epsilon_a(t)/(E_C/\Delta t)N_0$
10	$-$ 1,606
11	$-$ 1,158
12	$-$ 262
13	$+$ 1,530
14	$+$ 5,114
15	$+$ 12,282

Reference to Table 25.1-1 shows that the yearly energy output is positive three years after completion of initiation of construction for the assumed model. If we choose N_0 to be 10, then the annual energy output equals 15,300 times the annual energy charge for reactor development during the year with index 0 at a time corresponding to 13 years after program initiation and 3 years after initiation of construction has been terminated. Maximum output is reached 15 years after the year during which the program was initiated when the entire reactor-development program has been completed. Thereafter, the energy output remains constant until the year $t_\ell + \Delta t$ is reached, when the first completed reactors have attained their useful life span.

d.   Other Models and Concluding Remarks

The preceding calculations are easily generalized to arbitrary construction schedules. We conclude by noting the dominant importance of the net energy ratio R, of the reactor-construction time Δt, and of the reactor life time t_ℓ

in determining the magnitudes of available positive or negative energy charges that are incurred or released during the expansionary phase of an energy-resource development schedule. For fixed values of these parameters, a construction program may be invented that yields the desired energy return on a predetermined schedule provided the ratio $R(\Delta t / t_\ell)$ is sufficiently large. In the following Section 25.1C, we comment briefly on applicable values of R.

ii. Integrated Net-Energy Returns

The integrated energy investment at the beginning of the year t is

$$(E_C/\Delta t) \{N(t - 1) + 2N(t - 2) + 3N(t - 3) + \ldots \Delta t \, N(t - \Delta t)$$

$$+ \Delta t [N(t - \Delta t - 1) + \ldots N(0)]\} , \tag{25.1-9}$$

while the integrated energy return is

$$(E_n/t_\ell)[N(0)(t - \Delta t) + N(1)(t - \Delta t - 1) + \ldots + N(t - \Delta t - 1)] . \tag{25.1-10}$$

Hence the total or integrated net energy return is

$$E_T = (E_n/t_\ell)[N(0)(t - \Delta t) + N(1)(t - \Delta t - 1) + \ldots + N(t - \Delta t - 1)]$$

$$- (E_C/\Delta t)\{N(t - 1) + 2N(t - 2) + 3N(t - 3) + \ldots \Delta t N(t - \Delta t)$$

$$+ \Delta t [N(t - \Delta t - 1) + \ldots + N(0)]\} . \tag{25.1-11}$$

Thus the time t_r after which the entire enterprise becomes a net producer of energy is determined by the relation

$$(R\Delta t/t_{\ell})[N(0)(t_r - \Delta t) + N(1)(t_r - \Delta t - 1) + \ldots N(t_r - \Delta t - 1)]$$

$$= N(t_r - 1) + 2N(t_r - 2) + 3N(t_r - 3) + \ldots + \Delta t \, N(t_r - \Delta t)$$

$$+ \Delta t[N(t_r - \Delta t - 1) + \ldots + N(0)] . \qquad (25.1\text{-}12)$$

As in Section 25.1Ai, we shall now work out analytically simple but instructive examples referring to well-defined construction schedules.

a.   Uniform Scheduling of Reactor Development

For uniform scheduling of reactor development, all of the $N(t') \equiv N_0$ are equal. Hence Eq. (25.1-11) becomes

$$E_T/E_C = N_0[(R/t_{\ell})(t - \Delta t) + (t - \Delta t - 1) + \ldots + 1]$$

$$- (1/\Delta t)[1 + 2 + \ldots + \Delta t + \Delta t(t - \Delta t - 1)] . \qquad (25.1\text{-}13)$$

As an example, we consider the case $t_r = 7y$, $\Delta t = 5y$. In this case, the ratio $R\Delta t/t_{\ell}$ is determined by the relation $R\Delta t/t_{\ell} = 6.67$ or $R = 33.4$ for $t_{\ell} = 25y$. Similarly, for $t_r = 8y$, $\Delta t = 5y$, $R\Delta t/t_{\ell} = 4.17$ and $R = 21.8$ for $t_{\ell} = 25y$; for $t_r = 10y$, $\Delta t = 5y$, $R\Delta t/t_{\ell} = 2.33$ and $R = 11.7$ for $t_{\ell} = 25y$; for $t_r = 15y$, $\Delta t = 5y$, $R\Delta t/t_{\ell} = 1.09$ and $R = 5.45$ for $t_{\ell} = 25y$. Thus the net integrated energy break-even time that can be reasonably expected for a nuclear reactor with $R \simeq 30$, a five-year construction time and a twenty-five year life, for the special case of a uniform construction schedule, is a little longer than 7 years. Comparison with the analysis of Section 25.1Ai shows that t_r exceeds t_b by a little more than one year. For large values of R, this type of result is generally expected.

b. Sequential Doubling of Reactor Development for t_1 Years,
Followed by Termination of Construction

We now use again a model in which construction of N_0 reactors is initiated during the current year, $2N_0$ during the following year, $2^2 N_0$ during the year thereafter, etc. Equation (25.1-12) then takes the form

$$(R\Delta t/t_\ell) \left[(t_r - \Delta t) + 2(t_r - \Delta t - 1) + 2^{t_r - \Delta t - 1} \right]$$

$$= \{ [2^{t_1 - 1}(t_r - t_1) + \ldots 2^{t_r - \Delta t}(\Delta t - 1)]$$

$$+ \Delta t (2^{t_r - \Delta t - 1} + 2^{t_r - \Delta t - 2} + \ldots 1) \} \qquad (25.1-14)$$

for $t_1 \leq t_r \leq t_1 + \Delta t$; for $t_1 + \Delta t < t < t_\ell + \Delta t$, reactor development has been completed and maximum energy output occurs.

For $t_1 = 10y$ and $\Delta t = 5y$, we may then calculate from Eq. (25.1-14) the applicable values of $R\Delta t/t_\ell$ for selected values of t_r. The results of these computations are listed in Table 25.1-2, together with the required values of R for $\Delta t = 5y$, $t_1 = 10y$.

For a value of R = 10 referring to total energy production, or a value of R = 30 referring to electricity production, we find from Table 25.1-2 that the integrated energy return equals the integrated invested energy during the sixteenth or fourteenth year, respectively. Thus, for sequential doubling of nuclear-energy plants, the total energy investment will be returned during the sixth or fourth year after termination of construction, depending on the applicable value of R. For R = 30, the integrated energy return is positive before the last generation of nuclear reactors has been completed.

Table 25.1-2 Numerical values of $R\Delta t/t_\ell$ and of R, as computed from
Eq. (25.1-14) for selected values of $t_r - t_1$ with $t_1 = 10y$, and
$t_1 < t_r \leq t_\ell + \Delta t$.

t_r	$t_r - t_1$	$R\Delta t/t_\ell$	R(for $\Delta t = 5y$, $t_\ell = 25y$, $t_1 = 10y$)
10	0	34.72	173.6
11	1	33.03	165.2
12	2	23.82	119.1
13	3	15.29	76.45
14	4	9.09	45.45
15	5	5.11	25.55
16	6	2.51	12.55
17	7	1.67	8.35
18	8	1.25	6.25

C. Net Energy Ratios

We have considered methodological aspects in the calculation of net energy ratios elsewhere.[1,2] The term R in the following discussion is the same as R_1 in Refs. [1] and [2].

---

[1]UCSD/NSF(RANN) Workshop on Net Energy in Shale-Oil Production, report prepared by S. S. Penner; published as Section VIII in UCSD/NSF(RANN) Workshop on In Situ Recovery of Shale Oil, S. S. Penner, ed., U.S. Government Printing Office, Washington, D.C., 1975. A careful evaluation of net energy for nuclear reactors has been given by R. M. Rotty, A. M. Perry and D. B. Reister, "Net Energy from Nuclear Power," Report IEA-75-3, Institute for Energy Analysis, Oak Ridge Associated Universities, Oak Ridge, Tennessee, November 1975.

[2]See also, Volume II of this series of publications, Section 9.14.

Representative values of R for shale-oil recovery have been given by Clark and Varisco[3] and have been found to have values around 8.

Chapman[4] has estimated net energy ratios for a number of nuclear reactors. Energy charges for fuel processing (but not the intrinsic energy content of the fuel) were explicitly included for ores with uranium mass fractions of 3×10^{-3} and 7×10^{-5}, both for refueling and for initial fuel recovery. In Chapman's analysis, a 10^3 Mw_e plant produces a net output, after consideration of distribution and other losses, of 255.8 Mw_e or 5.602×10^{10} kwh_e over a life of 25y at a 62% average load factor when the lower grade ore is used; the corresponding values for the higher grade ore are 523.4 Mw_e and 1.14625×10^{11} kwh_e, respectively. The total energy costs for construction of the 10^3 Mw_e reactor are 2.2229×10^{10} kwh_t and 1.0162×10^{10} kwh_t for fuel processed from low-grade and high-grade ores, respectively. Thus, the energy ratios, defined as usable electrical energy produced divided by the thermal energy input, lie between 2.5 for low-grade ore and 11.4 for high-grade ore, respectively. Chapman has generally obtained energy ratios around 2 or 3 for low-grade ore and 10 for high-grade ore. Chapman notes[4] that other workers have given ratios as low as 3 even for high-grade ore. There is a basic criticism that may be raised against the conclusions that are implied by Chapman's numerical estimates, which is to some extent corrected by Chapman when he notes that the development of nuclear reactors may be viewed as an efficient procedure for converting fossil-fuel energy to electricity. We note that fossil fuel is used extensively today for the generation of electrical energy at an average conversion efficiency of about 33%. For this segment of the fossil-fuel industries,

---

[3] See the Appendix to Ref. [1] by C. E. Clark, Jr., and D. C. Varisco.

[4] P. Chapman, "The Ins and Outs of Nuclear Power," New Scientist, pp. 866-869, December 19, 1974.

the applicable energy ratio is therefore the ratio of electrical energy produced
by the nuclear reactor to the <u>electrical energy</u> which would be generated through
application of the fossil fuels used in the construction of nuclear reactors, i.e.,
the applicable energy ratios are effectively tripled for that portion of nuclear-
power generation which replaces electrical energy use that is normally produced from fossil-fuel plants. There is an obvious corollary to the preceding
critique, which refers to the idea expressed by some that nuclear energy will
ultimately become a primary source of electricity and a secondary source of
portable fuels. For applications of this type, the energy ratios of Chapman
must be <u>reduced</u> to allow for the loss of energy in the conversion of electricity
to portable fuels. For such sequential processes as electrolysis followed by
fuel-cell utilization for transportation, the overall energy loss would be expected to be perhaps 25% (if we assume that efficiencies for implementation of
transportation are comparable once we begin with either fossil fuels or portable
fuels in fuel cells).

W. K. Davis[5] has presented the following energy charges for an 1,100
Mw_e nuclear power plant operating at a 60% load factor: for plant construction,
0.74×10^9 kwh_e; for the initial fuel load, 1.00×10^9 kwh_e; for replacement
fuel, 0.50×10^9 kwh_e. Assuming a twenty-year plant life, the net energy ratio is seen to be

$$R = \frac{0.60 \times 9.64 \times 10^9 \text{ kwh}_e}{(1.74 \times 10^9/20) \text{ kwh}_e + 0.50 \times 10^9 \text{ kwh}_e} \simeq 9.9 \ .$$

―――――――――――――

[5]W. K. Davis, "Nuclear Power's Contribution to Energy Growth," paper presented at an Atomic Industrial Forum Conference on Accelerating Nuclear
Power Plant Construction, March 3, 1975, New Orleans, Louisiana.

This calculation involves pessimistic assumptions concerning the fuel-processing costs (i.e., no reduction in power consumption for diffusion plants and plutonium recycle nor allowance for anticipated fuel enrichment by centrifuging). The fuel is recovered from conventionally used, high-grade ore (2,000 to 2,500 ppm of U_3O_8 in rock).

For recovery from low-grade ore (e.g., Chattanooga shales) with 50 to 70 ppm of U_3O_8, the ore mining and milling costs have been estimated by A. Weinberg to be about 25 kwh_e/ton of ore.[5] For this low-grade ore, Davis then calculates an incremental energy charge of 0.32×10^9 kwh_e for the initial core and 0.14×10^9 kwh_e for annual fuel loading. The corresponding net energy ratio becomes

$$R = \frac{0.60 \times 9.64 \times 10^9 \text{ kwh}_e}{(2.06 \times 10^9/20)\text{kwh}_e + 0.64 \times 10^9 \text{ kwh}_e} = 7.7 \ .$$

Davis' net energy ratios are seen to be appreciably lower than the values which we derived from Chapman's data. However, Davis states that his numbers are uncertain by perhaps as much as a factor of 2.

i. Application of Chapman's R to a Reactor-Development Schedule

We conclude this brief commentary on energy ratios by noting that the rather large value of R = 30 may be considered to be consistent with the estimate for the ratio of _electrical_ energy derived from fission reactors to the _electrical_ energy which would have been derived from the fossil fuels used in the construction of nuclear reactors if they had instead been used in fossil-fuel power plants. Thus, some of our optimistic examples of energy return for nuclear-reactor development scenarios represent reasonable quantitative descriptions of possible implementation schedules. In particular, a ten-year development schedule with doubling each year of the installed number of reactors [see Section 25.1B(i)c and Table 26.2-1] will produce within 15 years a total supplementary annual electrical energy output of 2.5×10^{12} kwh$_e$ if the following number of 10^3 Mw$_e$ reactors N_0 were installed this year:

$$2.5 \times 10^{12} \text{ kwh}_e/y = 1.0235 \times 10^4 \times [10^6 \text{ kw}_e \times 8.76 \times 10^3 \text{ (h/y)}]$$

$$\times [(1/30) \text{ for the energy charge per reactor}] \times N_0$$

or $N_0 = 0.836$, i.e., 836 Mw$_e$ must be constructed during 1975 with a 5-year completion schedule and doubling of the capacity installed each year during the succeeding 10 years.   It should be noted that the preceding statements refer to the supplementary electrical energy that can be generated in excess of the electrical energy which would be produced if the fossil-fuels required for nuclear-reactor development were instead diverted for the direct production of electricity in fossil-fuel generating plants.

D.   Epilogue

The formal analysis presented in this section may be made more precise when the necessary quantitative data on energy requirements are available for all of the processes that are involved in the nuclear-fuel cycle.   A brief outline of the results obtained in a study of this type is presented in the following Section 25.2.

25.2 Net-Energy Analysis for Nuclear Reactors

A careful net-energy analysis for nuclear reactors has been performed recently using definitions related to those given for R_1 in Section 9.14 on shale-oil recovery.[1] Rotty et al[1] define the following net energy ratios:

$$R_1 = \frac{\text{electrical energy output (Mw}_e\text{h)}}{\text{thermal (fossil-fuel) energy required in Mw}_t\text{h} + 3.34 \text{ electrical energy required in Mw}_e\text{h}} \, ,$$

$$R_2 = \frac{\text{electrical energy output (Mw}_e\text{h)}}{\text{electrical energy required in Mw}_e\text{h} + (\text{thermal energy required in Mw}_t\text{h}/3.34)} \, ,$$

$$R_3 = \frac{\text{electrical energy output (Mw}_e\text{h)}}{\text{electrical energy required in Mw}_e\text{h} + \text{thermal energy required in Mw}_t\text{h}} \, ,$$

$$R_4 = \frac{\text{el. energy output(Mw}_e\text{h)} - \text{el. energy input (Mw}_e\text{h)}}{\text{thermal energy required in Mw}_t\text{h}} \, .$$

If the numerator in the ratio R_1 were multiplied by the reciprocal of the fossil-fuel-to-electricity conversion efficiency, then it would be identical with

[1]Ralph M. Rotty, A. M. Perry and David B. Reister, "Net Energy from Nuclear Reactors," Report IEA-75-3, Institute for Energy Analysis, Oak Ridge Associated Universities, Oak Ridge, Tennessee, November 1975.

the ratio R_1 defined in Eq. (9.14-1). It is apparent that R_1 measures the electrical energy produced per unit of fossil-fuel-energy-equivalent expended; R_2 is the ratio of electrical energy produced per unit of electrical-energy-equivalent expended; R_3 is the ratio of electrical energy output to total energy input regardless of type (i.e., whether electrical or thermal in origin); R_4 is the ratio of net electrical energy produced to thermal energy input. The ratios R_2 and R_3 are not related to the ratios R_2 and R_3 defined for shale-oil recovery in Eqs. (9.14-2) and (9.14-3), respectively, which include allowances for reprocessing of resources-in-place during recovery or for unrecovered resources. One might argue that the ratios R_1 to R_4 used by Rotty et al[1] are variations of parameters formed by considering only electrical energy outputs, electrical energy inputs, and fossil-fuel energy inputs.

Rotty et al[1] consider reactors with a power rating of 10^3 Mw$_e$, a 30-year life, and a 75% plant factor. They have evaluated the energy-input requirements for conventional uranium ores containing 0.176% of uranium and also for low-grade Chattanooga shales containing 0.006% of uranium. Net-energy analyses were performed for the following reactor types: PWR, BWR, HTGR, and the CANDU heavy-water-moderated reactor (HWR). Enrichment tails at levels of 0.3% or 0.2% were allowed for in the first three types of reactors.

The distribution of lifetime energy requirements for one of the proto-type systems is reproduced in Table 25.2-1 and clearly shows exceptional heavy electrical energy-input requirements for enrichment and of thermal energy-input requirements for plant operation during the 30-year service life.

Values for the net energy ratios R_1 to R_4 are reproduced in Table 25.2-2 for the different reactor types, ores and enrichment tails considered. For conventional ores, the CANDU HWR is seen to have an especially favorable

Table 25.2-1   Lifetime energy requirements for a 10^6 Mw_e reactor with a
30-year life and a 75% plant factor; reproduced from Ref. [1].

Process	$Mw_e h$	10^6 Btu_t
mining of 5,682 mt of U	110,100	3,006,000
milling of 5,682 mt of U	125,500	2,983,000
conversion of 5,682 mt of U	82,960	7,676,000
enrichment of 3.022×10^6 kg of U	8,533,000	2,412,000
fuel fabrication of 822 mt of enriched U	247,400	2,109,000
reprocessing of 822 mt of fuel	16,360	292,600
waste storage for 30 y	5,010	183,200
transportation of 5,682 mt of U and	597	81,930
of 822 mt of fuel	1,861	255,900

value of $R_1 \simeq 8.4$ whereas the HTGR yields the largest ratio of net production
of electrical energy to thermal energy input ($R_4 \simeq 22$).

A.   Dynamic Energy Analysis for a Single Reactor

We have considered in Section 25.1 the programming of reactor development in order to produce a desired level of electricity generation. To perform a task of this type in a realistic fashion, it is necessary to specify precise values for the energy requirements and energy outputs of each reactor.
The results obtained from a careful analysis for a particular reactor are reproduced in Fig 25.2-1.

Table 25.2-2 Lifetime net energy ratios R_1 to R_4 for 10^6 Mw_e reactors with a 30-year life and a 75% plant factor; reproduced from Ref. [1].

Reactor system	R_1	R_2	R_3	R_4
PWR; n.r.; c.o.; 0.3%	4.60	15.35	9.63	17.23
PWR; n.r.; c.o.; 0.2%	3.98	13.31	8.92	17.85
PWR, Pu r.; c.o.; 0.3%	6.05	20.20	12.10	20.38
PWR; n.r.; C.S.; 0.3%	2.19	7.30	3.40	4.14
PWR; Pu r.; C.S.; 0.3%	3.13	10.47	4.89	6.11
BWR; n.r.; c.o.; 0.3%	4.89	16.35	10.03	17.35
HTGR; c.o.; 0.3%	5.69	19.01	12.03	22.02
HTGR; C.S.; 0.3%	3.39	11.32	5.62	7.43
HWR; n.r.; c.o.; enrichment is not applicable	8.41	28.09	11.34	13.16

Legend: n.r. = no recycle; c.o. = conventional ores; percentages refer to levels of enrichment tails; Pu r. = plutonium recycle; C.S. = Chattanooga Shales.

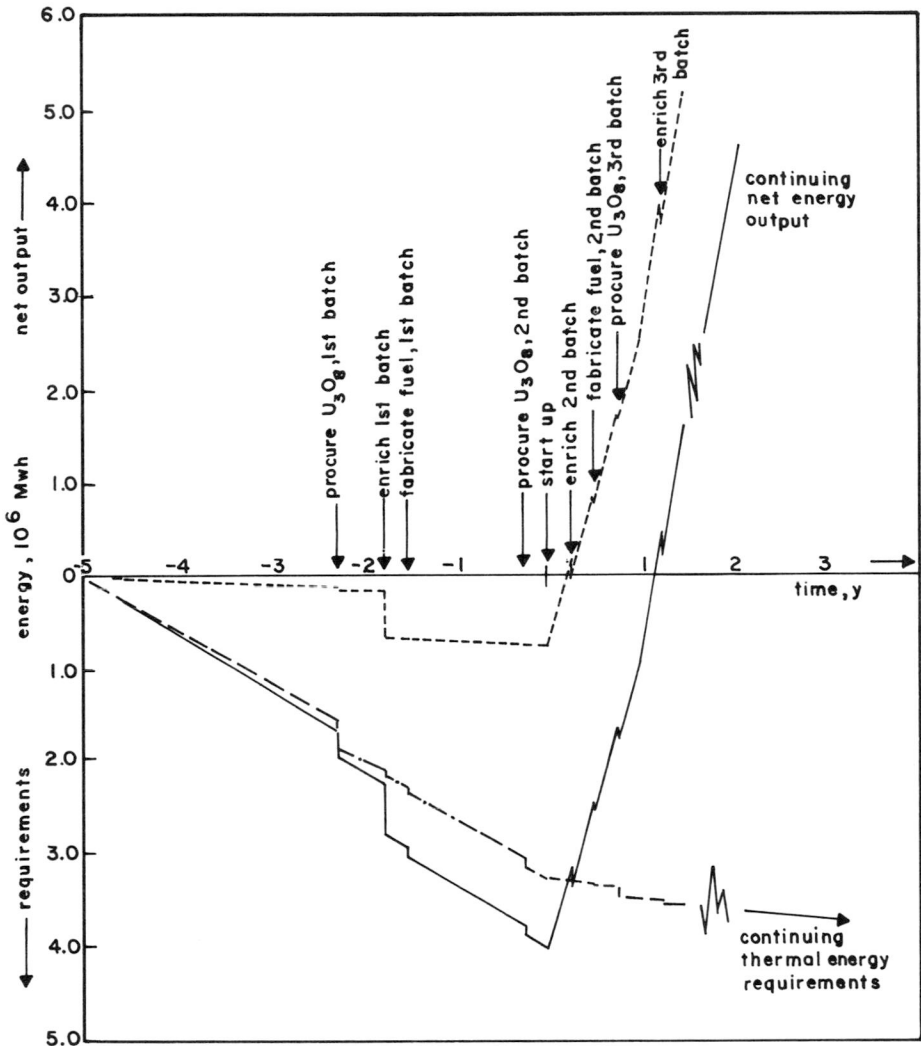

Fig. 25.2-1 The cumulative dynamic energy-requirements for a PWR rated at
             10^3 Mw$_e$ without fuel recycle, using conventional ores, enrich-
             ment tails of 0.3%, and a 75% load factor. Electrical-energies
             are shown by a dashed curve, thermal energies are indicated by
             a dot-dash curve, and total energies by a solid curve. The gross
             annual electrical energy output is 6.57×10^6 Mw$_e$h. This figure
             has been reproduced from Rotty et al.[1]

25.3 Nuclear Strategies

A. Introduction

Nuclear reactor strategy studies are concerned with the determination of the optimum reactor mix and its deployment to meet electrical energy-demand requirements with minimum cost. The major factors involved in this analysis are the capital and fuel-cycle components of the cost of electricity.

Nuclear reactor strategy studies have been conducted for over a decade.[1,2] The initial studies were primarily carried out under the auspices of the AEC (now ERDA) to examine the various nuclear options and their associated costs. The results of these studies have been used to justify R & D funding for nuclear development, particularly the LMFBR program.[3] More recently, the question of nuclear strategy analysis has been broadened to include the general problem of energy planning. Rose[4] and Häfele[5,6] have published extensively on this problem.

[1]USAEC, "Estimated Growth of Civilian Nuclear Power," WASH-1055, Washington, D.C., 1965.

[2]USAEC, Cost Benefit Analysis of the U.S. Breeder Reactor Program," WASH-1126, Washington, D.C., 1969.

[3]Paul W. MacAvoy, Economic Strategy for Developing Nuclear Breeder Reactors, MIT Press, Cambridge, Mass., 1969.

[4]D. J. Rose, "Energy Policy in the U.S.," Scientific American 230, 20-29 (1973).

[5]W. Häfele, "A Systems Approach to Energy," American Scientist 62, 438-447 (1974).

[6]W. Häfele and A. S. Manne, "Strategies for a Transition from Fossil to Nuclear Fuels," International Institute for Applied System Analysis, Report RR-74-7, Laxenburg, Austria, June 1974.

In the present section, the main aspects of nuclear strategy analysis will be outlined and two major techniques, linear programming and simulation analysis, will be discussed. Next, a simplified approach to nuclear strategy analysis will be described in detail.

B.   Nuclear Strategy Studies

In the present treatment, nuclear strategy analysis will be approximated by isolating and concentrating on the technical and economic aspects of the problem. The purpose of this analysis is threefold: (i) Shed further light on the nuclear option as compared with other energy options. (ii) Determine the major controlling elements of nuclear reactor implementation. (iii) Apply the results of the analysis to the question of funding allocation for the development of advanced nuclear reactor technology, i.e., how do development costs for an advanced reactor technology compare with the potential benefits of its implementation?

Nuclear reactor studies involve the following three major steps: (i) A set of reliable nuclear reactor parameters must be obtained. This preliminary step is a very crucial one since confidence in the results is directly related to the reliability of the input reactor parameters. (ii) A projected nuclear reactor installment-demand curve and associated costs must be obtained. This demand curve should be explicit and should indicate nuclear reactor introduction, retirement rate and date. (iii) The data described in (i) and (ii) are then used as inputs to a calculational model which determines a reactor mix satisfying the above-mentioned demand curve and additional constraints, e.g., the introduction rate for new reactor technology. (iv) The results of the various reactor mixes satisfying a given demand curve are organized to form a set of scenarios which can be compared in a systematic way, yielding associated costs and benefits.

The primary nuclear reactor-strategy analysis-effort to date has been centered around a linear programming model[7] of the U.S. power economy, in which various nuclear growth patterns were compared with fossil-fuel alternatives to the year 2020. This linear programming technique was further developed at the Hanford Engineering and Development Laboratory (HEDL) into a code known as ALPS.[8, 9]

In ALPS, an objective function (a general electrical energy cost function depending on reactor type, material costs and capital costs) is minimized. The reactor mix which provides the lowest costs is automatically selected by the code. After selecting the mix, ALPS may be used to calculate the material requirements and associated discounted costs. ALPS is a large code requiring large computer facilities and long running times. Since the output is highly sensitive to small cost variations,[10] which in their nature are very uncertain, the results must be examined with care.

At the Battelle Columbus Laboratory,[11, 12] a simulation technique, providing a simple means of carrying out a large number of sequential

[7] D. E. Deonigi, "A Simulation of the United States Power Economy," Proceedings of the American Power Conference 32, 105-115 (1970).

[8] R. W. Hardie, W. E. Black and W. W. Little, "ALPS - A Linear Programming System for Forecasting Optimum Power Growth Patterns," HEDL-TM 72-31, Hanford Engineering and Development Laboratory, Hanford, Washington, 1972.

[9] R. W. Hardie and R. P. Omberg, "An Economic Analysis of the Need for Advanced Power Sources," Trans. Am. Nucl. Soc. 22, 501-502 (1975).

[10] D. E. Deonigi, Ref. [7], p. 114.

[11] T. L. Willke, "NEEDS: Nuclear Energy Electrical Demand Simulation," Trans. Am. Nucl. Soc. 21, 257 (1975).

[12] W. M. Pardee, "An Evaluation of Alternative Nuclear Strategies," Ann. Nucl. Energy 2, 819-839 (1975).

calculations, is employed in nuclear strategy studies. NEEDS, the computer code developed for this purpose, is basically an accounting system which keeps track of the operating capacity of nuclear reactors, fuel-cycle requirements and the associated costs. Unlike the linear programming code, which requires a decision policy for selecting a reactor mix based on a priori criteria, in the simulation technique the user specifies the decision rules. NEEDS is used to assemble a nuclear scenario based on (i) reactor-type priority, (ii) maximum fraction of total additions to capacity, (iii) maximum fraction of total existing capacity, and (iv) maximum capacity in Mw_e.

C. Simplified Nuclear Strategy Studies

During the assessment of helium-cooled reactors (see Section 21.15), a need arose for carrying out simple nuclear strategy analysis. These efforts will now be discussed.

Following the general procedure for carrying out strategy studies outlined in Section 25.3B, we first tried to obtain an accurate and consistent set of nuclear-reactor parameters. A survey of previous strategy studies indicated the following problems in obtaining nuclear-reactor parameter data: (i) The units used are inconsistent. (ii) There is a lack of detailed references concerning the source of reactor parameters. In view of these difficulties and because of the primary importance of obtaining an acceptable nuclear-reactor parameter set, it was decided to generate this set independently. It was found useful to express the nuclear-reactor parameters in terms of initial inventory requirements (which refer to the material requirements for an initial critical reactor core, in kg of material per Mw_e) and of operating requirements (which refer to the net material requirements at steady-state conditions, in kg of material per Mw_e-y).

1. Nuclear-Reactor Parameters

The initial inventory requirements for material X (in t/Mw_e) are calculated from the following ratio: fraction of material X in the initial core $\div S\eta$, where S is the reactor specific power (Mw_t/t) and η is the thermal efficiency.

The operating requirements for material X are determined as follows: Consider a material X with a weight fraction of x and let B represent the burn-up (in Mw_t-d/t). For a 365-day year, at a load factor LF, the material X used (in t burned up per Mw_e-y) is given by $365\,LF\,x/B\eta$. Assuming an initial concentration x_i and a final concentration x_f, the net requirement of material X per Mw_e-y is given by $(365\,LF/B\eta)(x_i - \gamma x_f)$, where γ is the ratio of reactor core inventory at the end of the year to that at the beginning of the year.

Table 25.3-1 provides a summary of the reactor-parameter data-bank, which was generated in this search. Whenever possible, direct burn-up tables obtained from manufacturers were used to determine the reactor parameters. In Table 25.3-2, we have listed the associated source material, as well as a short description of reactor types.

In interpreting and applying the reactor parameters, it is important to keep in mind the following points: (i) The parameters refer to the fertile and fissile fuel requirements of the reactor. A negative sign implies breeding or production of the corresponding fissile material. (ii) The listed parameter values are averages referring to steady-state equilibrium conditions. (iii) The concept of net yearly requirements must be used with caution. Since the net yearly requirements do not take into account delays in fabrication and reprocessing of material, they are only useful for calculating total material utilization.

Table 25.3-1 Reactor parameters required for nuclear strategy analysis. The reactor types used in this table are defined in Table 25.3-2.

Reactor type	GCFR-Pu	GCFR-Th	LMFBR-EO	LMFBR-AO	PWR-U	PWR-Pu	PWR-Pu0	HTGR-1	HTGR-2	HTGR-3	HTGR-4
burner	Pu	Pu	Pu	Pu	^{235}U	Pu	^{235}U	^{235}U	^{233}U	^{235}U	^{233}U
producer	Pu	Pu, ^{233}U	Pu	Pu	Pu	Pu	Pu recycle	CR=0.66, ^{233}U recycle	CR=0.69, ^{233}U	CR=0.82, ^{233}U recycle	CR=0.84, ^{233}U
initial inventory											
^{238}U (kg/Mw$_e$)	62.9	30.1	44.8	41.0	78.4	81.9	78.4	N.A.	N.A.	N.A.	N.A.
Th (kg/Mw$_e$)	0.0	31.7	0.0	0.0	0.0	0.0	0.0	32.3	32.3	49.4	50.1
^{235}U (kg/Mw$_e$)	0.0	0.0	0.0	0.0	1.8	2.28	1.84	1.4	0.0	1.89	0.0
U$_3$O$_8$* (kg/Mw$_e$), 0.2% tails	0.0	0.0	0.0	0.0	383.0	487.3	383.0	320.7	0.0	434.8	0.0
^{233}U (kg/Mw$_e$)	0.0	0.0	0.0	0.0	0.0	0.0	0.0	0.0	1.01	0.0	1.29
Pu fissile (kg/Mw$_e$)	2.65	2.68	2.90	2.07	0.0	0.0	0.0	0.0	0.0	0.0	0.0
load factor	75.0	75.0	75.0	75.0	80.0	80.0	80.0	80.0	80.0	80.0	80.0
net yearly requirements											
^{235}U (kg/Mw$_e$-y)	0.0	0.0	0.0	0.0	0.618	0.0	0.467	0.324	0.0	0.200	0.0
^{233}U (kg/Mw$_e$-y)	0.0	-0.338	0.0	0.0	0.0	0.0	0.0	0.0	0.300	0.0	0.157
Pu fissile (kg/Mw$_e$-y)	-0.326	0.052	-0.141	-0.170	-0.168	0.320	0.003	0.0	0.0	0.0	0.0
U$_3$O$_8$* (kg/Mw$_e$-y), 0.2% tails	0.0	0.0	0.0	0.0	142.0	0.0	105.6	74.9	0.0	47.0	0.0
Th (kg/Mw$_e$-y)	0.0	0.51	0.0	0.0	0.0	0.0	0.0	0.520	0.550	0.660	0.670
^{238}U (kg/Mw$_e$-y)	2.50	0.74	1.06	0.97	1.00	0.0	1.0	N.A.	N.A.	N.A.	N.A.

*For fissile enrichment; N.A. = not applicable.

Table 25.3-2 Reactor types and references.

Type	Description	Source
GCFR-Pu	Pu burner, Pu breeder	R. Cerbone, "Physics Design of a Gas-Cooled Fast Breeder Reactor," pp. 625-650, in Advanced Reactors: Physics, Design and Economics, edited by J. M. Kallfelz and R. A. Karam, Pergamon Press, London, 1975.
GCFR-Th	Pu consumer, ^{233}U breeder	R. Cerbone, ibid.
LMFBR-EO	early oxide Pu burner, Pu breeder	WASH-1535, "Proposed Final Environmental Statement: Liquid Metal Fast Breeder Reactor Program," Vol. 4, U.S. AEC, Washington, D.C., 1974.
LMFBR-AO	advanced oxide Pu burner, Pu breeder	J. J. Taylor, "The Status of LMFBR Development," Ann. Nucl. Energy 2, 705-724 (1975).
PWR-U	^{235}U burner, Pu converter	Fuel burn-up information obtained by private communication from the Westinghouse Electric Co.
PWR-Pu	^{235}U is in the initial inventory, but the reactor runs on a Pu-cycle	Obtained from older runs of R. Graham, General Atomic Co., San Diego, Ca., 1973.
PWR-Pu0	complete Pu recycle (i.e., no net yearly Pu requirements) and hence lower ^{235}U yearly requirements	WASH-1297, "High Level Radioactive Waste Management Alternatives," U.S. AEC, Washington, D.C., 1974.
HTGR-1	CR = 0.66, complete ^{233}U recycle (i.e., no net ^{233}U yearly requirements) and hence lower ^{235}U yearly requirements	R. Brogli, "The High Conversion HTGR for Resource Conservation," Report GA-A13606, General Atomic Co., San Diego, Ca., 1975.
HTGR-2	CR = 0.69, pure ^{233}U burner	R. Brogli, ibid.
HTGR-3	CR = 0.82, complete ^{233}U recycle (i.e., no net yearly requirements) and hence lower ^{235}U yearly requirements	R. Brogli, ibid.
HTGR-4	CR = 0.84, pure ^{233}U burner	R. Brogli, ibid.

2.   Calculational Procedure

These nuclear-reactor parameters were used as input data to a computer program consisting of the following two parts: (i) A mass-flow program, which keeps track of the material mass flows in five-year intervals. (ii) A cost-flow program, which keeps track of the cost flows in five-year intervals.

The mass flows were calculated from the following expressions: a. The reactor-inventory requirements for the year N are given by the (number of Mw_e introduced in the year N minus the number of Mw_e retired in the year N) × (inventory requirements per Mw_e); b. The reactor operating requirements for the year N are given by (number of Mw_e operating through the year N) × (operating requirements per Mw_e-y) × (time interval in y).

Cost calculations were restricted to determining the present worth of the U_3O_8 requirements. These normally account[13] for 80% of the associated costs for a given scenario. They were determined by using a marginal cost curve obtained from WASH-1535.[13]

Using the anticipated demand and the rates of introduction of new reactors, calculations were done, for 30-y reactor life (see Fig. 25.3-1), by first adopting a total installed nuclear demand curve (from the HEDL[14] calculations) as the base-demand curve (see Fig. 25.3-2). Then a standard[13, 14] new reactor-introduction rate was adopted with the rate doubling every two years. Introduction dates for new reactor technologies were decided a priori (e.g., a breeder introduction date was set for 1993). Iterative calculations were then carried out to find the "best" reactor-mix combination

[13]WASH-1535, Vol. IV, "Liquid Metal Fast Breeder Program," U.S. AEC, Washington, D.C., 1976.

[14]Private communication from workers at HEDL.

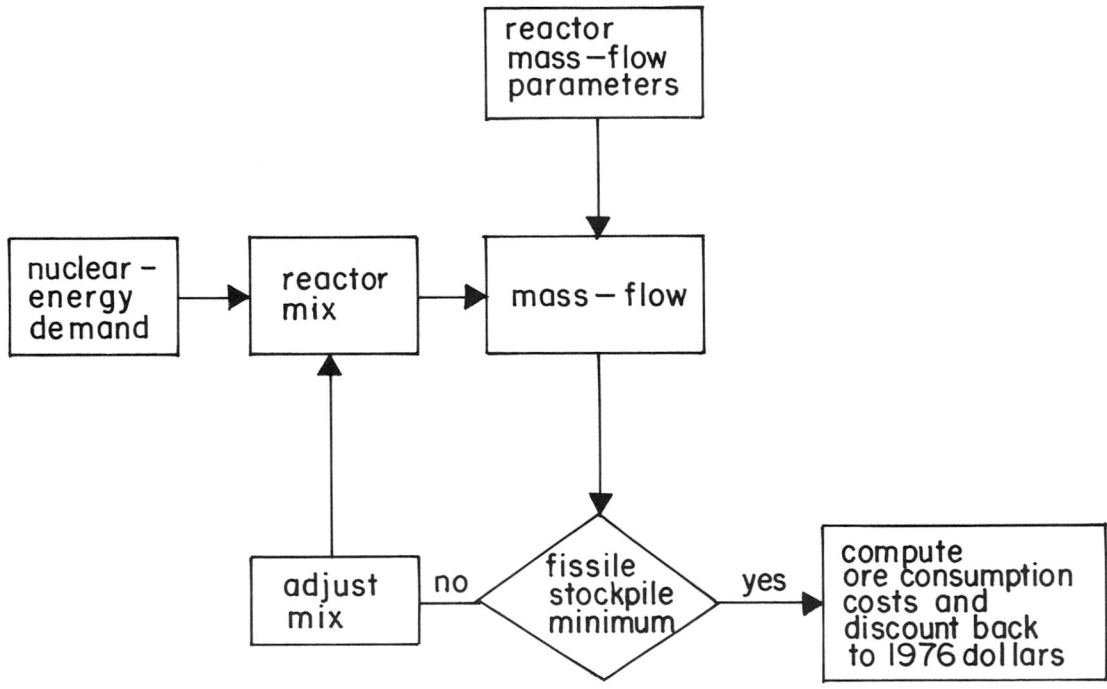

Fig. 25.3-1 Block diagram showing the procedure used for calculations.

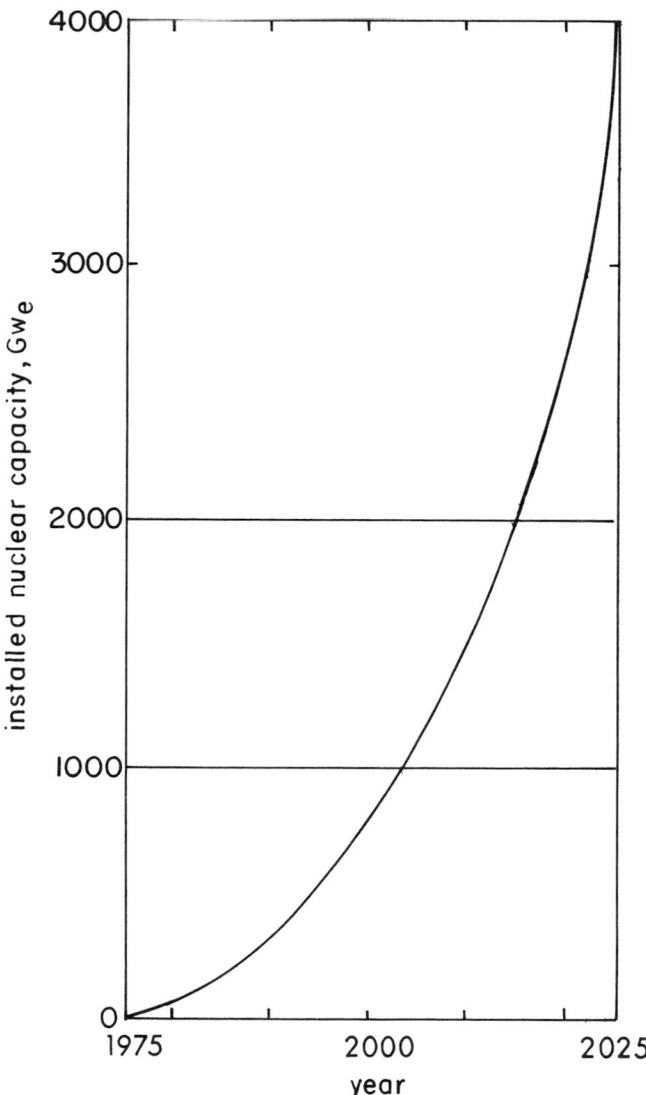

Fig. 25.3-2 Installed nuclear capacity in Gw_e as a function of time, adopted
from the HEDL[14] reference case.

in which U_3O_8 ore utilization was minimal, while balancing the Pu fissile and ^{233}U stockpiles to zero by the year 2025. This last criterion was adopted in order to avoid involving highly uncertain cost figures for these stockpiles. The study was restricted to resource utilization through the year 2025.

To test the overall adequacy of the method, a comparison was made with a reference calculation that incorporated adequate details regarding reactor types, dates and rates of introduction. A reasonably well documented case was found in the WASH-1535 cost-benefit studies.[13] Figure 25.3-3 shows the comparison between the results in WASH-1535 and those obtained by the use of our simpler method. We note good agreement. The differences in U_3O_8 requirements of up to 20% during the period 1990-2010 may be attributed to the lack of detailed information in our model or to the particular rate and date of introduction chosen for the PWR-Pu reactor.

3. Results

Having validated the overall method, a number of calculations were carried out by using the HEDL[14] scenarios as guidelines.

Case A - Case A is a base-case scenario (see Table 25.3-3) using only LWRs; the PWR-U, PWR-Pu0 and PWR-Pu were used. The PWR-Pu0 and PWR-Pu are new reactor types. The PWR-Pu0 is introduced in 1987 and the PWR-Pu is introduced at a later date when burning of plutonium produced by the PWR-U is needed.

Case B - For an LWR/HTGR scenario, the HTGR technology is introduced in 1983. The PWR-Pu is introduced in order to burn the plutonium produced by the PWR-U. Results are listed in Table 25.3-4.

Case C - Case C is an LWR/HTGR scenario, similar to Case B, but the HTGR-1 is replaced by the more efficient HTGR-3. Results are listed in Table 25.3-5.

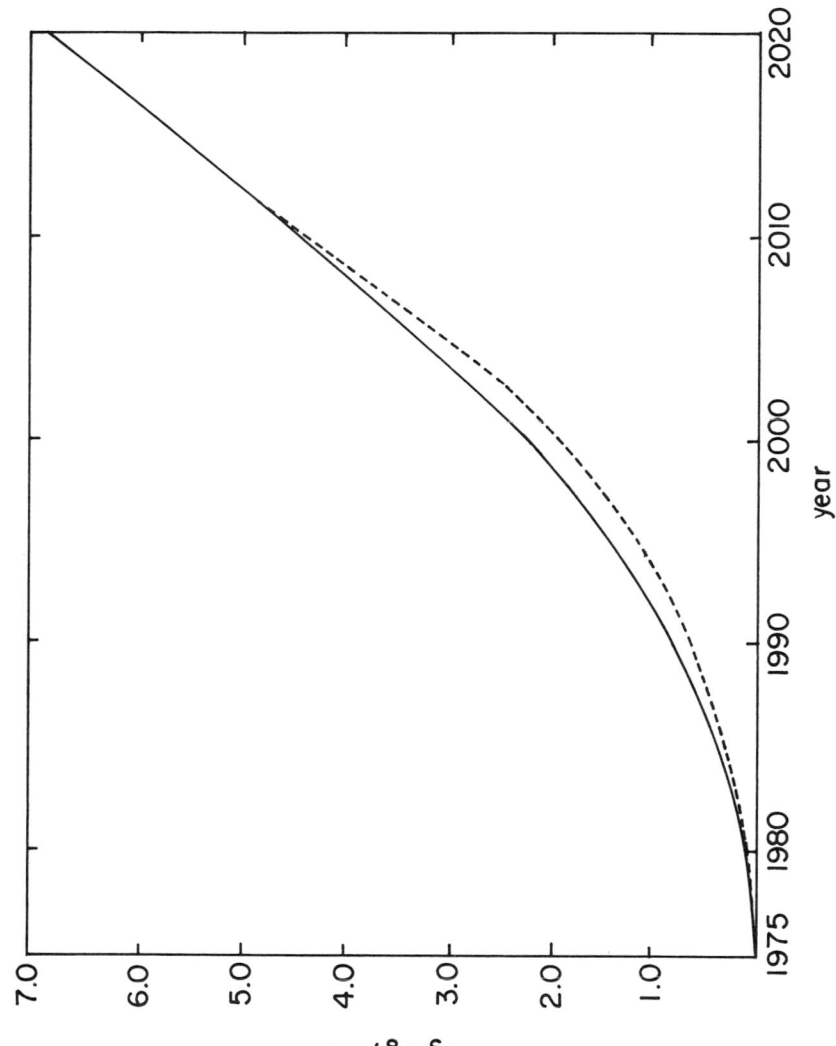

Fig. 25.3-3 Comparison of U_3O_8 ore requirements as a function of time in WASH-1535 (solid line) with our simpler calculations; 1 Mt = 10^6 tons.

Table 25. 3-3 Installed nuclear capacity in Gw_e as a function of time for
Case A.

Year	PWR-U	PWR-Pu0	PWR-Pu	Total
1975	43	0	0	43
1980	80	0	0	80
1990	365	0	0	365
2000	806	94	0	900
2010	806	944	0	1750
2020	806	1047	1077	2930
2025	806	1597	1630	4033

Table 25. 3-4 Installed nuclear capacity in Gw_e as a function of time for
Case B.

Year	PWR-U	PWR-Pu0	PWR-Pu	HTGR-1	Total
1975	43	0	0	0	43
1980	80	0	0	0	80
1990	349	0	0	16	365
2000	375	47	25	453	900
2010	475	47	255	1003	1750
2020	481	47	419	1983	2930
2025	481	47	555	2953	4033

Table 25. 3-5 Installed nuclear capacity in Gw_e as a function of time for
Case C.

Year	PWR-U	PWR-Pu0	PWR-Pu	HTGR-3	Total
1975	43	0	0	0	43
1980	80	0	0	0	80
1990	349	0	0	16	365
2000	375	47	25	453	900
2010	482	47	218	1003	1750
2020	482	47	418	1983	2930
2025	482	47	548	2956	4033

Case D - An LWR/HTGR/breeder scenario with the LMFBR-AO is used. The HTGR-1 is introduced in 1983 and the LMFBR-AO in 1993. Results are listed in Table 25.3-6.

Case E - An LWR/HTGR/breeder scenario is used similar to that in Case D, but the breeder is GCFR-Pu. This case illustrates the advantages of the GCFR-Pu. Results are shown in Table 25.3-7.

Case F - An LWR/HTGR/breeder scenario is used which shows how intensive implementation of helium-cooled technology will optimize the U_3O_8 resources. The GCRF-Pu and GCRR-Th are introduced in 1993. The HTGR-3 is introduced in 1990. In the year 2000, an HTGR on a direct cycle is introduced, thereby increasing the HTGR conversion efficiency by 25%. Case F also illustrates the concept of a symbiotic mix[15, 16] between GCFRs and HTGRs. Accordingly, the GCFR-Pu supplies Pu for the GCFR-Th which breeds ^{233}U for the HTGR-4. The results are shown in Table 25.3-8.

The preceding scenarios, their U_3O_8 requirements and the associated U_3O_8 costs discounted at 7.5% to 1976, as well as cost benefits, are summarized in Table 25.3-9. Figure 25.3-4 shows the U_3O_8 requirements for a number of cases as a function of time.

Table 25.3-9 and Fig. 25.3-4 show that the U_3O_8 requirements to the year 2000 are nearly independent of the scenarios adopted. The ore requirements are dictated by the total nuclear energy demand, the need of the PWR-U to supply Pu for the breeder, and the absence of new reactor technology which could

---

[15] P. Fortescue, "Fast Breeder and HTGR: A Profitable Partnership," Nuclear News 15, No. 4, 36-39 (1972).

[16] R. Brogli and G. Schlueter, "Fuel Economics of GCFR/HTGR Symbiotic Systems," GA-A13110, General Atomic Co., San Diego, Ca., 1974.

Table 25.3-6 Installed nuclear capacity in Gw_e as a function of time for Case D.

Year	LMFBR-AO	PWR-U	PWR-Pu	HTGR-1	Total
1975	0	43	0	0	43
1980	0	80	0	0	80
1990	0	349	0	16	365
2000	16	522	75	287	900
2010	512	662	188	388	1750
2020	1527	662	450	450	2930
2025	2554	662	301	516	4033

Table 25.3-7 Installed nuclear capacity in Gw_e as a function of time for Case E.

Year	GCFR-Pu	PWR-U	PWR-Pu	HTGR-1	Total
1975	0	43	0	0	43
1980	0	80	0	0	80
1990	0	349	0	16	365
2000	16	475	125	284	900
2010	512	475	390	373	1750
2020	1520	475	435	500	2930
2025	2554	475	488	516	4033

Table 25.3-8 Installed nuclear capacity in Gw_e as a function of time for Case F.

Year	GCFR-Pu	GCFR-Th	PWR-U	PWR-Pu	HTGR-3	HTGR-4	Total
1975	0	0	43	0	0	0	43
1980	0	0	80	0	0	0	80
1990	0	0	365	0	0	0	365
2000	16	0	710	126	48	0	900
2010	115	544	710	126	48	207	1750
2020	126	769	710	130	48	1147	2930
2025	126	779	710	130	48	2250	4033

Table 25. 3-9 Summary of the scenarios studies, their U_3O_8 requirements, and U_3O_8 costs and benefits.

Case	GCFR-Pu, Gw_e	GCFR-Th, Gw_e	LMFBR-AO, Gw_e	PWR-U, Gw_e	PWR-Pu, Gw_e	PWR-Pu 0, Gw_e
A	—	—	—	806	1630	1597
B	—	—	—	481	555	47
C	—	—	—	482	548	47
D	—	—	2554	662	301	—
E	2554	—	—	475	488	—
F	126	779	—	710	130	—

Case	HTGR-1, Gw_e	HTGR-3, Gw_e	HTGR-4, Gw_e	U_3O_8 requirements, Mt	U_3O_8 costs (discounted at 7.5% to 1976), $\$10^9$	Cost benefit, $\$10^9$
A	—	—	—	7. 89	67. 3	—
B	2950	—	—	6. 54	44. 7	22. 6
C	—	2956	—	5. 77	36. 7	30. 6
D	516	—	—	4. 11	25. 2	42. 1
E	516	—	—	3. 52	21. 7	45. 6
F	—	48	2250	3. 43	21. 3	46. 0

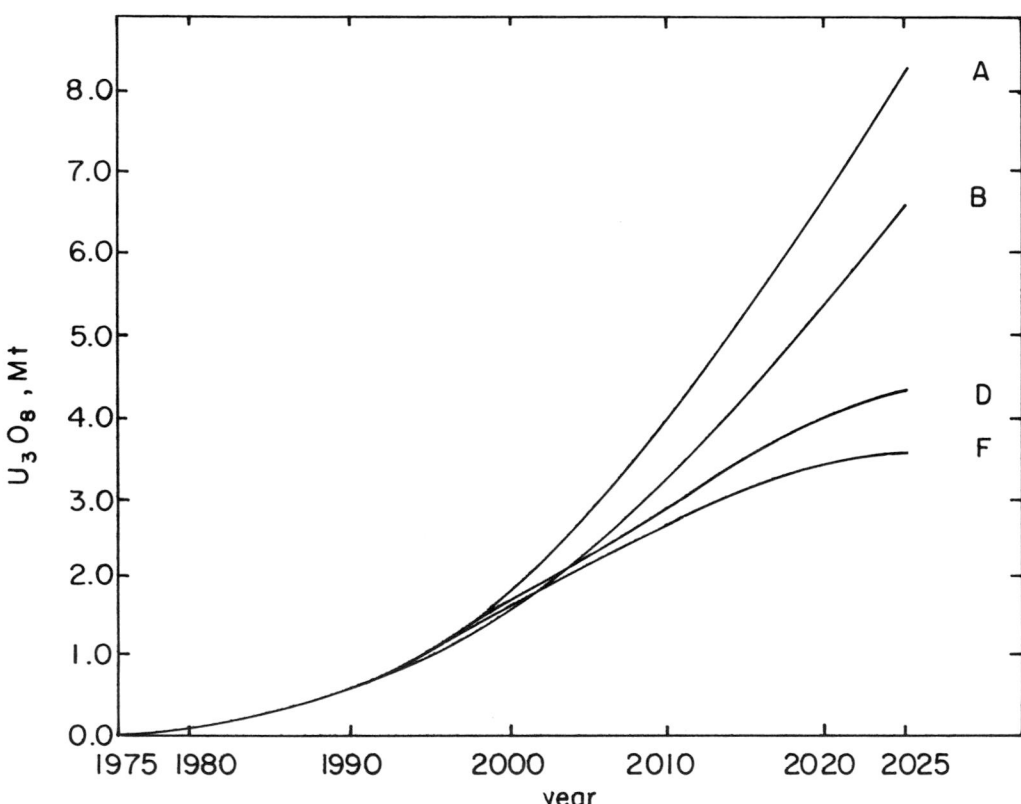

Fig. 25.3-4 U_3O_8 requirements in Mt as a function of time for the
major cases studied; 1 Mt = 10^6 tons.

supply a significant fraction of the total energy demand while minimizing the use of U_3O_8. Comparison of scenarios C and E with A indicates how LWR/HTGR and LWR/HTGR/GCFR implementation will reduce U_3O_8 demand by up to 4.47 Mt. The advantages of implementing helium-cooled reactor technology are illustrated by the following comparisons: (a) scenarios D and E indicate an advantage of 0.69 Mt in U_3O_8 requirement for implementing the GCFR-Pu in place of the LMFBR-AO; (b) scenarios D and F indicate how a symbiotic GCFR/HTGR system reduces the number of breeders needed by over a factor of two. This decrease in the number of breeders employed also reduces the plutonium mass flows, thus eliminating many of the risks involved in an intensive plutonium economy.

CHAPTER 26

ENERGY POLICIES

26.1 Introduction

Implicit in the population and per capita power consumption estimates listed in Table 1.2-2 are growth scenarios that may not be sustainable in view of limitations on the supplies of foods and raw materials (including energy sources) or of environmental loadings that are likely to be associated with the implied population and energy-use growth rates. A number of authors[1,2] have emphasized the importance of early and effective curtailment of population growth in the developed or "rich" countries, as well as in "poor" countries. For example, Holdren[2] has noted that a 1%/y growth rate of population in rich countries and a 2.5%/y population growth rate in poor countries, accompanied by a 5%/y growth rate in total energy consumption, will lead to a year 2020 world

---

[1] A. D. Sakharov, Progress, Coexistence, and Intellectual Freedom, W. W. Norton Co., New York, N.Y., 1968.

[2] J. P. Holdren, "Energy and Prosperity: Some Elements of a Global Perspective," paper presented at the 24th Pugwash Conference on Science and World Affairs, Baden, Austria, August 28 to September 2, 1974.

with 1.9×10^9 p in rich countries using 34.3 kw_t/p and 8.3×10^9 p in poor coun-tries using 1.66 kw_t/p; the corresponding worldwide rate of energy consumption is 80×10^9 kw_t or more than 10 times the 1972 use rate. Holdren considers this scenario to be unacceptable.

With heavy emphasis on the transfer of technical resources and capital from rich to poor countries, as advocated by Sakharov,[1] the following complex-ion for a year 2020 world may be achievable:[2] 1.4×10^9 p in rich countries using 8.03 kw_t/p (which is about equal to the 1960 U.S. consumption level) and 5.3×10^9 p in poor countries using 5.26 kw_t/p (which is about equal to the year 1972 use rate in Japan). The corresponding total power capacity is 28×10^9 kw_t or about four times the year 1972 use rate. Holdren[2] regards this scenario as desirable and achievable provided new energy-resource development is re-stricted to technologies that do not produce disastrous or irreversible ecologi-cal changes.

The definitions of global scenarios of the type defined in the preceding paragraphs and their practical realization represent political schedules which determine the making of energy policies. Before addressing basic issues in-volved in the design of a better world, we shall emphasize in our discussions im-portant factual information that must be considered in the definition of energy policies. Thus we address first the following problem areas: population growths and energy needs for improving worldwide standards of living, energy use in food production, monitoring the potential impact of growing energy use on the climates of the world, energy farms, and energy conservation. We con-clude with a discussion of energy policy in which we emphasize the U.S. ener-gy-policy plan as defined in ERDA-48.

26.2 Population Growths and Energy Needs

Population growth projections for the U.S. have been discussed briefly in Section 8.1.   More detailed projections made at the U.S. Bureau of the Census during 1972 are listed in Table 26.2-1 and refer to fertility levels of 1.8, 2.1, 2.5, and 2.8 with annual net immigration of 400,000 p.   Here the fertility rate is defined as the number of births per woman at the end of childbearing.   The U.S. Bureau of the Census median estimates for the 1985 and 1995 U.S. population are 231×10^6 p and 246×10^6 p, respectively (see p. 338 in Volume I). Comparison of these data with the entries in Table 26.2-1 shows that they correspond to the lowest listed mean fertility rate of 1.8.

World population projections to the years 1980, 2000 and 2020 are listed in Table 26.2-2 for mean annual growth rates of 1% and 2%, starting from known population levels in 1971.   During the period 1963 to 1971, annual percent increases in population[1] were 0.4%/y in the U.K.; 0.8%/y in Germany, Italy and Poland; 0.9%/y in France; 1.1%/y in the U.S.S.R., USA, Japan and Spain; 1.8%/y in the People's Republic of China; 2.1%/y in Pakistan; 2.2%/y in India and Korea; 2.5%/y in Nigeria, Turkey and Egypt; 2.7%/y in Thailand; 2.8%/y in Indonesia and Brazil; 3.0%/y in the Phillipines; 3.2%/y in Mexico.   During these same years, the average world population growth rate was 2%/y.   Thus, an average growth rate of 2%/y for the world represents a continuation of trends during the recent past while a growth rate of 1%/y represents a reduction of 50% below recently applicable levels.   Although we cannot predict with a high degree of certainty what the population of the world will be in the future, it appears likely that the upper values (6.58×10^9 p during the year 2000 and

---

[1]U.S. Bureau of the Census, Pocket Data Book, USA 1973, p. 38, Table 4, U.S. Government Printing Office, Washington, D.C., 1973.

Table 26.2-1 U.S. population (in 10^6p) projections during 1972 to the year 2020 for selected fertility levels (in number of births per woman at the end of childbearing) and annual net immigration of 400,000 p. Based on data published by the U.S. Bureau of the Census.[1]

Year	Births per woman			
	2.8	2.5	2.1	1.8
Estimates				
1960	181	181	181	181
1970	205	205	205	205
1972	209	209	209	209
Projections				
1975	216	215	214	213
1980	231	229	224	222
1985	249	244	236	231
1990	266	259	247	239
1995	283	272	256	246
2000	300	286	264	251
2005	321	301	273	255
2010	344	318	282	259
2015	368	335	290	263
2020	392	351	298	265

9.78×10^9 p by 2020) will apply unless a very drastic reorientation in social values concerning population growth is achieved in the absence of major natural or man-made calamities.

During 1970, the average continuous thermal power consumption by the population of the world as a whole was 2 kw_t/p, while the corresponding U.S. consumption amounted to 10 kw_t/p (see Table 1.2-2). We have listed the average continuous thermal power consumption for the entire world population in Table 26.2-3 for assumed annual growth rates of 1%/y, 2%/y, 4%/y, and 6%/y. At a 4%/y rate of increase annually, the average world consumption will reach

Table 26.2-2 World population (in 10^6p) during 1971 and projected populations at various growth rates. The 1971 data are taken from a publication by the Statistical Office of the United Nations, New York.

Country, 1971 population	Populations in 10^6p at 1 and 2% annual growth rates		
	1980	2000	2020
People's Republic of China, 787	861; 941	1,050; 1,398	1,280; 2,077
India, 550	602; 657	734; 977	895; 1,451
U.S.S.R., 245	260, 293	327, 435	399; 647
USA, 207	226; 247	302; 368	337; 546
Indonesia, 125	156; 149	208; 222	203; 330
Pakistan (including Bangladesh), 117	137; 140	156; 208	190; 309
Japan, 105	110; 125	140; 186	171; 277
Brazil, 95	104; 114	127; 169	155; 251
German Federal Republic, 59	65; 71	79; 105	96; 156
Nigeria, 57	62; 68	76; 101	93; 150
United Kingdom, 56	60; 67	75; 99	91; 148
Italy, 54	59; 65	72; 96	88; 142
France, 51	56; 61	68; 91	83; 135
Mexico, 51	56; 61	68; 91	83; 135
Philippines, 38	42; 45	51; 67	62; 100
Turkey, 36	39; 43	48; 64	59; 95
Thailand, 35	38; 42	47; 62	57; 92
Spain, 34	37; 41	45; 60	55; 90
Egypt, 34	37; 41	45; 60	55; 90
Poland, 33	36; 39	44; 59	54; 87
Republic of Korea, 32	35; 38	43; 57	52; 84
World, 3,706	4,053; 4,429	4,946; 6,581	6,035; 9,779

Table 26.2-3 Projected equivalent world-average continuous uses of thermal
power to the year 2020 for selected compound growth rates. A
year 1970 base of 2 kw_t/p is assumed.

Year	Compound continuous equivalent power growth rate			
	1%/y	2%/y	4%/y	6%/y
1975	2.10	2.21	2.43	2.67
1980	2.20	2.44	2.96	3.58
1985	2.32	2.69	3.60	4.89
1990	2.44	2.97	4.38	6.41
1995	2.56	3.28	5.33	8.58
2000	2.70	3.62	6.49	11.5
2005	2.83	4.00	7.89	15.4
2010	2.98	4.42	9.60	20.6
2015	3.13	4.88	11.7	27.5
2020	3.29	5.38	14.2	36.8

the U.S. level of 1970 about 2011; at a 6%/y rate of annual growth in per capita
energy consumption for the population of the world, the U.S. level of 1970 will
be reached during 1998.

We may postulate that an asymptotic limit of 30 kw_t/p will assure the
complete comfort of a terminally blissful society. The peoples of the world
will reach this state around 2017 at a compound energy-consumption growth
rate of 6%/y from 1970; the population in the U.S. will reach this same value
at a 3%/y increase in energy consumption above the 1970 value by the year 2007.

If we assume that the average world population growth rate will be reduced by 50% to 1%/y and that the terminal required equivalent continuous power consumption is 30 kw_t/p, then it follows that the year 2020 world energy

needs become 6.035×10^9 p $\times$ 30 $(kw_t/p) \times (8.76 \times 10^3 \ h/y) \times (0.949 \ Btu/$

$2.78 \times 10^{-4} \ kwh_t) \times (Q/10^{18} \ Btu)$ = $5.41Q/y$ or about 22.6 times the energy

used during 1970. If the required terminal value is only 20 kw_t/p, the year

2020 world energy consumption is reduced to $3.6Q$.

It is not difficult to invent alternative population and energy-consumption

schedules. For example, at the total 1970 world energy-consumption level of

$0.24 \ Q/y$, a population reduction to about 400×10^6 p will yield an average con-

tinuous thermal power consumption of 20 kw_t/p, i.e., the world population

would have to be reduced to about half of that of the People's Republic of China

during 1975; correspondingly, the year 1970 total world energy consumption

would accommodate about 800×10^6 p (i.e., about the 1975 population of the

People's Republic of China) at the average 1970 U.S. continuous thermal power

consumption level of 10 kw_t/p.

Population growth, economic well being, and energy use are intimately

connected, as we have emphasized in Chapter 8. If we insist on maintaining

historically applicable relations between economic well being and energy con-

sumption (see Chapter 4 for details), then we may relate the average per capita

continuous power consumption to average per capita income. In this type of

correlation, we may differentiate between countries and regions by allowing for

variations in the nature of industrial activities that contribute to income (com-

pare Section 4.1). When the indicated analysis is properly performed, the ter-

minally blissful society with adequate energy supply will show divergencies in

energy-consumption that are consistent with local technological idiosyncracies.

We shall not elaborate on this hypothesis here although its quantification may

be readily accomplished by using available data on current production, gross

national product, and energy consumption as reference bases.

26.3 Energy Use and Food Production

A serious determination of energy policy must allow for possible limitations in food production associated with curtailment of energy availability. Preferred allocation of energy resources to the food sector is mandatory up to the
level where adequate production for world needs occurs.

A. Arable World Land Areas

In 1965, The world had 7.9×10^9 ac* of arable land, of which about
3.4×10^9 ac, corresponding to 44%, were actually used as cropland.[1, 2, 3]
Since the best land is used first, expansion of land use will be costly. About
1.75×10^9 ac have low mineral content and consist of soils classified as Podzol,
Red-Yellow Podzolic, Latosol, and Ando;[1] these soils require very heavy fer-
tilization for use as cropland.

About 0.85×10^9 ac are potentially usable after irrigation. Of this total,
about 0.40×10^9 ac were being irrigated in 1965 and perhaps 0.50×10^9 ac by
1975.[1] Arkley[1] has estimated the cost for irrigation at about \$500/ac, which
amounts to $\$175 \times 10^9$ for a cropland expansion of 0.350×10^9 ac. These land-
development charges do not include annual water expenses involved in land use

---

*1 ac $= 43,560$ ft$^2 = 4,840$ yd$^2 = 4,046.873$ m$^2 = 0.4046873$ hectare; 1 hectare
$= 1$ square hectometer $= (100$ m$)^2 = 10^4$ m$^2 = 2.471044$ ac.

[1]R. J. Arkley, "Energy Costs of World Agriculture," paper presented at a
Conference on Limits to Non-Growth, University of California/San Diego,
La Jolla, Calif., October 10-12, 1974.

[2]"The World Food Problem," Report of the President's Science Advisory Commission (PSAC) 2, 5-135 (1967).

[3]R. Revelle, "Population and Resources," Research Paper No. 5, Center for
Population Studies, Harvard University, Cambridge, Mass., July 1974.

and crop production.   Irrigated land is generally dedicated to intensive agriculture and requires heavy application of fertilizers and correspondingly great energy consumption.

The potentially available crop areas are being continuously reduced by erosion and by urbanization.   In 1965, the ratio of world-cropland area ($\simeq 3.4 \times 10^9$ ac) to world population was very close to 1 ac/p.

B.   Arable U.S. Land Areas

In 1958, the potentially usable U.S. arable land area was estimated to be 807×10^6 ac.[4]   This estimate allowed for destruction of about 50×10^6 ac by erosion* and 40×10^6 ac loss by urbanization (including construction of cities, highways, airports, etc.).   Erosion had also removed about half (7 in.) of the topsoil from an additional 50×10^6 ac,[5] which remained arable but less fertile. Of the 807×10^6 ac, 34×10^6 ac had been allocated for non-agricultural farm use (farmsteads, country roads, etc.) while 161×10^6 ac were only usable intermittently.[1]   The remaining arable 612×10^6 ac were allocated as follows: 367×10^6 ac (60%) for cropland and 245×10^6 ac (40%) for forest and pasture.

The 1958 ratio of cropland use in the U.S. (367×10^6 ac) to the 1958 population ($\simeq 174 \times 10^6$ p; compare Table 8.1-2 in Volume I) is seen to be 2.11 ac/p. The loss of arable land in the U.S. per person, prior to 1958, amounted to -90×10^6 ac/174×10^6 p $\simeq -0.52$ ac/p lost from all causes and -40×10^6 ac/

---

[4] Basic Statistics of the National Inventory of Soil and Water Conservation, U.S.D.A. Statistical Bulletin No. 317, Washington, D.C., 1967.

* The U.S.D.A. Soil Conservation Service (SCS) was formed to help alleviate this acute problem.   As of 1958, it was estimated that only 31% of U.S. cropland was adequately protected against erosion.

[5] H. H. Bennett, Elements of Soil Conservation, 2nd ed., McGraw Hill Book Co., New York City, N.Y., 1955.

174×10^6 p = -0.23 ac/p lost as the result of urbanization.[1]

Between 1958 and 1975, 15.8×10^6 ac were urbanized in the U.S., of which 12.6×10^6 ac (79.7%) were arable land. The corresponding population growth was 44×10^6 p (compare Table 8.1-2 in Volume I). Hence the cropland loss per person associated with urbanization amounted to -0.29 ac/p for this period. Accelerated loss of cropland with population growth appears to be occurring in the U.S.

The conversion of cropland to land used for pastures or forests has been estimated[1,4] to be 12.1×10^6 ac during the seventeen-year period between 1958 and 1975. This loss of 0.71×10^6 ac/y has been attributed[1] to the utilization of tractors and other power equipment, which is making land more vulnerable to erosion. Nevertheless, Arkley[1] has estimated that the ratio of cropland area in ac to U.S. population will be maintained close to 2 ac/p for another 50 y.

In California, about 25% of the total potentially available arable land had been converted for urbanization by 1975.

C. Energy Costs in Food Production

Ennis et al[6] have noted that the crop yield per acre is strongly affected by minimal use of tractor horsepower (hp), fertilizers and pesticides, each of which represents a significant energy charge. An an example, the substitution of 0.5 to 0.8 tractor hp for human and animal power is needed for optimal crop production.[7] Similarly, the 1966 use of 6.15×10^6 mt of fertilizers in devel-

---

[6]W. B. Ennis, Jr., I. T. Ellis, L. L. Jansen, and L. D. Newsom, "Inputs for Pesticides. The World Food Problem," PSAC Report 3, 130-175 (1967).

[7]G. W. Giles, "Agricultural Power Requirements," pp. 175-216 in Ref. [2].

oping countries would have had to be increased nearly ten times in order to double crop production.[8] More than half of the fertilizer requirement is of the form of nitrogen-containing compounds which are highly energy-intensive; the remaining chemicals used for fertilizers contain potassium or phosphate compounds with energy costs of 12 to 18% per unit mass of that required per unit mass of nitrogen-containing fertilizer (see Table 26.3-1).[9] For primary crop production, Pimentel et al,[9] Revelle[10] and others have noted that the ratio

Table 26.3-1 Energy costs in U.S. agriculture.[9]

Service or material	Energy cost, in $(bbl)_e$ at 5.82×10^6 Btu/$(bbl)_e$
typical farm laborer working 1 year, 40 h/week, 50 weeks/y	0.74
nitrogen-containing fertilizer per ton of N	11.4
phosphorus-containing fertilizer per ton of P	2.06
potassium-containing fertilizer per ton of K	1.43
insecticides or herbicides per ton	14.9
one ton of corn	0.87

[8] L. B. Nelson and R. Ewell, "Fertilizer Requirements for Increased Food Needs," pp. 95-117 in Ref. [2].

[9] D. Pimentel et al, "Food Production and the Energy Crisis," Science 182, 443-449 (1973).

[10] R. Revelle, "Food and Population," Scientific American, No. 231, pp. 161-170, September 1974.

of primary energy to food energy lies in the range 0. 3 to 0. 6. On the other

hand, for the entire U.S. food system, the ratio of primary energy to the

energy content of food products[*] has increased from about 4 in 1940, to about

5. 5 in 1950, to about 5. 7 in 1960, to about 6. 5 in 1970.[11, 12]

We have noted in Table 3. 2-6 of Volume I that the average 1971 dollar

cost of the fuel used, per dollar value of food shipped in the U.S., was about

$0.004. Figure 4.7-1 in Volume I contains a listing of the 1974 dollar values

for various food items, as well as a summary of food-product energy costs in

$/$10^6$ Btu; these entries are easily converted to energy costs per unit mass

for selected food items but are only rough estimates.

Referring to Fig. 4.7-1 of Volume I, we note that corn at $3.00/bushel

corresponds to about 33×10^6 Btu or 9.09×10^4 Btu/bushel. But one bushel

in agricultural trade refers to 56 lbs. Hence the energy cost (from Fig. 4.7-1)

is 1.62×10^3 Btu/lb or 0.56 bbl_e/t, which is appreciably smaller than the value

0.87 bbl_e/t listed in Table 26.3-1. The corresponding energy cost for boneless

Swiss steak is 0.30 bbl_e/t while that for cheddar cheese at $1.20/lb and $650/

10^6 Btu is 0.63 bbl_e/t. Energy costs of other food products may be estimated

similarly from the compilation given in Fig. 4.7-1.

The primary energy per unit of food product used in the U.S. for agri-

cultural purposes is larger than in undeveloped countries by factors of 3 to 7.

Odum[12] has referred to this U.S. practice as energy-subsidized industrial

agriculture. As the use of tractor hp increases, the use of farm man-hours

decreases with a correlation coefficient of -0.974 in the U.S.[1] A composite

---

[*]Energy values refer to the heat of complete combustion.

[11]P. R. Stout, "Agriculture's Energy Requirements," paper presented at the
65th Annual Meeting of the American Society of Agronomy, Las Vegas,
Nevada, 1973.

[12]H. T. Odum, "Energetics of World Food Production," pp. 55-94 in Ref.
[2].

diagram showing farm man hours, energy use, and persons supplied with food products in the U.S. as functions of time is shown in Fig. 26.3-1. Reference to Fig. 26.3-1 indicates a significant increase in worker productivity, which has been relatively greater than the increasing energy costs that have occurred with time. For example, between 1940 and 1970, the number of persons supplied per food worker increased by about a factor of 4 (from about 12 to about 47) while the total energy input into the food <u>system</u> tripled and the <u>per</u> <u>capita</u> energy input did not quite double [it increased from $\sim 0.75 \times 10^{15}$ kcal/y $\div 131.7 \times 10^6$ p $= 5.7 \times 10^6$ kcal/py $= 3.87$ (bbl)$_e$/py in 1940 to $\sim 2.3 \times 10^{15}$ kcal $\div 209 \times 10^6$ p $= 11 \times 10^6$ kcal/py $= 7.48$ (bbl)$_e$/py in 1970].

The total 1970 energy use for agriculture in the U.S. is seen to represent 2.2×10^{15} kcal/y $= 8.7 \times 10^{15}$ Btu/y or about 12.5% of the total 1970 energy use of 7×10^{16} Btu/y. It should be noted that this entry for energy charges in the food <u>system</u> includes not only energy costs for production but also for harvesting, processing, packaging, distribution, etc. By contrast, the entries in Table 3.2-3 show that 3.2×10^{11} kwh$_t$ were used in food production, exclusive of beverages, during 1968 while industrial energy consumption corresponded to 18.3×10^{12} kwh$_t$, or 1.75% of industrial energy use was allocated to food production during 1968. Furthermore, as is shown in Table 3.2-7, 35.4×10^9 kwh$_e$ of electrical energy were used in the U.S. during 1971 in the manufacture of food and kindred products; this represented 6.83% of the total electrical energy $(517.7 \times 10^9$ kwh$_e$) used in all industries; similarly (see Table 3.2-6), 266.8×10^9 kwh$_t$ were used in the manufacture of food and kindred products during 1971, corresponding to 8.01% of $3,332.4 \times 10^9$ kwh$_t$ used in all industries.

The estimate of about 13% of total energy used in the food system during 1970 is consistent with calculations by Steinhard and Steinhart,[13] whose

[13] J. S. Steinhard and C. E. Steinhart, "Energy Use in the U.S. Food System," Science <u>184</u>, 307-316 (1974).

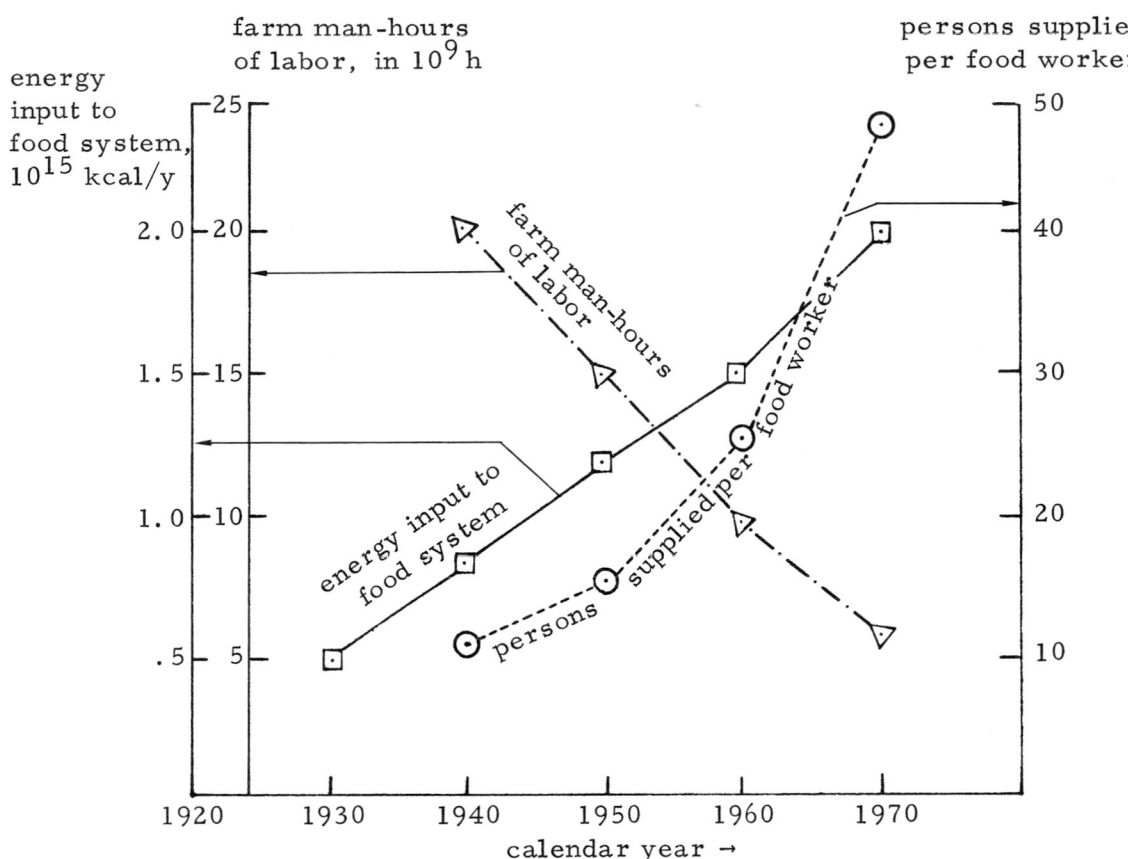

Fig. 26.3-1 Composite representation of energy input, man-hours of labor, and persons supplied per food worker as functions of time; reproduced from A. B. Cambel and R. C. Warder.[14]

results are shown in Table 26.3-2; similar values have been given by Hirst.[15]

---

[14]A. B. Cambel and R. C. Warder, "Energy Resource Demands of Food Production," Energy (in press in Volume 1, 1976).

[15]E. Hirst, "Food-Related Energy Requirements," Science 184, 134-138 (1974).

Table 26.3-2 Energy use in the food sector during 1970; reproduced from
Ref. [14] and based on data given in Ref. [13].

Process component	Consumption in 10^{12} kcal/y	Percentage of total required in food processing	Percentage of total U.S. consumption
food growing on farms	526.1	24	3.11
industrial food processing	841.9	39	4.97
food storage and preparation for commercial and home use	804.0	37	4.76
totals	2,172	100	12.84

D. Energy Savings in Food Production

A compilation for 1964 of tractor, animal and human labor in agriculture
is reproduced from Arkley[1] in Table 26.3-3. The discrepancy shown between
tractor use in developed countries and in the world as a whole has not changed
greatly in recent years. Fuel consumption may be estimated roughly by assuming a utilization rate of 0.1 gallon per tractor-hp per 400 h. Optimal crop production requires about 0.5 tractor-hp per hectare or 0.202 tractor-hp/ac.
Giles has estimated that this goal will be reached by 1998 in Latin America and
Oceania, while Asia and Africa will fall short by 45% and 34%, respectively. Increasing use of tractor-hp in food production will lead to higher total energy consumption and to reduced energy use per unit of product. Extreme examples of
energy consumption per unit area are provided by the U.S. with 0.48×10^6
kcal/ac for crop growth and 1.24×10^6 kcal/ac in the food system and by
Tsembaga with 0.09×10^6 kcal/ac in the food system. [14, 16]

----

[16] R. A. Rappaport, "The Flow of Energy in an Agricultural Society," Scientific American, No. 225, p. 117, September 1971.

Table 26.3-3 Sources of agricultural power during 1965; reproduced from Arkley.[1]

Power source	Use in developed countries:		Use in developing countries:	
	10^6 hp	% of total	10^6 hp	% of total
4-wheel tractor	333	90.4	56	36.2
garden tractor	17.5	4.8	17.5	11.3
draft animal	13	3.5	51	33.0
human	4.8	1.3	30.2	19.5
total	368.3	100	154.7	100

Among obvious conservation measures are reduced populations and reduced per capita food consumption, especially in affluent countries where people habitually overeat.

Cambel and Warder[14] have discussed possible uses of solar and wind energy as primary sources, solar drying, farm-waste utilization, optimization of farm implements, and increased use of manual labor. An authoritative evaluation of worldwide political and technical problems associated with food production and nutrition has been published recently.[17] Pimentel et al[18] have projected 1985 energy consumption for food production at 3.4×10^{15} kcal (11.9 quads) while between 9.0×10^{15} and 12.0×10^{15} kcal (35.6 to 47.5 quads) will be used in the entire food system (20 to 30% of total world-energy consumption by 1985).

[17]Special issue on Food, Science 188, 503-653 (1975).

[18]D. Pimentel et al, "Energy Use in World Food Production," Report 74-1, Environmental Biology, Cornell University, Ithaca, N.Y., November 1974.

26.4 <u>Monitoring the Global Impact of Energy Use</u>[*]

The current and anticipated thermal loadings associated with world-wide energy use represent a potentially significant influence on local, synoptic and world-wide meteorological conditions, as was discussed in Chapter 6 of Volume I. In this discussion, we review first our limited knowledge of the environmental impacts of thermal plumes in an effort to define energy-use levels for which significant disturbances are expected to occur. Next, we consider the definition of a world-wide satellite network and the sensors which will be needed to provide the basic data input on which quantitative predictions of ultimately tolerable thermal loads must be based. The utilization of an adequate monitoring service is required for the development of a meaningful, long-range predictive model dealing with the geophysical implications of escalating energy use, especially insofar as important long-term global climate modifications are concerned. The principal reasons why this predictive geophysical analysis is not currently well under way are associated with the extreme conceptual and analytical difficulties involved in performing the required studies. However, we can no longer run the risk of potentially catastrophic consequences resulting from a significant upset in climatic balance and must therefore proceed with the difficult task at hand as best we know how. A useful input will be provided by careful measurements of the thermal energy balance of the earth. In a program of this type, we may regard the current and planned collection of data as providing a reasonable approximation to a baseline. Even small deviations from past average behavior may indicate incipient instabilities and, possibly, major climatic alterations.

---

[*] This Section is based on a more extensive analysis by S.S. Penner, "Space Monitoring of the Thermal Impact of Energy Use," Acta Astronautica <u>2</u>, 755-769 (1975).

In summary, an immediate problem of first-order importance is associated with increasing world-wide energy utilization and the long-term possibility of inducing climatic changes not only on the mesoscale but also on the synoptic or global scales. The preceding statement applies independently of the primary energy source used and includes, in particular, all solar options (e.g., solar heating and cooling, photovoltaic power conversion using either land-based or satellite-based collectors, the solar sea power plant, wind energy), fusion energy, geothermal sources, etc. Although the feedback processes which are responsible for large-scale climate modifications are not sufficiently well understood to make quantitative predictions at the present time, it is appropriate to emphasize the relatively short time periods (one to two centuries) which have been claimed by some to be involved in the initiation or termination of ice ages.[1]

During 1970, average world-wide, man-made power densities amounted to only 0.054 w/m^2, as compared with an averaged solar-energy input of 340 w/m^2 into the outer atmosphere and about 159 w/m^2 initially retained from the sun in the atmosphere or initially absorbed on the surface of the earth (see Chapter 6 in Volume I for the source of these results). However, man-made power densities associated with energy use have exceeded the average normal solar power input in some metropolitan areas for more than a decade, in some cases by substantial margins (e.g., in Moscow by about a factor of 3 and in Manhattan by more than a factor of 6 during the 1965-68 period). By the year 2000, such a heavily industrialized area as the Ruhr Valley in Western Germany will produce thermal loads over a wide region which exceed the normal solar input by about a factor of 10 while the new generation of nuclear farms, which is currently being considered for implementation, will be responsible for thermal loads that

---

[1]R. A. Bryson, "A Perspective on Climatic Change," Science 184, 753-760 (1974).

are larger than "normal" by more than a factor of 100.[2]

A global monitoring system should include total infrared emission measurements from earth-based regions with characteristic lengths between a few miles, tens of miles, and hundreds of miles. A prototype design should involve sensors of a type widely employed for other purposes. The final system would be expected to be more extensive in infrared coverage and more selective in regional coverage than the system used in the current NASA earth-radiation budget experiments.[3] Hopefully, the coupling of this data output with an evolutionary world-climate model can be achieved sufficiently rapidly to provide an interpretive output that is believable and will serve, if necessary, as a guide to mandatory world-wide limits on energy utilization.

A.   Projections of World-Wide Energy Use and Regional Energy
     Densities

Using data from Chapter 6 in Volume I, we have compiled in Table 26.4-1 a list of representative world-wide average power densities and total power outputs and of the approximate land-surface areas to which the listed power outputs refer. Energy sources are, of course, not uniformly distributed over the source areas. However, in first approximation and in the absence of better data inputs, we shall treat the data listed in Table 26.4-1 as input information into a normal thermal plume model in order to assess the magnitudes of normal atmospheric penetration in the absence of cross winds, using the simplified plume model described in the following Section 26.4B. Reference to the data listed in Table

[2]W. Häfele, "Energy Systems," in Proceedings of IIASA Planning Conference on Energy Systems, pp. 9-78, International Institute for Applied Systems Analysis, Schloss Laxenburg, Austria, July 17-20, 1973.

[3]J. R. Hickey and A. R. Karoli, "Radiometric Calibrations for the Earth Radiation Budget Experiment," Applied Optics 13, 523-533 (1974).

Table 26.4-1 Man-made average power densities associated with energy use; computed from data given in Ref. [2] and in Chapter 6 of Volume I.

Reference area, km^2	Time period	Energy sources	Average man-made	
			power density w/m^2	total power Mw$_t$ [a]
World land area, 1.4885 × 10^8	1970	all sources	0.054	8.03 × 10^6
Continental U.S.A., 9.4 × 10^6	1970	all sources	0.26	2.44 × 10^6
West Germany, 2.5 × 10^5	1970	all sources	0.96	2.40 × 10^5
Ruhr area in West Germany, 6.5 × 10^3	1970	all sources	16.60	1.08 × 10^5
World land area, 1.4885 × 10^8	2000	all sources	0.47*	7.00 × 10^7*
World land area, 1.4885 × 10^8	2050	all sources	1.34*	1.99 × 10^8*
Continental U.S.A., 9.4 × 10^6	2000	all sources	0.57*	5.35 × 10^6*
West Germany, 2.5 × 10^5	2000	all sources	4.80*	1.20 × 10^6*
Ruhr area in West Germany, 1 × 10^4	2000	all sources	100*	1 × 10^6*
Nuclear energy farm generating 30,000 Mw$_e$, [b] 3.5	1990	thermal energy associated with electricity production	2 × 10^4	7 × 10^4

[a] Mw$_t$ = total thermal power in megawatts.

[b] Mw$_e$ = megawatts of electrical power.

*Projected values based on 1973 estimates of future energy use.

26.4-1 shows that the average continuous power levels over concentrated thermal sources lie between 7×10^4 Mw over an area of 3.5 km^2 (for a 30,000 Mw$_e$ nuclear farm) and $\sim 8 \times 10^6$ Mw (in 1970) for the entire world land area. While total power levels of $\sim 2.5 \times 10^6$ Mw were reached in 1970 only over large land areas (e.g., the entire land area of the Continental U.S.A. $\simeq 9.4 \times 10^6$ km^2), projected energy-growth rates to the year 2000 indicate that total continuous power sources of this magnitude will occur in heavily industrialized regions (e.g., the Ruhr area in West Germany) with much smaller land area ($\sim 1 \times 10^4$ km^2). Currently considered large nuclear farms for generating electricity may produce a continuous thermal power output of $\sim 10^5$ Mw over land areas of a few km^2 as early as 1990 (see Table 26.4-1).

B.   Atmospheric Disturbances Produced by Thermal Plumes

Plume aerodynamics has been studied extensively, beginning with investigations by Tollmien[4] and Schmidt[5] on convective plumes above a steady source. Recent work on plume rise and plume spreading in stagnant air is largely based on an entrainment model first described by Morton, Taylor and Turner.[6] We refer to the literature for an outline of the applicable theoretical

[4]W. Tollmien, "Berechnung der turbulenten Ausbreitungsvorgänge," Z. Angew. Math. Mech. 4, 468-478 (1926).

[5]W. Schmidt, "Turbulente Ausbreitung eines Stromes erhitzter Luft," Z. Angew. Mathem. 21, 265-278, 351-353 (1941).

[6]B. R. Morton, G. I. Taylor, and J. S. Turner, "Turbulent Gravitational Convection from Maintained and Instantaneous Sources," Proc. Roy. Soc. London 234A, 1-23 (1956).

considerations and content ourselves here with two references to reviews[7, 8]
on buoyant plumes in air, without and with cross winds and including considerations of stratified atmospheres.   Important work on plume rise over industrial
chimneys and the dispersal of pollutants has been published by Fay, Hoult and
their collaborators.[9, 10]

i.   Meteorological Changes Produced by Thermal Plumes

The observation of cumulus and cumulonimbus clouds over man-made
heat sources associated with power stations has been documented for some
time.[11]   These are local disturbances for which no significant coupling with
synoptic meteorological development has been demonstrated.   Waste heat and
water vapor discharges tend to be coupled over cooling towers.   For a given
set of environmental conditions, it is not known whether wet or dry cooling towers favor cloud formation more strongly at a fixed thermal energy input.[12]
Augmentation of precipitation is generally associated with cooling-tower dis-

[7] G. A. Briggs, Plume Rise, U.S. Atomic Energy Commission, Division of
Technical Information Extension, Oak Ridge, Tennessee, November 1969.

[8] B. E. Boyack and D. W. Kearney, "Plume Behavior and Potential Environ-
mental Effects of Large Dry Cooling Towers," Report Gulf-GA-A12346,
Gulf General Atomic Company, P. O. Box 81608, San Diego, California 92138,
February 1973.

[9] D. P. Hoult, J. A. Fay and L. J. Forney, "A Theory of Plume Rise Compared
with Field Observations," J. Air Pollution Control Association 19, 585-590
(1969).

[10] J. A. Fay, M. Escudier and D. P. Hoult, "Comments on Plume Rise at Industrial Chimneys," J. Atmospheric Environment 3, 311-313 (1969).

[11] G. J. Jefferson, "Man-Made Cumulus," Meteorology Magazine 89, 16-17
(1960).

[12] F. A. Huff, "Potential Augmentation of Precipitation from Cooling Tower
Effluents," Bulletin of the American Meteorological Society 53, 639-644
(1972).

charges, which are often viewed primarily as the carriers of noxious combustion products and air pollutants. Downwind precipitation, as well as augmentation of snowfalls, rain showers and thundershowers, have been described.[12] The possibility of using thermal plumes for the penetration of inversion layers and dispersal of air pollutants has been suggested.[13, 14]

ii. Comments on the Influence of Energy Use on the Creation of Thermal Plumes

The rise and spreading of a buoyant plume in the atmosphere depend on atmospheric turbulence and on self-induced turbulence which causes entrainment of the ambient air. The theory of Morton, Taylor and Turner[6] has been found to be in good agreement with observed data; it involves the following basic postulates: (1) The rate of entrainment of ambient air in the plume edge is proportional to the velocity difference between the plume fluid and the ambient air. (2) The spacial dependencies of the mean velocity and mean buoyancy are similar. (3) Density fluctuations are small compared with the absolute value of the ambient or reference density. Here the buoyancy is directly proportional to the total power of the heat source associated with the creation of the thermal plume. There are two characteristic heights associated with buoyant plumes, one of which refers to the altitude at which the vertical velocity component vanishes while the other refers to the altitude where the buoyancy vanishes. Thermal plumes will generally rise to the higher level before spreading horizontally.

Using the measured value[6] $\alpha = 0.093$ for the ratio of the entrained

---

[13] "Cooling Tower Designed for Smog Dispersal," Chemical and Engineering News, Vol. 49, No. 16, p. 57, April 19, 1971.

[14] E. J. Tschupp, "Effects on Regional Pollution of Alternative Power Plant Stacks," General Electric-Tempo, Center for Advanced Studies, Santa Barbara, California, November 1971.

radial to axial plume velocities and a standard tropospheric temperature gradient of 6.5°C/km, the following approximate relation is obtained between plume height, H in meters, and source power, Q in megawatts, in a stationary atmosphere:

$$H \simeq 281 \, Q^{1/4}. \qquad\qquad (26.4\text{-}1)$$

An excellent empirical verification of Eq. (26.4-1) appears in Fig. 5.7 of Ref. [7]. In Table 26.4-2 we list some representative data calculated from Eq. (26.4-1).

   Reference to the literature and to the data shown in Table 26.4-2 leads to the important conclusions that the thermal loads provided by large nuclear farms are larger than those which were involved in the creation of whirlwinds over a burning oil tank[15] and are within a factor of three of the estimated thermal source magnitude in the eruption of the Surtsey Volcano[16] for which extensive and prolonged atmospheric disturbances, including whirlwinds, were observed. What is particularly noteworthy is the fact that the total thermal energy output over the Ruhr Valley (with an area of about 1×10^4 km^2) will probably exceed that of the Surtsey eruption by a factor of five during the year 2000.

   Our knowledge of atmospheric physics is not sufficiently well developed to allow a confident statement about the climatological impact of the projected year 2000 energy use in the Ruhr Valley because we do not understand the important role of energy density in sufficient detail. The simplified plume models

---

[15] I.N. Hissong, "Whirlwinds at Oil Tank, San Luis Obispo, California," Monthly Weather Review 54, 161-164 (1926).

[16] S. Thorarinsson and B. Vonnegut, "Whirlwinds Produced by the Eruption of Surtsey Volcano," Bull. Am. Meteorol. Soc. 45, 440-444 (1964).

Table 26.4-2   Representative calculated plume heights over selected thermal
              sources in a standard atmosphere with a tropospheric lapse rate
              of 6.5°C/km.

Description of thermal source	Total thermal power release, Mw	Plume height, m
household chimney for coal burning at the rate of 4 lbs/hour, with half of the combustion energy wasted in the chimney	8×10^{-3}	~ 84
bonfire (2 ft^3 of wood/hour)	0.45	~ 230
annual bonfires at Texas A & M University	1×10^2	~ 890
large power-station chimney	2×10^3	~ 1880
burning forest (1000 tons/hour)	5×10^3 to 4×10^4	~ 2360 - 3980
burning town (250 to 500 houses/ hour, each containing 10 to 20 tons of combustible material)	2.5×10^4	~ 3530
petroleum fire, 5×10^6 bbl in 5 days[15]	4×10^4	~ 3980
large nuclear farm (see Table 26.4-1)	7×10^4	~ 4570
eruption of Surtsey Volcano,[*] during the first 10 days[16]	2×10^5	~ 5940
Ruhr area of West Germany in the year 2000 (see Table 26.4-1)	1×10^6	~ 8890[**]

[*] For an extensive compilation of data on volcanoes, see G.A. MacDonald,
Volcanoes, Prentice Hall, New York, N.Y., 1972.

[**] This source is not sufficiently concentrated to justify use of the simplified
plume model.   F. K. Moore (Cornell University report, June 1975) has
concluded that the regional climatic impact of 50,000 Mw$_e$ stations operating
by the year 2000 will be negligibly small.

are only applicable to intense thermal sources with initially small spacial dimensions. However, the case appears to have been made that the thermal plume
problem with large initial dimensions must be studied with appropriate urgency.
Furthermore, it is reasonable to conclude that the construction of nuclear farms
with a total power output of 30,000 Mw$_e$ and air cooling should be implemented
in stages with appropriate monitoring of environmental impact during assembly.

C. Energy Consumption and Climatological Changes

Following Budyko,[17, 18] H.-J. Bolle[19] has reviewed currently employed methods for zonal and global climate prediction with emphasis on the
required measurement precision for such parameters as the solar constant,
the earth surface albedo, average CO_2 concentration, particle concentrations,
etc. We note two important aspects of Bolle's review: (a) the governing factor
in the ocean-atmosphere system is the radiation balance and (b) very small (of
the order of one percent) changes in a parameter such as the solar constant
could be responsible for the creation of ice ages.[20] Some examples[19, 20]
of estimated mean earth-surface temperature changes, ΔT_o (in ^{o}C), for changes
in the solar constant, ΔS_o (in %), are the following: -5°C for a change in ΔS_o

[17] M. I. Budyko, Atlas of the Heat Balance of the Earth, Moscow, U.S.S.R.,
    1963.

[18] M. I. Budyko, "The Effect of Solar Radiation Variations on the Climate of
    the Earth," Tellus 21, 611-619 (1969).

[19] H.-J. Bolle, "Possibilities of Remote Sensing to Determine Parameters
    Needed for Climate Change Studies," Acta Astronautica 1, 1399-1413 (1974).

[20] W. D. Sellers, "A New Global Climate Model," J. Applied Meteorology 12,
    241-254 (1973).

of -0.9%, + 12°C for a change in ΔS_o of + 7.0%. Changes in CO_2-concentration are now believed[19,20,21] to be responsible for relatively small variations in T_o; thus ΔT_o has been estimated to lie between 0.1 and 2.36°C for a doubling of the CO_2 concentration and between -1.0 and -2.28°C for halving of the CO_2 concentration.

Noting that a global mean value for the absorbed solar radiation which is converted to heat energy on the earth is about 160 w/m^2, it is readily seen that world-averaged, man-made heat additions are expected to approach the 1% level of S_o by 2050 and that large regions (e.g., averages over the entire continental U.S.) will provide man-made energy inputs above the 1%-level by the early part of the next century. Thus, if the conclusion is valid that the solar flux must be known with an accuracy of 1%, then it is also true that man-made energy sources will influence global climatological predictions by the early part of the twenty-first century to a significant extent.

D.   Monitoring World-Wide Energy Use*

A first step in the independent monitoring of world-wide energy use has been made by NASA with the introduction of satellite observations of the earth radiation budget.[3] Kontratyev et al[22] have discussed remote sensing to determine the presence of natural formations (e.g., forests, various food crops, agricultural land, water surfaces, etc.). However, a detailed global program

---

[21]S. Manabe and R. T. Wetherald, "Thermal Equilibrium of the Atmosphere with a Given Distribution of Relative Humidity," J. Atmos. Sci. 24, 241-259 (1967).

*Drs. H. G. Wolfhard and O. Kosovych have greatly contributed to this discussion.

[22]K. Ya. Kontratyev, O. B. Vasilyev, O. M. Pokrovsky, and G. A. Ivanyan, "Remote Sensing of Natural Formations from Measurements of Radiance Coefficients," Acta Astronautica 1, 1415-1426 (1974).

for the quantitative measurement of heat dissipation from thermal sources remains to be defined.   We address this problem in the following discussion.

Local heat release will lead to the following climatic changes: (a) An increase in surface temperature in areas ranging from about 1 km × 1 km to 100 km × 100 km. (b) A change in the vertical temperature profiles of the atmosphere above heated areas. (c) Local disturbances of the general patterns of air movements (winds) in the vicinities of the thermal sources. (d) The formation and/or dissolution of clouds.

Before making specific recommendations as to how to measure these possible climatic changes, we note a list of measurement programs that are currently in progress or are being planned for later years.   This information has been primarily compiled from Refs. [23] to [25] and is summarized in the paper by Penner.[26] The reader should note that future missions of spacecrafts are listed in this paper as planned in 1975 and will be subject to considerable change with time.

High-resolution multispectral scanners have very good spatial resolution and are suitable for monitoring individual streets and large buildings.   The

---

[23]N. W. Stoldt et al, Compendium of Meteorological Satellites and Instrumentation, National Space Science Data Center, July 1973, NSSDC 73-02, and National Technical Information Service, U.S. Department of Commerce, Washington, D.C., N74-20540.

[24]R. R. Drummond, Digest of NASA Earth Observations Sensors, December 1972, National Technical Information Service, U.S. Department of Commerce, Washington, D.C., N73-15482.

[25]Advanced Scanners and Imaging Systems for Earth Observations, NASA Report of Working Group, Cocoa Beach, Florida, December 1972.

[26]S. S. Penner, "Space Monitoring of the Thermal Impact of Energy Use," Acta Astronautica 2, 755-769 (1975).

ERTS 1 (launched in 1972) and ERTS B (to be launched in 1975) may be used to differentiate clouds with still smaller characteristic lengths. The orbital altitudes used are about 900 km; thus, the satellites may be employed to observe clouds relative to the ground for at least 400 seconds (assuming nadir angles of ±60°). During this time, clouds moving with a speed of 10 m/sec will cover distances of 4.0 km, which is orders of magnitude above the available spatial resolution. Thus, by observing clouds relative to the earth at and around the locations of interest, one may follow locally-induced cloud movements.

ERTS C will have a long-wavelength infrared scanner with a resolving power of 240 m and may be used to measure thermal emission near 10μ and, therefore, also the ground temperature. The thematic mapper on the EOS will permit surface-temperature measurements with an accuracy of $0.5°C$.

An important problem with the current program of measurements involves the management of enormous numbers of data that will be collected. Furthermore, long-term instrument stability must be maintained so that the data derived from different satellites over long periods of time may be compared and correctly evaluated.

Intermediate-resolution radiometers have advantages over high-resolution instruments. First of all, more frequent coverage of locations of interest is achieved because the total scanned field of view is of the order of a kilometer. Surface temperatures may be determined accurately provided the ground is not totally or partially obscured by clouds. Large clouds can still be tracked and local disturbances observed. In a representative system, the temperature of the ground, cloud cover and cloud movement will be determined twice a day or, in the case of synchronous satellites, every 20 minutes.

The low-resolution radiometers are of lesser value to our objectives.[26] The THIR instrument may be useful for observations of the Ruhr Valley since its resolution fits the extent of this particular area. Also, it has been designed for measurements of cloud cover, surface or cloud temperatures, and relative humidity.   The instrument is, however, unsuitable for monitoring the climatic impact of nuclear energy farms.   Unfortunately, the earth radiation budget instrument is also totally unsuitable since its field of view is much too coarse and surface temperatures cannot be measured.   This instrument has been designed to measure the balance of the incoming solar radiation and outgoing earth radiation, thus fulfilling our stated objectives superficially.   However, the measurements will be done with such poor resolution, both spatially and spectrally, as to be of little use.

Microwave radiometers will also be employed to measure thermally emitted radiation, but are generally of too coarse spatial resolution to provide useful data.

Infrared sounders are probably the most important instruments for our objectives.   They will be used to measure vertical temperature profiles in the atmosphere, a property which is most profoundly disturbed by the thermal plumes that are expected to form over areas with great heat release.   The basic instrument consists of many channels within the 15μ CO_2 absorption-emission band and allows us, by inversion of the data, to reconstruct the temperature profile to an altitude of 30 km.   The additional improvements planned in infrared instrumentation are probably not needed for our purposes.   The principal drawback of the sounders is the very large field of view, which may be suitable for observations of such regions as the Ruhr Valley but is too coarse for studies on more concentrated energy-release areas.   GOES-D may have a sufficiently small field of view with the AASIR instrument, but this mission is not well defined as yet and has a proposed launch date during the nineteen eighties.[26]

E. Summary and Recommendations on Monitoring the Climatological
    Impact of Energy Use[26]

We have seen that surface temperatures, cloud movements and cloud
formations may be studied in great detail through available and planned satellite missions. The sounders which will be used to measure temperature profiles have insufficient spatial resolving power. However, a redesign of existing
instruments to a field of view of about 1 km × 1 km does not seem to be an in-
surmountable task. The ERB instrument could equally well be redesigned to
a smaller field of view in order to provide information on the overall thermal
balance of a given area. We note that an instrument with small field of view
is unsuitable as the primary instrument for measurement of radiant-energy
transfer and for measurements of the transport of thermal energy through wind,
condensation, or vaporization.[26]

We now address the question of how to monitor a nuclear energy farm
over long periods of time (10 to 20 years). A costly answer is to design a satellite of say 2 to 3 years lifetime and to replace it thereafter. This satellite
would have to be provided with high- as well as low-resolution scanners in the
visible and also at about 10μ. Furthermore, one would use sounders and earth-
radiation-budget instruments of appropriate spatial resolution. For continuity
of measurements, one would want a synchronous-satellite altitude for this monitoring task, On the other hand, the need for high spatial resolution and the desire to maintain low cost suggest the use of low-altitude observations. More
realistically, we propose utilization of the extensive instrumentation (perhaps
with some upgrading in spatial resolving power) which will become available
on a great variety of satellites and at a number of orbital altitudes.[26]

The greatest difficulties will be associated with data management. Every satellite mission that has been planned serves a primary purpose that is

distinct from the objectives stated in this section.   The actual design objectives
of the satellite missions may, however, not be contradictory to our purposes.
Really detailed data will be required at a relatively small number of locations
(i.e., the locations of the primary heat-release areas and of corresponding
areas without artificial heat release for comparison).   Often data are not stored
or transmitted for all instruments at all times.   Therefore, there is a need for
mission-program managers to integrate monitoring of thermal-load effects with
existing primary missions.[26]

  Data derived from different satellites at different times are difficult to
compare and to utilize in the overall data-management process.   Calibration
procedures for different instruments and long-term stability of the calibration
must be known.   It will clearly be necessary to conduct some ground-truth experiments whereby locally well known conditions (obtained by the use of sounding
balloons, ground or airborne instrumentation, etc.) will be measured by the instruments involved in the satellite-measurements program.   These ground-truth
experiments must be available over sufficiently long periods of time to provide
adequate verification of measurement procedures.[26]

  Analysis of the data presently being taken in the Ruhr Valley should be
started at once since this is the only region where sounders can now be utilized.
This schedule would permit a more sophisticated assessment of what present
and future satellite instrumentation can materially contribute to the analysis of
meteorological changes induced by large artificial heat releases.   An overall
framework for data collection, reduction and correlation may be constructed
and initial attempts for a meaningful theoretical treatment can then be initiated.[26]

26.5 Energy Farms

The energy plantation has been discussed briefly in Chapter 14 (Volume II, pp. 327-329). The idea of developing a balanced, steady-state farming schedule for energy production (using selected crops, farm wastes, or a combination of crops and wastes) constitutes an especially appealing solution to our "energy needs". The scale on which energy farms can be built is clearly limited. An optimistic practical objective is 5 to 8% of total needs by the year 2,000. Ideally, implementation of energy farms will reestablish the earth albedo at a value characteristic of earlier, "unspoiled" times while providing for CO_2 sinks of adequate scale to remove accumulation of this pollutant as a potential long-term causative factor in undesirable climate modifications. These advantages are well understood. Unfortunately, the economic performance of energy farms will not be well defined until the performance of a large-scale energy plantation has been studied carefully for some years.

The production of biomass for energy use has recently been reviewed by Alich and Inman,[1] who have discussed the types of vegetations that are best suited for energy production. This study also includes an overview of the logistics and economics of energy crops.

Representative biomass yields for selected crops are reproduced in Table 26.5-1.[1] We note large variations in energy yields, depending on the plant species involved and on the locations where they are grown. In California, experimental sugarcanes yield more than 30 t/ac-y while species of eucalyptus produce between 13 and 24 t/ac-y and fresh-water algae between 8 and 39 t/ac-y. In Florida, average sugarcanes currently produce 17.5 t/ac-y

---

[1] John A. Alich, Jr., and R. E. Inman, "Energy from Agriculture," Energy (in press, Volume 1, 1976).

Table 26.5-1 Representative dry biomass yields of selected plant species or complexes; abstracted from Alich and Inman, Ref. [1].

Species	Location	Yield, t/ac-y
Annuals		
exotic forage sorghum	Puerto Rico	30.6
forage sorghum (irrigated)	Kansas	12
silage corn	Georgia	6-7
kenaf	Florida	20
kenaf	Georgia	8
Perennials		
water hyacinth	Florida	16
sugarcane (state average)	Florida	17.5
sugarcane (10-year average)	Hawaii	26
sugarcane (5-year average)	Puerto Rico	15.3
sugarcane (6-year average)	Philippines	12.1
sugarcane (experimental)	California	32
sugarcane (experimental)	California	30.5
sudangrass	California	15-16
alfalfa	New Mexico	8
bamboo	South East Asia	5
abies sacharinensis (dominant species) and other species	Japan	6
cinnamomum camphora (dominant species) and other species	Japan	6.8
fagus sylvatica	Switzerland	4.3
picea abies (dominant species) and other species	Japan	5.5
red alder (1-14 years old)	Washington	10
eastern cottonwood (8 years old)		3
eucalyptus sp.	California	13.4
eucalyptus sp.	California	24.1
eucalyptus sp.	Spain	8.9
eucalyptus sp.	India	17.4
eucalyptus sp.	Ethiopia	21.4
eucalyptus sp.	Kenya	8.7
eucalyptus sp.	South Africa	12.5
eucalyptus sp.	Portugal	17.9
Miscellaneous		
algae (fresh-water pond culture)	California	8-39
tropical rainforest complex (average)		18.3
subtropical deciduous forest complex (average)		10.9
world's oceans (primary productivity)		6

while corresponding average values for the Philippines are 12.1 t/ac-y and for Puerto Rico 15.3 t/ac-y. Species of eucalyptus yield from 8.9 t/ac-y in Spain to 17.4 t/ac-y in India to 24.1 t/ac-y in California.

Biomass crops mature more rapidly than seed crops, thus facilitating multiple cropping of annual species. The gains derivable from the use of this procedure have been allowed for in the compilation of Table 26.5-1.

Alich and Inman[1] have examined a representative plantation and have estimated energy requirements for farming operations such as applications of herbicides, fertilizer, cutting, hauling, etc. They arrive at an energy-output to input ratio of 18:1. The energy charges in Btu/ac-y were found to be distributed as follows: 7.2×10^6 for farming operations (for three crops), 16.8×10^6 for the manufacture of farm chemicals, 0.412×10^6 for manufacture of the farm machinery used, 0.073×10^6 for seed production. The total energy yield was 450×10^6 Btu/ac-y.

In 1973, the cost of producing and harvesting 30 t/ac-y of dry biomass was estimated to be between $9.50 and $10.00/t with fertilizer accounting for 23% of this cost. Updated costs[1] for 1975 fell in the range of $13.00 to $16.00/t.

Electricity-generation costs have been compared by Alich and Inman[1] with coal costs. The results are reproduced in Fig. 26.5-1. At a cost equivalent to 1975 oil costs of about $2.00/$10^6$Btu, biomass production appears to be a cost-effective procedure. We conclude that energy plantations deserve to be built and studied to ascertain what the operational costs really are.

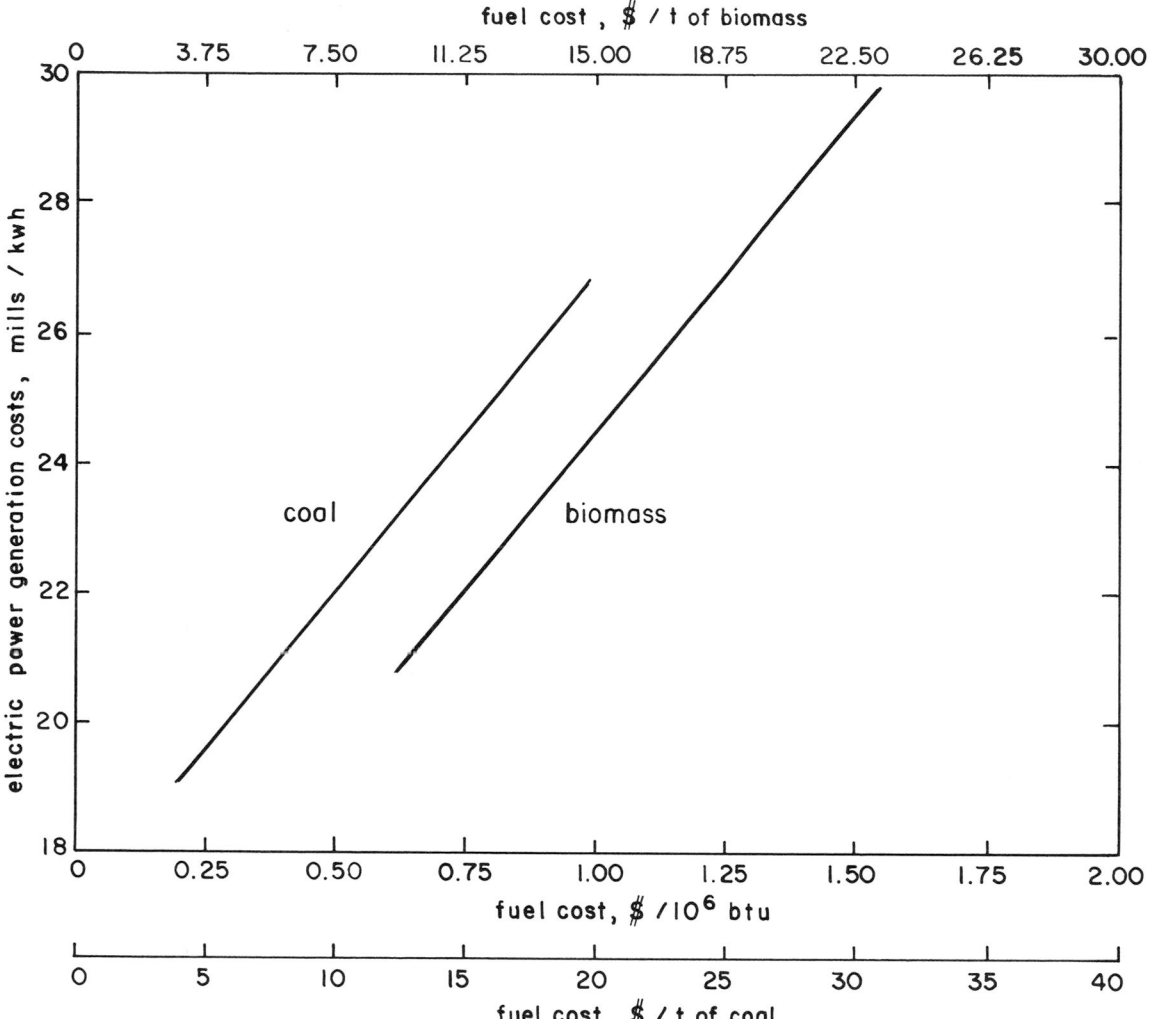

Fig. 26.5-1 An estimate of power-generation costs as a function of fuel costs;
reproduced from Alich and Inman, Ref. [1].

26.6 Energy Conservation

The seemingly obvious implementation of energy conservation contains subtle implications that can be understood only in the context of a careful input-output analysis for the entire economic complex that is affected by energy-conservation measures. As we have seen, urgent motivation for energy conservation is provided by a number of generally accepted facts, viz. (a) limited and decreasing availability of cheap conventional energy sources and (b) undesirable environmental impacts which may ultimately place absolute limits on such factors as the total tolerable man-made heat loads; (c) long lead times and heavy capital costs are required for effective conversion to major alternative energy-supply bases (e.g., solar or fusion energy); (d) dependence on imported energy sources may produce trade imbalances and may be affected by political pressures or even by outright embargoes; (e) responsible people who are convinced that the energy-supply base is in jeopardy want to help and believe that their most obvious and significant immediate contribution can be made by limiting consumption.

Ideally, we might argue that reduced per capita energy consumption is equivalent to an improvement in life style or at least to the development of an improved life style. Unfortunately, this idealistic definition is not compatible with the composite effect of reduced energy consumption on our economic system or on the economic well-being of a majority of our citizens. Hannon[1] has articulated several undesirable aspects of reduced energy consumption. These are listed below.

1. A decrease in energy consumption per dollar value of GNP will

---

[1]Bruce Hannon, "Energy Conservation and the Consumer," Science 189, 95-102 (1975).

generally lead to a change from highly paying jobs for workers to employment at reduced remuneration. This conclusion is associated with the fact that jobs tend to be replaced by mechanization when wages increase relative to costs. Thus, reduced energy consumption implies reduced mechanization and hence reduced wages per unit product.

2. Since it has been shown that energy consumption increases nearly linearly with family-income levels above sustenance earnings, at least up to 20×10^3/y,[1] incremental dollar expenditures for all goods and services require essentially the same energy. Thus, dollars saved by reductions in direct energy consumption tend to be disbursed on other items that require essentially equivalent energy consumption. Our per capita GNP is maintained by total expenditures which, in turn, are closely related to total energy consumption (compare Chapter 4 in Volume I). A significant shift can only occur if an essential restructuring of our society is accomplished or if fundamental alterations occur in the types of goods which we manufacture and use.

Reference to the appropriate data in Chapter 4 of Volume I shows that large disparities in energy consumption per dollar of GNP produced correspond to different types of industrial activities and lifestyles. The U. S. per capita GNP could probably be maintained at high levels, under conditions of reduced energy consumption, through changes of the following types: increased production of high-cost goods that have relatively low energy costs (e. g., computers, watches, perfumes, champagnes); increased participation in highly labor-intensive activities such as advanced education and personal grooming or in costly recreational endeavors such as sailing, operatic productions, and horse racing.

Energy conservation is a universally acceptable goal; its judicious implementation remains to be defined and accomplished.

An interesting approach to energy conservation has been taken by a panel of physicists.[2] These authors note that the efficiencies of existing devices are customarily described by using the first law of thermodynamics and that far more efficient energy-conversion systems with higher first-law efficiencies can be constructed, in principle, by admitting a priori only the general limitations imposed by the basic laws of thermodynamics. Examples of this approach are worked out[2] for house heating and cooling, utilization of low-quality waste heat, improved transportation systems, and industrial processing. The physical ideas that are implicit in the preceding statement are readily understood. Thus, a gas furnace with a flame temperature above $2,000°$K is an intrinsically inefficient device for heating a house a few degrees. A perfectly operating gas heater (first law efficiency $\simeq 100\%$, i.e., all of the energy released by combustion is ultimately transferred to heating of the air) has a low second-law efficiency of 6 to 12%. It is an existing device for the intended application and it is wasteful of energy in performing its function. A much lower total energy requirement would result if we had a low-temperature heat pump. Similarly, significant energy economies are possible in principle by applying small, efficient engines in the transportation sector with boosters to meet peak demands or by using appropriate physical phenomena in selected industrial processing.

Second-law efficiencies are device-independent limiting efficiencies. They provide an indication of what can be done over the long term after required

---

[2]Efficient Use of Energy: A Physics Perspective, sponsored by the American Physical Society, January 1975. Copies of this report were obtainable from Robert H. Socolow, Center for Environmental Studies, Engineering Quadrangle D-325, Princeton University, Princeton, N. J. 08540.

inventions have been made. They have no general bearing on conservation measures for existing energy-conversion systems. Second-law efficiencies help to define desirable long-term goals which may be attained if as yet undefined inventions that can be marketed at competitive costs meet wide consumer acceptance. It is not reasonable to confound these long-term goals with strategies that can now be implemented to conserve energy and replace alternative energy sources.

A. Available Work Associated with the Transfer of Heat

We proceed to illustrate the methodology used in Ref. [2] by referring without detailed verification to the applicable general laws of thermodynamics. As an example, we choose the home heater (see Fig. 26.6-1) which, ideally, involves the transfer of heat from a uniform, isothermal high-temperature reservoir at temperature T_1 and pressure p (i.e., the burner) to the ambient atmosphere which we represent ideally as a uniform, isothermal reservoir at temperature T_0 and the same constant pressure p (see Fig. 26.6-1). Heat transfer occurs through regions of progressively lower temperature, the last of which we denote by T_2.

A useful version of the combined first and second laws of thermodynamics for analysis of this problem is the following:[3]

$$\mathrm{d}\varphi = dH - TdS - Vdp.$$ (26.6-1)

Here $\mathrm{d}\varphi$ is the __useful__ or __available__ work (__not__ including pressure-volume work

---

[3]S. S. Penner, Thermodynamics, Eq. (6.3-2) on p. 71, Addison-Wesley Publishing Co., Reading, Mass. 1968.

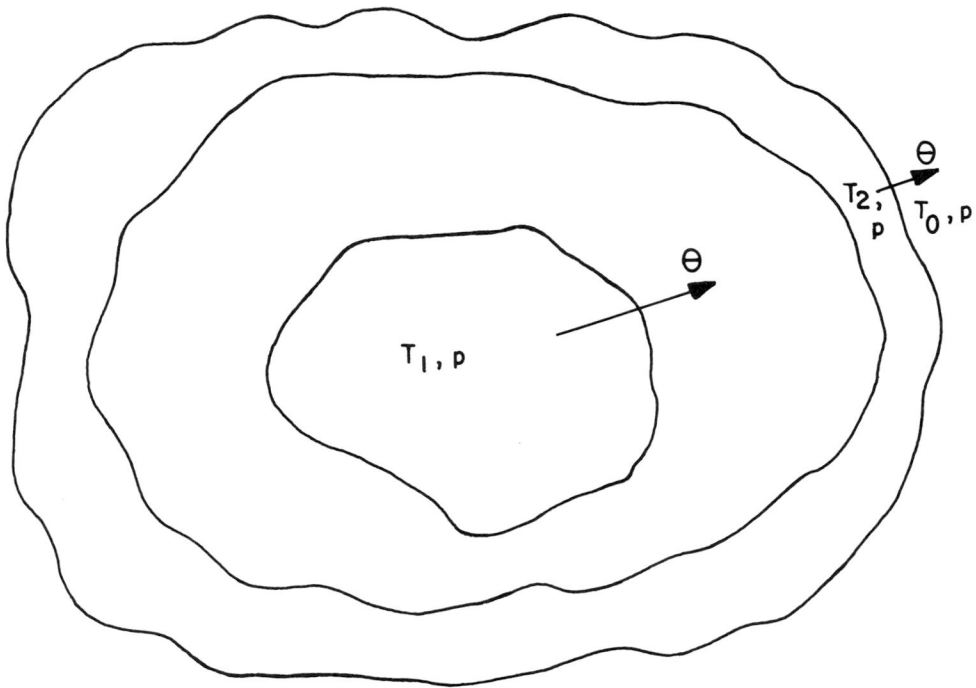

Fig. 26.6-1 An idealized representation of the home heater (burner temperature T_1, pressure p) as a source of heat for the ambient atmosphere (temperature T_0, pressure p). The heat Q is transferred
(e.g., by convection and radiation) from the hot combustion products to progressively cooler regions. The region in contact with
ambient air is symbolically represented as being isothermal at
T_2 and p.

done on the surrounding atmosphere), dH is the enthalpy change, T the temper-

ature, dS the entropy change, V the volume, and dp the pressure change. It

follows from Eq. (26.6-1) for the postulated constant-pressure process that the

total available work for a thermodynamic system at the initial temperature T_i

and a cooler system at T_f is

$$\varphi = (\Delta H_i - T_i \Delta S_i) + (\Delta H_f - T_f \Delta S_f) =$$

$$(\Delta H_i + \Delta H_f) - T_i \Delta S_i [1 + (T_f/T_i)(\Delta S_f/\Delta S_i)], \qquad (26.6\text{-}2)$$

where a symbol preceded by Δ denotes the incremental change in the succeeding thermodynamic state variable. For the transfer of heat Q at constant pressure, $\Delta H_i = -Q$ and $\Delta H_f = +Q$ {compare Eq. (3.2-16), p. 30, of Ref. [3]}, where the negative sign indicates heat lost from the system and the positive sign indicates heat gained by the system. The corresponding entropy change for the home heater is $\Delta S_i = -Q/T_i$ {see Eq. (4.1-8), p. 38, Ref. [3]}. The maximum value of ΔS_f is $-\Delta S_i$ and is attained for an idealized, reversible transfer of heat between the burner and the ambient atmosphere, which is isentropic such that $\Delta S_i + \Delta S_f = 0$. The condition $\Delta S_i = -\Delta S_f$ leads to an upper limit for φ and corresponds to the result

$$\varphi = Q[1 - (T_f/T_i)] \qquad (26.6\text{-}3)$$

for the available work.

An upper limit for the heat transfer Q is the total heat of combustion Q_{comb}. Therefore, for $T_f/T_i \ll 1$, $\varphi \simeq Q_{comb}$. This value of the available work measures the best use that can be made of the energy released by combustion processes in the home heater. The maximum available work is thus given by the relation

$$\varphi_{max} = Q_{comb}. \qquad (26.6\text{-}4)$$

B.   Definitions of First- and Second-Law Efficiencies

The device-dependent efficiency η of Ref. [2] is the normally used expression for the efficiency derived from the first law of thermodynamics, viz.,

$$\eta = \frac{\text{energy transfer of the desired kind achieved in a given device}}{\text{required energy input for the given device to achieve the desired energy transfer}} . \qquad (26.6\text{-}5)$$

It is apparent that

$$\eta = \frac{Q}{\text{energy available in the combustible gases when the heat transfer } Q \text{ was achieved}} . \qquad (26.6\text{-}6)$$

The denominator in Eq. (26.6-6) is the heat of combustion, Q_{comb}. The first-law efficiency η defined by Eq. (26.6-6) is generally high, e.g., 0.6 (60%).

A device-independent efficiency may be defined[*] as follows:

$$\epsilon = \frac{\text{available work achievable in an optimally designed device}}{\text{maximum of the available work which can be achieved in principle}} . \qquad (26.6\text{-}7)$$

In view of Eq. (26.6-4), the denominator is the same in Eqs. (26.6-6) and (26.6-7); also, $\eta \leq 1$, $\epsilon \leq 1$.

---

[*]Our definition of ϵ appears to differ from that used in Ref. [2] according to the descriptive statements used. However, at least for the home heater, our definition leads to results that are equivalent to those obtained in Ref. [2].

Referring now to the home heater shown schematically in Fig. 26.6-1, we note that the numerator in Eq. (26.6-7) is given by Eq. (26.6-3) with $Q = Q_2$, $T_f = T_0$, and $T_i = T_2$. Thus, for the home heater represented in Fig. 26.6-1,

$$\epsilon = \frac{\varphi}{\varphi_{max}} = \frac{Q_2[1 - (T_0/T_2)]}{Q_{comb}} = \eta[1 - (T_0/T_2)]. \qquad (26.6-8)$$

In a practical furnace system, we may operate with $T_0 = 275°K$ and $T_2 = 300°K$. Under these conditions,

$$\epsilon = \eta(1 - 0.92) = 0.048$$

if $\eta = 0.6$. We conclude that the home furnace with its very high combustion temperature of about $2,000°K$ is an exceedingly inefficient device for space heating since only a very small operative temperature rise is actually used. While its first-law efficiency is high, a different device could be constructed which operates at much lower temperatures (e.g., a fuel cell with a resistive heater element) or else a combustion system could be designed with efficient waste-heat utilization.

The difficulties of inventing highly efficient devices are exemplified by recent efforts to achieve theoretical first-law (Carnot) efficiencies in the combustion of low-Btu gases by introducing a regeneratively heated "Swiss role" heat-exchanger-combustor.[4]

---

[4]S. A. Lloyd and F. J. Weinberg, "Limits to Energy Release and Utilization from Chemical Fuels," Nature 257, 367-370 (1975).

C. Cost Evaluation for an Efficient House Heater

We now turn to the problem of defining a device that is suitable for house heating at an efficiency approaching that achievable for an ideal thermodynamic system. An ideal fuel cell coupled to a resistive heater constitutes a good approximation. We have seen (see Volume II, Section 13.3) that $H_2(g)$-air and $H_2(g)$-oxygen fuel cells have been built with (first-law) energy conversion efficiencies of 55 to 60%. An ideal fuel cell (see Table 13.3-1) with heat transfer from the surroundings has a theoretical efficiency greater than unity for the graphite-oxygen and propane-oxygen cells (see Table 13.3-1). If the fuel cell is operated in such a manner that an externally connected resistive element is maintained at a temperature just above the desired room temperature, the overall efficiency achievable in convective room heating should be very close to the combustion-energy-to-electricity conversion efficiency of the fuel cell. Thus, we know today how to build house-heating units with values of ϵ that are substantially larger than those achieved in a gas burner, viz. $0.55 \leqslant \epsilon \leqslant 1$. Why then are we still using gas-fired burners and blowers instead of fuel cells coupled to resistive elements in our house-heating units?

The answer to the question posed in the preceding paragraph is, as usual, based immediately on economic considerations (which, in turn, are generally linked to hidden energy charges when total energy costs of manufacture, operations and maintenance are properly accounted for).

The current installed cost for a gas-fired house-heating unit with a capacity of 1×10^5 Btu/h is about $500 in new construction. This rated capacity corresponds to 29.3 kw of thermal power. The equivalent capacity for a fuel-cell-resistor combination with an overall "second-law" conversion efficiency of 50%, which is higher by a factor of 10 than that of the gas heater, is then 2.93 kw_e. Current fuel cell costs are about $500/kw_e$ and we assume that this value applies also to an NG-air fuel cell.   The fuel-cell itself will require a

capital investment of about $1,500 while costs for the entire system will pro-

bably exceed the cost of the gas unit by about $3,200.   At a 15% annual capital

charge, we see that installation of the fuel cell would have to save about $480

per year to be competitive.   With total heating costs averaging about $300/y

of which about $270/y could be saved if the efficient fuel-cell-resistor unit

were installed, we must conclude that the more efficient fuel-cell-resistor

system is not now an economically viable alternative to the gas-fired home

burner.   Actual costs for the fuel-cell-resistor use will, of course, be sub-

stantially higher than we have estimated because of operations and mainte-

nance costs which are likely to be about 12%/y or $450/y.

The invention of efficient and economical alternatives to the house

heater remains to be accomplished for new construction.   The costs for retro-

fitting existing structures with inventions will always be higher than the intro-

duction of these devices into new units.   The replacement times for houses

are long, with a half life probably exceeding 40 years.   For this reason, con-

servation measures involving space heating should preferably be addressed

to improvements in combustion efficiency for existing furnaces such as intro-

duction of interruptible pilots (which can be built for $10 to $20 and hopefully

installed for $100 to $200 per unit).

D.   Indefensible Economic Policies that Deter Energy Conservation[5]

G. S. Gill[5] has discussed economic policies that deter implementation

of energy-conservation measures.   Among "perverse" incentives, he notes the

following:

---

[5]G. S. Gill, "Perverse Economic Incentives and Energy Conservation,"
Energy, Volume 1 (in press, 1976).

(a) Promotional pricing associated with declining block-rate structures by electric and gas utilities. When this type of pricing is replaced by a shift to higher costs during peak loads, a significant incentive will have been created for energy conservation.

(b) Depletion allowances of about 13% for oil and gas and about 1% for coal have distorted fossil-fuel prices and consumption patterns.

(c) Failure to charge societal costs for coal use (see Chapter 7 of Volume I) and for other fossil-fuel applications to the consumers directly have created artificially low prices and have provided disincentives for energy conservation.

(d) Regulatory agencies have been responsible for waste-inducing programs and practices. Examples are public subsidies for road construction and trucking when intrinsically more efficient rail transport should have been supported instead. Social costs of shipping by one means or another have not been considered by the Interstate Commerce Commission. In air travel, this practice has led to restrictions on competition, excessive capacity, relatively small load factors, and fuel waste.

(e) Pricing policies for natural gas (see pp. 53-58 of Volume I) seem to have been designed to exhaust known resources while preventing new exploration.

At issue are the complex problems of price-use elasticities for energy resources, "windfall" profits for large companies, and the reluctance of successful politicians to be identified with logically defensible policies the most visible effects of which are price rises over the near term.

   E.   Energy-Demand Projection with Proper Allowance for
        Conservation[6]

Bullard and Foster[6] have emphasized that exponential growth of
energy use will inevitably be resumed, after an intermediate period while
technical improvements are implemented, unless lifestyle changes are introduced as well. Using a disaggregated model of 106 sectors (in place of the 9
sectors employed in the Ford Foundation project) and an appropriate computational program, these authors have investigated the impacts of a number of
scenarios. With unrestricted growth, energy demand in the U.S. will increase
to 0.185 Q by the year 2000 from 0.059 Q in 1967. If post-embargo prices are
maintained, technological innovations will be introduced for economic reasons
and the year 2000 demand will be reduced to 0.147 Q. With the addition of
important lifestyle changes (e.g., a nationwide program of home insulation,
a 50% reduction in gasoline consumption by the private sector allocated to bus
and rail transportation, etc.), a further reduction to 0.126 Q is considered to
be feasible. If, in addition, zero-population growth is achieved linearly from
the current level of 1.5% by the year 2030, only 0.116 Q will be needed for the
year 2000.

      All of the previously specified energy-conservation measures are voided
if a real compound GNP per capita growth rate of 4% per year is to be achieved
to the year 2000. This last statement serves to reemphasize the close connection between energy use and economic well being.

_____

[6]Clark W. Bullard, III, and Craig Z. Foster, "On Decoupling Energy and
    GNP Growth," Energy, Volume I (in press, 1976).

26.7 Facts and Fancy Bearing on Energy Policies

Factual data alone do not determine the makings of energy policies, which are highly sensitive to political factors. Energy-policy implementation schedules may be decided by social objectives, such as the "desirability" for solar-energy development because it is "clean", does not require imports of fossil fuels, is "obviously cheaper" over the long term than other energy sources, etc. The predetermination of social objectives of this type, independently of careful cost and environmental-impact analyses, may well lead to attempted implementation of new technologies that are premature, not cost-effective over the long term, require rapidly escalating capital investments that have been grossly underestimated initially, etc.

A major difficulty in the intelligent definition of energy policies is attributable to the fact that responses to well defined questions are not usually well defined answers, as will now be illustrated by a number of representative examples.

A. Are Solar Heating or Implementation of Energy Plantations
Justifiable Economic Developments at the Present Time?

To illustrate that policy responses to well defined questions are usually not well defined answers, we ask the following question about the implementation of a solar energy collection system: what is the maximum cost in dollars per square foot at which the solar-collection device becomes competitive with fossil fuel costing $4.00/10^6$ Btu? In the absence of a developed technology, we can only find a parametric response. Thus, at a location where the solar flux density is 200 w/m^2, we collect in one year

$$200\,(\text{w-y/m}^2\text{-y}) \times (0.949 \times 10^{-3}\ \text{Btu}/2.78 \times 10^{-4}\ \text{wh}) \times (8.76 \times 10^3\ \text{h/y}) \times \epsilon$$

$$= 5.98 \times 10^6\ \epsilon\,(\text{Btu/m}^2\text{-y})\,,$$

where ϵ is the solar-energy collection efficiency. The dollar value of this collected solar energy is \$23.92 ϵ/m^2-y. Assuming next a 15%/y return on investment, the solar collector may cost at most \$159 ϵ/m^2 if we disregard such major items of expense as energy-storage systems. Thus, at a collection efficiency ϵ of 10%, we are allowed at most \$1.47/ft^2 for the collector; at a collection efficiency ϵ of 20%, the allowed cost becomes \$2.95/ft^2; etc. Our policy answer becomes then that we should support solar-energy development provided collector costs for achievable efficiencies are sufficiently low when fuel costs have risen to \$4.00/10^6 Btu.

In the Appendix to this section, we present a more adequate treatment of the complex issues involved in an economic analysis of solar-energy costs for space and water heating.

Next let us consider the Btu-bush for a location at which the solar-power density is 170 w/m^2. Proceeding as in the preceding paragraph, we find a Btu-bush crop of $5.08 \times 10^6\ \epsilon\,(\text{Btu/m}^2$-y). At a 1% conversion efficiency, the dollar value of the crop is then $5.08 \times 10^4\,(\text{Btu/m}^2$-y) $\times$ (\$4/10^6 Btu) $\times$ ($4.0469 \times 10^3\ \text{m}^2$/acre) = \$822/acre-y. Again requiring a 15%/y return on investment, we conclude that the total allowed cost becomes \$5,482/acre-y, which appears to be a very adequate allowance for the project under discussion. But suppose the achievable solar-energy conversion efficiency is only 0.2% instead of the optimistic value of 1%. Then the allowable cost becomes \$1,096/acre-y, which would be fully absorbed by the annual rental or purchase value of land at \$7,307/acre with a 15%/y return. Energy-farm management costs are uncertain and not known for the Btu-bush. Our conclusion regarding the energy plantation must be that we should build it if the land values are sufficiently low, if the solar-energy conversion efficiencies are sufficiently high, if farm management is sufficiently inexpensive, etc. Again, a straightforward energy-policy question allows a categoric response only when a developed, large-scale technology is at hand for proper assessment.

B. Can the Development of Controlled Thermonuclear-Fusion Reactors Be Obviated by Using Fusion Bombs for Energy Production?

The development of controlled thermonuclear fusion reactors is currently a high-priority goal. There are means for replacing this program entirely. An example is provided by the concept of "pacer power" (see Fig. 26.7-1). The pacer-power proposal involves application of a fission reaction to ignite a fusion reactor in an underground cavity (see Fig. 26.7-1). The released thermal energy may then be used for steam production and electricity generation. This system contains an attractive application for the disposal of unwanted fusion bombs. Without extensive analysis and field testing, it is not certain that the pacer-power scheme is more costly or more damaging to the environment than a fully developed, successful, controlled thermonuclear reactor. However, there is significant popular antagonism to the use of fission and fusion bombs. It is therefore not unreasonable to eliminate this plan as a serious contender for replacement of conventional energy sources.

C. Is the Catastrophic Failure of a Nuclear Reactor Comparable in Impact with an Explosion Following the Spilling of an LNG Supertanker?

As we have discussed in Chapter 24, nuclear reactors pose, in principle, three types of hazards: hazards associated with accidental or catastrophic failure of the power plant during operation or immediately after

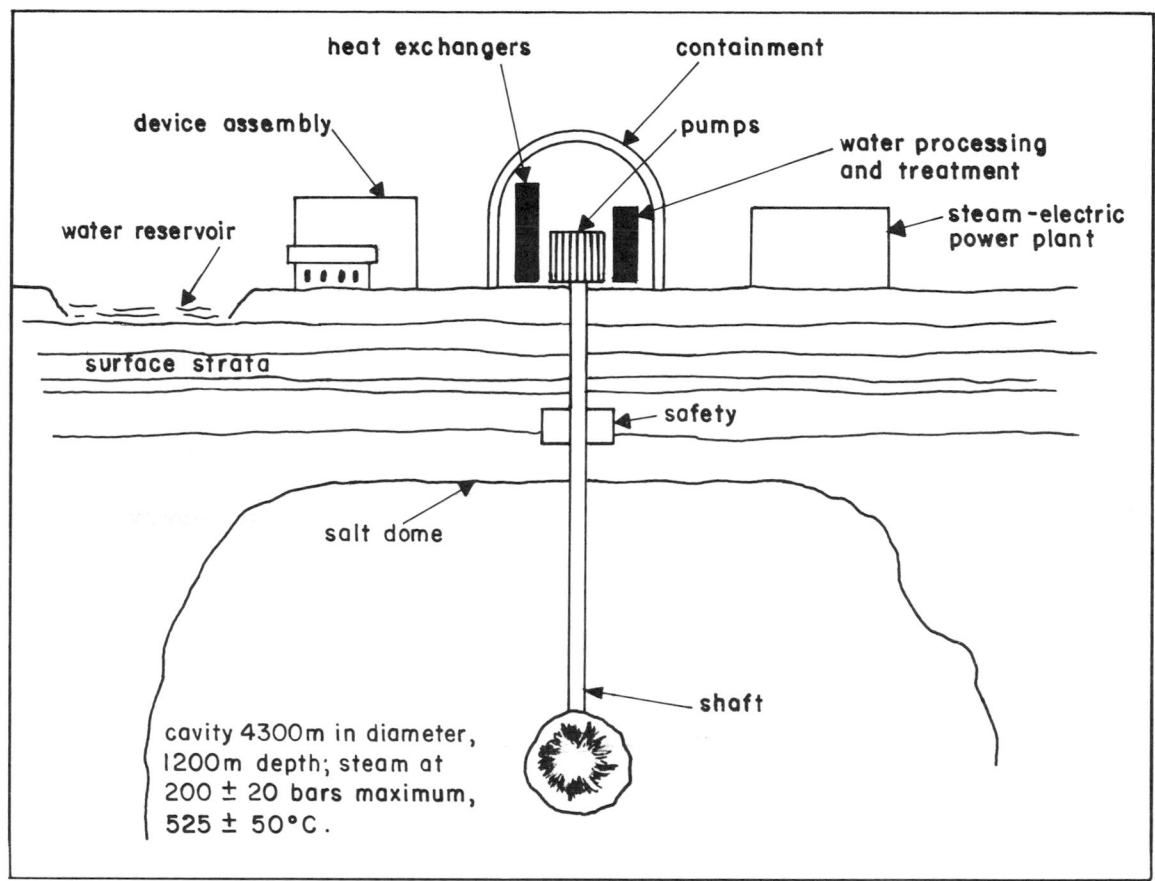

Fig. 26.7-1 Schematic diagram showing the concept of pacer power, as proposed by workers at the Los Alamos Scientific Laboratory of the University of California, Los Alamos, New Mexico; reproduced from Energy/Environment, University of California, Vol. 50, No. 27, February 18, 1975. Explosives and water are injected and steam is recovered through the shaft.

shutdown, hazards associated with the disposal and storage of nuclear-waste materials, and hazards associated with the deliberate misuse of power plants or waste materials for the purpose of implementing nuclear blackmail or subversion. In contrast to these hypothetical dangers, the number of actual deaths produced by fossil-fuel use has been significant. We have discussed the social costs of coal use in Chapter 7. Starr[1] stated in 1964 that coal and oil were responsible for 19,000 deaths per year. We have presented in Chapter 24 an extensive comparison of hazards associated with different energy sources.

The energy contents of supertankers carrying LNG or petroleum are so great that a very real potential for a major catastrophic event cannot be ruled out. While we are preoccupied with nuclear-hazard assessments,[2] relatively little work has been done on the potential dangers of LNG-based explosions.[3]

Consider a supertanker carrying 10^5 m^3 of LNG. The liquified natural gas is mostly methane with a boiling point of $112°K$ and a liquid density of 0.42 g/cm^3. According to Fay,[3] the following events will occur after spilling of the load on water in the absence of winds: the LNG will float on the water while boiling vigorously. The heat for evaporation (138 cal/g) is supplied primarily by the warmer water layers below the LNG which will thereby be cooled and freeze. The density of methane in the vapor phase at its boiling point exceeds that of normal air by about 34%. The layer of gaseous LNG will accordingly tend to lie on the water surface and spread laterally. Large spills evaporate at a relatively faster rate than small spills.[3]

The NG-vapor cloud will tend to drift downwind. Wind-generated turbulence will cause mixing with air above the cloud. The processes of spreading and mixing are accompanied by heating until the vapor cloud reaches

---

[1]C. Starr, "Radiation in Perspective," Nuclear Safety 5, 325-335 (1964).

[2]N. C. Rasmussen, "An Assessment of Accident Risks in U.S. Commercial Nuclear Power Plants," U.S. Atomic Energy Commission, Washington, D.C., August 1974.

[3]J. A. Fay, "Unusual Fire Hazards of LNG Tanker Spills," Combustion Science and Technology 7, 47-49 (1973).

ambient conditions. A very large area is thus covered very quickly with a potentially ignitable or explosive mixture of NG and air. Fay has estimated that the 10^5 m^3 of LNG will create a vapor cloud with a radius of 3,800 m in 15 min.

Since the vapor-to-liquid density ratio is about 254, the spill of a supertanker holding 10^5 m^3 of LNG will produce a gaseous volume of about 2.5×10^7 m^3 or 8.83×10^8 ft^3. With an energy content of 10^3 Btu/CF, the potential heat release in combustion or explosion is 8.83×10^{11} Btu or 2.21×10^5 ton of TNT$_e$. By contrast, explosion of the dirigible Hindenburg was associated with an energy content equal to about 10^{-4} of that of the LNG spill while the nuclear bombs at Hiroshima (12 to 14 kt) and Nagasaki (about 22 kt) released energy equivalent to less than 10^{-1} of the hypothetical LNG spill followed by explosion. It is thus not difficult to contemplate a major disaster as the result of an LNG spill near a populated coastal region. The potentials and likelihood of this type of disaster must be carefully weighed relative to those associated with nuclear-reactor developments. Large spills are also expected to constitute a major hazard because of asphyxiation produced by replacement of air with NG. We conclude that nuclear-reactor development and the use of LNG-carrying supertankers are closely-linked topical problems.

An extensive risk assessment associated with delivery of LNG at the equivalent rate of 4×10^9 SCF/d at Point Conception near Los Angeles has recently become available.[4] The LNG is assumed to be delivered in cryogenically cooled tankers and is then transferred to four shore-based storage tanks through a cooled pipeline. The following initiating events for catastrophic systems failure were considered: natural events (severe storms, tsunamis,

---

[4]"LNG Terminal Risk Assessment Study for Point Conception, California," Science Applications, Inc., 1200 Prospect St., La Jolla, Calif. 92037, January 23, 1976.

earthquakes, meteorites), ship collisions, air crashes, missile impacts, internal failures of the tanker, and failures during ship-to-tank transfers. The
analysis of hazard assessment included consideration of vapor-cloud dispersal
in view of expected meteorological conditions, ignition probability as a function
of time, and flame propagation with allowance of radiative and convective
energy transport.

The results are presented in terms of contour maps showing estimated
fatalities per person per year averaged over all possible spill locations. Within one and one third miles of the spill location, the fatality probability is about
7×10^{-8} per person per year and decreases to 1×10^{-10} per person per year
at 2 miles for the given population density. These estimates include the assumption that all of the people encountering burning plumes would become fatalities; for persons in shelters, the percentage-fatality estimate should actually be reduced to about 20%.

The methodology of fault-tree and event-tree analysis employed in the
Rasmussen study on accident-probability assessment for nuclear reactors (see
Section 24.3ii) was employed in the LNG accident probability assessment given
in Ref. [4].

D.   What Are the Likely Economic Impacts of Secondary and Tertiary
     Oil and Gas Recoveries?

Primary oil recovery yields between 13 and 46% of the original resource
in place, depending on the nature of the geological strata involved.[5] The
mean recovery efficiency in the U.S. has been estimated to be 32%.[5] Secondary recovery is performed by injecting water or gas under pressure into the
oil-bearing rocks, with supplementary yields amounting to perhaps 10% of the

---

[5]Chemical and Engineering News, December 16, 1974. p. 27.

original resource. Workers at the FEA have concluded[5] that tertiary recovery methods (e.g., injection of surfactant materials or partial combustion
after air injection) will yield between 30×10^9 and 60×10^9 bbl of oil from
existing U.S. wells.

Primary recovery of NG is believed to yield about 50% of the original
resource in place.[5] Secondary recovery techniques generally require stimulation by the use of explosives. Workers at the FEA have estimated that 300
to 600×10^{12} SCF of NG are recoverable from existing U.S. gas wells.[5]

Field implementations of secondary and tertiary recovery techniques
have been performed only to a limited extent. Workers at the Marathon Oil
Corporation have used the "Maraflood" process,[5] which involves injection
of an emulsion of water, hydrocarbons, surface-active agents, and alcohols
into an oil deposit as a plug or buffer ahead of water injection. The composition of the emulsion is chosen in such a manner that it readily displaces the
oil. Recovery efficiencies for the tests were estimated to be about 45% of the
oil in place.

Carbon dioxide injection at pressures around 1,000 psi and temperatures above 90°F has been used to dissolve some of the hydrocarbons in the
oil which, in turn, changes the physico-chemical properties of the remaining
oil in such a manner that water flooding may increase oil yields by 10 to 30%.
The technique of partial, underground combustion to stimulate oil recovery
has not yet been adequately explored.

Encouraging results have been obtained with nuclear stimulation of gas
wells, both in the U.S. and in the U.S.S.R. Commercialization of these techniques has been impeded by the usual constraints that relate to applications of
nuclear explosions. It is not inconceivable that widespread use of nuclear

stimulation will provide the most economical supplementary source of conventional fossil fuels. Unfortunately, we cannot really be certain that it will come out this way before performing extensive field studies.

E. Should We Attempt to Recover Electric Energy from Lightning?

We have the following factual information concerning lightning strokes:[6, 7] The energy dissipation is about 10^5 j/m and the average length of a stroke is 3 km. Hence the energy per stroke is 3×10^8 j = 3×10^8 w-sec, while the average duration is 30×10^{-6} sec. Thus, the average power per stroke is 10^{13} w. On the average, 100 lightning flashes, each containing 4 strokes per flash, occur per sec near the surface of the earth.[8] Thus the equivalent continuous electrical power dissipation is 4×10^{15} w and the corresponding annual energy dissipation* becomes 3.5×10^{16} kwh/y = 1.2×10^2 Q/y, a value which is nearly 20 times larger than the estimated year 2050 world-wide energy-use level.

Is it worthwhile to develop a major effort to harness some of this enormous resource? The answer must be negative until an ingenious utilization scheme has been evolved by someone.

---

[6] R. D. Hill, "Lightning Research," Naval Research Reviews, pp. 1-14, October 1975; published by the Office of Naval Research, Dept. of the Navy, Arlington, Va. 22217.

[7] E. P. Krider, G. A. Dawson and M. A. Uman, "Peak Power and Energy Dissipation in a Single-Stroke Lightning Flash," J. Geophys. Res. 73, 3335-3339 (1968).

[8] E. T. Pierce, "Charge Transferred to Earth by a Lightning Flash," J. of the Franklin Institute 286, 353-354 (1968).

*The listed number is larger by a factor of 3.5×10^7 than the value given in Ref. [6].

F.   Should We Develop Salinity Power?

Power may be obtained by mixing fresh and ocean waters.[9-11] The difference in salinity between fresh water and seawater (which contains about 35×10^3 ppm of dissolved salts) corresponds to a salinity potential of approximately 22.4 atm.   Dilution of $1 \text{ m}^3/\text{sec}$ of fresh water in a large volume of seawater will generate 2.25 Mw of salinity power, which is probably convertible to electrical power at an efficiency of 15 to 50%.   The worldwide generation of salinity power is obtained by multiplying 2.25 Mw by the worldwide river discharges to the oceans ($\sim 1.08 \times 10^6 \text{ m}^3/\text{sec}$) and yields a total salinity power of 2.43×10^6 Mw, which is seen to be roughly comparable with the total power dissipation by ocean waves ($\sim 2.5 \times 10^6$ Mw, see p. 465 in Volume II) and tides ($\sim 3 \times 10^6$ Mw, see p. 442 in Volume II).

There are no known, economically defensible, techniques for the utilization of salinity power.   Capital-cost estimates are too high by factors of 10 to 10^2 as compared with other electric power generators.   However, if fresh water were available in areas such as the Dead Sea, power generation would be competitive since the very high salt content of the Dead Sea leads to augmentation of the power output per m^3/sec of fresh water mixed by about a factor of 20.

Practical utilization of salinity power must await inventions and large reductions in component costs.   The utilization of salinity power is an interesting research program at the present time.

---

[9] J. N. Weinstein and Frank B. Leitz, "Electric Power from Differences in Salinity: The Dialytic Battery," Science 191, 557-559 (1976).

[10] R. S. Norman, "Water Salination: A Source of Energy," Science 186, 350-352 (1974).

[11] S. Loeb, "Osmotic Power Plants," Science 189, 654-655 (1975).

G.   Will Population Growth Be Limited?

The following, surprisingly accurate relation[12] was presented in 1960 for the growth of world population N(t) between 1750 and 1960:

$$N(t) = \frac{1.79 \times 10^{11}}{(2026.87-t)^{0.99}},$$

(26.7-1)

where the time t is measured in years A.D.   The 1975 estimate for N(t) is 3.65×10^9 p as compared with the actual value of 3.97×10^9.[13]   According to Eq. (26.7-1), N(t) will become infinite during the year 2026 which would then represent a special form of doomsday.[14]   What policies must be implemented to prevent this ultimate catastrophe from occurring?

H.   Comments on the U.S. National Energy-Policy Plan

The 1975 version of the U.S. national energy plan is defined in the document labeled ERDA-48.   The Office of Technology Assessment (OTA) of the U.S. Congress has recently released a document[15] entitled "An

---

[12] H. von Foerster, P. M. Mora, and L. W. Amiot, "Doomsday: Friday, 13 November, A.D. 2026," Science 132, 1291-1295 (1960).

[13] P. F. Myers, L. F. Bouvier, and J. R. Echols, 1975 World Population Data Sheet, Population Reference Bureau, Washington, D.C., 1975.

[14] J. Serrin, "Is Doomsday on Target?," Science 189, 86-88 (1975).

[15] U.S. Government Printing Office, Washington, D.C., 1975.

Analysis of the ERDA Plan and Program. "[*] Under the column News and Comment, Philip M. Boffey has given a review of the salient features of the OTA evaluation in Science.[16] Boffey[16] has emphasized the adversary aspects of the OTA study. A more personal commentary on ERDA-48 is presented here.

a. Basic Constraints Implicit in the Preparation of ERDA-48[**]

There is a general concensus that Volume I of ERDA-48 represents a significant articulation of U.S. energy-policy objectives while the implementation plan in Volume II is incomplete and, in some instances, inadequate to achieve the objectives specified in Volume I.

Of particular importance in understanding ERDA-48 are five statements of national policy goals, which include (i) maintenance of "security and policy independence of the Nation," (ii) maintenance of a healthy economy allowing "fulfillment of economic aspirations (especially in the less affluent parts of the population)," (iii) provision of adequate energy for future needs, (iv) contribution to world stability through cooperative efforts, and (v) protection of the Nation's environmental quality.

Much of the criticism of ERDA-48 relates to implicit refusal by reviewers to accept the implications of the third stated policy goal. This reads in full as follows: "This Plan recognizes five national goals as a focus for energy policy" including the objective "To provide for future needs so that life

---

[*] The author served as Chairman of the OTA Panel on Environment and Health and as a member of the Overview Panel, whose job it was to prepare a summary assessment of the ERDA plans and programs.

[**] The 1976 version of the ERDA plan is described in ERDA-76, which is generally similar to ERDA-48.

[16] Science 190, 535-537 (1975).

styles remain a matter of choice and are not limited by the unavailability of energy." This stated policy goal must properly be viewed as having been mandated for ERDA by higher authority (e. g. , by the President or other superior levels of the Administration). Once we accept the doctrine that energy availability must not limit choices, we are inexorably committed to the primary task of developing new and promising technologies. It is then not useful to argue that ERDA is "too hardware oriented."

A statement of "National Energy R, D and D (Research, Development and Demonstration) Goals" follows inevitably from a policy statement that energy availability must not limit choices of lifestyles. The identified Energy R, D and D Goals are listed below.

"I.   Expand the domestic supply of economically recoverable energy producing raw materials.

II.   Increase the utilization of essentially inexhaustible domestic energy resources.

III.   Efficiently transform fuel resources into more desirable forms.

IV.   Increase the efficiency and reliability of the processes used in the energy conversion and delivery systems.

V.   Transform consumption patterns to improve energy utilization.

VI.   Increase end-use efficiency.

VII.   Protect and enhance the general health, safety, welfare and environment related to energy.

VIII. Perform basic and supporting research and technical services related to energy."

It is interesting to note that R, D and D Goals III, IV, V, and VI are directly related to aspects of energy conservation while VII emphasizes the importance of environmental and health factors.

b.   The Policy Plan in Volume I of ERDA-48

The policy plan of Volume I has been constructed after considering the implications of five scenarios with widely differing import implications.   These scenarios include extreme alternatives.   As an illustration of scenario implications, we have calculated the year 2000 import requirements in $(bbl/d)_e$ (barrels per day equivalent of petroleum) and trade deficits in dollars per year at a cost of \$10 per bbl of energy-equivalent.   These are listed in Table 26.7-1. The detailed scenario inputs that are required to arrive at the listed estimates are not especially transparent.   But we may safely conclude that we cannot afford to live without new initiatives, that energy-conservation measures of all types are of primary importance, and that we must, in fact, implement all environmentally and economically acceptable technologies which conserve energy or enhance the domestic sources of supplies.

Given limited capital and manpower resources, the order of priority for development must reflect assessments of promise, ready availability and low cost.   The ERDA listing here simply represents educated expectations of viable accomplishments.   In the "Highest Priority Supply" category, one must agree that the proper near-term major energy systems are coal, nuclear reactors, and oil and gas.   Among the new sources of liquids and gases for the mid-term, we would favor oil shale for liquids and coal for gases, whereas the ERDA listing refers simply to "gaseous and liquid fuels from coal and oil shale."   As "inexhaustible" sources for the long term, the ERDA listing identifies breeder reactors, fusion reactors and solar-electric options.   Likely implementation schedules for these technologies differ widely and the economic and environmental implications for successful developments are not at all comparable.

The ERDA document stresses correctly, under the category of "Highest Priority Demand," near-term efficiency (conservation) technologies.   For

Table 26.7-1 Implications of five ERDA scenarios for the year 2000.

Scenario	Import requirement, $(bbl/d)_e$	Annual trade deficit at $\$10/(bbl)_e$, $\$$
0: no new initiatives	2.8×10^7	100×10^9
I: improved efficiences in end use	9.4×10^6	34×10^9
II: synthetics from coal and shale	8.5×10^6	31×10^9
III: intensive electrification	12×10^6	43×10^9
IV: limited nuclear power	9.0×10^6	33×10^9
V: combination of all technologies[*]	-2.4×10^6	-8.6×10^9

[*]A negative sign indicates exports or a trade surplus.

"Other Important Technologies," we find appropriate listings of "under-used mid-term technologies" (geothermal, solar heating and cooling, waste-heat utilization). Under "technologies supporting intensive electrification," the following are emphasized: electricity conversion efficiency, power transmission, distribution, and storage. For the long-term, biomass conversion and the production of hydrogen-based fuels are identified as promising technologies.

In the "National Ranking of R, D and D Technologies," there may well be an overemphasis on electrification and failure to recognize properly the important role that will be played in our society by the personal transportation vehicle. Thus, preferred emphasis on the development of portable fuels for transportation would have been in order.

We now come to identified "five major changes"..."needed in the nature and scope of the Nation's energy R, D and D program." These are (i)

emphasis on accelerated development of coal supplies and nuclear reactors,
(ii) focus on conservation efforts, (iii) accelerated gaseous and liquid fuel production from coal and oil shale, (iv) high priority for solar-electric development, (v) increased attention on under-used technologies. Effort (iv) may be
premature while (v) is a platitude.

The following additional important problem areas have not been adequately assessed in ERDA-48:

(a) The effect of price on energy demand.

(b) Removal of constraints to commercialization, including the development of incentives.

(c) Working relations with State and local governments and their
participation in ERDA-sponsored programs.

(d) Availability of resources, especially when multiple, competing
uses are involved.

(e) Detailed consideration of global trade balances and energy uses
and their interactions with U.S. developments.

The energy-policy goals described in this section will form useful and
acceptable guidelines. Where we go from here falls in the implementation
schedule and Volume II. Where we go from here is not clear, not so well
thought out, not easy to define, and will, in any case, require continuous re-
assessments and modifications.

c. Policy Implementation

It is possible that the best policy-implementation plan for Volume I
of ERDA-48 is the abolition of ERDA, coupled with preferred reliance on profit
motives, the private sector, and a free economy in which the government role
is that of reducing encumbrances, defining required environmental and health
constraints, providing low-cost loans to industry for the development of new
energy technologies, and perhaps taxing unreasonable gains which may accrue

to anyone as the result of total price deregulation. Supporting research on energy technologies under this scheme could still be partially funded through one or more federal agencies.

For good or bad reasons, the decision seems to have been made that only the "Government" is big enough, smart enough, and rich enough to guide the Nation's energy future. An immediate and, at the present time, most visible result of this decision is a significant growth in the number of functionaries dealing with the energy problem at all levels of government.

Volume II of ERDA-48 is more technical in nature than Volume I. For each energy technology, there is an identified implementation schedule. These schedules are characterized by the following features: <u>decision</u> on program initiation, program <u>start</u>, completion of <u>pilot</u> stage, completion of <u>demonstration</u> plant. For ocean-thermal electricity generation, program start is listed as 1977 and completion of demonstration as 1988; for fusion, demonstration plants producing 1 to 10 Mw-sec pulses are to be built by 1982; a near-commercial liquid-metal-fast-breeder-reactor is to be completed after 1985; a controlled environment agricultural pilot plant to produce fuels from biomass is scheduled for 1984-85; etc. It is neither difficult nor productive to find fault with one or more of the schedules for energy-development technologies. Depending on the point of view, each one is too optimistic, too slow, too costly, not economically viable at this time, environmentally unacceptable, a health hazard, urgently needed, premature, etc. Energy-development technologies must be reassessed as they are being investigated. The more urgent the implementation schedule is, the larger the price tag and the more wasteful the program will be. Here we content ourselves with a few detailed comments.

Oil- and gas-recoveries from conventional on-shore and off-shore resource areas do not figure prominently in the ERDA plan. This omission is intentional and appropriate. New technologies are not at issue here. Further-more, the existence of federal agencies with overlapping authorities and mis-sion objectives may have dissuaded ERDA planners from giving these resources a major play. Oil- and gas-price regulations fall under other jurisdictions (e.g., the Federal Power Commission, the Federal Energy Administration, the Department of the Interior, State Lands Commissions, utility-regulating agen-cies, etc. etc.). The plethora of overlapping authorities is a major deterrent to progress. It is a conspicuous ailment of an overpowering bureaucratic complex.

As we have seen, energy conservation in all its forms appears as a major policy objective. The ERDA implementation plan is notably deficient in concrete prescriptions on how the desired measures of conservation may be achieved. Even a well conceived public-education activity is absent. The tech-nology and science that are of basic importance in implementing fuel-conserva-tion measures are covered by combustion research. This vital topic appears in various boxes (e.g., conservation and atomic and molecular physics) as a subsidiary item. One might argue that, over the near term, its importance is commensurate with, or even superior to, that of solar-energy research for which we are now laboring under a Congressional mandate to establish a major national Solar Energy Research Institute (SERI). A national combustion labora-tory has also been proposed and will probably appear in some form at one of the national, in-house ERDA laboratories. The basic question of in-house ver-sus out-house R, D and D reappears as a major unresolved issue.

The emphasis by ERDA on "hard science" as a legacy of its derivation from the AEC (Atomic Energy Commission), OCR (Office of Coal Research), BM (Bureau of Mines) is often deplored. No energy program will ever be im-mune from criticism of this type. Energy use is environmentally hazardous, detrimental to human and animal health, an aberration on the natural order, fraught with social and political consequences.

Under the stress of technological innovation, health hazards remain a persistent problem. The basic difficulty is that the time scales are not matched: it usually takes too long to find out what the real health costs are. Here we need improved methodology, better understanding, more comprehensive approaches. The deficiencies in the ERDA plan are deficiencies in knowledge that will only be corrected by careful, long-term, basic studies of the type best done at the universities.

The role of ERDA with respect to university support is undefined and in urgent need of clarification. Effective means for technology transfer and implementation remain to be constructed. ERDA's role in industrial development remains unconvincing. Such questions as the following must be answered: who owns the patents if ERDA shares in the costs of technological innovations?

The proposed RD and D 1977 budget of about 2. 4 billion dollars is a primary input into the program plan of ERDA-76. The costs were not first determined for program needs and then added to arrive at the ERDA operating budget. Are 2. 4 billions of dollars per year too much or too little? What could be done for one billion dollars per year and what for four billions of dollars? It is unfortunate that no indication to the proper answer to these questions can be found in the ERDA-76 document.

As time goes on and ERDA proceeds beyond restructuring itself and functions, hopefully effectively, the policy-implementation schedule should become real, better defined, and ultimately a tool for U.S. energy sufficiency.

     d.   What Must Be Done to Proceed Beyond the Preliminary Planning Stage?

We have already alluded to critical, largely political problem areas.

These are repeated here in order of our estimate of their decreasing importance.

(i) How large will the federal energy bureaucracy be allowed to become? What fraction of the total budget may be spent on in-house research and on management? What is the correct number of people for efficient management of the ERDA programs?

(ii) What total or fraction of the R, D and D moneys should be spent within the ERDA-operated government laboratories on each activity?

(iii) How are problems of overlapping jurisdictions between federal agencies to be adjudicated in the energy area?

(iv) How large should the ERDA budget be to give the taxpayers an optimal return on their investment in energy R, D and D?

A26.7 Appendix: Solar Heating and Cooling

We present in this section a description of an economic analysis of solar-assisted space- and water-heating units, as well as quantitative cost estimates.

A.   Financial Analysis

In order to evaluate the net annual cost of a supplementary solar-heating system, a number of concepts in financial analysis must be understood.

An annuity is a series of equal payments made at equal intervals of time.   Equal annuity payments do not correspond to equal reductions in loan equity during the life of the annuity.   The magnitude of each payment is a periodic payment which is more simply called a rent and should not be confused with the amount paid to a landlord for a property rental.   The amount of the annuity is the sum of all the payments made in the time interval between the beginning of the first rent period and the end of the last rent period when all accrued interest has been paid.   The time interval between the first and last rent periods is called the term of the annuity.   The first rent period corresponds to the beginning of the contract before accumulation of interest; at the end of n rent periods, interest has been paid for n - 1 rent periods.

We shall now evaluate the amount of an annuity with a term of n periods and a periodic rent of $1.00 when payments are accumulated at a constant interest rate of i per period.   We assume that the payment and interest periods coincide.   The payment at the end of the first period (before interest accumulation begins) is $1; the accumulated balance at the end of the second period is $1 + $1 + $i = $1 + $1(1 + i); the balance at the end of the third period (in dollars) is $1 + (1 + i) + (1 + i)^2 = 1 + [1 + (1 + i)](1 + i)$; etc. Thus, if the accumulated balance at the end of k periods is given by x_k, then the balance at the end of k + 1 periods is

$$x_{k+1} = 1 + x_k (1 + i).\qquad\qquad\text{(A26.7-1)}$$

The accumulated balance at the end of n periods may be written as

$$x_n = 1 + (1 + i) + (1 + i)^2 + \cdots + (1 + i)^{n-2} + (1 + i)^{n-1}.\quad\text{(A26.7-2)}$$

Sequential terms on the right-hand side of Eq. (A26.7-1) correspond[*] to the first n terms of the modified binomial series

$$x_n = [(1 + i)^n - 1]/i, \; n \geq 1,\qquad\qquad\text{(A26.7-3)}$$

which is thus equal to the amount (sum of all payments including interest) of an n-period annuity of \$1.00 per period accumulated at an interest rate of i per period.

---

[*] In order to verify Eq. (A26.7-3), we begin with the following binomial series: $(1 + i)^n = 1 + ni + \dfrac{n(n-1)}{2!} i^2 + \dfrac{n(n-1)(n-2)}{3!} i^3 + \cdots$, from which it follows that $x_n \equiv [(1 + i)^n - 1]/i = n + [n(n-1)/2]i + [n(n-1)(n-2)/3!]i^2 + \cdots$. For n = 1, x_n = 1; for n = 2, x_n = 2 + i = 1 + (1 + i); for n = 3, $x_n = 3 + 3i + i^2 = 1 + (1 + i) + (1 + i)^2$; etc.

The present value (PV) of an annuity[*] is the sum of the present values of all the payments of the annuity. Since the amount x_n of an n-period annuity with a rent of \$1.00 is given by Eq. (A26.7-3), it follows that the present value of this annuity, $PV(x_n)$, is obtained by dividing x_n by $(1 + i)^n$, i.e.,

$$PV(x_n) = \frac{[(1 + i)^n - 1]/i}{(1 + i)^n} = [1 - (1 + i)^{-n}]/i. \qquad (A26.7-4)$$

It is apparent from Eq. (A26.7-4) that the present value of an annuity, $PV(x_n)$, is \$1 if the periodic rent of the annuity of \$1 is replaced by

$$rent = i/[1 - (1 + i)^{-n}], \qquad (A26.7-5)$$

where the right-hand side of Eq. (A26.7-5) is called a capital-recovery factor and is often used in the financial analysis of investments.

---

[*] The concept of present value accounts for the time value of money (i.e., a dollar tomorrow, even in the absence of inflation, is not worth as much as a dollar today because invested money will earn some interest by tomorrow). Investing one dollar today at an annual interest rate of i, which is assumed to represent the maximum possible interest rate obtainable from a riskless venture, yields $(1 + i)$ dollars one year from now. Thus, one dollar received today is equivalent to $(1 + i)$ dollars received one year from now. In general, the present value of an amount y to be received n periods in the future (i.e., discounted over n periods) and discounted at the rate i is $PV(y) = y/(1 + i)^n$. We refer to Appendix 9-A for further discussion of this concept.

B. Consumer Costs for an Incremental Investment of One Dollar[1]

The annual cost (C) for an incremental investment with a present value of $1.00 is

$$C = T + I + R, \qquad (A26.7\text{-}6)$$

where T is the annual property-tax rate in $/$-y = $/($ of appraised value)-y, I is the insurance payment in $/$-y, and R is the annual mortgage payment obtained from Eq. (A26.7-5) in $/$-y with i replaced by the mortgage interest i_m, viz.

$$R = i_m / [1 - (1 + i_m)^{-N}]. \qquad (A26.7\text{-}7)$$

Here N is the mortgage lifetime in years and we have assumed for simplicity that a consumer makes annual payments rather than following the more traditional schedule of monthly payments.

Since the annual property-tax and annual mortgage-interest payments are tax-deductible, the actual net annual cost to the borrower for an incremental investment of $1.00 in a solar-energy utilization system is less than the cost calculated from Eq. (A26.7-6). The proper analysis of this cost reduction will now be described.

---

[1]The economic analysis presented in this Section follows closely the paper by F. Kreith and J. F. Kreider, "Preliminary Design and Economic Analysis of Solar Energy Collectors for Heating and Cooling of Buildings," Energy 1, (in press, 1976).

The mortgage interest paid at the end of the kth year, for an incremental investment with a present value of $1.00, is

$$M_k = i_m P_k,$$ (A26.7-8)

where P_k is the principal remaining on the $1.00 present investment during the kth year. We note that the remaining principal during the kth year is given by the relation

$$P_k = (1 + i_m)^{k-1} - R[\{(1 + i_m)^{k-1} - 1\}/i_m].$$ (A26.7-9)

The first term on the right-hand side of Eq. (A26.7-9) represents the sum that would be owed after $k - 1$ years on the mortgage if no payments had been made at all while the second term corresponds to the amount of a $(k - 1)$-year annuity of $R per year accumulated at an interest rate of i_m per year, as evaluated from Eq. (A26.7-3).

The present value at the beginning of the first year of the mortgage-interest payment M_k at the end of the kth year is

$$PV(M_k) = M_k/(1 + r)^k,$$ (A26.7-10)

where r is the applicable interest rate corresponding to the return earned by the consumer if the incremental investment in a solar-energy utilization system were placed in an alternative interest-earning venture (e.g., a personal savings account). The interest r will generally be less than the interest i_m. The total

present value of all of the interest payments is the sum of the present values
of the payments, i.e.

$$\sum_{k=1}^{N} PV(M_k) = \sum_{k=1}^{N} M_k/(1 + r)^k. \tag{A26.7-11}$$

It is now convenient to evaluate the annual rent M for an N-year annuity corresponding to the total present value given by Eq. (A26.7-11). The annual rent
payment is found from Eq. (A26.7-5) by replacing i by r, n by N, and multiplying by $\sum_{k=1}^{N} PV(M_k)$ since Eq. (A26.7-5) refers to an annuity of $1.00 rather
than to an annuity of $\sum_{k=1}^{N} PV(M_k)$; thus,

$$M = \{r/[1 - (1 + r)^{-N}]\} \times [\sum_{k=1}^{N} M_k/(1 + r)^k]. \tag{A26.7-12}$$

The net annual cost for an incremental investment of $1.00 in a solar-
energy utilization system becomes then

$$R_T = R + T + I - t(T + M), \tag{A26.7-13}$$

where t is the applicable total income-tax rate to the borrower. The last term
on the right-hand side of Eq. (A26.7-13) represents the total annual income-
tax savings associated with a $1.00 investment in a solar-energy utilization
system and may be an important factor in reducing the actual cost of the system to the consumer, particularly if repayment of the loan must be made over

a long period of time. For example, with a mortgage-interest rate of i_m = 0.09 (9%), an alternative investment rate of r = 0.06 (6%), a property-tax rate of T = 0.03 (3%), an insurance rate of I = 0.005 (0.5%), and a 30-year mortgage paid annually, the annual cost per investment dollar is found from Eq. (A26.7-6) to be $0.1323. On the other hand, if the tax credit for a consumer in a 30% (t = 0.3) composite tax bracket is properly included [see Eq. (A26.7-13)], the net annual cost becomes $0.1011, which is a 23.6% reduction.

In the future, consumers may desire to retrofit a dwelling with a solar-energy system. At the present time, lending institutions will not ordinarily allow the cost of home improvements to be added to the existing mortgage and will instead require the borrower to finance the retrofit by a home-improvement loan. Under these circumstances, with an interest rate of i_m = 0.12 (12%), an alternative investment rate of r = 0.06 (6%), a property-tax rate of 0.03 (3%), an insurance rate of 0.005 (0.5%), and a 5-year loan paid annually, the annual cost of a solar energy-utilization system during the 5-year mortgage period, per investment dollar, is $0.3124. If the tax credit for a consumer in the 30% (t = 0.3) tax bracket is included, the net annual cost becomes $0.2794, which is a 10.6% reduction.

Long-term federal loans for retrofitting of supplementary solar heaters will stimulate introduction of these devices.

C. Cost Minimization of Solar-Assisted Energy Systems

Optimal design of a solar-heating unit involves determination of the sizes and methods of integration of the collector, storage system, distribution and transport system, control system, and auxiliary energy system. For a specified heating or cooling load, the optimal combination of the solar and auxiliary systems is defined to be the integrated unit which provides energy at minimum cost. In the construction of the storage, distribution and transport, and control systems of a solar installation, collector size is the single most important design parameter. The detailed configurations of the storage, distribution and transport, and control systems have a second-order effect on solar-system cost and performance.

The total net annual cost (C_S), per unit of energy delivered for a solar-energy utilization system requiring a capital investment of P_S dollars is

$$C_S = (P_S R_T + C_{OM,S})/Q_S,$$
                                                                             (A26.7-14)

where $P_S = C_C A_C + C'_S$ (with C_C = cost per unit of collector area; A_C = total collector area; C'_S = total capital cost for the storage, distribution and transport, and control systems) and $C_{OM,S} = P_S C_P + C_{M,S} + C_{\ell,S}$. Here the net annual cost for an incremental investment of \$1.00, R_T, is given by Eq. (A26.7-13); $C_{OM,S}$ is the sum of the annual operating and maintenance costs; Q_S is the average annual energy delivered by the solar-energy system; P_S is the annual power requirement for operation of the solar-energy system; C_P is the unit power cost; $C_{M,S}$ is the annual charge to maintain structures and materials; $C_{\ell,S}$ is the annual labor cost for maintenance.

The total net annual cost per unit of energy delivered by an auxiliary energy system may be written as

$$C_a = (P_a R'_T + C_{OM,a})/Q_a,$$
                                                                             (A26.7-15)

where P_a is the capital cost of the auxiliary energy system; R'_T is the net annual cost for an investment of \$1.00 in an auxiliary energy system;

$C_{OM,a} = E_a C_f + P_a C_P + C_{M,a} + C_{\ell,a}$ is the sum of the annual operating and maintenance costs; E_a is the annual energy (fuel) requirement of the auxiliary energy system and C_f is the corresponding unit fuel cost; P_a is the annual power requirement of the auxiliary energy system and C_P is the corresponding unit power cost; $C_{M,a}$ is the annual cost for maintenance of structures and materials in the auxiliary power system; $C_{\ell,a}$ is the annual labor cost for maintenance; Q_a is the average annual energy delivered by the auxiliary energy system.

The total net annual cost of the solar-assisted energy system depends on the fraction f_S of the total load that is supplied by the solar system. We may write, in general,

$$C_T = Q_T[C_S f_S + C_a(1 - f_S)], \tag{A26.7-16}$$

where Q_T is the total annual energy requirement for heating and cooling and $1 - f_S$ is the fraction of the load supplied by the auxiliary energy system.

Determination of the total annual energy needs for heating and cooling of a building requires knowledge of the heat-transfer characteristics of the building walls, ceiling, roof, windows, doors, and of other heat-flow paths for a given geographic location and climate. The average annual energy available from the solar-energy system is

$$Q_S = I_S A_C \eta_C \eta_S, \tag{A26.7-17}$$

where I_S is the average annual solar insolation, A_C is the collector area, η_C is the collector efficiency, and η_S is the composite efficiency for the storage, distribution, and transport systems.

Substitution of Eq. (A26.7-17) into Eq. (A26.7-14) yields

$$C_S = \frac{(C_C A_C + C_S')R_T + P_S C_P + C_{M,S} + C_{\ell,S}}{I_S A_C \eta_C \eta_S}$$

(A26.7-18)

after explicit introduction of the defining relations for P_S and $C_{OM,S}$. The required collector area A_C depends only on the collector efficiency η_C for given Q_S, i.e., given insolation and fixed systems performance. Therefore, the change in required collector cost per unit area with collector efficiency may be estimated from Eq. (A26.7-18) by treating all of the other terms in this relation as independent of η_C. With this assumption, we find that

$$dC_C/d\eta_C = C_S I_S \eta_S/R_T$$

(A26.7-19)

or, in terms of discrete changes ΔC_C and $\Delta \eta_C$,

$$\Delta C_C = (C_S I_S \eta_S/R_T)\Delta \eta_C.$$

(A26.7-20)

Assessments of the economic implications of installing a more efficient and expensive collector with two glass covers rather than a cheaper (single glass cover) model, or of implementing a costly collector-surface treatment to improve the optical properties, may be made by starting from Eq. (A26.7-20).

D. Determination of Systems Performance

As we have stated previously, in considerations of economic constraints on the engineering design of solar-assisted energy systems, the key problem is the determination of optimal collector size. Appropriate values for storage capacity, piping, pumps, heat exchangers, flow rates, controls, and other ancillary equipment may then be matched to the specified collector size. It has been found that in moderate, mid-latitude climates the optimal capacity of a well-insulated thermal-storage system for a water-cooled flat-plate collector is about 10 to 15 lb of water per ft^2 of collector area.[1,2]

The value of the optimal collector size is a function of the energy requirements for heating and cooling of a structure. The energy loads may be calculated[3] by employing standard engineering methods, e.g., by using thermal response or degree-day techniques for heat loads and peak cooling hour or cooling degree-day methods for cooling loads. The final quantitative design of solar-energy systems requires detailed hour-by-hour simulation (by digital computers) of systems performance for specified collectors and auxiliary systems. Each of the following systems parameters must be known: collector-cover transmittance, collector-surface absorptance, collector and system thermal-loss coefficients, fluid flow rates, heat-exchanger effectiveness. Systems performance must be matched to requirements determined by structural factors and climatic conditions.

A useful approach has been developed by Klein, Beckman, and Duffie,[4] which allows the designer to approximate the results obtained from an hour-by-hour digital simulation by using a parametric design chart which requires

[2]G. O. G. Löf and R. A. Tybout, "Cost of House Heating with Solar Energy," Solar Energy 14, 253-278 (1973).

[3]See, for example, ASHRAE Handbook of Fundamentals, American Society of Heating, Refrigeration, and Air-Conditioning Engineers, New York, N.Y., 1972.

[4]S. A. Klein, W. A. Beckman, and J. A. Duffie, "A Design Procedure for Solar Heating Systems," University of Wisconsin, Madison, Wisconsin, 1975.

only knowledge of collector properties and of average monthly climate and load data described by two dimensionless parameters. Klein et al[4] define the following solar-insolation parameter:

$$P_{solar} = F'_R A_C S(\overline{\tau\alpha})/Q_L, \qquad (A26.7-21)$$

where F'_R is the collector heat-exchanger efficiency factor, A_C is the collector area, S is the total monthly insolation incident on the plane of the collector, $(\overline{\tau\alpha})$ is the time-average product of the collector-cover transmittance and absorber-plate absorptance, and Q_L is the total monthly energy load. The collector heat-exchanger efficiency factor F'_R is given by

$$F'_R = F_R/[1 + F_R U_L A_C/(\dot{m}C_p)_C]\{[(\dot{m}C_p)_C/(\dot{m}C_p)_{min} \varepsilon_C] - 1\}, \qquad (A26.7-22)$$

where F_R is the collector heat-removal factor, U_L is an overall energy-loss coefficient for the collector, $(\dot{m}C_p)_C$ is the rate at which the collector fluid is capable of absorbing heat energy (collector-fluid capacitance rate), $(\dot{m}C_p)_{min}$ is the smaller of the fluid-capacitance rates in the collector and storage-tank heat exchangers, and ε_C measures the heat-exchanger effectiveness in the collector storage tank. The collector heat-removal factor F_R may be determined[5] from

$$F_R = [(\dot{m}C_p)_C/A_C U_L]\{1 - \exp[F'U_L A_C/(\dot{m}C_p)_C]\}, \qquad (A26.7-23)$$

---

[5] J. A. Duffie and W. A. Beckman, Solar Energy Thermal Processes, John Wiley and Sons, New York, N.Y., 1974.

where F' is the collector geometry efficiency factor, which is a function of the collector construction, and may be determined by methods described by Bliss.[6]

The energy-collection rate of a flat-plate collector may be expressed[7] as

$$Q = A_C[S(\overline{\tau\alpha}) - U_L(T_p - T_a)], \tag{A26.7-24}$$

or

$$Q = F_R A_C[S(\overline{\tau\alpha}) - U_L(T_i - T_a)], \tag{A26.7-25}$$

where T_p is the instantaneous absorber-plate temperature, T_i is the instantaneous fluid temperature at the collector input, and T_a is the ambient air temperature. Expressions for the efficiency of a flat-plate collector may be determined from either Eq. (A26.7-24) or Eq. (A26.7-25), as follows:

$$\eta_C = Q/A_C S = (\overline{\tau\alpha}) - U_L(T_p - T_a)/S \tag{A26.7-26}$$

or

---

[6] R. W. Bliss, "The Derivation of Several 'Plate-Efficiency Factors' Useful in the Design of Flat-Plate Solar Heat Collectors," Solar Energy 3, 55-64 (1959).

[7] H. C. Hottel and A. Whillier, "Evaluation of Flat-Plate Solar Collector Performance," paper presented at the International Conference on the Use of Solar Energy, Tucson, Arizona, 1955; transactions published by the University of Arizona Press, Tucson, Arizona, 1958.

$$\eta_C = F_R[(\overline{\tau\alpha}) - U_L(T_i - T_a)/S]. \tag{A26.7-27}$$

The collector heat-removal factor may be determined empirically from relations between collector efficiency and the ratio of temperature differences and solar-insolation values. Empirical values may be obtained for the parameters $(\overline{\tau\alpha})$, U_L, and F_R for use in Eqs. (A26.7-26) and (A26.7-27) from plots of measured values for η_C as functions of $(T_p - T_a)/S$ and of η_C as functions of $(T_i - T_a)/S$.

Klein et al[4] have defined the thermal-loss parameter

$$P_{loss} = F'_R A_C U_L (T_{ref} - \overline{T}_a)\Delta t/Q_L, \tag{A26.7-28}$$

where T_{ref} is a reference temperature equal to the boiling point of water, $\overline{T}_a$ is the average monthly ambient temperature, and Δt is the number of hours in the specified month. The dimensionless parameters given in Eqs. (A26.7-21) and (A26.7-28) have the following physical interpretations. The solar-insolation parameter P_{solar} is the product of the collector heat-exchanger efficiency factor and the ratio of the total energy absorbed on the collector-plate surface to the total monthly energy load. The thermal-loss parameter is the product of the collector heat-exchanger efficiency factor and the ratio of the maximum collector-plate energy losses to the total monthly energy load. Typical values for the collector heat-exchanger efficiency factor are 0.86 to 0.90.[4] By using Fig. A26.7-1 for known monthly average values of P_{solar} and P_{loss}, we may determine the optimal value f_S of the fraction of the energy load that is to be supplied by the solar-energy system each month.

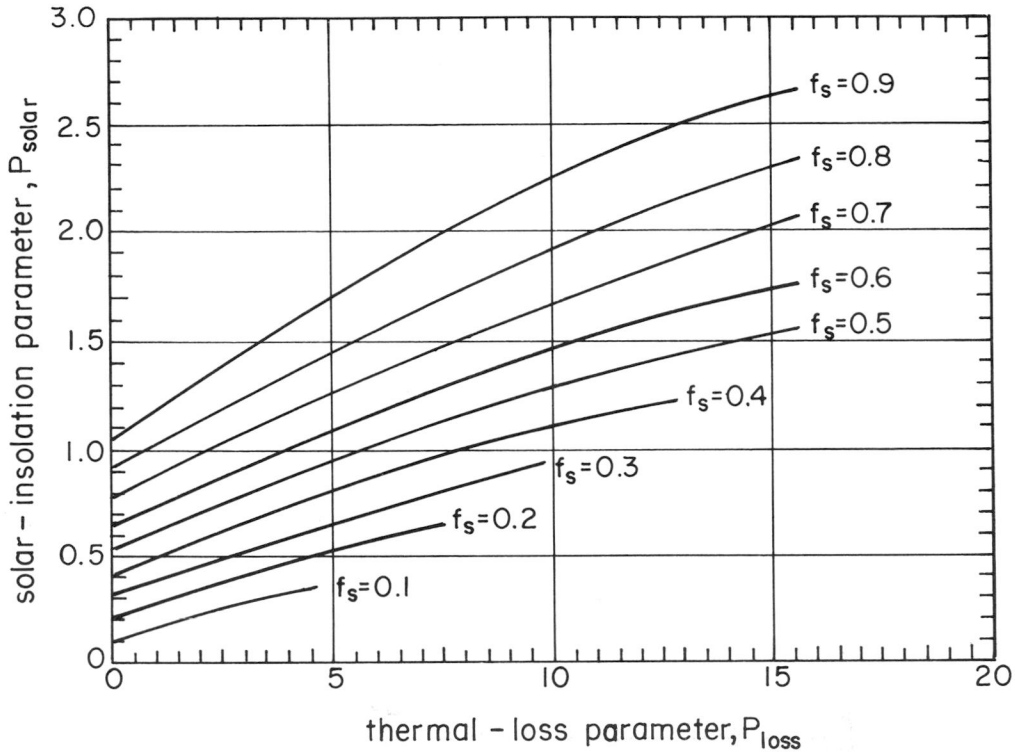

Fig. A26.7-1 Correlations of the dimensionless design parameters P_{solar} and P_{loss} for determination of the fraction of the monthly energy load that is to be supplied by the solar-energy system. The parameters P_{solar} and P_{loss} are defined in Eqs. (A26.7-21) and (A26.7-28), respectively; f_s is the fraction of the monthly load supplied by the solar-energy system. A simple procedure for estimating the relative performance of systems with various collector sizes involves construction of a straight line through the origin and a point corresponding to a specified collector area (e.g., 1,000 ft^2), which is calculated from Eqs. (A26.7-21) and (A26.7-28) for constant and known values of the other parameters. Along this line, the meteorological conditions for the month and all systems-design parameters except the collector area are constant. This line may then be interpreted as the "collector-area axis" and may be suitably scaled to facilitate the construction of a plot showing the variation of collector area with the fraction of the monthly heating load supplied by solar energy. Reproduced from Ref. [4].

The validity of the empirically determined form of Fig. A26.7-1 has been validated by a computer simulation model using eight years of meteorological data recorded at half-hour intervals for Madison, Wisconsin, in more than 300 simulation runs. The standard error of the differences between the simulated and empirically estimated (from Fig. A26.7-1) yearly average values of the fraction of energy supplied by the solar-energy system for four collector sizes has been judged[4] to be substantially lower than the errors inherent in the simulation model and the recorded meteorological data.

Although Fig. A26.7-1 was developed from empirical data for Madison, Wisconsin, it will also yield good results for other locations. Klein et al[4] have compared simulated performance of solar systems in Massachusetts, North Carolina, New Mexico, and Colorado with Fig. A26.7-1. In each location, four system designs were chosen to provide from 25 to 90% of the heating load. In all cases, the standard errors of the differences between the simulated and estimated performance characteristics were equal to or less than the corresponding standard errors incurred for the Madison, Wisconsin, data.

The energies delivered <u>monthly</u> by the solar-energy system and by the auxiliary energy system are given, respectively, by

$$Q'_{solar} = f_s Q_L, \tag{A26.7-29}$$

$$Q'_{auxiliary} = (1 - f_s) Q_L, \tag{A26.7-30}$$

where f_s is the fraction of the <u>monthly</u> load supplied by the solar-energy system. For a specified collector area and known collector properties, monthly averages of ambient temperature and energy load, the dimensionless parameters P_{solar} and P_{loss} may be calculated from Eqs. (A26.7-21) and (A26.7-28).

These values may then be used for the determination of the fraction of the energy load f_s that is to be supplied by the solar-energy system each month (from Fig. A26.7-1). These monthly data, in turn, yield the total net annual cost of the solar-assisted energy system when Eq. (A26.7-16) is used and the monthly averages are summed. Various values of the collector area may be chosen and tested in order to determine optimal collector area corresponding to minimum total net annual cost for supplying the annual energy requirements by a combined solar and auxiliary energy system.

E. Effects of Price Escalation of Alternative Fuels

In the preceding analysis, the unit cost of fuels for the auxiliary energy system has been assumed to be constant with time. During a period in which the unit costs of fuel increase rapidly, the annual fuel bill for operating the auxiliary system will, of course, rise substantially. A projected escalation in fuel cost may be readily accounted for by modification of the capital recovery factor [see Eq. (A26.7-7)]. Kreider and Kreith[8] have shown that, if the annual fuel bill is expected to increase at an annual percentage rate j, the mortgage-interest rate i_m in Eq. (A26.7-7) may be replaced by an effective interest rate which properly accounts for fuel-cost escalation and is given by

$$i_{effective} = [(1 + i_m)/(1 + j)] - 1 \qquad\qquad (A26.7-31)$$

which, for j << 1, reduces to

---

[8] J. F. Kreider and F. Kreith, Solar Heating and Cooling, McGraw-Hill, New York, N.Y., 1975.

$$i_{effective} \approx i_m - j. \qquad\qquad (A26.7-32)$$

Thus, when an economic analysis of a solar-assisted energy system is performed during a period when the fuel costs for the auxiliary system escalate rapidly, the effective interest rate allowable on the investment for the solar system is approximately equal to the difference between the actual mortgage-interest rate and the rate of fuel-price escalation. With current, variable interest rates, it is unrealistic to expect that $i_{effective}$ will be much smaller than i_m.

F. Examples of Optimization of Solar-Assisted Energy Systems

Tybout and Löf,[9] Löf and Tybout,[2] and Kreith and Kreider[1] have performed hour-by-hour digital simulations of solar-assisted heating-system performance and cost-minimization analyses based on a simplified version of Eq. (A26.7-16), which may be written as

$$C_T/Q_T = f_S(C_C A_C + C_S')R_T/Q_S + (1 - f_S)E_a C_f/Q_a, \qquad (A26.7-33)$$

where the values of $C_{OM,S}$, $P_a R_T'$, $P_a C_P$, $C_{M,a}$, and $C_{\ell,a}$ have been assumed to be negligible.

Löf and Tybout[2] have established a range of optimal values of design parameters (e.g., collector tilt, number of glass covers over the collector, and storage area per unit collector area) for use with the major design parameter of collector area (i.e., the fraction of the total annual heating load to

---

[9]R. A. Tybout and G. O. G. Löf, "Solar House Heating," Natural Resources Journal 10, 268-326 (1970).

be supplied by the solar system) in the determination of minimum-cost solar
systems for residences located in eight U.S. cities with different climatic conditions. The results obtained by Löf and Tybout[2] for an assumed collector
cost of $2/ft^2, an equipment cost of $375 + $0.45/ft^2 of collector area, a stor-
age cost of $0.05/ft^2 of collector area, a mortgage-interest rate of 6%, and a
mortgage lifetime of 20 years are summarized in Figs. A26.7-2 and A26.7-3.
The 1961 costs given in Figs. A26.7-2 and A26.7-3 refer to the energy delivered by the solar system only. In every city except Seattle and Miami, the
solar-heat costs were less than electric space-heating costs. In Santa Maria
and Albuquerque, solar-assisted gas and oil heating systems will have a lower
total delivered energy cost than conventional systems. The optimum combination of solar and conventional heating is found when the marginal, or incremental, costs of solar and conventional heat are equal. Reference to Figs.
A26.7-2 and A26.7-3 indicates that the minimum-cost systems represent a
wide range of solar-system sizes. For example, in Santa Maria (see Fig.
A26.7-2), there is a cost differential of only about $0.10/10^6 Btu when the
range of the water- and space-heating load supplied by the solar system is
changed from 65 to 85%; in Boston, the same cost differential occurs over a
range of 35 to 65%. Although some cost minima are narrower (e.g., Omaha
and Charleston in Fig. A26.7-3), the collector size is not of critical importance within a reasonable range of minimum-cost designs.

Kreith and Kreider[1] have performed a similar economic optimization
for a ten-family apartment complex with many energy-conserving features in
Pueblo, Colorado. The results of this analysis are given in Fig. A26.7-4 for
several auxiliary fuels. Reference to Fig. A26.7-4 indicates that the minimum-
cost system with fuel oil as the auxiliary fuel has a collector size that is approximately 50% less than the optimal system using electricity as the auxiliary
fuel.

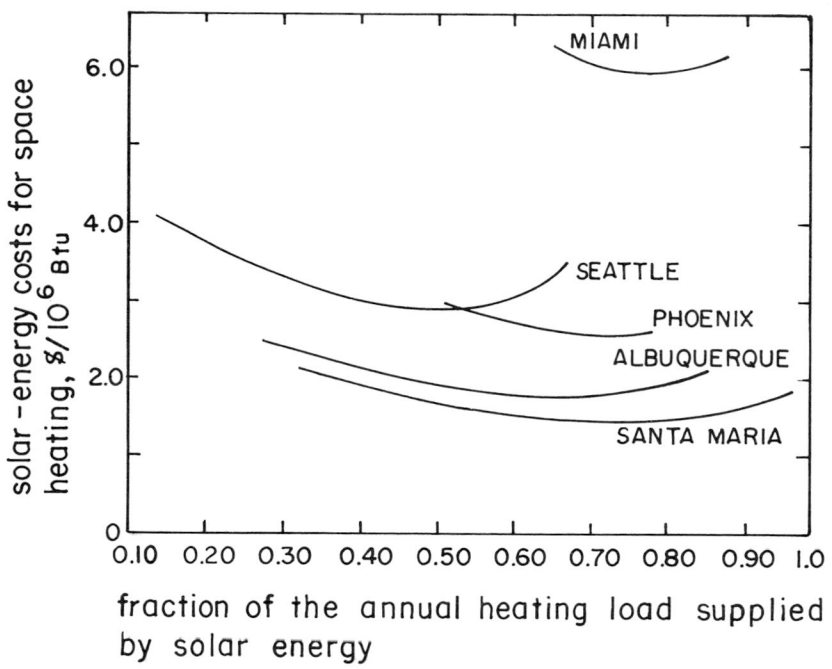

Fig. A26.7-2 Unit solar-energy costs (1961) for water heating and space heating residences in five areas of the United States as functions of
the fraction of the annual heating load supplied by the solar-
energy system. All residences are assumed to have a 1.5×10^4
Btu/degree-day space-heating demand. The average hourly hot-
water demand is assumed to be equal to one degree-hour of
space-heating demand. The large difference between the values
of the fraction of annual heating load supplied by the solar-energy
system or, equivalently, the optimal collector sizes corresponding to the minimum unit energy costs in Miami and Seattle should
be noted. As has been emphasized in the text, the data refer to
highly optimistic collector costs of $2/ft^2. Reproduced with
modifications from Ref. [2].

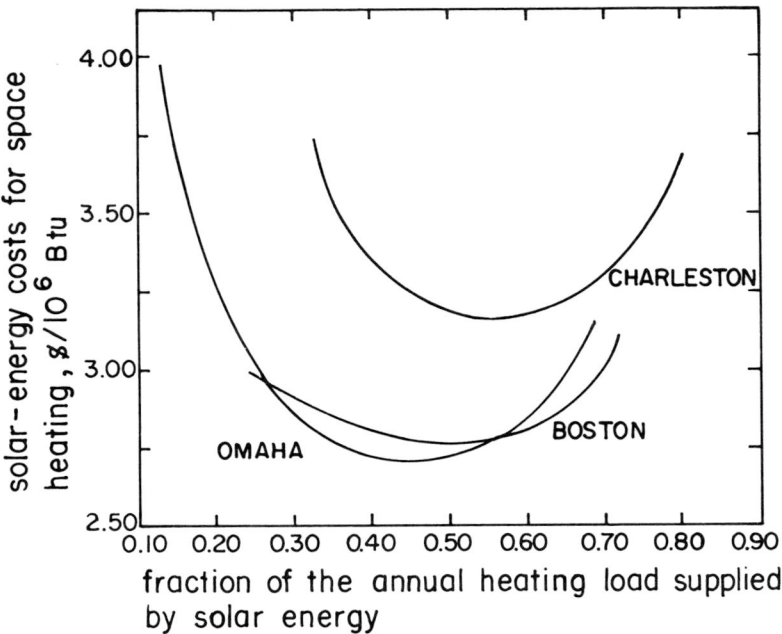

Fig. A26.7-3 Unit solar-energy costs (1961) for water heating and space heating residences in three areas of the United States as functions
of the fraction of the annual heating load supplied by the solar-
energy system. All residences are assumed to have a 1.5×10^4 Btu/degree-day space-heating demand. The average hourly
hot-water demand is assumed to be equal to one degree-hour of
space-heating demand. As has been emphasized in the text, the
data refer to highly optimistic collector costs of $2/ft^2. Repro-
duced with modifications from Ref. [2].

Although different cost assumptions, alternative system designs, various space- and water-heating demands, and different climatic conditions will
lead to changes in the optimal fractions of the annual heating load supplied by

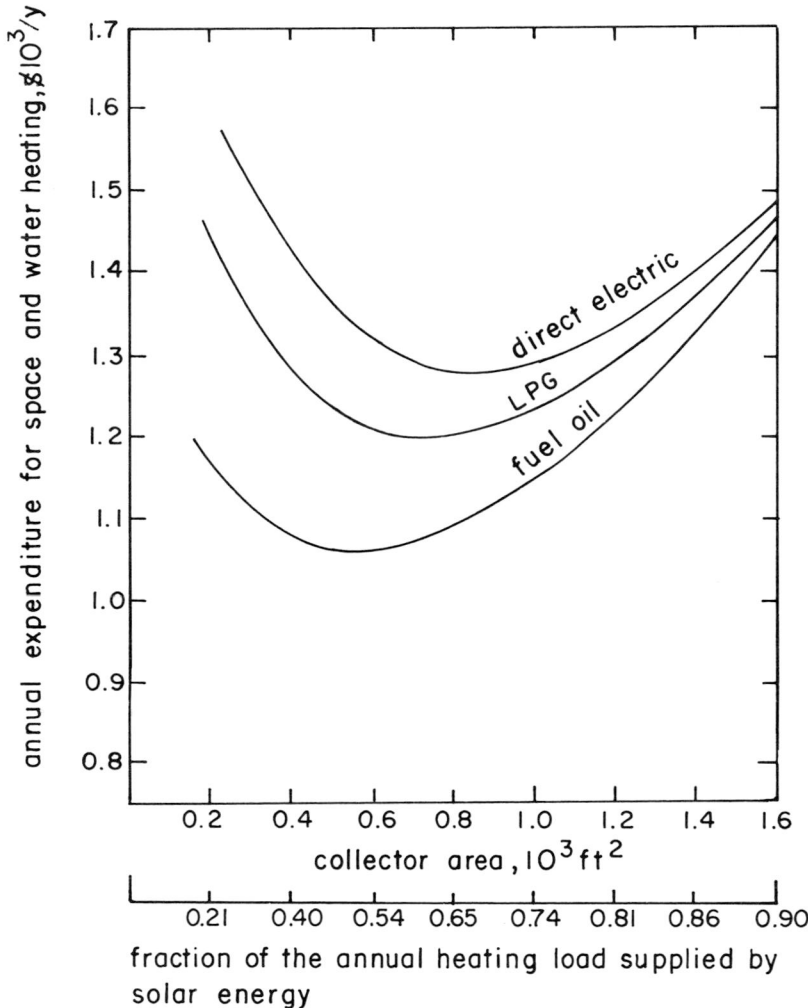

Fig. A26.7-4 Annual expenditure for space and water heating of a ten-family apartment complex in Pueblo, Colorado, as a function of the area of the solar-energy collector and, equivalently, the fraction of the annual heating load supplied by the solar-energy system. A collector cost of $8/ft^2, a heated floor area of 1,200 ft^2 per family unit, and a hot-water demand of 15 gallons per person per day have been assumed. The applicable auxiliary fuel costs are: electricity, $4.60/10^6 Btu; LPG, $3.94/10^6 Btu; fuel oil, $3.06/10^6 Btu. All cost data are given in early 1975 dollars. Reproduced with modifications from Ref. [1].

the solar-energy system and to the minimum costs of delivered energy, the results of cost-minimization analyses will generally be similar to those shown in Figs. A26.7-2 to A26.7-4.

G.   Some Commonly Used Units in Solar Energy Applications

In addition to units listed in Table 1.2-2, the following photometric units and conversion factors are often employed in solar energy applications:

$$1 \text{ langley} = 1 \text{ ly} = 1 \text{ cal/cm}^2,$$
$$1 \text{ ly/min} = 221 \text{ Btu/ft}^2\text{-h}.$$

PROBLEMS

Problems for Chapter 21

1.   Estimate the maximum number of grams of uranium that will be con-
        sumed per year when the largest number of power plants listed in
        Fig. 21.1-1 are all operating, if their average efficiency is 32% and
        they are in use 75% of the time. How much ^{236}U will be produced?

2.   Identify a technology, other than nuclear fission power, that may be said
        to have originated from pure science. Identify a field of science that
        grew out of technology.

3.   Using the definition of the ev and the definition and value of the Faraday
        ($\equiv 96,500$ coulombs/6.03×10^{23} electrons), convert the energy per
        mole (gram atomic weight) of ^{235}U from ev to j.

4.   Calculate from Eq. (21.4-1) the most probable energy of neutrons from
        the fission of ^{235}U, assuming $E_o = 1.29$ Mev.

5.   It has been found that a fission chain reaction occurred naturally in
        a uranium deposit in Gabon, Africa, about 1.8×10^9 years ago, pro-
        bably at the bottom of a body of water. Estimate the fraction of ^{235}U
        in the uranium deposit at that time. The half lives of ^{235}U and ^{238}U
        are 7.1×10^8 y and 4.5×10^9 y, respectively. What was the relative
        abundance of the two isotopes at the time of the formation of the earth
        about 4.55×10^9 y ago? Hint: Calculate the number of half lives in each
        time interval for each isotope and compare the results of the number of
        doublings of the amounts of each nuclide.

6.   Estimate η for ^{239}Pu for neutrons having energy equal to the resonance
        peak in Fig. 21.7-1 (using approximate data read from the plots in the
        figure) and compare the result with the appropriate entry in Table 21.5-1.

7. A collimated beam of neutrons with a flux of 10^{13} n/cm^2-sec from a reactor falls on a slab of paraffin (with the approximate formula $C_{20}H_{42}$) 1 cm thick and of density 0.8 g/cm^3. Estimate the flux in a collimated beam emerging from the other side of the slab. Assume that the scattering cross sections of hydrogen and carbon are 55 b and 4.8 b, respectively. What fraction of the neutrons is scattered by the carbon atoms?

8. One of the fission products is ^{135}Xe. It occurs in a mass chain that has a yield from fission of 6.4%. A portion of this chain is $^{135}I \xrightarrow[6.7h]{\beta} {}^{135}Xe \xrightarrow[9.16h]{\beta} {}^{135}Cs \xrightarrow[2.3 \times 10^6 y]{\beta} {}^{135}Ba$. In addition, this nuclide has the largest known capture cross section for thermal neutrons, $\sigma_c = 2.7 \times 10^6$ b ($\bar{v} = 2200$ m/s). Formulate relations that yield the steady-state number densities of the ^{135}Xe nuclei in a reactor of known power output per unit mass of fissile uranium (specific power). Discuss briefly what effect this fission product might have on reactor controls.

9. Using data in Table 21.8-1, compare the mean free path for scattering of neutrons at room temperature in H_2O, D_2O, BeO, and graphite.

10. Using the temperatures listed in Fig. 21.13-1, estimate the flow of water (in kg/sec) in the primary loop for a PWR producing 1000 Mw_e with a heat-power efficiency of 32%. Use density and specific heat data from a standard handbook. Estimate the flow rate of water (vaporized to steam) in the steam-power loop.

11. Using typical dimensions for the PWR and BWR pressure vessels, compare the hoop stresses in the walls of the two vessels and in the mid-sections. Hint: For an approximate comparison, assume that the thickness of the wall is small compared to the diameter of the vessel

and that no nozzles are located nearby and no changes occur in the diameters or thicknesses.

12. Assuming that an LWR fuel pellet has a diameter and length of one cm, a density of 10 g/cm^3, the chemical formula UO_2, and contains 2% of ^{235}U, calculate the fission power per unit length (w_t/cm) at a uniform thermal flux of neutrons = 10^{14} n/cm^2-sec with an average velocity of 2200 m/s. What is the heat flux per cm^2 from the fuel rod that corresponds to this power per unit length?

13. Approximately 1800 ppm of boron may be present in the primary coolant of a PWR at start-up to compensate for the initial excess of ^{235}U that will be consumed. If approximately one half of the coolant is in the reactor core, compare the approximate, relative rates of removal of ^{235}U and boron by neutron reactions in a 1000 Mw_e power plant having a heat-power efficiency of 32% (assume that the neutron flux is uniform and use room-temperature cross sections). The capture cross section of naturally occurring boron is 760 b and the atomic weight is 10.811 (19.8% ^{10}B, $\sigma_c = 3.84 \times 10^3$ b and 80.2% ^{11}B, $\sigma_c = 5 \times 10^{-3}$ b). If one supposes that boron is removed only by neutron capture, estimate the ratio of the boron isotopes after a fuel burn-up of 10 $Mw_t d/kg$.

14. If the average burn-up of the fissile material in a PWR is 30 Mw_t/kg, what is the average atom fraction of fission products that must be present in a fuel element on removal from a reactor?

15. Compare the working Rankine cycle efficiency of an LWR with the ideal Carnot cycle in which heat is transferred at the maximum steam and minimum (condenser) water temperatures ($21°C$), respectively. Make a similar calculation for the HTGR and compare the two reactor concepts.

16.   Estimate the cooling water evaporated per second and per day in an
       evaporative cooling tower for two 1000 Mw_e power plants operating at
       efficiencies of 32% and 40%, respectively.

17.   Assuming that the heat-power efficiency of a power plant is a constant
       fraction of the Carnot efficiency, compare the heat discharged in the
       two cases in problem 16 with a cooling system per $kw_e h$ supplied to an
       electrical distribution network.

18.   Estimate the fractional reduction in the amount of coal to be mined if
       H_2 were produced in an idealized process using reaction (21.17-1).
       Nuclear heat should be compared with coal as primary energy source.

Problems for Chapter 22

1.   Calculations have shown that the use of fuel elements containing urani-
      um-plutonium carbides instead of the oxides in an LMFBR may yield a
      maximum power per unit length of fuel rod of 1000 w_t/cm at a breeding
      ratio of 1.33. Assuming that the amount of plutonium in the reactor
      core is about the same for both the carbide and the oxide fuels, esti-
      mate the improved doubling time afforded by using the carbide elements.

2.   A heavy-water-moderated reactor (HWR, as is used in Canada) fueled
      with natural uranium might have a conversion ratio of 0.85 and a heat-
      power efficiency of 32%. How much UO_2 would the Canadians have to
      produce per year by mining to meet a nuclear power demand of 10 Gw_e?

3.   Discuss briefly why fuel rods for the FBR have smaller diameters than
      those for the LWR.

4.   If the volume expansion of the stainless-steel cladding material, in the central region with maximum fast-neutron flux in an LMFBR, is 10% at the end of life, estimate the final diameter of an 8 mm diameter rod located in this region.

5.   Present briefly your ideas concerning the effect of a given increase in the cost of uranium on the cost of electric power produced from both the LWR and the FBR.

6.   Using the data in Table 22.2-1, estimate the number of HTGRs that could be fueled with ^{233}U produced by one GCFR with a radial blanket of ThO_2, which is operated at the same thermal power level. Hint: Note that, if production of Pu for additional reactors is not required, the breeding ratio for Pu in the core and axial blanket may be unity and that the excess number of neutrons may then be allowed to produce ^{233}U in the radial blanket.

Problems for Chapter 23, Section 23.1

       Problems 1 to 3 have been constructed to show how cross section data may be used to obtain rate constants and ignition temperatures and to obtain Lawson's criterion for a fusion reaction. Problem 9 shows how the incremental radiative loss produced by an impurity of 0.1% Mo may be approximated and how the ignition temperature is changed.

1.   Arnold et al[6] have measured cross-sections for the $He^3(d, p)He^4$ reaction (i.e., for $D + He^3 \rightarrow p + \alpha$). Here 1 millibarn = 10^{-3} barn = 10^{-27} cm^2. Determine the best fit to $\sigma = (A/E)\exp - (\gamma/\sqrt{E})$, where E is the relative energy of collision between D and He^3, A is in barn-kev and E in kev; find A and γ. Hint: (a) Change E_D to E; (b) by

Deuteron energy (E_D), kev	σ, millibarn(s)	Deuteron energy (E_D), kev	σ, millibarns
36.0	0.124	60.0	2.32
40.0	0.258	67.0	3.84
46.0	0.605	73.0	5.65
53.0	1.28	80.0	8.45
		93.0	16.2

plotting $E\sigma$ on semi-log paper against $1/\sqrt{E}$ (Gamov plot), obtain A and γ; or (c) by plotting $\ell n(E\sigma)$ against $1/\sqrt{E}$ on regular graph paper, obtain the slope γ and the intercept $\ell n A$; or (d) by taking $y = \ell n(\sigma E)$ and $x = 1/\sqrt{E}$, obtain a least-squares-fit to $y = a - bx$. Compare the results with values given in Table 23.1-5.

2.   Using results from Problem 1, (a) deduce the rate constant for the same reaction, i.e., find β and Λ in $\mathcal{K} = \beta T_i^{-2/3} \exp(-\Lambda T_i^{-1/3})$ [see Eq. (23.1-50)] and (b) plot $\mathcal{K}$ against T_i on log-log paper.

3.   Using results from Problem 2 and assuming that the only fusion reaction is $D + He^3 \rightarrow p + \alpha$, obtain (for $T_e = T_i$) (a) the ignition temperature, (b) the Lawson curve of $N_e \tau$ against T_i (with only radiative energy loss by bremsstrahlung), and (c) discuss how the feasibility of this reaction compares with those of the DT and DD reactions. Hint: Use Eq. (23.1-55) for the ignition temperature and Eq. (23.1-69) for the Lawson curve (omit the last term involving cyclotron radiation).

4.   In addition to the rate equations in Eq. (23.1-82), add a rate equation for the tritium reaction using Li^6. Obtain a steady state-equation for the concentration of tritium. (a) Use only the DT reaction in Eq. (23.1-82). (b) Use both the DD and DT reactions in Eq. (23.1-82). Discuss the results.

5.   Consider the reaction $D + D \rightarrow T + p$. (a) Calculate from the mass de-
          fect the total reaction-energy release and the energy release per deu-
          teron. (b) Calculate the kinetic energy carried by tritons and by pro-
          tons.

6.   Derive k [Eq. (23.1-50) from Eq. (23.1-48)], including the applicable
          numerical factors. Hint: To evaluate the integral approximately, ex-
          pand the exponent under the integral around $W = B^{2/3}$.

7.   Derive Eq. (23.1-51), including the numerical coefficient.

8.   Calculate the energy released in the carbon cycle in stars and show
          that the overall reaction corresponds to $4H \rightarrow \alpha + 2e^{+}$.

9.   Estimate the radiative energy loss from a plasma containing 0.1% of a
          molybdenum impurity; assume, for $10 \leqslant T_e$, kev $\leqslant 50$, that Mo^{40+} is the
          most abundant Mo-species. Using the approximate relations in Eqs.
          (23.1-93) to (23.1-95), extend Fig. 23.1-16 to find the new ignition tem-
          perature. Compare the results with those depicted in Fig. 23.1-17,
          obtained using different approximations.

Problems for Chapter 23, Sections 23.2 through 23.6

1.   (a) Find the length of a straight θ-pinch system needed to produce a
               reactor if the available magnetic field is 100 kG.
          (b) If the plasma in this device is required to be at least 50 r_L in radius,
               what is the minimum power level?
          (c) What must the wall radius be to keep classical diffusion losses at
               one tenth of the end losses?
          (d) What must the wall radius be if the wall loading is to be kept at less
               than 1 Mw/m^2?

(e) What are the magnetic and plasma energies stored in the system?

2. Using the MHD equilibrium equation, prove the assertion that a purely toroidal field cannot confine a plasma.

3. Using drift orbits, prove the statement made in Problem 2. Establish the magnitude of the parallel current in a Tokamak that is required for equilibrium.

4. Show that if a magnetized plasma is disturbed by a potential Φ, the $\delta E_j = \langle e \vec{\Delta} \cdot \nabla \Phi \rangle$ where Δ is the displacement due to the drift velocity.

Problems for Chapter 23, Sections 23.7 through 23.13

1. Calculate the reaction rate in DT at a number density of $10^{26}/cm^3$ and 50 kev temperature; estimate the time for 50% burn-up of the DT. What are the final temperature and pressure if expansion does not occur?

2. If the available reaction time is about equal to the sound transit time across the fuel, what fuel radius is required to give the burn-up time estimated in Problem 1?

3. The laser radiation pressure in the plasma is approximately equal to $2\varphi/c$ in units of the laser flux. What is the pressure at a flux of 10^{16} w/cm^2? At a laser wavelength of 10 microns, what is the critical density? What plasma temperature is required to balance the radiation pressure?

4. How does the black-body radiation in the solar interior at $20 \times 10^{6}\,°K$ compare with the laser flux at 10^{16} w/cm^2?

Problems for Chapter 24

1.   What are your views on the implications of the data listed in Tables
     24. 1-1 and 24. 1-2 concerning nuclear-reactor implementation on an
     accelerated basis? Do you consider the differences between risks
     associated with voluntary and involuntary activities (see Fig. 24. 1-1)
     as justifiable?

2.   Determine radiation doses from various man-made and natural sources
     and discuss their likely effects on human health.

3.   Summarize salient aspects of the engineered safety features that have
     been incorporated in nuclear reactors. How might they fail?

4.   What is meant by fault-tree analysis? Do you consider this procedure
     as adequate for estimation of nuclear-reactor safety? What would you
     do to assure greater reliability in reactor-safety assessments?

5.   Discuss the relation between nuclear-fuel reprocessing and subversion.

6.   What are some important options in high-level waste management?

7.   Compare the likely environmental hazards associated with large-scale
     uses of fusion and breeder reactors.

8.   Verify the estimates given in Section 24. 5F for the environmental and
     health effects of waste disposal.

Problems for Chapter 25

1.   Verify Eqs. (25.1-13) and (25.1-14).

2.   Compare the definitions given for net-energy ratios in Sections 25.2 and 9.14.

3.   Verify the entries listed in Table 25.2-1 by a careful study of available mining, milling and processing data.

4.   Discuss the relation to uranium-fuel availability and costs of staged development of fission and breeder reactors.

5.   (a) Using Table 25.3-1, calculate the fissile requirements and fissile production for each reactor, assuming a 30-y reactor life.

      (b) Compare the PWR-U and PWR-Pu0 with the HTGR-1 and HTGR-3. Calculate fissile material-inventory requirements and total operating fissile requirements. Which reactor minimizes U_3O_8 consumption?

6.   (a) What is meant by a symbiotic reactor system?

      (b) Consider the following two symbiotic systems: LMFBR-AO/PWR-Pu and GCFR-Pu/GCFR-Th/HTGR-4. Starting with 100 Gw_e of breeders, how many Gw_e of thermal reactor capacity (assuming a 30-y life) could be supported by each of the two symbiotic systems? Hint: Read Ref. [15] in Section 25.3.

Problems for Chapter 26

1. Estimate world-wide energy use in the year 2100 if population and energy-use growths are limited to 0.5% annually during the period 1970 to 2000.

2. If all of the energy requirements determined in problem 1 are met by electricity production from biomass at an overall conversion efficiency of 1% and the applicable average insolation corresponds to 200 w/m^2, what land area will be dedicated by the year 2100 to energy plantations for erg-growths?

3. What is your estimate of arable land-area requirements to feed the world population during the year 2100? Discuss the environmental implications of providing for world energy and food needs according to the scenario specified in problems 1 to 3.

4. Estimate the required number and discuss the likely environmental impacts of fusion-reactor farms generating 50×10^3 Mw_e each by the year 2100 to supply the energy requirements estimated in problem 1.

5. Develop procedures for implementing energy-conservation measures that will reduce world-wide energy requirements for the year 2100 (see problem 1) by 50%.

6. If you were given authority to develop long-range solutions to the world's energy-supply problems, what would be your percentage allocation for RD and D funds during the current year for each of the following technologies: development of fusion reactors, construction of breeder reactors, shale-oil recovery by in situ techniques, in situ coal gasification, wind-power systems, ocean-thermal technology, photovoltaic power

generation, utilization of salinity differences between river and ocean waters, waste utilization, uncommitted exploratory research? What percentage of your total budget would you allocate to environmental and health studies or to socio-political research? Justify your preferred allocations in a carefully constructed essay defining your views on how to make a better world by the year 2100.

INDEX

INDEX

705